PAVEMENT ENGINEERING

Principles and Practice

PAVEMENT ENGINEERING

Principles and Practice

Rajib B. Mallick and Tahar El-Korchi

CRC Press
Taylor & Francis Group
Boca Raton   London   New York

CRC Press is an imprint of the
Taylor & Francis Group, an **informa** business

CRC Press
Taylor & Francis Group
6000 Broken Sound Parkway NW, Suite 300
Boca Raton, FL 33487-2742

© 2009 by Taylor & Francis Group, LLC
CRC Press is an imprint of Taylor & Francis Group, an Informa business

---

Library of Congress Cataloging-in-Publication Data

Mallick, Rajib Basu, 1966-
 Pavement engineering : principles and practice / Rajib B. Mallick, Tahar El-Korchi.
  p. cm.
 Includes bibliographical references and index.
 ISBN 978-1-4200-6029-4 (hardback : alk. paper)
 1. Pavements--Design and construction. I. El-Korchi, Tahar. II. Title.

TE251.M3245 2009
625.8--dc22   10 05549916   2008025434

Visit the Taylor & Francis Web site at
http://www.taylorandfrancis.com

and the CRC Press Web site at
http://www.crcpress.com

Dedication

*This book is dedicated to the
past, present, and future generations of
civil engineers*

Contents

Preface

This book is written for undergraduate and graduate students of civil engineering, as well as practicing engineers. The authors have been greatly influenced by two classic books in this area: E. J. Yoder and M. W. Witczak, *Principles of Pavement Design*, 2nd ed. (John Wiley & Sons, 1975); and Yang Huang, *Pavement Analysis and Design* (Prentice Hall, 1993). These two books lay down most of the important principles of pavement engineering, and it is difficult, if not impossible, to find any alternative for these two books.

However, it is our belief that most students at the undergraduate level can spend only a few terms or semesters on pavements, and for this reason they need a textbook that presents a broad but concise view of the more relevant topics in pavement engineering. The need is the same for many practicing engineers, who can use a reference textbook that is current and practical.

For graduate students, it is important that they get exposed to a broad overview of different topics in pavement engineering before starting in-depth research in any specific area. This can be helpful when they are focusing on a narrow topic of research to view the relevance of their work in the larger scope of pavement engineering.

For students coming back to school from the profession, it is often the case that they gain professional experience in either a specific area of pavement engineering or some other area (such as structures), and they need a broad view of the subject before starting in-depth study. This book can be used for providing such a broad overview on pavement engineering for graduate students.

The Internet has changed both teaching and learning forever. In most cases, the use of a single textbook is just not enough. A vast treasure-house of online documents is accessible on the Internet. When possible, this textbook has referenced numerous websites with a wealth of information and documentation relevant to pavement engineering. Students can supplement their textbook with the relevant online documents. This is important because students can now view, read, and understand "real-world" case studies, research reports, and specifications from pavement professionals, and are no longer restricted to the knowledge confined within a single textbook. And since learning through multiple sources has become easy and accessible through the Internet, students are encouraged to practice it more frequently. This book makes reference to Internet resources wherever they are appropriate, and encourages students to download freely available design guides/charts and procedures to work out the practice problems provided at the end of each chapter.

The objective of this book is to explain the basic principles of the various design and construction procedures, with relevant examples. The required resources in the form of charts, tables, and weblinks have been provided wherever appropriate for such examples. Note that different agencies have developed different types of pavement design and construction methods over the years, and have published comprehensive manuals. The authors have purposefully not provided the full range of charts or tables from any specific organization in this textbook for two reasons. First, the manuals published by the different organizations are complete by themselves and are the best source of information, and they should be consulted for design or construction. Secondly, it is the objective of this book to assist the user in self-learning about design and construction, and not to actually be a design and construction manual or guide. We firmly believe that while learning can be accomplished with a textbook, the actual application of knowledge requires the utilization of resources provided by the industry and pavement agencies. As such, students are encouraged to use appropriate guidelines from these sources.

Which unit system should be used in this book? We have debated this question at length, since both the English/U.S. and International System of Units (SI)/metric systems continue to be used in pavement engineering. While most people use the SI units outside the United States, many of the original methods/charts that are still used today were developed using English/U.S. units. One perfect example is the California Bearing Ratio (CBR). A good approach would be to utilize both systems by putting one or the other units in parentheses. However, the authors feel that there are two drawbacks of this approach. First, this system does not encourage the students to learn conversions from one system to the other, and, second, equations, tables, numbers, and figures often become cluttered and clumsy with the use of a double-unit system. It is important for students in pavement engineering to be conversant in both systems of units. With these considerations in mind, the authors have used both systems, but in many cases, not together. The units corresponding to the original models/equations have been retained. A conversion guide is provided in the Appendix to assist the user.

This book is offered as a textbook. Obviously, it draws information from a number of sources—notes, research reports, specifications, and manuals. The importance of these documents cannot be overstated, and students are encouraged to review them, either online or in school libraries. Some of the more relevant ones are as follows:

- Yoder, E. J. and M. W. Witczak. *Principles of Pavement Design*, 2nd ed. New York: John Wiley, 1974.
- Huang, Yang. *Pavement Analysis and Design*. Englewood Cliffs, NJ: Prentice Hall, 1993.
- American Association of State and Highway Transportation Officials. *MDM-SI-2, Model Drainage Manual, 2000 Metric Edition*. Washington, D.C.: AASHTO, 2000.
- American Association of State and Highway Transportation Officials. *AASHTO Guide for Design of Pavement Structures*. Washington, D.C.: AASHTO, 1993.
- ACI Committee 211. *Standard Practice for Selecting Proportions for Normal, Heavyweight and Mass Concrete, ACI 211. 1-91*. Farmington Hills, MI: American Concrete Institute, 1991.
- ACI Committee 214. *Recommended Practice for Evaluation of Strength Test Results for Concrete, ACI 214-77*. Reapproved 1997. Farmington Hills, MI: American Concrete Institute, 1977.
- Federal Aviation Administration. FAA Advisory Circular AC No. 150/5320-6D. Washington, D.C.: U.S. Department of Transportation, 1995.
- Roberts, Freddy L., Prithvi Kandhal, E. Ray Brown, Dah-Yinn Lee, and Thomas W. Kennedy. *Hot Mix Asphalt Materials, Mixture Design, and Construction*. Lanham, MD: NAPA Education Foundation, 1996.
- Read, John and David Whiteoak. *Shell Bitumen: The Shell Bitumen Handbook*, 5th ed. 2003. London: Thomas Telford.
- Barksdale, R. D., Ed. *The Aggregate Handbook*. Washington, D.C.: National Stone Association, 1991.
- ASCE. *Introduction to Mechanistic-Empirical Pavement Design: Reference Manual*. Seminar. Baltimore: American Society of Civil Engineers, 2004.
- Coduto, Donald P. *Geotechnical Engineering: Principles and Practices*. Upper Saddle River, NJ: Prentice Hall, 1999.
- Federal Highway Administration. *Pavement Deflection Analysis*, NH1 Course No. 13127, Participant Workbook, FHWA Publication No. FHWA-HI-94-021, 1994.
- Ullidtz, Per. *Pavement Analysis*, New York: Elsevier, 1987.

In addition, students are encouraged to view the Federal Highway Administration (FHWA; www.fhwa.gov) and the Federal Aviation Administration (FAA; www.faa.gov) websites for their pavement-related resources.

Acknowledgments

This book is the collective result of many years of learning and experience. Countless people have contributed to this book, and a list would certainly fill up the rest of this book. But then there are some without whom this book would not have been possible.

Ray Brown of the US Army Corps of Engineers, and Prithvi Kandhal, Frazier Parker, and Dan Brown of Auburn University

Rick Bradbury, Dale Peabody, Brian Marquis, and Wade McClay of Maine Department of Transportation

Sudhanshu Bhattacharjee

Robert Pelland of the Massachusetts Port Authority (Massport)

Mike Marshall of Wirtgen GmbH

Ed Kearney

Wouter Gulden of the American Concrete Pavement Association (ACPA-SE Chapter)

Maureen Kestler of the U.S. Forest Service

Frances Wychorski, Agata Lajoie, and Don Pellegrino of Worcester Polytechnic Institute

The authors thank them from the bottom of their hearts.

Writing a book requires a significant amount of support and encouragement from the family. Special thanks go to Sumita and Urmila, and to Marya, Miriam, and Youssef, for their patience, help, and encouragement.

This book would not have been possible without the encouragement of the authors' parents, Bimalangshu Kumar Basu Mallick and Monika Basu Mallick, and M'hamed El-Korchi and Bahija El-Bouchti El-Korchi.

Last, but not least, the authors thank the ever helpful members of the Taylor & Francis Group, especially Joe Clements, acquiring/contact editor, Andrea Dale, editorial assistant, Jill Jurgensen, production coordinator, Nadja English, marketing manager, and Michele Dimont, project editor. It has been a pleasure to work with the group.

About the Authors

Dr. Rajib Mallick has extensive experience with asphalt materials, mixture design, construction, pavement design, recycling, and nondestructive testing. From 1992 to 1998, Dr. Mallick worked as a research assistant and as a senior research associate at the National Center for Asphalt Technology (NCAT), and is currently working as an associate professor of civil and environmental engineering at Worcester Polytechnic Institute (WPI). He has completed numerous research projects for several departments of transportation, the Federal Highway Administration (FHWA), the Massachusetts Port Authority (Massport), the National Science Foundation (NSF), the Federal Aviation Administration (FAA), and several private practitioner organizations. Dr. Mallick has coauthored more than 60 papers for journals and conference proceedings, as well as several practical reports, manuals, and state-of-the-practice reports for federal, state, and local highway agencies. He has taught professional courses on asphalt technology, presented courses on  recycling for the Federal Highway Administration, as well as lectured in the National Workshop on Recycling and Other Pavement Rehabilitation Methods at the Indian Institute of Technology (IIT), Kanpur, India. He has presented a series of seminars on hot mix asphalt paving for the Baystate Roads Programs for the Massachusetts Highway Department. With three awards from the National Science Foundation (NSF), he has introduced several innovations in his courses. He is a member of the Transportation Research Board, Committee AFK30; Association of Asphalt Paving Technologists; American Society for Testing and Materials, Committee D04, Road and Paving Materials; ASCE Highway Construction and Maintenance Technical Committee; Bituminous Mix Testing and Evaluation Consortium for Massport's Logan International Airport; and Sigma Xi. He is a registered professional engineer (PE) in Massachusetts. Dr. Mallick has served as a consultant in several projects for practitioner organizations.

Dr. Tahar El-Korchi is a professor and interim head of civil and environmental engineering at Worcester Polytechnic Institute. He has been at WPI since 1987. He teaches courses at the undergraduate and graduate levels in pavement analysis, design and management, construction materials testing and evaluation, cement and concrete materials and construction processes, structural analysis and design, and high-performance structural materials. He advises undergraduate projects and graduate theses. He has served at numerous WPI Global project centers, including Puerto Rico; Washington, D.C.; Morocco; and Costa Rica. Dr. El-Korchi was granted  the prestigious Presidential Young Investigator Award by the National Science Foundation in 1991. His research funding has been generated from federal and state sources, including the National Science Foundation, the U.S. Army Waterways Experiment Station (WES), the U.S. Geological Survey's Columbia River Research Laboratory (CRRL), the Small Business Innovation Research

(SBIR), the Federal Highway Administration, the New England Transportation Consortium, the Massachusetts Highway Department, and numerous industries. Dr. El-Korchi was an invited scientist at the Turner Fairbanks Highway Research Center (Federal Highway Administration), Washington, D.C. (1993–1994). He has conducted research on the durability of concrete pavements. Research included analytical and experimental aspects of concrete freeze-thaw characterization for pavement applications. He also conducted research at the National Institute of Standards and Technology using the cold neutron research facility to characterize and quantify the freezable water in hardened Portland cement. Dr. El-Korchi is also a consultant to several major corporations. He provides recommendations for the design and evaluation of pavements, materials, and structures.

1 Introduction and Description of Pavements

1.1 IMPORTANCE

Pavements are an essential part of our life. We use them as roads, runways, parking lots, and driveways. Pavements are engineered structures and are important for our everyday life, commerce and trade, and defense. Surface transportation is the most widely used mode of transportation in the world, and a country's development is often measured in terms of its total paved road mileage. The construction of road is and will continue to be a major industry in developing countries, and as the infrastructure matures, it will be a major industry in developed countries as well.

Like any other engineered structure, pavements are expected to be adequately strong and durable for their design life. They are expected to function properly by providing a smooth traveling surface for the traffic under various conditions of the environment. In order to ensure this, pavements must be designed, constructed, maintained, and managed properly.

Pavements can be broadly classified into asphalt (or flexible) and concrete (or rigid) pavements (Figure 1.1). Pavements consist of different layers, more so in the case of asphalt pavements than concrete ones. From the bottom up, these layers are known as the subgrade, subbase, base, and binder and/or surface. There are certain pavements with asphalt surface layers on top of concrete layers.

In the United States, there are about 4 million miles of roads, of which approximately 2.5 million are paved. The federal government and state departments of transportation (DOTs) spend a significant portion of their budget on maintaining and managing existing pavements and rehabilitating old pavements. More than 90% of commodities are transported on highways in the United States. Roads in poor condition end up costing the DOTs a lot of money for repairs, as well as to the users for repairing damaged vehicles. These roads are also unsafe for travel—for example, more than 30% of traffic fatalities in the state of Massachusetts in the United States have been reported to be due to poor road conditions.

1.2 FUNCTIONS

The most important function of the pavement is to withstand the load applied from a vehicle such as a truck or an aircraft, without deforming excessively. The layered structure of the pavement is meant for ensuring that the load is spread out below the tire, such that the resultant stress at the bottom layer of the pavement, the subgrade, is low enough not to cause damage. The most significant load applied to a pavement surface comes from a truck or an aircraft tire. The approach in a flexible pavement is to spread the load in such a way that the stress at the subgrade soil level is small enough so that it can sustain the stress without any major deformation. When the existing soil is not stiff enough to support the relatively small stress, then there is a need to improve the soil. There is also a need to improve the soil if it is susceptible to moisture. Such a problem can be solved by treating the soil with an additive, such as lime and a Portland cement.

Since pavements are exposed to the environment, a very important factor in the design of pavements is the consideration of water, which could be coming from rain/snow (surface water) and/or from the ground (ground/subsurface water). Since water can be detrimental to a pavement, a basic necessity of designing a proper pavement is to provide adequate drainage for both surface and subsurface water. Standing water on a pavement can cause hydroplaning, skidding, and accidents. There is a need to make sure that water from precipitation is drained away quickly and effectively and that there is no depression on the roads to collect water. Water present in frost-susceptible soils

FIGURE 1.1 Asphalt and Concrete Pavements.
Courtesy: Mike Marshall, Wirtgen, GmbH, and Wouter Gulden, ACPA-SE Chapter.

in the subgrade can freeze, causing heaving and failure of the pavement. Therefore, frost susceptible materials should be avoided. If this is not possible, then the pavement structure above the subgrade should be thick enough to prevent the freezing front from reaching the frost susceptible soil. Similarly, as one expects some water to make its way through random cracks and joints, proper subsurface drainage must be provided, and the material within the pavement structure should be made resistant to the actions of water—otherwise the aggregates, for example, would be washed away due to repeated traffic-induced pressure or freeze–thaw pressures.

In most cases it is not possible to completely avoid water flowing inside a pavement. Such ingress of water can physically remove materials from inside a pavement structure, and also freeze and cause deformations in the pavement. To prevent this, the pavement material must be selected properly and, if needed, modified.

1.3 DESIGN AND CONSTRUCTION

Generally the layers in a pavement improve in quality as one goes up from the bottom to the surface layer. The surface layer, which can be asphalt or concrete, is the most expensive and stiff/durable layer in the entire pavement structure. Components of this layer are mostly naturally occurring materials, for example, asphalt binder is a by-product of the petroleum distillation process and aggregates are obtained from rock quarries or riverbeds. These materials are combined and used in different proportions to produce the final material that is used in the pavement. For example, asphalt binder is mixed with aggregates to produce hot mix asphalt (HMA) for asphalt pavements, while Portland cement is mixed with aggregates in Portland cement concrete (PCC) pavements. In both cases, the mixing must be conducted in the correct proportions to ensure adequate quality of the mixture. It is important to find out whether the resultant mix has the adequate strength and stiffness through testing. Such *testing* is generally conducted in the laboratory during the mix design process. During this process, loading and effect of environmental conditions can be simulated in the laboratory. If the responses to this testing do not meet our *expectations*, then the mix needs to be redesigned and/or the materials need to be reselected. These "expectations" are specifications that have been developed on the basis of experience and research. State DOTs or agencies can use their own specifications and/or use specifications from organizations such as the American Association of State Highway and Transportation Officials (AASHTO) or the Federal Aviation Administration (FAA). Similarly, testing is also conducted according to standards, which again can be from state DOTs or the AASHTO. Whether or not the material can be placed and compacted is also determined during the mix design process.

In most cases pavement engineers are restricted to using locally available materials, with or without some modifications, because of economic and practical reasons. With these available materials it is important to determine what thickness of each layer, and hence the entire pavement, is required to carry the loads under different environmental conditions without any problem. This step, known as the structural design, makes sure that the pavement structure as a whole can withstand traffic for its design life, even though the traffic might increase and the properties of the layer might change cyclically and/or progressively during its design life.

Generally several layers are present in an asphalt pavement. From the bottom up, the layers are known as the subgrade, subbase, base, and binder and/or surface. Generally, the bottommost layer is soil; the subbase and/or base layers can be granular soil, or aggregates or asphalt-aggregate mixtures (mixes); and the binder and surface are asphalt mixes. While designing, adequate thickness to each layer is assigned, so as to obtain the desirable properties in the most cost-effective way. Concrete pavement may not have as many layers, and in many cases the concrete slab rests on a stabilized subgrade, which consists of soils modified with some additives.

Once the pavement materials/mix and structure are designed, it must be constructed properly. To ensure this, the material must be laid down and cured (if needed) and compacted in the proper way so that it has the desirable qualities, such as density and/or stiffness. While selecting the materials and designing the mix and the structure, it is important to keep workability in consideration, since the best-designed mix would be worthless if it cannot be constructed properly. Furthermore, quality control must be carried out during construction to ensure strict adherence to specification and hence uniformly good quality over the entire project duration.

1.4   MAINTENANCE AND REHABILITATION

Starting from day one after construction, a pavement starts deteriorating in quality. Even though properly designed and constructed pavement will not deteriorate so as to cause total failure within its design life, if no maintenance is performed, the entire pavement will become totally worthless at the end of its design life. Furthermore, ingress of water through random openings such as surface cracks, and well-defined openings such as joints, can lead to quick deterioration of the quality of the pavement. The best approach is to regularly perform maintenance operations, similar to any engineered structure or product, through actions such as joint filling, crack repair or filling, or pothole patching. Note that these operations do not specifically increase the design life but do prevent its rapid deterioration. For pavements such as those on highways and airport runways, proper maintenance is critical for the safety of the vehicles and their occupants.

In most cases, once built, a pavement can be "recycled" at the end of its design life almost infinite times, by reusing the existing materials solely or in combination with new materials. Off-course rehabilitation of pavements is a costly process. Hence, for any pavement network, a proper inventory of the condition of different pavements must be kept and utilized effectively to determine the time/order in which the different pavement sections should be rehabilitated. This is important because of two reasons. First, there is never enough funding for rehabilitation of all of the roads in a network at the same time, and, secondly, different roads deteriorate at different rates and hence are in different conditions at a specific time. It is important to "catch" the pavement at the most "appropriate" condition such that the rehabilitation can be done economically—a totally damaged pavement will need too much money to rehabilitate/replace it. This process of keeping an inventory of condition and selecting pavements for rehabilitation/reconstruction is called *pavement management*. This step includes the use of tools for determining the condition of existing pavements. The fastest growing method of such detection is nondestructive testing and evaluation (NDT and NDE). One example of a widely used form of NDT is the falling weight deflectometer (FWD) that works on the principle of evaluating deflections of pavements under known loads and making an assessment of the structure and the condition of the pavement.

1.5   IMPORTANT ISSUES

Traffic keeps on increasing, while the costs of materials and methods keep on climbing, as budgets dwindle everywhere. The only way to keep up and still have good roads is through learning and applying good principles, implementing proven new concepts and technologies, and continually researching for better materials and methods.

Therefore, to summarize:

1. Drainage is needed to drain water away from the pavement.
2. The materials must be evaluated and selected properly so that they can withstand the effects of the traffic and the environment.
3. The mix must be designed properly such that it can withstand traffic and environmental factors.
4. The structure should be designed properly such that it has adequate thickness to resist excessive deformation under traffic and under different environmental conditions.
5. The pavement must be constructed properly such that it has desirable qualities.
6. The pavement must be maintained/managed properly through periodic work, regular testing, and timely rehabilitation.
7. Generation of knowledge through research is critical for ensuring good pavements in the future.

This book aims at providing the relevant principles and practical concepts of pavement engineering for both students and currently practicing engineers.

1.6   FUNCTIONAL REQUIREMENTS

A pavement's primary purpose is to provide a functional surface for a specific transportation need. The basic function is to withstand load, under different seasonal environmental conditions, without deforming or cracking, since either of these distress conditions would reduce the functionality of the pavement. The function of the different layers in the pavement is to spread out the load on the surface and reduce its intensity with depth, such that the pressure on the subgrade is much less than the pressure on the surface, and can be tolerated by the subgrade without undergoing excessive deformation.

A pavement consisting of asphalt mixes (and aggregate and soil layers) only is referred to as a flexible pavement, since the pavement layers deflect under a traffic load. The typical applied concept of a flexible (or asphalt) pavement is that a layered structure (Figure 1.2) with better materials near the top would distribute the load in such a way that the resulting stress in the bottommost layer will be small enough so as to cause no significant deformation of the layer. The bottommost layer is the existing layer or the existing layer modified with some materials. The materials and the thicknesses of the different layers will be such as to be able to withstand the different effects of temperature and moisture due to changes in season in a specific location. The subbase, in addition to providing structural support, may also serve as a platform for constructing the base and prevent the fine materials from the subgrade from contaminating the base layer. If the subgrade is of frost-susceptible material, then the subbase could be made up of non-frost-susceptible materials to prevent frost-related damage.

Rigid (or concrete) pavements, which deflect very little under traffic loads, behave differently than flexible pavements under loads. The wearing layer, which is in contact with the traffic, is a Portland cement concrete slab that ranges in thickness between 5 and 12 inches depending on traffic loading. The thicker pavements are typical of heavier and more repetitive loads. The slab, due to its higher stiffness as characterized by elastic modulus, usually distributes the loading across a large pavement area. This in turn reduces the stresses experienced by the underlying base and

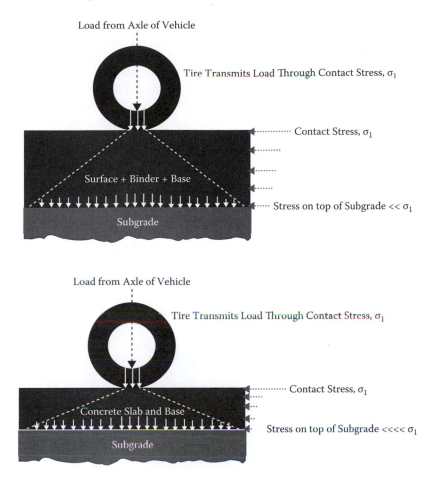

FIGURE 1.2 Flexible Pavement (top) and Rigid Pavement (bottom).
Note: The function of the pavement is to decrease the tire contact stress on the subgrade to a tolerable level.

subgrade layers. Rigid pavements may or may not have a base or subbase layer, and could be placed directly over the subgrade. However, in high-performance pavements, a base or subbase is typically included. Besides providing a wearing course and a contact surface with traffic, the slab provides friction, drainage, smoothness, noise control, and waterproofing for the underlying layers.

1.7 TYPES AND USES OF PAVEMENTS

Under the broad definition, there exist several different types of pavements with specific functions. Let us review the more important types.

1. *Pavement for roads.* There are different types of roads ranging from high-traffic-volume interstate highways to low-traffic-volume local roads. These roads have different types and volumes of traffic. Accordingly, they are designed and constructed in different ways. Thicker pavements are constructed for both heavier and high-volume traffic, of which trucks are the major load applicators. Thinner pavements are for low-volume roads, although in many cases low-volume roads may carry heavy trucks such as log trucks.

2. *Pavement for airports.* Pavements are required in airports in aircraft holding/terminal areas, taxiways, and runways. Just as road pavements are subjected to a wide range of vehicles, airport pavements are subjected to a wide range of aircrafts—a small general aviation airport may have only light aircrafts (< 30,000 lbs., for example), while a large hub/major airport would have major aircrafts. The different pavement areas are runways, taxiways, and aprons. Special considerations such as protection from fuel are required in many cases.

3. *Pavement for parking lots.* Parking lots are essential features in cities and towns, and are commonly found adjacent to business/office buildings, including those near hospitals, schools, and airports.

4. *Loading and unloading areas in ports and other areas.* Heavy-duty pavements are constructed to support equipment and materials unloaded from ships, rail, and trucks. These areas may also require special protections such as those from fuel droppings/spillage.

Figure 1.3 shows a comparison of thickness of different types of pavements. Note that in addition to thicker layers, pavements carrying heavier and more traffic also consist of better materials. For example, a > 300 mm thick full depth asphalt pavement is found in an airport, whereas a 63 mm

>300 mm Thick Airport Pavement

200 mm Thick Highway Pavement

130 mm Thick Low Traffic Volume Pavement

FIGURE 1.3 Thicker Pavements Are Provided for Heavier Loads.

hot mix asphalt layer over a 75 mm aggregate base course is sufficient for a low-traffic-volume local road.

1.8   DIFFERENT FEATURES OF TYPICAL ASPHALT PAVEMENTS

The bottom layer on which the pavement is built is called the *subgrade*. The other layers, upwards in order, are the subbase, base, binder, and surface layers. In many cases the pavement can be a full depth asphalt pavement, in which case all of the layers above the subgrade are composed of hot mix asphalt constructed in several layers (or lifts). In other cases the subbase and base may be combined to form one single layer, as far as materials are concerned, but constructed in multiple layers. And finally, above the surface layer of HMA, there may be a very thin wearing course of specialty material such as open graded friction course (OGFC), to provide better friction and drainage of water.

Generally, the materials become costlier as one moves up from the subgrade layer. On a per unit weight cost basis, therefore, the surface layer is probably the most expensive layer. This is because, as layers at the surface and near it are subjected to the direct application of traffic loads as well as environmental effects, they need to be "fortified" to resist the traffic and environmental effects. These "fortifications" come mostly in the form of higher asphalt binder content, which is the more costly component. Furthermore, stricter specifications on size, shape, and quality of aggregates are also responsible for higher prices of the surface and near surface layers.

A binder layer is provided because it would be difficult to compact the surface and the binder layers in one step, and also the binder layer part can be made up of larger aggregates and with lower asphalt content, thus costing less.

The base layer could be designed as a permeable base layer to let any water inside the pavement flow down to the sides for drainage. This water can be coming from the top—rain and snow through open voids or cracks in the surface—or from groundwater. The surface layer should be compacted to a high density, and cracks should be filled as soon as possible to prevent the flow of rain and snow water inside. On the other hand, groundwater level should be reduced before construction of the pavements by providing interceptor drains. However, there could still be some water coming inside the pavement which needs to be drained out.

The layers in the pavement provide different functions. While the wearing layer in an OGFC may provide friction and channels for quick draining of water, the binder and layers below provide the stiffness to prevent excessive deflection and hence cracking and structural rutting. Good performing surface and binder layers should have sufficient stiffness to prevent rutting related to poor materials. Furthermore, the layer (surface or wearing) exposed to the environment should not be too stiff at low temperatures (during the winter months), to avoid cracking caused by a drop in temperature. There could be a design in which the surface layer is stiff, followed by less stiff binder, base, and subbase layers. There could also be a design in which the bottom parts are less stiff but made with high asphalt contents to prevent cracking, whereas the top parts are stiffer with low asphalt content to prevent rutting, and the surface is sealed with a thin layer of mix with high asphalt content to provide a smooth and dense wearing layer. The point is that the materials and the structure can be engineered to provide a pavement with desirable qualities.

The density of the different layers is crucial for the proper functioning of an asphalt pavement. Soil/aggregate layers are compacted to maximum density using an optimum additive content. In the case of the HMA layers, they are compacted to a specific density (or voids) that has been found to be optimum for that specific mix (for example, 6–8% air voids for dense graded mixes), using a design asphalt content. There is a subtle difference between the two approaches that are used for the selection of the optimum additive content for the layers other than HMA and the one that is used for selecting the design asphalt content for HMA. For non-HMA layers, the optimum additive content is that which produces the maximum density, whereas for HMA layers, the design asphalt content is one that produces 4% void in compacted mixes in the laboratory.

The base course and subbase course are sometimes made up of different materials since the sub-base can be made up of less costly material. If the base is open graded, then the subbase can serve as a filter with fine graded material, between the base course and the subgrade. Even if no outside material is added, before starting construction of the pavement the top 6 inches of the existing subgrade are usually scarified and compacted to a high density using optimum moisture content.

One important requirement of an asphalt pavement is that there must be sufficient bonding between the different layers so as to prevent any slippage at the interface between the layers. Tack coatings of asphalt emulsions, for example, are applied between two HMA layers to achieve this. A tack coat is used between two asphalt layers or between an old PCC layer and a new asphalt layer to form a good bond. The spraying must be uniform and of the right amount per unit area, and if emulsion is used it must reach the desirable condition before the application of the upper layer. A prime coat is used between a granular material layer and an asphalt mix layer. The requirement is that the prime coat material must penetrate the voids and seal them up in the granular material.

In a full depth asphalt pavement, all layers starting from the base layers and up are made of asphalt mix. This type of pavement has a higher initial cost, but if designed and constructed properly, it provides several advantages over layered structures with different materials, such as little or no ingress of moisture and reduction in time of construction. Such pavements may deteriorate only at the surface with time, and hence rehabilitation work will be far less than that in pavements with different materials.

1.9   DIFFERENT FEATURES OF TYPICAL CONCRETE PAVEMENTS

Concrete pavements also consist of different layers (Figure 1.4). Based on the concrete slab, these pavements can be classified into three broad types: jointed plain concrete pavement (JPCP), jointed reinforced concrete pavement (JRCP), and continuously reinforced concrete pavement (CRCP).

Jointed plain concrete pavements are the most common type of rigid pavements. Contraction joints are constructed every 3.7 m (12 ft) to 6.1 m (20 ft) apart to control cracking. In JPCP, no reinforcement is used except for dowel bars at transverse joints or tie bars at longitudinal joints. Changes in temperature and moisture in the concrete will induce stresses that may cause cracking between joints; hence, reinforcing steel or a steel mesh is used to hold these cracks tightly together. Dowel bars are typically used at transverse joints to enhance load transfer, while the reinforcing steel or wire mesh assists in load transfer across cracks.

FIGURE 1.4 Different Layers in a Rigid Pavement.
Courtesy: Wouter Gulden, ACPA-SE Chapter.

Jointed reinforced concrete pavements are similar to JPCP, where contraction joints are used to control cracks in the concrete slab. Structural longitudinal reinforcement is used within the slab, which allows contraction joints to be spaced further apart, approximately 15 m (50 ft).

Continuously reinforced concrete pavement does not use contraction joints but uses continuous reinforcing steel within the slab for crack control. However, hairline cracks do occur in CRCP. Cracks typically appear every 1.1–2.4 m (3.5–8.0 ft) and are held tightly together by the underlying reinforcing steel. The reinforcing steel is present in typically 0.6–0.7% of the cross-sectional area of the concrete, and consists of typically No. 5 or 6 bars of grade 60. According to a 1999 survey, at least 70% of the state highway agencies in the United States use JPCP. About 20% of the states build JRCP, about six or seven state highway agencies build CRCP, and a couple of states reported that they do not build concrete pavements (American Concrete Pavement Association, 1999).

The cost of CRCP is higher than that of JPCP or JRCP. CRCP also takes more attention during construction, and could be difficult to rehabilitate. However, CRCP may prove cost-effective in high-traffic-volume roadways due to its superior long-term performance compared to the other types of concrete pavements.

1.10 RESEARCH ON PAVEMENTS

Continuous research is needed to produce better materials, processes, and designs to make pavements safer, durable, and more cost-effective. Such research can be basic or applied, for solving specific problems. In the United States, the benefits of the Strategic Highway Research Program (SHRP) have been estimated to be a savings of up to $785 million annually (for an investment of $50 million in asphalts), not considering the benefits to the motorist of a reduction in delays and vehicle wear and tear (Halladay, 1998). Benefits of investments in research can be quantified in terms of reduction of road agency costs also. The benefit-to-cost ratios are dependent on the inflation-free discount rate and have been reported to range from 4 to 5. In addition, unquantifiable benefits, such as boosting innovation and improved cooperation of researchers and practitioners, have also been noted (Rose and Bennett, 1994).

Since pavement engineering involves innumerable different disciplines, research potentials are abundant in this field. In a broad way, research on pavement engineering can be grouped into three major categories—materials, structures, and construction related. Material-related research work involves the development of innovative materials, mixtures, and additives, and chemical and material characterization. Structures-related research involves the development and use of advanced techniques, sensors, equipment, and models for understanding the pavement structure and its different components. Research related to construction includes activities such as the development of new procedures and equipment for faster and better construction, construction under difficult conditions, and/or the utilization of recycling/marginal materials, as well as those related to better quality control and specifications. Some of the emerging areas of research include those related to the environment, maintenance materials, microscale characterization, the use of nondestructive and accelerated testing, and recycling.

Research on pavements is conducted by industry and government organizations all over the world. In the United States, pavement research is conducted at local, state, and federal levels. The National Cooperative Highway Research Program (NCHRP), which brings together the AASHTO, FHWA, U.S. Department of Transportation (USDOT), and National Academies, is the primary sponsor of national-level pavement research in the United States. Similarly, the Federal Aviation Administration sponsors pavement-related research through the Asphalt Airfield Pavement Technology Programs (AAPTP) and the Innovative Pavement Research Foundation (IPRF). The FAA and FHWA also conduct research on material characterization as well as accelerated loading and testing, through both in-house as well as academia-based centers.

Research on pavements requires basic knowledge of principles, formulation of problems, development and application of appropriate theory, utilization of appropriate equipment and tools, and proper inference of results. A thorough literature review is the key to successful research. The Transportation Research Information System (TRIS; http://ntlsearch.bts.gov/tris/index.do) is a very powerful database that can be accessed and queried online to conduct a literature review on pavement engineering. Information is also available from state DOT websites (see the following Google site: http://www.google.com/coop/cse?cx=006511338351663161139%3Acnk1qdck0dc), as well as industry websites such as those of the National Asphalt Pavement Association (www.hotmix.org) and the Portland Cement Association (www.cement.org).

A number of general and specialty conferences are held every year or on alternate years all over the world on pavement-related research. One of the biggest assets in pavement research (as in any other case) is the teamwork/cooperation of industry/academic and government agencies, as well as the cooperation between researchers in different countries. Researchers and students should take opportunities to learn about new things as well as to network with fellow students and researchers in different schools/industries and countries. So wide is the area of pavement research that persons from a wide variety of backgrounds can find their niche areas of research. However, in general, knowledge on numerical techniques, mechanics, basic physics, and chemistry is essential.

QUESTIONS

1. What are the main functions of an interstate highway pavement and an airport taxiway pavement?
2. Why are these pavements generally built in layers?
3. Why are not all pavements very thick or very thin?
4. What happens if fuel such as kerosene or diesel drops in large amounts on asphalt pavements?
5. Why are roads and airport pavements important?
6. What are the steps in building a stable and long-lasting pavement?
7. Why is maintenance of pavements important?
8. How can roads be improved?
9. While designing a road pavement, are both weight and number of vehicles important?
10. How is the repetitive loading taken into account?
11. What are the steps in designing and constructing a road?
12. How can the road be rehabilitated in the most cost-effective way?
13. Is there a difference between airport and road pavement?
14. Are there special requirements for airport pavements?
15. How can roads be made safer for travel?
16. Can pavement be designed and built in an environment-friendly way?
17. What is the best way to rehabilitate an old pavement?
18. What are the different types of rigid pavements?
19. What are the advantages and disadvantages of the different types of rigid pavements?
20. What are the considerations for the selection of the type of pavement—flexible or rigid?
21. Why is research on pavements important?
22. What are the general areas of research on pavements?
23. Conduct an Internet search and list the primary pavement-related research being carried out in the different parts of the world.

2 Principles of Mix and Structural Design, and Construction of Asphalt Pavement

2.1 OVERVIEW

Flexible/asphalt pavements are generally made up of layers. Each layer is made up of a combination of materials—generally aggregate, binder, and any other additive. The process of designing the optimum combination of materials is called *mix design*. Once the structural properties of each layer are known (or estimated accurately), the optimum structure of the pavement is determined, that is, the thickness of each layer is determined. This process is called *structural design*, which also involves structural analysis—determination of stresses, strains, and deformations in different layers.

Note that mix design and structural design are dependent on each other, for example, a structurally better mix may only require a thin layer of the material to sustain a given load. However, the mix design is conducted in such a way that locally available (and hence relatively low-cost) materials are utilized for economic reasons. Hence this can be viewed as one of the most important "constraints" in the overall design of an asphalt pavement. The next logical consideration is that of the locally available materials; one would utilize the relatively less costly materials more (that is, in thicker layers) than the more costly components. Hence, for a given condition of traffic and environment, and for a given period of time for which the pavement is expected to be usable (design period), it is the most judicious combination of materials and thicknesses that results in the optimum design of a pavement structure. In this process, the two factors, material quality *and* material (as well as construction) cost, are balanced. Note that a pavement with very cheap materials would probably be able to sustain traffic loads in the short term but would require frequent maintenance, which could be costly in terms of both money and user delay.

2.2 TRAFFIC AND LOAD DISTRIBUTION CONCEPT

One of the basic purposes of providing a good mix and structural design is that the pavement can withstand traffic loads without deforming or deteriorating to a degree that it becomes unusable within the design period. Heavy freight truck traffic is the main contributor of load on highways, since their weight is significantly higher than that of cars, small pickup trucks, or other passenger vehicles. Similarly, the load from big multiengine aircrafts is significantly higher than that coming from small single aircrafts on airport pavements.

The load (from a vehicle/aircraft) is transferred to the pavement through load-bearing axles and pressurized tires. The resulting pressure or stress on the pavement, at any depth, is dependent on many factors, such as total load, the number of axles and tires, and the condition of the tires. The stress on the surface of the pavement just below any tire is concentrated in the tire contact area, often assumed as a point for structural analysis. Theoretical concepts and validations have shown that this

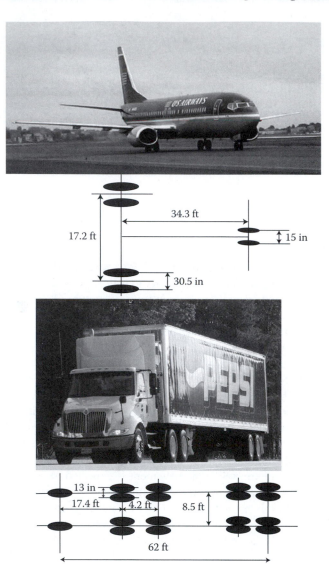

FIGURE 2.1 Gear/Axle Layout of a Boeing 737 Aircraft and a Semi-Tractor Trailer.

stress gets distributed in an inverted V form from the surface downwards. In other words, the stress intensity decreases along the depth of the pavement. However, note that if the stress on the surface is coming from two tires placed, say, 13 inches center to center (or even axles placed close to each other), due to the stress distribution, at some point below the surface, the two stresses might overlap, causing a stress at that point to be higher than each of the two stresses on the surface. Obviously, whether or where (in the pavement) this would happen depends on the tire/axle configuration and the thickness of the pavement. Since the pavement needs to sustain the stresses, one needs to take into consideration not only the gross load but also relative positions of the axles and the gears during the structural analysis part. Figure 2.1 shows the gear/axle layout of a Boeing 737 aircraft and a semi-tractor trailer, and Figure 2.2 illustrates the concept of distribution of stress in the inverted V form.

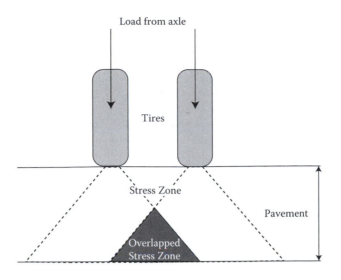

FIGURE 2.2 Concept of Distribution of Stress in the Inverted V Form.

2.3 MATERIALS AND LAYERS

The layer on the surface of the pavement has to withstand the maximum stress and bear the changing conditions of the environment. Therefore, this surface layer usually consists of the "best" and most costly materials. Also, this layer is always "bound"—that is, mixed with a "binder," in this case asphalt binder—to prevent raveling of materials under traffic, as well as to provide a dense surface to prevent ingress of water, unless it is an open graded friction course. Therefore, the surface layer has two major components—asphalt binder and aggregates, with relatively small amounts of additive, if any. The mixture (referred to as *mix*) of this layer needs to be designed properly to give it adequate stiffness, strength, and durability. This mix is usually prepared by combining hot aggregates and asphalt binder (hot mix asphalt, or HMA).

The next layer could be either binder or base. Generally, the binder layer is almost similar to the surface layer, except that it may consist of aggregates of larger size. The base may be made up of a bound layer or an unbound aggregate layer. The base layer has to be sufficiently strong in shear as well as bearing capacity, but need not be as good as the binder and surface course, since the stresses at this level are much lower. However, the base needs to be stiff enough to provide an overall stiffness to the pavement structure such that it does not deflect too much under load. How much is too much? That depends on the specific mode of failure we are concerned with, but it would suffice to say here that if the pavement can sustain repetitive recoverable deflection without cracking or having permanent deformation within the design period, then the deflection is acceptable. One of the most important criteria for designing the base course is to provide adequate density. If it is a bound layer, it may be with aggregates and asphalt binder or Portland cement, and if it is an unbound layer, it is most likely with aggregates only, with water used as an additive for aiding compaction during construction. The function of the subbase is almost similar to that of the base, except that it could be of lower quality materials, since the stress is even less at this level. Generally, it is made up of aggregates only.

The subgrade is made up of the existing soil or the soil mixed with some additives to enhance its properties. It serves as the foundation of the pavement and should be of such quality as to resist excessive deflection under load. One important requirement is to make sure that the subgrade is capable of resisting the detrimental effects of water. Generally, the presence of excessive amounts of fine materials such as clays and silts is not desirable.

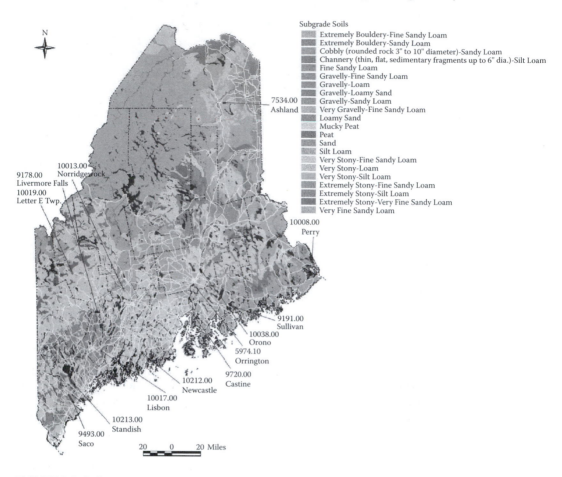

FIGURE 2.3 Soil Type Map of the State of Maine, United States.
Courtesy: Brian Marquis, Maine Department of Transportation.

2.3.1 Soils

Soils or aggregates, which can be defined as processed soils, are present in every layer, either mainly by themselves, such as in unbound layers, or as a major component, such as in a bound layer. In the subgrade, the soil generally consists of unprocessed native soil, compacted to a maximum achievable density, using an optimum moisture content and roller compaction.

If the soil is incapable of resisting the detrimental effects of the environment or is too weak in shear strength and stiffness, it is generally "stabilized" with additives and/or adequate compaction. Stabilization is the process of improving the quality of the soil (in most cases, reducing its susceptibility to the effect of moisture) with the addition of some other materials (known as *additives*).

Soils can vary widely over a region or a state (for example, see Figure 2.3). To determine whether the native soil is adequate for the proposed pavement (and if not, what kind of stabilization process is needed), one needs to test and characterize the soil. These tests are mostly conducted to evaluate its physical properties and the effect of moisture on some of those properties. Strength/stiffness under different stress levels is also determined. Based on the results of these tests, one decides whether the soil can be compacted and used as it is, or should be mixed with an additive and then compacted. Note that some compaction is almost always necessary to improve its bearing capacity to sustain the stress from the weight of the construction equipment and layers of the pavements. In general, since the subgrade is the lowest layer, the stresses from these two sources are much greater than the stress from a vehicle. One important factor is the consideration of the thickness of the subgrade layer. That is, to be adequate, the subgrade must have desirable properties throughout a specific depth.

2.3.2 AGGREGATES

Mostly naturally occurring aggregates are used in pavements. Aggregates are soil/rock, processed to produce materials of a certain range of sizes (and other properties). *Aggregates* generally refer to soil particles whose major portion is of a size greater than 0.075 mm. These particles are tested for their physical properties and characterized to determine their suitability. Unlike existing soil in the subgrade, the aggregates used in bound courses are actually the result of a combination of different proportions of different sized particles—the combination is done as part of the mix design process.

Therefore, not only each aggregate particle but also the blend of the different size particles should be of desirable quality. Furthermore, such aggregates, when used in combination with asphalt, must also be of such (physical as well as chemical) quality as to resist cracking, and retain the coating of asphalt binder under the action of traffic and harsh environmental conditions, such as repetitive pore pressure due to the presence of moisture. Thus the characterization of aggregate involves a host of testing and evaluation steps that range from those used for determination of size and shape to the action of moisture under different conditions.

2.3.3 ASPHALT

Paving asphalt, also known as bitumen, is a product that is mostly obtained from crude petroleum through a series of refining steps. Asphalt can also be used in other forms, such as in emulsified form in asphalt emulsions. With reference to paving mixes, asphalt is commonly referred to as *asphalt binder*, since its basic purpose is to "bind" the aggregate particles together. In general, asphalt binders are semisolid or solid at room temperature and liquid at a relatively high temperature. Different "grades" of asphalt binders are produced by changing the source (of crude petroleum) as well as the refining conditions to meet different paving demands, mainly arising from differences in environmental and traffic conditions. Asphalt binders are classified into different "grades" with the help of characterization tests.

The properties of asphalt binders are affected significantly by the temperature and time of loading. Thus, characterization of asphalt binders is required to determine the effect of temperature and stress on the relevant engineering properties—mainly for two basic applications. The first application is the actual mix design process, where the asphalt should be of such grade as to produce a mix with aggregates that would be able to sustain the effects of traffic and the environment for the design life of the pavement. The second application of the characterization test results is for the construction process, where the asphalt binder must be transported from a refinery, stored in a mixing plant, pumped through pipes and mixed with aggregates, transported to the job site, and laid down and compacted at specific temperatures that would allow the completion of all those steps in the most convenient way. Therefore, the characterization tests of asphalt binders involve similar tests at multiple temperatures, using specific loading rates.

2.4 ENVIRONMENT

Since pavements are exposed to the environment, proper consideration of environmental factors must be made. There are three major factors—temperature, moisture, and the effect of temperature on moisture. These factors affect the properties of the pavement materials significantly. For example, a higher temperature decreases the stiffness of the asphalt materials, whereas a lower temperature increases its stiffness. A higher moisture content would reduce the strength/stiffness of unbound materials, whereas a frozen moisture would increase the overall stiffness of the pavement. In addition to one time effect, the environmental factors affect the properties of the pavement

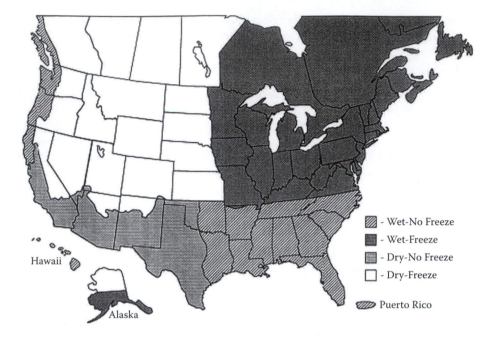

FIGURE 2.4 Freeze-Thaw Regions in the United States.
Source: From Hanna et al., 1994, reprinted with kind permission of the Transportation Research Board.

materials through their repetitive/cyclic nature. Frost-susceptible soils, with enough capacity to draw capillary water, can form ice lenses during the winter and result in frost heaves, and as the ice lenses melt in spring (spring thaw), voids form underneath the top layers of the pavement, weakening it and ultimately resulting in destruction of the pavement under traffic load. Figure 2.4 shows the different regions of the United States with respect to freeze thaw.

To consider these factors in the appropriate way, three steps are taken. First, the effects of the different factors on the pavement materials are evaluated and those materials which are proven to be susceptible to the detrimental effects are eliminated from consideration. Second, to use available materials, and to avoid excessive costs, if it is not possible to eliminate all such materials, the available materials are modified with additives to make them resistant to the detrimental effects of the environment. Third, during the structural analysis and design, the effect of the environmental conditions on the relevant parameters (such as stiffness) is duly taken into consideration, and any pavement layer/material needed to eliminate the detrimental effect of the environment (such as frost action) is provided.

One of the important pavement-related issues directly related to the environment is providing an adequate surface and subsurface drainage system in the pavement. No matter what, some rain/snow water will always find its way into the pavement, and some groundwater will always seep inside the pavement. There should be a proper system to divert this water away from the pavement into side ditches/drains.

2.5   MIX DESIGN

The basic purpose of mix design is to select and combine the different components in such a way as to result in a mix that has the most optimum levels of all relevant properties so as to produce the best mix possible, within the constraints of available materials and funds available to produce the mix.

Therefore, the process starts with the selection of the most appropriate standard specification (or the development of one, if needed). The specification should clearly spell out the requirements for

Consider Traffic and Environmental Factors

Select Appropriate Materials

Prepare Mixes with Different Percentages of Asphalt Binders
Compact samples Using Appropriate Laboratory Compactor
to Simulate Field Compaction

Test Samples for Volumetric Properties

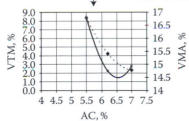

Analyze: Plot Volumetric Properties against Asphalt Content
and Select Optimum Asphalt Content on the Basis
of Desirable Volumetric Properties

Compact Samples at Optimum Asphalt Content and
Check for Mechanical Properties Such as Resistance Against
Moisture Damage After Appropriate Conditioning

Finalize Design Asphalt Content and Recommend Job Mix Formula

1. Aggregate Information
2. Asphalt Amount and Type
3. Temperature Information
4. Limits and Target Values

FIGURE 2.5 Key Steps in Asphalt Mix Design.

the materials in terms of properties to be determined by specific tests, details of the tests, and acceptable/not acceptable levels of results that are to be used. Then, according to the specification, aggregates, blends of aggregates, and asphalt binder are selected. Next, different blends of aggregates are tested for their properties. Then a selected blend is tested with different asphalt contents, and the optimum asphalt content is determined. Finally, the mix (design mix) with the selected gradation and the optimum asphalt content is tested at appropriate levels of controllable properties (such as air voids or density) to make sure that it is resistant to the detrimental effects of the environment. Furthermore, one needs to estimate the relevant structural design parameters for the designed mix.

Depending on the importance of the project, and the levels of sophistication employed in testing and evaluation of the materials, the mix design process can range from a simple process to a fairly complex one. In any case, it is always a trial-and-error experimental process, involving the testing of multiple samples for different test properties. Figure 2.5 shows the key steps in the mix design of an asphalt mixture.

2.6 STRUCTURAL DESIGN

Structural design of a pavement is conducted to determine the thickness of the different layers to prevent the occurrence of problems or distresses in the pavement due to traffic loading and the environment. Such distresses may include rutting or depressions in the wheel path and cracking. Figure 2.6 shows the effect of different stresses and strains on distress or problems in asphalt pavements.

The desirable thickness of the soil layers is determined from the consideration of their stiffness/resistance to deformation properties (such as with the use of the test property California Bearing Ratio, or CBR), as well as the change in the properties that can be expected due to a change in the environmental conditions, primarily moisture and the effect of temperature on moisture (Figure 2.7).

A Flexible Pavement "Flexes" Under a Load; Repeated Loadings
Cause Repeated Compressive Strains As Well As Tensile Strains

FIGURE 2.6 Major Distress-Causing Responses in a "Flexible" or Asphalt Mix Pavement.

Test Samples of Materials As Designed in Mix Design

Determine Pavement Material Response and Effect of Temperature and Loading Period

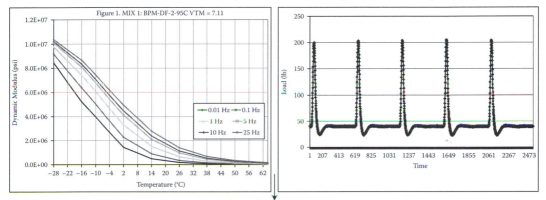

Consider Traffic and Environmental Factors for Project

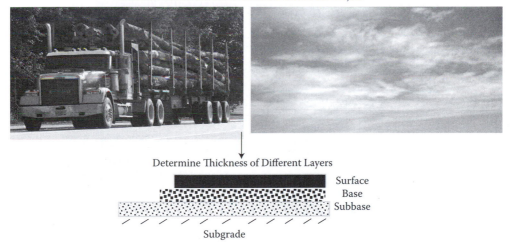

Determine Thickness of Different Layers

FIGURE 2.7 Basic Steps in Structural Design.

However, note that in a pavement structure, the stress in the subgrade or subbase/base is governed by the imposed stress (weight of layers above them and vehicle load) as well as the thickness of the layers above them. Therefore, if the subgrade/subbase or base is known to be relatively weak, it should be protected by a proper selection of binder/surface mix and layer thickness. It follows that the structural design process in pavements is an iterative process that balances the availability of materials and the cost of the total pavement structure.

The basic purpose of the structural design process is to combine the different layers in such a way so as to result in the most cost-effective functional pavement structure. This can be achieved by primarily two different techniques: (1) by using empirical methods—that is, charts and equations developed from experimental studies carried out with a set of traffic, environment, and pavements; or (2) by using a mechanistic method, in which concepts of mechanics are used to predict responses and performance of the pavement. Note that a purely mechanistic approach is not possible at this time—the responses can be predicted by employing concepts of mechanics, but the performance has to be predicted by empirical models. Hence, it is more appropriate to say that pavements can be designed either by using the empirical approach or by using the mechanistic-empirical approach (ME).

There are obviously differences in steps and levels of sophistication between the two processes. In the empirical approach, considering the traffic and environment, and the available subgrade, an appropriate "chart" may be used to "read off" the required base course and/or the asphalt mix layer thicknesses. Such methods can be simple or fairly sophisticated, utilizing different factors such as axle numbers, drainage parameters, concepts of reliability, and performance-related properties of the pavement. Figure 2.8 shows an example of an empirical procedure for designing airport

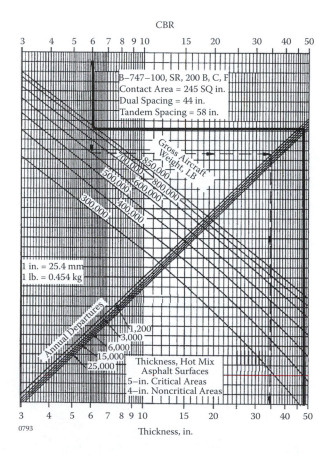

FIGURE 2.8 Example of an Empirical Procedure for Designing Airport Pavements for the Boeing 747 Aircraft.
Source: Adapted from Federal Aviation Administration, *FAA Advisory Circular No. AC 5320-6D* (Washington, D.C.: U.S. Department of Transportation, 1995), Figure 3–7.

pavements for the Boeing 747 aircraft. Note that in this case, the CBR of the soil is used, along with aircraft information, to determine the thickness of the pavement (subbase + base + surface) required on the subgrade.

The ME process is more of an iterative nature, and can be divided into three basic steps. It starts with the consideration of traffic, design period, projected total traffic (expected over the design period), and environmental conditions. In the second step, considering the load as well as environment-related properties of the available materials, a trial structure with different layers is designed, and stresses/strain/deflections at different critical levels are determined. Finally, the stresses and strains are utilized in the empirical models to predict the performance of the pavement over the design period. If the performance is not found to be satisfactory, the process loops back to the second step and starts with a different trial structure. This process continues until a pavement structure with satisfactory performance over the design period is obtained.

Inherent in the design processes are the very important considerations of economics and the cost of the pavement over its entire "life" (referred to as *life cycle cost*, or LCC). Considerations of both initial cost and regular maintenance and salvage costs must be made during the design process, particularly for large projects (for example, for pavement projects costing more than $1 million).

2.7 LINK BETWEEN MIX AND STRUCTURAL DESIGN

Mix design is generally conducted on the basis of volumetric properties. This process produces the job mix formula (JMF), according to which the mix to be used for paving is produced. This mix is tested for its structural properties, and, using the structural properties, the structural performance of the pavement with the specific mix is predicted. Therefore, it is important to note that the structural performance is very much dependent on the mix design, even though structural properties may not be generally considered explicitly in the mix design process.

An example is shown in Figure 2.9. It shows the effect of two parameters on the rutting performance of a mix under medium-volume traffic. The two parameters are the type of asphalt and the modulus of the base. The type of binder and the asphalt content of the base are selected during mix design. The years-to-failure chart shows that a "stiffer" asphalt binder takes a longer time to fail by rutting. That is, it has a higher resistance to structural rutting. At the same time, a mix with a high modulus base shows more resistance to structural rutting than the one with a low modulus.

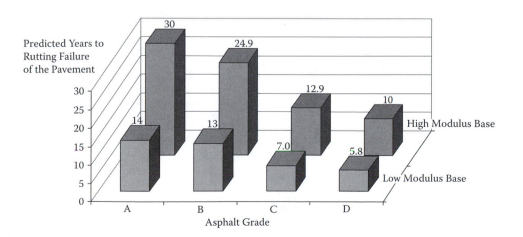

FIGURE 2.9 Effects of Mix and Structural Design on Performance of Pavement.

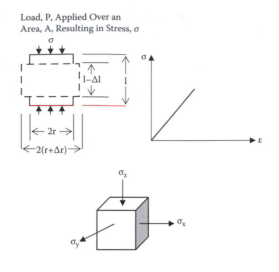

FIGURE 2.10 Schematics of Stress and Strains.

2.8   THEORETICAL CONSIDERATIONS FOR STRUCTURAL DESIGN

Before proceeding to structural design, it is necessary to review some of the basic mechanistic techniques that enable us to compute stresses and strains due to loads.

2.8.1   HOOKE'S THEORY OF ELASTICITY

Ratio of stress over strain is a constant, Young's Modulus, E (modulus of elasticity). This constant
relates stress to strain: $\sigma = E\varepsilon$. Poisson's ratio (μ) is the ratio of radial and longitudinal strains.
Figure 2.10 explains the different parameters:

$$\varepsilon_l = \frac{\Delta l}{l}; \quad \varepsilon_r = \frac{\Delta r}{r}; \quad \sigma = \frac{P}{A}; \quad E = \frac{\sigma}{\varepsilon_l}; \quad \mu = \frac{\varepsilon_r}{\varepsilon_l}$$

For a three-dimensional case,

$$E\varepsilon_x = \sigma_x - \mu\sigma_y - \mu\sigma_z$$
$$E\varepsilon_y = \sigma_y - \mu\sigma_x - \mu\sigma_z$$
$$E\varepsilon_z = \sigma_z - \mu\sigma_x - \mu\sigma_y$$

Note that for pavement materials, E and μ are not constants but are affected by different factors such
as temperature, moisture content, and stress conditions.

Example Problem

An elastic bar is subjected to the force shown below. Determine the total elongation of the bar.
Area through which load is applied = 113 sq inch; modulus = 22,000 psi.

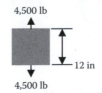

Stress: σ = load divided by cross-sectional area = 4,500/113 = 39.8 psi.

Strain: ε = elongation divided by original length = $\Delta/12$.

By definition, modulus of elasticity, E = stress / strain = σ / ε.

Therefore, $\frac{39.8}{\Delta/12} = 22,000$.

Or, $\Delta = 0.022$ inch.

What happens if E was exactly double, that is, 44,000 psi?

$\Delta = 0.011$ inch (half of the original elongation).

2.8.2   Boussinesq's Method

This method provides a way of determination of stresses, strains, and deflections of homogeneous, isotropic, linear elastic, and semi-infinite space under a point load. Consider the schematic shown in Figure 2.11.

Vertical stress:

$$\sigma_z = \frac{-3Pz^3}{2\pi R^5}$$

Radial stress:

$$\sigma_{rz} = \frac{P(1+\mu)}{2\pi R^2}\left[\frac{-3r^2z}{R^3} + \frac{(1-2\mu)R}{R+z}\right]$$

Vertical deformation below the surface:

$$u_{zr} = \frac{P(1+\mu)}{2\pi E}\left[\frac{2(1-\mu)}{R} + \frac{z^2}{R^3}\right]$$

Surface (that is, at z = 0) vertical deflection:

$$u_r = \frac{(1-\mu^2)P}{\pi ER}$$

where:

r = radial distance from the point load

z = depth

$$R^2 = r^2 + z^2 = x^2 + y^2 + z^2$$

Deflection along the centerline of a rigid circular load, with radius, a:

$$u_r = \frac{(1-\mu^2)P}{2Ea}.$$

For a distributed load under a circular area, the responses can be found out by numerical integration. Along the centerline of the load,

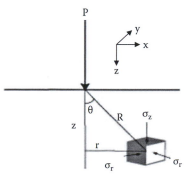

FIGURE 2.11 Schematic for Boussinesq's Method.

Vertical stress at depth z:

$$\sigma_z = \sigma_o \left[1 - \frac{1}{\{1 + (a/z)^2\}^{3/2}} \right]$$

Radial and tangential stress at depth z:

$$\sigma_r = \sigma_t = \sigma_0 \left[\frac{1 + 2\mu}{2} - \frac{1 + \mu}{\sqrt{1 + \left(\frac{a}{z}\right)^2}} + \frac{1}{2\left[\{1 + \left(\frac{a}{z}\right)^2\}^{3/2}\right]} \right]$$

Vertical strain at depth z:

$$\varepsilon_z = \frac{(1 + \mu)\sigma_0}{E} \left[\frac{z/a}{\{1 + (z/a)^2\}^{3/2}} - (1 - 2\mu)\left\{ \frac{z/a}{\sqrt{1 + (z/a)^2}} - 1 \right\} \right]$$

Deflection at depth z:

$$d_z = \frac{(1 + \mu)\sigma_0 a}{E} \left[\frac{1}{\sqrt{1 + (z/a)^2}} + (1 - 2\mu)\left\{ \sqrt{1 + (z/a)^2} - z/a \right\} \right]$$

where:
  σ_0 = stress on the surface
  E = elastic modulus
  a = radius of the circular area of the load
  z = depth below pavement surface
  μ = Poisson's ratio

For flexible pavement design, layers are often simplified as homogeneous, isotropic, linear elastic materials. However, in reality the traffic load is not applied at a point but spread over an area; for practical purposes, the equations for a point load can be used for a distributed load at distances equal to or greater than 2r from the centerline of the load. From Boussinesq's equations, note that modulus and deflection are inversely related. When the modular ratio of the pavement (that is, the combination of all layers above the subgrade) and the subgrade is close to unity, the equations for one-layer elastic solutions for subgrade stress, strain, and deflection are applicable. Tests show that in most cases, stresses and deflections predicted by this method are larger than measured values.

2.8.3 APPLICATION, EXTENSION, AND REFINEMENT OF BOUSSINESQ'S METHOD

Foster and Ahlvin developed solutions for vertical and horizontal stress and vertical elastic strains due to circular loaded plate, for $\mu = 0.5$. Ahlvin and Ulery refined Foster and Ahlvin's charts to develop solutions for complete pattern of stress, strain, and deflection at any point in the homogeneous mass for any value of μ. The solutions are as follows. Examples of coefficients in equations are presented in Table 2.1.

$$\sigma_z = p[A + B]$$

$$\sigma_r = p[2\mu A + C + (1 - 2\mu)F]$$

TABLE 2.1
Example Values of Coefficients

r/a	z/a	A	B	C	D	E	F	G	H
0	0	1	0	0	0	0.5	0.5	0	2
	0.1	0.9005	0.09852	–0.04926	0.04296	0.45025	0.45025	0	1.80998
	0.5	0.55279	0.35777	–0.17889	0.17889	0.27639	0.27639	0	1.23607
	1	0.29289	0.35355	–0.17678	0.17678	0.14645	0.14645	0	0.82843
	2	0.10557	0.17889	–0.08944	0.08944	0.05279	0.05279	0	0.47214
	5	0.01942	0.03772	–0.01886	0.01886	0.00971	0.00971	0	0.19805
0.2	0	1	0	0	0	0.5	0.5	0	1.97987
	0.1	0.89748	0.10140	–0.05142	0.04998	0.44949	0.44794	0.00315	1.79018
	0.5	0.54403	0.35752	–0.17835	0.17917	0.27407	0.26997	0.04429	1.22176
	1	0.28763	0.34553	–0.17050	0.17503	0.14483	0.14280	0.05266	0.85005
	2	0.10453	0.18144	–0.08491	0.09080	0.05105	0.05348	0.02102	0.47022
	5	0.01938	0.03760	–0.01810	0.01950	0.00927	0.01011	0.00214	0.19785
1	0	0.5	0	0	0	0.5	0	0.31831	1.27319
	0.1	0.43015	0.05388	0.02247	0.07635	0.39198	0.03817	0.31405	1.18107
	0.5	0.28156	0.13591	0.00483	0.14074	0.21119	0.07037	0.26216	0.90298
	1	0.17868	0.15355	–0.02843	0.12513	0.11611	0.06256	0.18198	0.67769
	2	0.08269	0.11331	–0.04144	0.07187	0.04675	0.03593	0.07738	0.43202
	5	0.01835	0.03384	–0.01568	0.01816	0.00929	0.00905	0.00992	0.19455
2	0	0	0	0	0	0.12500	–0.12500	0	0.51671
	0.1	0.00856	–0.00845	0.01536	0.00691	0.11806	–0.10950	0.00159	0.51627
	0.5	0.03701	–0.02651	0.05690	0.03039	0.09180	–0.05479	0.03033	0.49728
	1	0.05185	–0.01005	0.05429	0.04456	0.06552	–0.01367	0.06434	0.45122
	2	0.04496	0.02836	0.01267	0.04103	0.03454	0.01043	0.06275	0.35054
	5	0.01573	0.02474	–0.00939	0.01535	0.00873	0.00700	0.01551	0.18450

$$\sigma_t = p[2\mu A + D + (1 - 2\mu)E]$$

$$\Delta_z = \frac{p(1+\mu)a}{E_1}\left[\frac{z}{a}A + (1-\mu)H\right]$$

$$\tau_{rz} = \tau_{zr} = PG$$

$$\theta = \sigma_z + \sigma_r + \sigma_t$$

$$\varepsilon_z = \frac{p(1+\mu)}{E_1}[(1-2\mu)A + B]$$

$$\varepsilon_\theta = \varepsilon_z + \varepsilon_r + \varepsilon_t$$

$$\tau_{zt} = \tau_{tz} = 0$$

$$\varepsilon_r = \frac{p(1+\mu)}{E_1}[(1-2\mu)F + C]$$

$$\sigma_{1,2,3} = \frac{(\sigma_z + \sigma_r) \pm \sqrt{(\sigma_z + \sigma_r)^2 + (2\tau_{rz})^2}}{2}$$

$$\varepsilon_t = \frac{p(1+\mu)}{E_1}[(1-2\mu)E - D]$$

$$\tau_{max} = \frac{(\sigma_1 - \sigma_3)}{2}$$

2.8.4 Burmister's Method for Two-Layer Systems

Burmister developed solutions for stresses and displacements in two-layer systems: $\mu = 0.5$ for each layer, in homogeneous, isotropic, and elastic materials, with the surface layer infinite in extent in the lateral direction but of finite depth, the underlying layer infinite in both horizontal and vertical directions, layers in continuous contact, and the surface layer free of shearing and normal stress outside loaded area.

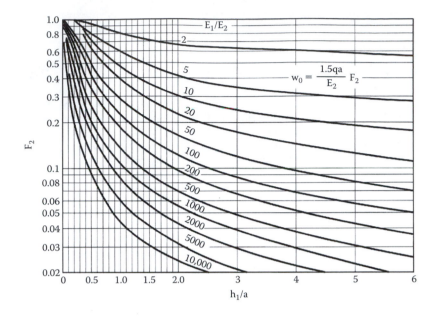

FIGURE 2.12 Values of F_2.
Source: From Huang, Yang H., Pavement Analysis and Design, 2nd edition, © 2004, pg. 60. Reprinted by permission of Pearson Education, Inc., Upper Saddle River, NJ.

Total surface deflection Δ_t for a two-layer system:
Flexible plate:

$$\Delta_t = 1.5 \frac{pa}{E_2} F_2$$

Rigid plate:

$$\Delta_t = 1.18 \frac{pa}{E_2} F_2$$

where:
  p = unit load on circular plate
  a = radius of plate
  E_2 = modulus of elasticity of lower layer
  F_2 = dimensionless factor depending on the ratio of moduli of elasticity of the subgrade and pavement as well as the depth-to-radius ratio, as shown in Figure 2.12.

Note that stress and deflections are influenced by the ratio of the modulus of the pavement (everything above subgrade) and subgrade, there is significant effect of the layers above the subgrade, and there is significant difference in stress gradients obtained from Boussinesq's and Burmister's theory.

2.8.5 Odemark's Method Equivalent Layers

Odemark developed a method whose principle is to transform a system consisting of layers with different moduli into an equivalent system where all the layers have the same modulus, and on which Boussinesq's equation may be used. This method is known as the method of equivalent thickness (MET).

$$\text{Stiffness} = \frac{IE}{(1 - \mu^2)} \quad \text{where, } I = \text{moment of inertia}$$

For stiffness to remain constant, this expression must remain constant, from which we can say the following:

$$\frac{h_e^3 E_2}{1 - \mu_2^2} = \frac{h_1^3 E_1}{1 - \mu_1^2}$$

$$h_e = h_1 \left[\frac{E_1}{E_2 \left(\frac{1-\mu_2^2}{1-\mu_1^2} \right)} \right]^{1/3}$$

where h_e = equivalent thickness. Note that this is an approximate method, and a correction factor is used to obtain a better agreement with elastic theory.

In many cases, Poisson's ratio may as well be assumed to be the same for all materials. Then the above equation becomes the following:

$$h_e = fh_1 \left[\frac{E_1}{E_2} \right]^{1/3}$$

$$[\mu_1 = \mu_2]$$

where f = correction factor that depends on layer thickness, modulus ratio, Poisson's ratio, and number of layers in the structure. Frequently used values of f are as follows: 0.9 for a two-layer system, and 0.8 for a multilayer system, except for the first interface, where it is 1.0. A factor, f, of $1.1 (\frac{a}{h_1})^{0.3}$ can be used for the first interface when $a > h_1$.

For the multilayer system shown above, the equivalent thickness of the upper $n - 1$ layers with respect to the modulus of the layer n may be calculated as follows:

$$h_{e,n} = f * \sum_{i=1}^{n-1} \left\{ h_i * \left[\frac{E_i}{E_n} \right] \right\}^{1/3}$$

Note that to use the MET, moduli should be decreasing with depth, preferably by a factor of at least 2 between consecutive layers, and the equivalent thickness of a layer should preferably be larger than the radius of the loaded area.

2.8.6 FOX AND ACUM AND FOX'S SOLUTIONS

Fox and Acum and Fox developed exact solutions for boundary stresses in the centerline of a circular, uniformly distributed load acting on the surface of a three-layer half-space, $\mu = 0.5$ for all layers. They provided a tabular summary of normal and radial stresses in three-layer systems at the intersection of the plate axis with the layer interfaces; Jones and Pattie expanded Fox and Acum and Fox's solutions to a wider range.

To find stresses, four parameters are needed:

$$K_1 = \frac{E_1}{E_2} \qquad A = \frac{a}{h_2}$$

$$K_2 = \frac{E_2}{E_3} \qquad H = \frac{h_1}{h_2}$$

The procedure to calculate the different stresses is as follows. For vertical stresses, obtain, from graphs, ZZ1 and ZZ2 (see Figure 2.13 and Figure 2.14 for examples of values) and use the

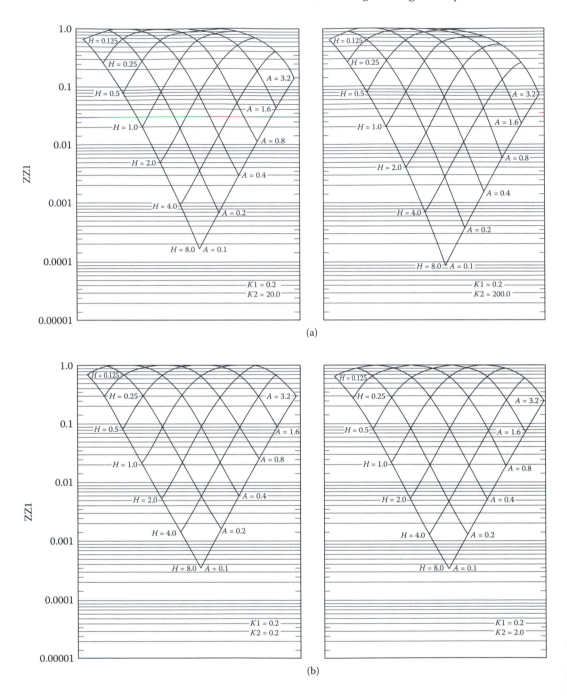

FIGURE 2.13 Examples of Values of ZZ1.
Source: From Yoder, E.J., and M.W. Witczak (1975) Principles of Pavement Design, 2nd edition © 1975.
Reprinted with permission of John Wiley & Sons, Inc.

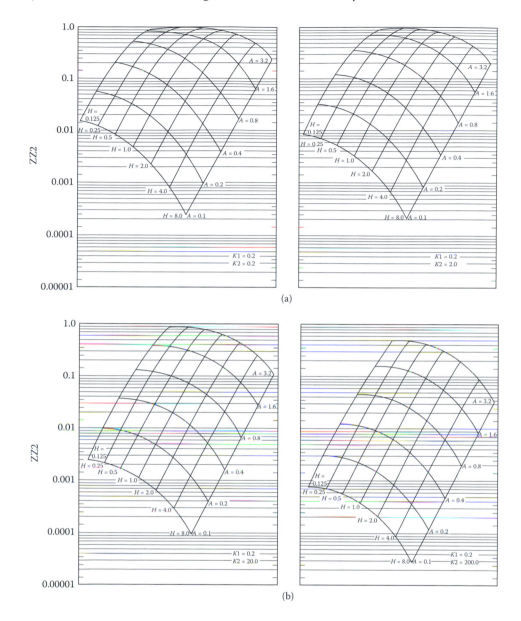

FIGURE 2.14 Examples of Values of ZZ2.
Source: From Yoder, E.J., and M.W. Witczak (1975) Principles of Pavement Design, 2nd edition © 1975. Reprinted with permission of John Wiley & Sons, Inc.

following:

$$\sigma_{z1} = p(ZZ1)$$
$$\sigma_{z2} = p(ZZ2)$$

For horizontal stresses, first obtain the following from a table (see Table 2.2 for an example of values):

[ZZ1-RR1]
[ZZ2-RR2]
[ZZ3-RR3]

TABLE 2.2
Example Values of ZZ1-RR1, ZZ2-RR2, and ZZ3-RR3

a1	H = 0.125, k₁ = 0.2 (ZZ1-RR1)	(ZZ2-RR2)	(ZZ2-RR3)	H = 0.125, k₁ = 2.0 (ZZ1-RR1)	(ZZ2-RR2)	(ZZ2-RR3)	H = 0.125, k₁ = 20.0 (ZZ1-RR1)	(ZZ2-RR2)	(ZZ2-RR3)	H = 0.125, k₁ = 200.0 (ZZ1-RR1)	(ZZ2-RR2)	(ZZ2-RR3)
			$k_2 = 0.2$			$k_2 = 0.2$			$k_2 = 0.2$			$k_2 = 0.2$
0.1	0.12438	0.00332	0.01659	0.71614	0.00350	0.01750	1.80805	0.00322	0.01611	2.87564	0.00201	0.01005
0.2	0.13546	0.01278	0.06391	1.01561	0.01348	0.06741	3.75440	0.01249	0.06244	7.44285	0.00788	0.03940
0.4	0.10428	0.04430	0.22150	0.83924	0.04669	0.23346	5.11847	0.04421	0.22105	15.41021	0.02913	0.14566
0.8	0.09011	0.10975	0.54877	0.63961	0.11484	0.57418	3.38600	0.11468	0.57342	9.70261	0.08714	0.43568
1.6	0.08777	0.13755	0.68777	0.65723	0.13726	0.68630	1.81603	0.13687	0.68436	7.02380	0.13705	0.68524
3.2	0.04129	0.10147	0.50736	0.38165	0.09467	0.47335	1.75101	0.07578	0.37890	2.34459	0.06594	0.32971
			$k_2 = 2.0$			$k_2 = 2.0$			$k_2 = 2.0$			$k_2 = 2.0$
0.1	0.12285	0.01693	0.00846	0.70622	0.01716	0.00858	1.81178	0.01542	0.00771	3.02259	0.00969	0.00485
0.2	0.12916	0.06558	0.03279	0.97956	0.06647	0.03324	3.76886	0.06003	0.03002	8.02452	0.03812	0.01906
0.4	0.08115	0.23257	0.11629	0.70970	0.23531	0.11766	5.16717	0.21640	0.10820	17.64175	0.14286	0.07143
0.8	0.01823	0.62863	0.31432	0.22319	0.63003	0.31501	3.43631	0.60493	0.30247	27.27701	0.45208	0.22604
1.6	−0.04136	0.98754	0.49377	−0.19982	0.97707	0.48853	1.15211	0.97146	0.48573	23.38638	0.90861	0.45430
3.2	−0.03804	0.82102	0.41051	−0.28916	0.84030	0.42015	−0.06894	0.88358	0.44179	11.87014	0.91469	0.45735
			$k_2 = 20.0$			$k_2 = 20.0$			$k_2 = 20.0$			$k_2 = 20.0$
0.1	0.12032	0.03667	0.00183	0.69332	0.03467	0.00173	1.80664	0.02985	0.00149	3.17763	0.01980	0.00099
0.2	0.11787	0.14336	0.00717	0.92086	0.13541	0.00677	3.74573	0.11697	0.00585	8.66097	0.07827	0.00391
0.4	0.03474	0.52691	0.02635	0.46583	0.49523	0.02476	5.05489	0.43263	0.02163	20.12259	0.29887	0.01494
0.8	−0.14872	1.61727	0.08086	−0.66535	1.49612	0.07481	2.92533	1.33736	0.06687	36.29943	1.01694	0.05085
1.6	−0.50533	3.58944	0.17947	−2.82859	3.28512	0.16426	−1.27093	2.99215	0.14961	49.40857	2.64313	0.13216
3.2	−0.80990	5.15409	0.25770	−5.27906	5.05952	0.25298	−7.35384	5.06489	0.25324	57.84369	4.89895	0.24495
			$k_2 = 200.0$			$k_2 = 200.0$			$k_2 = 200.0$			$k_2 = 200.0$
0.1	0.11720	0.05413	0.00027	0.67488	0.04848	0.00024	1.78941	0.04010	0.00020	3.26987	0.02809	0.00014
0.2	0.10495	0.21314	0.00170	0.85397	0.19043	0.00095	3.68097	0.15781	0.00079	9.02669	0.11136	0.00056
0.4	−0.01709	0.80400	0.00402	0.21165	0.71221	0.00356	4.80711	0.59391	0.00297	21.56482	0.43035	0.00215

0.8	0.00765	1.53070	41.89873	0.00979	1.95709	1.90825	0.01163	2.32652	−1.65954	0.01340	2.67934	−0.34427
1.6	0.02284	4.56707	69.63157	0.02626	5.25110	−5.28803	0.03133	6.26638	−6.47707	0.03680	7.35978	−1.21139
3.2	0.05710	11.42045	120.95931	0.06225	12.45058	−21.52546	0.07128	14.25621	−16.67376	0.08114	16.22830	−2.89282
$k_2 = 0.2$												
0.1	0.00451	0.00090	0.86644	0.01011	0.00202	0.61450	0.01384	0.00277	0.28658	0.01370	0.00274	0.05598
0.2	0.01784	0.00357	2.71354	0.03964	0.00793	1.76675	0.05377	0.01075	0.72176	0.05302	0.01060	0.12628
0.4	0.06824	0.01365	6.83021	0.14653	0.02931	3.59650	0.19211	0.03842	1.03476	0.18722	0.03744	0.14219
0.8	0.23118	0.04624	13.19664	0.43854	0.08771	4.58845	0.51687	0.10337	0.88833	0.49196	0.09839	0.12300
1.6	0.52955	0.10591	13.79134	0.70194	0.14039	2.31165	0.70510	0.14102	0.66438	0.69586	0.13917	0.10534
3.2	0.43037	0.08608	2.72901	0.37934	0.07587	1.24415	0.49020	0.09804	0.41539	0.55569	0.11114	0.05063
$k_2 = 2.0$												
0.1	0.00203	0.00407	0.96553	0.00481	0.00962	0.63215	0.00677	0.01353	0.28362	0.00704	0.01409	0.05477
0.2	0.00806	0.01611	3.10763	0.01891	0.03781	1.83766	0.02639	0.05278	0.70225	0.02742	0.05484	0.12136
0.4	0.03110	0.06221	8.37852	0.07079	0.14159	3.86779	0.09589	0.19178	0.96634	0.09890	0.19780	0.12390
0.8	0.10930	0.21860	18.95534	0.22355	0.44710	5.50796	0.27605	0.55211	0.66885	0.28019	0.56039	0.06482
1.6	0.29277	0.58553	31.18909	0.45058	0.90115	4.24281	0.47540	0.95080	0.17331	0.48108	0.96216	−0.00519
3.2	0.44595	0.89191	28.98500	0.46627	0.93254	1.97494	0.44695	0.89390	−0.05691	0.43610	0.87221	−0.02216
$k_2 = 20.0$												
0.1	0.00043	0.00861	1.08738	0.00096	0.01930	0.65003	0.00136	0.02728	0.27580	0.00156	0.03116	0.05192
0.2	0.00171	0.03421	3.59448	0.00381	0.07623	1.90693	0.00536	0.10710	0.67115	0.00611	0.12227	0.11209
0.4	0.00668	0.13365	10.30923	0.01454	0.29072	4.13976	0.01996	0.39919	0.84462	0.02275	0.45504	0.08622
0.8	0.02457	0.49135	26.41442	0.04928	0.98565	6.48948	0.06328	1.26565	0.21951	0.07214	1.44285	−0.07351
1.6	0.07692	1.53833	57.46409	0.12762	2.55231	6.95639	0.14743	2.94860	−1.22411	0.16850	3.37001	−0.40234
3.2	0.18048	3.60964	99.29034	0.23812	4.76234	6.05854	0.24494	4.89878	−3.04320	0.25503	5.10060	−0.71901
$k_2 = 200.0$												
0.1	0.00007	0.01311	1.19099	0.00014	0.02711	0.65732	0.00019	0.03814	0.26776	0.00024	0.04704	0.04956
0.2	0.00026	0.05223	4.00968	0.00054	0.10741	1.93764	0.00075	0.15040	0.63873	0.00093	0.18557	0.10066
0.4	0.00103	0.20551	11.96405	0.00207	0.41459	4.26004	0.00285	0.57046	0.71620	0.00353	0.70524	0.04248
0.8	0.00388	0.77584	32.97364	0.00735	1.46947	6.94871	0.00963	1.92636	−0.28250	0.01203	2.40585	−0.24071
1.6	0.01320	2.63962	82.77997	0.02183	4.36521	8.55770	0.02680	5.35936	−3.09856	0.03412	6.82481	−1.00743
3.2	0.03801	7.60287	189.37439	0.05468	10.93570	10.63614	0.06322	12.64318	−9.18214	0.07730	15.45931	−2.54264

Source: From Yoder, E.J., and M.W. Witczak (1975) Principles of Pavement Design, 2nd edition © 1975. Reprinted with permission of John Wiley & Sons, Inc.

Then, use the following to calculate horizontal stresses from vertical stresses calculated earlier:

$$\sigma_{z1} - \sigma_{r1} = p[ZZ1 - RR1]$$
$$\sigma_{z2} - \sigma_{r2} = p[ZZ2 - RR2]$$
$$\sigma_{z2} - \sigma_{r3} = p[ZZ2 - RR3]$$

Knowing σ_{z1} and σ_{r1}, the horizontal strain at the bottom of layer 1, ε_{r1} can be calculated as follows:

$$\varepsilon_{r1} = \frac{\sigma_{r1}}{E_1} - \mu_1 \frac{\sigma_{t1}}{E_1} - \mu_1 \frac{\sigma_{z1}}{E_1}$$

2.8.7 COMPUTER PROGRAMS

Computer programs have been developed on the basis of the theory discussed above for calculating stresses, strains, and deflections of a layered elastic system. They have gradually became more sophisticated in capability to handle linear elastic materials, nonlinear elastic granular materials, vertical as well as horizontal loads, elastic multilayer systems under multiple wheel loads, and stress-dependent material properties, and in application of finite element linear and nonlinear analysis.

Examples include the following: BISTRO and BISAR (from Shell), ELSYM5 (from Chevron), ALIZEIII (LCPC) and CIRCLY (from MINCAD, Australia), DAMA (from the Asphalt Institute), SAPIV and ELSYM5 (from the University of California, Berkeley), PDMAP (Finn et al., 1986), the IILI-PAVE computer program (Raad and Figueroa, 1980), MICH-PAVE (Harichandran et al., 1989), and Everstress (from the Evercalc suite of software from the Washington State DOT [1995]).

The reader is advised to acquire any of the above software for the practice problems. One suggested software is Everstress (http://www.wsdot.wa.gov/biz/mats/pavement/pave_tools.htm; Washington State Department of Transportation, 2007).

2.9 PRINCIPLES OF GOOD CONSTRUCTION

Pavement construction is a complex process involving many factors—each of which has the potential of affecting the quality of the constructed pavement in a very significant way. The process of good construction should start with the identification/adoption or creation of an appropriate specification. This specification should lay out details of steps, determine tests required to make sure that the quality of the product is acceptable, and monitor the key parameters during construction, such that errors can be identified and rectified quickly.

Paving mixes are produced in plants, which can be of different types. The basic processes are the same. Aggregates of different sizes are mixed in specific proportions, dried, mixed with asphalt at a high temperature, and then stored in silos or transported to the job site through insulated trucks. During the mixing process, the steps are monitored closely to identify any malfunction. Since the mixing process is conducted at a specific range of high temperature to make sure that the aggregates are dried and the asphalt is of a sufficiently fluid nature to allow coating and mixing of aggregates, and because the proportioning of different components is conducted in terms of weight, temperature and weight measurements need to be monitored very closely during the process.

Trucks take the mix to pavers at the job site. Pavers are self-propelled machines that lay down asphalt mix at a specific depth and provide the initial compaction. A paver is a complex piece of machinery and has a number of moving parts, including an auger and conveyors to mix the mix and send it to the back from the front hopper where it receives the mix from the trucks. The screed at the back of the paver maintains a specific depth of the mix and must be controlled carefully.

Right behind the paver come the rollers. Different rollers have different functions. Steel wheel vibratory drum rollers help in initial compaction, and rubber-tired and static steel-wheel rollers provide intermediate and final compaction. The rollers must be of appropriate weight, and frequency and amplitude of vibration (for vibratory rollers) must be set at the appropriate level. The rollers should also move at the correct speed and in the correct pattern to ensure the best compaction possible.

Finally, althrough the production and laydown operation, an appropriate quality control procedure must be adopted to monitor key properties and correct, if necessary, any errors as quickly as possible. There may also be a quality assurance process to make sure that the quality control process is actually working.

2.10 PUTTING EVERYTHING TOGETHER

The process of pavement construction actually begins during the mix design process and ends with compaction. While selecting materials for mix design, one needs to use experience and judgment, not only from the point of view of the optimum mix but also from the consideration of construction. In many cases, a mix that can be produced in small amounts in the laboratory during mix design may not be produced in large batches in the plant because of practical reasons—for example, due to the lack of locally available materials in sufficient amounts, the lack of appropriate fixtures in the plant machinery, or even the lack of sufficient paver and compaction equipment.

Hence, looking at the big picture is necessary from the very beginning. Consideration should be given to the type of project, environmental conditions expected during construction, and availability of plant and field equipment.

The mixing process in the plant involves batching or proportioning a variety of aggregates of different sizes as well as liquid asphalt binder. Any deviation in these parameters would be reflected in the constructed pavement—the key is to catch it through the use of appropriate tests in the plant and in the field, and to correct it quickly.

One important consideration is the availability of the fleet of the machinery required for transporting mix from the plant to the job site. The whole operation should be coordinated in such a way that the pavers and rollers keep moving smoothly at the appropriate speed. If required, a material transfer vehicle may be used to receive and store mix from the trucks, such that the paver can keep working even if the flow of mix through trucks is hampered for a specific time interval due to any reason.

One significant factor that affects the quality of the finished pavement is its air voids or density. Given that the mix design is accurate and the mixing process in the plant is working fine, four things have the most significant effect on the density of the pavement. These are the temperature of the mix, the ambient temperature (and wind), the availability of required rollers in appropriate numbers, and the thickness of the layer. It is crucial that appropriate considerations are given to these factors to ensure that for the given thickness and the existing environmental conditions at the site, within the allowable range of mixing temperature, there is sufficient time to compact the mix in the field with the use of the available equipment.

QUESTIONS

1. What is the purpose of mix design?
2. How does structural design differ from mix design?
3. What are the most important considerations for construction of an asphalt pavement?
4. Are mix design, structural design, and construction all related to a certain extent? Can you explain this schematically?

5. Determine the vertical and radial stresses at nine points for a 9000-lb point load on a homogeneous, isotropic, linear elastic, semi-infinite space. Consider a Poisson's ratio of 0.3.

Point	z, inch	r, inch	Point	z, inch	r, inch	Point	z, inch	r, inch
1	0	0	4	6	0	7	12	0
2	0	6	5	6	6	8	12	6
3	0	12	6	6	12	9	12	12

6. If the deflection at the center of a rigid plate is found out to be 0.03 in. from a load on a subgrade with Poisson's ratio of 0.35, what is the estimated modulus of the subgrade? Consider a load of 9,000 lb and plate radius of 6 inches.

7. Use any layered elastic analysis program to compute the vertical and radial stresses and strains directly below the load at a depth of 149 mm in a full depth 150 mm thick asphalt pavement with a modulus of 3500 MPa and Poisson's ratio of 0.35 for the following loading conditions. In each case, half of a standard 20 KN axle (only the main load-bearing axles, not including the steering axle) has been indicated. Can you sketch the axle/wheel configuration of the entire vehicles? For the subgrade consider a modulus and Poisson's ratio of 100 MPa and 0.4, respectively. Consider all layers to be stress insensitive
    A. Loads of 20 KN, with coordinates in cm (x,y): (0,0) (33,0); (0,122) (33,122); tire pressure of 690 kPa
    B. Loads of 20 KN, with coordinates in cm (x,y): (0,0) (33,0); (0,122) (33,122); (0,244) (13,244); tire pressure of 690 kPa

8. Use a layered elastic analysis program to determine the radial stresses at the bottom of the surface layer directly under the load for a pavement with three layers as follows:

Layer	Modulus (psi)	Poisson's Ratio	Thickness (in.) Full Friction between All Layers
1	435,113	0.35	10.63
2	21,755.7	0.4	20.08
3	7251.9	0.4	Infinite

Consider all layers to be stress insensitive.
Consider three different cases of loads, as follows:
A. Single axle with dual tires

Tire #	X (in)	Y (in)	Load (lb)	Pressure (psi)
1	0	0	5000	100
2	13.5	0	5000	100

B. Tandem axle with dual tires

Tire #	X (in)	Y (in)	Load (lb)	Pressure (psi)
1	0	0	5000	100
2	13.5	0	5000	100
3	13.5	54	5000	100
4	0	54	5000	100

C. Tridem axle with dual tires

Tire #	X (in)	Y (in)	Load (lb)	Pressure (psi)
1	0	0	5000	100
2	13.5	0	5000	100
3	13.5	54	5000	100
4	0	54	5000	100
5	0	108	5000	100
6	13.5	108	5000	100

3 Principles of Mix and Structural Design of Concrete Pavement

Overview

3.1 MIX DESIGN

Portland cement concrete (PCC) consists of Portland cement, aggregates, water, and admixture, and may contain entrained air. Concrete mix design is the process of determining the proportions of the different components, and achieving the desirable mix characteristics based on the most economical and practical combination of readily available materials. The produced concrete mix should satisfy the performance requirements, including workability, ease of placement, strength, durability, appearance, and economy. A mix design leads to a concrete specification that imposes limits on used materials and properties.

The steps in mix design of concrete consist of the selection of coarse and fine aggregates; cement and pozzolanic materials such as fly ash, slag, silica fume, and natural pozzolans; and a proportion of water to cement and admixtures, if any.

Prior to proportioning the ingredients, the desirable concrete properties must be determined based on the intended use of the concrete to handle load and environment stresses. For example, these include resistance to imposed stresses due to traffic load, freezing and thawing conditions, and salt or aggressive chemical exposure. The selection of the appropriate water-cementitious ratio (W/CM) influences the quality of the water-cement paste and the mix. More importantly than strength, a concrete must be designed for durability, impermeability or watertightness, wear, and abrasion resistance. Once these characteristics are achieved, usually the required strength is satisfied too.

3.1.1 HYDRATION, STRENGTH, AND MATERIALS

When water is added to cement, an exothermic reaction called *hydration* occurs, resulting in the formation of a number of compounds, the main one being calcium silicate hydrate. This process and the resulting chemical compounds provide the glue to bind the components of the mix together and provide the strength to hardened concrete.

The water-cementitious ratio is the mass of water divided by the mass of all cements, blended cements, and pozzolanic materials such as fly ash, slag, silica fume, and natural pozzolans. In most literature, *water-cement ratio* (W/C) is used synonymously with W/CM.

Assuming that a concrete is made with clean sound aggregates, and that the cement hydration has progressed normally, the strength gain is inversely proportional to the water-cementitious ratio by mass. The paste strength is proportional to the solids volume or the cement density per unit volume. Differences in strength may also be from the influence of aggregate gradation, shape, particle size, surface texture, strength, and stiffness; or factors associated with cement materials (i.e., different sources, chemical composition, and physical attributes), the amount of entrained air, and effects of admixtures and curing. The strength of concrete is measured in terms of compressive

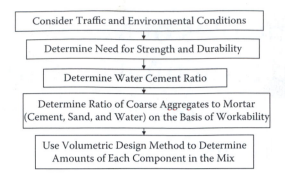

FIGURE 3.1 Basic Steps in Concrete Mix Design.

strength of cylinders, and the strength gain is evaluated by checking strengths at different periods of time from the mixing of the concrete.

Aggregates have a significant effect on the workability of fresh concrete. The aggregate particle size and gradation, shape, and surface texture will influence the amount of concrete that is produced with a given amount of paste (cement plus water). The selection of the maximum size aggregate is governed by the thickness of the slab and by the closeness of the reinforcing steel. The maximum size aggregates should not be obstructed and should flow easily during placement and consolidation.

The purposeful entrainment of air in concrete provides tremendous protection against freezing and thawing action and against deicing salts. Hydraulic pressure is generated when water in the paste pore structure freezes and pushes against the unfrozen water. The tiny entrained air bubbles act as relief valves for this developed hydraulic pressure.

Fresh concrete must have the appropriate workability, consistency, and plasticity suitable for construction conditions. Workability is a measure of the ease of placement, consolidation, and finishing of the concrete. Consistency is the ability of freshly mixed concrete to flow, and plasticity assesses the concrete's ease of molding. If the concrete is too dry and crumbly, or too wet and soupy, then it lacks plasticity. The slump test is a measure of consistency and indicates when the characteristics of the fresh mix have been changed or altered. However, the slump is indicative of workability when assessing similar mixtures. Different slumps are needed for different construction projects.

The cementing materials content is usually determined based on the specified water-to–cementitious materials ratio. However, usually a minimum amount of cement is also specified to ensure satisfactory durability, finishability, and wear resistance of slabs even though strength needs may be satisfied at lower cement contents. For durability, the type of cement used is also important. And finally, admixtures are used in concrete to enhance the desirable characteristics in the fresh and hardened concrete.

A flow chart showing the basic steps in mix design of concrete is shown in Figure 3.1.

3.2 STRUCTURAL DESIGN

Structural design of concrete pavements is based on the concept of limiting stresses to prevent excessive damage and deterioration of the pavements. The different distresses are summarized in Figure 3.2.

Distresses are caused by stresses in rigid pavements. These stresses are imposed by traffic through wheel loads, temperature and moisture variations (causing warping, and expansion and shrinkage stresses), and volume changes in the base and subgrade.

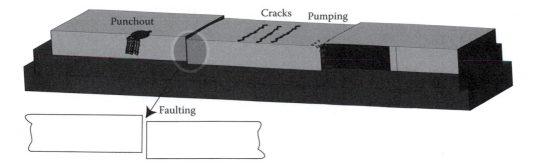

FIGURE 3.2 Distress Conditions in Portland Cement Concrete (PCC) Pavements.

For analysis of stresses, a concrete pavement is idealized as a rigid slab resting on a spring-like foundation. Theories, as explained in the next section, are used to develop equations relating stresses to material and structural properties, and then analytical methods are used to predict these stresses on the basis of properties such as modulus and Poisson's ratio and stiffness of the subgrade, as well as temperature fluctuations. The design process essentially consists of consideration of traffic and environmental factors and the use of the relevant properties (through material characterization tests) to ensure that the slab can sustain the stresses for the duration of its life, without failing. The basic steps in structural design are shown in Figure 3.3.

3.3 THEORETICAL CONSIDERATIONS

A rigid pavement is basically a slab resting on a subgrade or base. The slab is much stiffer than the supporting base or foundation material, and therefore carries a significant portion of induced stresses. The load-carrying mechanism is similar to beam action, although a concrete slab is much wider than the beam and should be considered as a plate. Westergaard (1926b) developed stress equations for rigid pavement slabs supported on a Winkler or liquid foundation, which is a conceptual model that considers the foundation as a series of springs. When the slab is loaded vertically

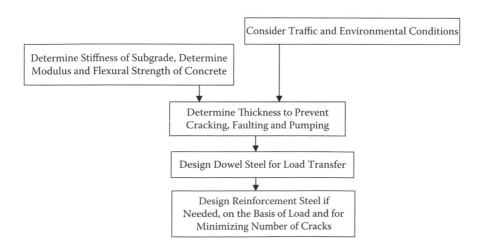

FIGURE 3.3 Basic Steps in PCC Pavement Design.

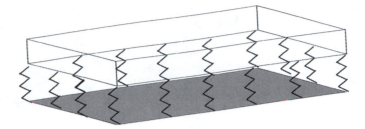

FIGURE 3.4 Slab on Winkler Foundation.

down, the springs tend to push back; when the enviroment-related loads are pulling up on the slab, the springs tend to pull down toward the foundation (Figure 3.4).

The induced stresses and their effects are controlled by a number of factors, such as restrained temperature and moisture deformation; externally applied loads; volume changes of the supporting material, including frost action; continuity of subgrade support through plastic deformation; or materials loss due to pumping action.

Mathematical modeling of the true stress state of a PCC slab on a soil foundation is very complex due to the nonelastic behavior of the soil foundation. Models can be developed once certain assumptions are made.

3.3.1 STRESSES DUE TO CURVATURE AND BENDING CAUSED BY LOADS

If we consider that the pavement slab is a beam on an elastic foundation, then the reactive pressure, p, can be related to the deformation, Δ, through the equation $p = k\Delta$, where k is the modulus of subgrade reaction, which is the ratio of the pressure applied to the subgrade using a loaded area divided by the displacement experienced by that loaded area. It is assumed that the modulus of subgrade reaction k is a constant and that the behavior is elastic. This assumption is only true over very small deformations and is greatly affected by numerous conditions such as soil type, density, moisture content, and stress state.

A concrete pavement slab will deform under load. The resistance to deformation due to loading depends upon the characteristics of the foundation and the stiffness of the slab. The moment due to bending in a beam is given by the following equation:

$$M = EI \frac{d^2 y}{dx^2}$$

where E is the modulus of elasticity, and I is the moment of inertia of the slab.

The stiffness term in the beam equation is expressed by the EI term. For an infinite thin plate, the moment equation is given by the following equation:

$$Mx = \frac{Eh^3}{12(1 - \mu^2)} \left(\frac{\partial^2 \omega}{\partial x^2} + \mu \frac{\partial^2 \omega}{\partial y^2} \right)$$

Or

$$Mx = D \left(\frac{\partial^2 \omega}{\partial x^2} + \mu \frac{\partial^2 \omega}{\partial y^2} \right)$$

where the stiffness term for the plate is given by the following equation, and is represented by the letter D.

$$D = \frac{Eh^3}{12(1-\mu^2)}$$

Westergaard (1927) applied plate theory to a finite PCC slab and supporting foundation, and developed the relative stiffness expression, which is called the *radius of relative stiffness*.

$$l = \sqrt[4]{\frac{Eh^3}{12(1-\mu^2)k}}$$

where:
  l = radius of relative stiffness (in.)
  E = modulus of elasticity of the pavement (psi)
  h = thickness of the pavement (in.)
  μ = Poisson's ratio of the PCC
  k = modulus of subgrade reaction (pound per cubic inch, pci)

3.3.2 STRESSES DUE TO CURLING AND WARPING CAUSED BY THE ENVIRONMENT

A concrete slab will undergo volume changes due to changes in temperature and moisture. During the day, as the air temperature and sun increase the surface temperature of the concrete slab, the top of the slab will tend to expand relative to the neutral axis and the bottom of the slab will tend to contract. However, the weight of the slab will prevent it from contraction or expansion, and compressive stresses will be induced in the top slab layer while tensile stresses will be induced in the bottom layer. The opposite will occur at night. The top of the slab will be cooler compared to the bottom and will tend to contract. The slab weight will prevent the upward curling, and therefore tensile stresses will develop in the top of the slab while compressive stresses will be induced in the bottom of the slab (Figure 3.5).

The approach used to explain curling stress is using plate theory and the concept of a liquid or Winkler foundation (Figure 3.4). When the surface temperature is greater than the bottom temperature, the top of the slab/plate tends to expand and the springs react to this movement. The outside springs are compressed and push the plate outward, and the inner springs are in tension and pull the plate inward. This condition induces compressive stresses in the top of the slab and tensile stresses in the bottom. When the surface temperature of the slab is lower than the bottom temperature, the

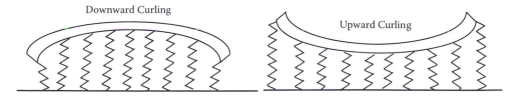

FIGURE 3.5 Slab Curling Due to Temperature Variations.

top tends to contract and the bottom tends to expand. This produces an upward curvature of the slab (such that it can virtually hold water). The springs attached to the exterior edge of the slab are now in tension and pull the slab down, and the springs attached to the interior of the slab are in compression and push the slab outward, inducing tensile stresses in the top of the slab and compressive stresses in the bottom of the slab.

Westergaard (1926a) developed equations for approximating the curling stresses in concrete pavements based on plate theory. These equations are very complex and will not be discussed in this book. However, based on Westergaard's curling stress equations, Bradbury (1938) developed a simple method for determining curling stresses due to temperature in a finite slab. The maximum total stresses at the interior of the slab, edge, and corner warping are given by the following equations:

$$\text{Interior stress}, \sigma_t = \frac{E\alpha\Delta t}{2}\left[\frac{C_x + \mu C_y}{1 - \mu^2}\right]$$

$$\text{Edge stress}, \sigma_t = \frac{CE\alpha\Delta t}{2}$$

where:
   E = modulus of elasticity of concrete
   μ = Poisson's ratio of concrete
   α = coefficient of thermal expansion
   C = C_x and C_y are correction factors for the finite slab and are dependent on L_x/l and L_y/l; l is defined as the radius of relative stiffness and is derived based on plate theory
   L_x L_y = finite slab length along the x-axis and y-axis, respectively (Figure 3.6 explains the dimensions.)
   Δt = change in temperature

Figure 3.7 shows the variation of warping stress with dimensions of the slab.

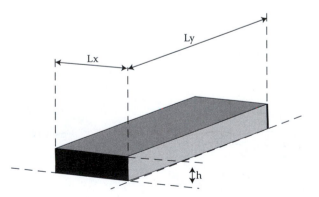

FIGURE 3.6 Slab Dimensions.

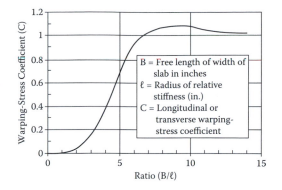

FIGURE 3.7 Variation of Warping Stress with Slab Dimensions (B = L_x or L_y).
Source: From Yoder, E.J., and M.W. Witczak (1975) Principles of Pavement Design, 2nd edition © 1975. Reprinted with permission of John Wiley & Sons, Inc.

Example

Consider a concrete slab 30 ft (9.14 m) by 12 ft (3.66 m) and 8 in. (203 mm) thick, subjected to a temperature differential of 20°F (11.1°C). Assuming k = 200 pci (54.2 MN/m³) and α = 5 × 10⁻⁶ in./in./°F (9 × 10⁻⁶ mm/mm/°C), determine the maximum curling stress in the interior and at the edge of the slab.

Solution

$$l = \sqrt[4]{\frac{Eh^3}{12(1-\mu^2)k}}$$

$$l = \sqrt[4]{\frac{4 \times 10^6 \times 8^3}{12(1-0.15^2)200}} = 30.4 \text{ in (772.04 mm)}$$

$$L_x/l = 11.84$$
$$L_y/l = 4.74$$

From Figure 3.7, $C_x = 1.04$
$$C_x = 0.63$$

$$\text{Interior stress, } \sigma_t = \frac{E\alpha\Delta t}{2}\left[\frac{C_x + \mu C_y}{1-\mu^2}\right]$$

$$\sigma_t = \frac{4 \times 10^6 * 5 \times 10^{-6} * 20}{2}\left[\frac{1.04 + 0.15 * 0.63}{1-0.15^2}\right]$$

$$\text{Interior Stress, } \sigma_t = 232 \text{ psi}$$

$$\text{Edge stress, } \sigma_t = \frac{CE\alpha\Delta t}{2}$$

$$\sigma_t = \frac{1.04 * 4 \times 10^6 * 5 \times 10^{-6} * 20}{2}$$

$$\text{Edge stress, } \sigma_t = 208 \text{ psi}$$

QUESTIONS

1. What are the basic steps in PCC pavements' mix and structural design?
2. What are the primary distresses against which PCC pavements are designed?
3. Consider a concrete slab 20 ft by 12 ft and 6 in. thick, subjected to a temperature differential of 20°F. Assuming k = 200 pci and $\alpha = 5 \times 10^{-6}$ in./in./°F, determine the maximum curling stress in the interior and at the edge of the slab, and the corner warping stress.

4 Standards

4.1 IMPORTANCE OF STANDARDS

Standards are documents provided by expert and recognized organizations, which provide guidelines or characteristics of common activities to ensure consistency, quality, and safety. Standards are developed on the basis of research and experience (and are subsequently revised, as needed) through consensus of groups of experts. Generally, standards used in pavement engineering are those that are developed by traditional formal organizations such as the American Society for Testing and Materials (ASTM International; www.astm.org), the American Association of State and Highway Transportation Officials (AASHTO; www.transportation.org), the Asphalt Institute (www.asphaltinstitute.org), the American Concrete Institute (ACI; http://www.aci-int.org), and the British Standards Institution (BSI; www.bsi-global.com). They are developed through a very open and transparent process, ensuring the representation of all stakeholders. Standards on safety can be obtained from the Occupational Safety and Health Administration (OSHA) website (http://www.osha.gov/comp-links.html).

Each standard is an accumulation of years of experience in that specific topic. Use of appropriate standards can help the proper utilization of that combined experience; make the process (being used) rational, practical, and up-to-date; and, as required, ensure, to the maximum extent at that point of time, safety. Therefore, standards ensure uniformity and consistency, reliability, and safety in products and processes. Use of standards also ensures compatibility between different products, and the best use of moneys. This is important since it means, for pavement engineering, the best use of the taxpayers' dollars for road construction.

For pavement engineering, there are various types of standards—such as those for conducting tests, for interpreting results, as well as for writing specifications. Every test process is described in detail, with mentions of relevant other standards, steps in inference, and methods of reporting the results. If a new test is developed, standards provide guidance for conducting a round-robin study to determine the variability of results from that specific test. Once conducted, the results are considered and transferred into inter- and within-laboratory tolerance values. This ensures that in the future, if the tests are conducted, the variability in the results would point out inconsistencies or greater than expected errors, if any.

4.2 THE AMERICAN SOCIETY OF TESTING AND MATERIALS

A typical ASTM standard on a test consists of the following items:

1. *Scope.* The scope defines the purpose of the standard and provides relevant details related to specific applications of the test/procedure.
2. *Referenced documents.* This section lists the other standards that are used as part of this standard.
3. *Summary of test method.* The summary describes briefly the steps, ranging from sample preparation (for example) to reporting of results.
4. *Significance and use.* This section relates the results of the test/process that is described in this standard to practical problems/solutions and applications.
5. *Apparatus.* The apparatus section describes, in detail, each and every apparatus or piece of equipment that is used. Wherever applicable, appropriate dimensions/capacities of apparatus are provided.

Standard Test Method for

CBR (California Bearing Ratio) of Laboratory-Compacted Soils

FIGURE 4.1 Example of American Society for Testing and Materials (ASTM) Standard Designation.

6. *Sample.* This section describes how a sample, from which test specimens are to be prepared, is to be handled or modified, if required.
7. *Test specimen.* This part provides a detailed description of the process of preparing test specimens. If the test consists of determination of multiple parameters, it describes the steps needed for the preparation of samples for testing of the different parameters.
8. *Procedure.* This section describes the actual test procedure.
9. *Calculation.* This part presents the steps needed to calculate the different parameters, and provides steps to analyze the data, make corrections, and plot the data, as required.
10. *Report.* The report section points out what needs to be indicated in the document that reports the results. Specific items include those such as other relevant properties of the test specimens, any special sample preparation procedures, results, and any other tests conducted as part of this.
11. *Precision and bias.* This part provides the maximum variation that can be expected between two tests conducted in the same laboratory and between tests conducted in different laboratories, as well as between the results obtained from this test and those from more accurate tests, considered as "true" values.
12. *Keywords.* The last part presents a list of keywords against which this standard has been indexed.

Other than tests, some of the other relevant ASTM standards include those defining different terminology, sampling materials, conducting ruggedness tests for detecting sources of variation in test methods, and establishing precision of test methods.

The ASTM standards are generally designated as a two-part number, the first number being the number of the standard, and the second number indicating the year in which it was first adopted (or, if revised, the year of the last revision). A letter (for example, D) indicating the group that deals with the subject matter under which the specific topic presented in the standard falls is also mentioned in the designation. For example, see Figure 4.1.

With the advent of global business and economy, in more and more cases organizations are working on globalization of standards such that materials and process become compatible regionally, nationally, and internationally.

4.3 THE AMERICAN SOCIETY OF STATE HIGHWAY AND TRANSPORTATION OFFICIALS

AASHTO standards are developed either separately or, if appropriate, on the basis of existing ASTM standards by the AASHTO Highway Subcommittee of Materials, with representatives of transportation department officials in the United States, associate members, and international members, such as those from the Canadian provinces and Korea. The set of standards is made available in two parts—Part 1 contains specifications for materials, and Part 2 contains methods of tests and specifications for testing equipment.

A typical standard test method from the AASHTO consists of the following:

1. *Scope*. This part states the objective of the test method.
2. *Referenced documents*. Relevant other AASHTO and ASTM standards are mentioned here.
3. *Terminology*. This part defines the different terms used in this standard.
4. *Significance and use*. In this section, the practical applications of the results of this test are indicated.
5. *Apparatus*. All of the required apparatus are listed and described briefly in this part.
6. *Sample*. The procedure of sampling or preparation of sample is described here. Required sample sizes are indicated.
7. *Procedure*. This section describes the test procedure along with required warnings and cautions.
8. *Calculation*. The equations and expressions for calculating the results are presented in this section.
9. *Precision and bias*. The precision section indicates the within- and between-laboratory standard deviations, and hence guidelines for maximum allowable differences in results obtained by tests conducted by the same operator and those conducted in different laboratories. The bias section presents data on differences between results obtained from this test and more accurate tests (considered as a "true" value).
10. *Keywords*. The index keywords are listed in this part.

4.4   THE USE OF STANDARDS IN MATERIALS SELECTION, MIX DESIGN, AND STRUCTURAL DESIGN

Standards are used in selecting materials such as aggregates and asphalt, conducting tests, and interpreting results in mix design and structural design of pavements. The selection part involves using appropriate project/application conditions to choose specific types of materials and avoid those that could be detrimental for such conditions. For example, for selecting aggregates, AASHTO Specification M 283 could be used. In the design step, standards are used for conducting tests, which involve sample preparation, fabricating test specimens, conducting tests, analyzing data, and reporting results. An example is AASHTO T 27 for sieve analysis of fine and coarse aggregates.

4.5   THE USE OF STANDARDS IN QUALITY CONTROL IN CONSTRUCTION

In quality control and testing of pavements, the different steps involve developing a plan for quality control, sampling materials from the plant and the field, conducting either field or laboratory tests, and interpreting the results as evidence against which a pavement section could be accepted or rejected by the owner. These steps could involve the use of a series of standards, for example, ASTM D3665 (Standard Practice for Random Sampling of Construction Materials) and ASTM D295 (Determination of Density of Pavement Using the Nuclear Method).

4.6   IMPORTANT SPECIFICATIONS

Two lists of important specifications are provided in Table 4.1 and Table 4.2. The reader is encouraged to visit the ASTM (www.astm.org) and AASHTO (www.aashto.org, www.transportation.org) websites to learn more about the standards.

TABLE 4.1
List of Relevant ASTM and AASHTO Standards for Flexible Pavement

Item	AASHTO	ASTM
Specification: coarse aggregate	M 283	
Specification: fine aggregate	M 29	
Specification: mineral filler	M 17	
Specification: aggregate gradations		D 3515
Tests: sampling aggregates	T 2	D 75
Random sampling		D 3665
Reducing field samples to test size	T 248	C 702
Sieve analysis of fine and coarse aggregates	T 27	C 136
Amount of material finer than 0.075 (no. 200) sieve	T 11	C 117
Soundness of aggregates by freezing and thawing	T 103	
Soundness of aggregates by sodium or magnesium sulfate method	T 104	C 88
Resistance to abrasion of small-size coarse aggregate by LA abrasion	T 96	C 131
Resistance to abrasion of large-size coarse aggregate by LA abrasion		C 535
Insoluble residue in carbonate aggregate		D 3042
Specific gravity and absorption of fine aggregate	T 84	C 128
Specific gravity and absorption of coarse aggregate	T 85	C 127
Moisture in aggregate by drying	T 255	C 566
Coating and stripping of aggregate mixtures	T 182	C 1664
Mechanical analysis of extracted aggregate	T 30	C 136, C 117
Effect of water on cohesion of compacted mixtures	T 165	D 1075
Resistance of compacted mixtures to moisture-induced damage	T 283	
Effect of moisture on asphalt concrete paving mixtures		D 4867
Asphalt content by nuclear method	T 287	
Compressive strength	T 167	
Specification: mixing plants for hot mix asphalt	M 156	
Hot asphalt paving mixtures		D 3515
Plant inspection	T 172	D 290
Random sampling		D 3665
Sampling paving mixtures	T 168	D 979
Degree of particle coating	T 195	D 2489
Marshall tests	T 245	D 1559
Hveem tests	T 246	
Bulk-specific gravity and density of compacted mixtures	T 166	D 2726
Bulk-specific gravity and density of compacted mixtures, paraffin coated	T 275	D 1188
Density of pavement, nuclear method		D 2950
Degree of compaction	T 230	
Theoretical maximum specific gravity of mixtures	T 209	D 2041
Thickness of compacted mixtures		D 3549
Percentage air voids of compacted mixtures	T 269	D 3203
Moisture or volatiles in mixtures	T 110	D 1461
Extractions	T 164	D 2172

TABLE 4.2

List of Relevant ASTM and AASHTO Standards for Rigid Pavement

Item	AASHTO	ASTM
Resistance to freezing and thawing	T 161, T 103	C 666, C 682
Resistance to disintegration by sulfates	T 104	C 295, C 3398
Gradation	T 11, T 27	C 117, C 136
Fine aggregate degradation		C 1137
Uncompacted void content of fine aggregate	T 304	C 1252
Bulk density (unit weight)	T 19	C 29
Fine aggregate grading limits	M 6	C 33
Compressive and flexural strength	T 22, T 97	C 39, C 78
Definitions of constituents		C 125, C 294
Aggregate constituents—maximum allowed of deleterious and organic materials	T 21	C 40
Resistance to alkali reactivity and volume change	T 303	C 227
Sizes of aggregate for road and bridge construction	M 43	D448
Bulk density ("unit weight") and voids in aggregate	T 19	C29
Coarse aggregate for Portland cement concrete	M 80	—
Organic impurities in fine aggregate for concrete	T 21	C40
Portland cement	M 85	C150
Compressive strength of cylindrical concrete specimens	T 22	C39
Making and curing concrete test specimens	T 23	C31
Sodium chloride	M 143	D632
Obtaining and testing drilled cores and sawed beams of concrete	T 24	C42
Calcium chloride	M 144	D98
Quality of water to be used in concrete	T 26	C1602
Liquid membrane-forming compounds for curing concrete	M 148	C309
Effect of organic impurities in fine aggregate on strength of mortar	T 71	C87
Ready-mixed concrete	M 157	C94
Air-entraining admixtures for concrete	M 154	C233
Resistance to degradation of small-size coarse aggregate by abrasion and impact in the Los Angeles machine	T 96	C131
Chemical admixtures for concrete	M 194	C494
Flexural strength of concrete (simple beam with third point loading)	T 97	C78
Lightweight aggregate for structural concrete	M 195	C330
Fineness of Portland cement by the turbidimeter	T 98	C115
Epoxy protective coatings	M 200	A884
Soundness of aggregates by freezing and thawing	T 103	—
Soundness of aggregates by use of sodium sulfate or magnesium sulfate	T 104	C88
Chemical analysis of hydraulic cement	T 105	C114
Use of apparatus for the determination of length change of hardened cement paste, mortar, and concrete	M 210	C490
Compressive strength of hydraulic cement mortar (using 50-mm or 2-in. cube specimens)	T 106	C109
Clay lumps and friable particles in aggregate	T 112	C142
Lightweight pieces in aggregate	T 113	C123
Slump of hydraulic cement concrete	T 119	C143
Mass per cubic meter (cubic foot), yield, and air content (gravimetric) of concrete	T 121	—
Blended hydraulic cements	M 240	C595
Making and curing concrete test specimens in the laboratory	T 126	C192
Concrete made by volumetric batching and continuous mixing	M 241	C685

QUESTIONS

1. Why are standards important?
2. List five different standard-authoring organizations for pavements.
3. How are standards developed?
4. Contact any pavement industry in your area and ask them which standards they use regularly. List these standards; critically review the different parts of any commonly used standard.

5 Traffic

5.1 DIFFERENT TYPES OF HIGHWAY TRAFFIC

Different types of vehicles use roadways, and different types of aircrafts use airport pavements. For roads, the main destructive effect comes from trucks, since other vehicles such as passenger cars are significantly lighter than trucks. Again, actual truck traffic on pavements consists of a variety of loads and axles. Trucks can be of the single-unit type or of the tractor-semitrailer or trailer type.

In the United States, combination trucks with 53-foot trailers are the most common. The gross maximum permissible weight is dictated by road agencies (such as the states in the United States), and depends on the number of axles that the truck consists of. Gross weights can range from 70,000 lb to 164,000 lb. There are specific guidelines regarding maximum weights per axle. For example, for highways other than interstates in the United States, the single-axle maximum is 18,000 lb and the tandem-axle maximum is 32,000 lb. The nominal spacing between a pair of axles in a tandem is 4 ft. Axles can be spread apart to accommodate different weights. The federal bridge formula in the United States specifies the maximum weights as follows:

$$W = 500 \left[\frac{LN}{N-1} + 12N + 36 \right]$$

where:
  W = maximum weight in lbs carried by any group of two or more axles
  L = the distance in feet between the extremes of any group of two or more axles
  N = the number of axles under consideration

Definitions of truck dimensions as well as some typical values are shown in Figure 5.1 and Figure 5.2.

5.2 MEASUREMENT OF TRAFFIC LOADS

For highway pavements, with the help of a weigh-in-motion (WIM) system, the load on each and every axle of each and every vehicle can be determined. In a mechanistic or mechanistic-empirical structural design process, such data can be used directly to determine the damage on the pavement by linking the stress/strain caused by such loads and the stress/strain versus performance relationships.

ASTM Standard E1318 describes WIM as a system to measure the dynamic tire forces of a moving vehicle and estimate the tire loads of the static vehicle. WIM is used to collect traffic data and classify it according to different times, such as days and weeks. The specific type of WIM that is used depends on the intended applications of the data, for example, collection of traffic data for design and/or weight enforcement. There are three primary types of WIM systems in use—load cell, bending plate, and piezoelectric sensor. In the load cell system, the load is directly measured with a load cell on scales placed on the pavement. In the bending plate system, strain gages attached at the bottom of a plate (that is, inserted in the road) record the strain from a moving vehicle. The strain

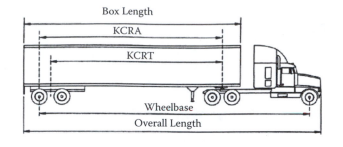

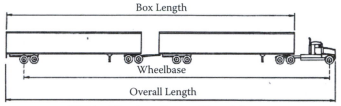

KCRA – Kingpin to Center of Rear Axle
KCRT – Kingpin to Center of Rear Tandem

Definition of Truck Dimensions

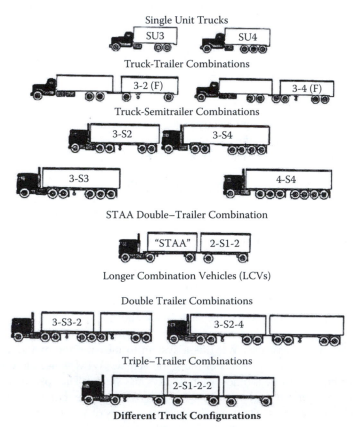

Different Truck Configurations

FIGURE 5.1 Truck Dimensions and Some Truck Configurations.
Source: From Harwood et al., 2003, Review of Truck Characterization as Factors in Roadway Design, NCHRP Report 505, Reprinted with kind permission of the Transportation Research Board.

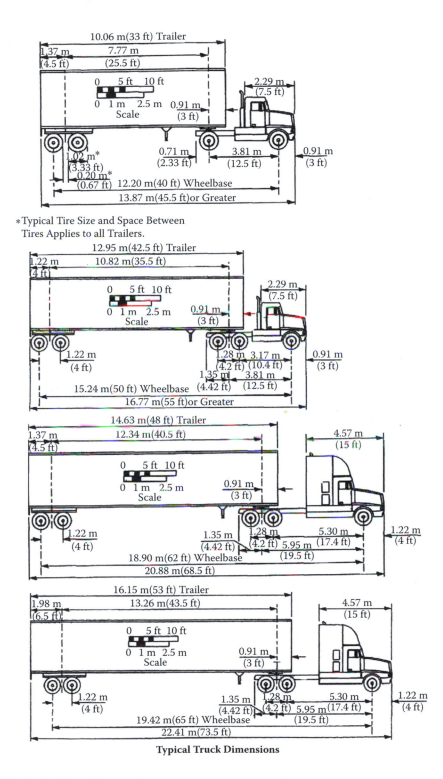

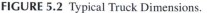

Typical Truck Dimensions

FIGURE 5.2 Typical Truck Dimensions.
Source: From Harwood et al., 2003, Review of Truck Characterization as Factors in Roadway Design, NCHRP Report 505, Reprinted with kind permission of the Transportation Research Board.

data is used to calculate the dynamic load and then the static load (using calibration constants). Piezoelectric sensors operate on the basis of generation of electricity in quartz-piezo sensors as a result of application of pressure on them. The quartz units in a sensor (placed in the pavement) are usually inserted in epoxy-filled aluminum channels. The electric charge generated from the sensor is detected, and the dynamic and hence the static load of the axles are calculated.

The data from the WIM system are processed by a data acquisition system and software at the site to generate time and traffic information (speed, weight of axle, and classification, for example), and then are transmitted to the office computer as data files (such as in ASCII format). These data files can then be processed, using separate software, to generate detailed reports, such as those required for checking and calibration of the WIM, and classification of the traffic according to days or weeks. Figure 5.3 shows the commonly used piezoelectric sensor, installation on the pavement, the on-site data acquisition system, the software output, and an example of traffic classification.

As noted in Figure 5.3, the traffic can be grouped into different types. For example, Table 5.1 shows the FHWA classification system.

In mechanistic-empirical processes, data from WIM and automatic vehicle classification (AVC) systems can be used to determine the number of axle applications for each axle type and axle load group (full axle-load spectrum) over the design period. Generally, vehicles in the range of classes 4 to 13 (FHWA classification system) are considered.

5.3   LOAD EQUIVALENCY FACTOR AND EQUIVALENT SINGLE-AXLE LOAD

The mixed stream of traffic is often considered in empirical design processes by converting the different axles with different loads to an equivalent number of standard axles, for example, the 18,000-lb- or 18-kip-load single axle. The equivalency is based on the assumption that the load equivalency factor (LEF; also known as the equivalent single-axle load factor, or ESAL factor) of a specific load/axle combination is the ratio of damage caused by one pass of the load/axle to a single pass of a standard 18-kip single axle. The damage can be represented in different methods—each design method may use a different parameter. The AASHTO design process uses the loss in serviceability, which is represented in terms of the Present Serviceability Index (PSI), as the selected parameter.

In the AASHTO design process, the terminal serviceability index, p_t, and the rigidity of the pavement structure (denoted by structural number, SN, for a flexible pavement and slab thickness, D, for a rigid pavement) are utilized, along with the specific axle/load information to determine a load equivalency factor. The equations that were used to generate the LEFs for different axle loads and configurations (single, tandem, and extended for tridem) are as follows (given in the 1986 AASHTO Guide, vol. 2, Appendix MM).

5.3.1   Flexible Pavements

$$\log_{10}\left[\frac{w_{t_x}}{w_{t_{18}}}\right] = 4.79 * \log_{10}(18+1) - 4.79 * \log_{10}(L_x + L_2) + 4.33 * \log_{10} L_2 + \frac{G_t}{\beta_x} - \frac{G_t}{\beta_{18}}$$

$$G_t = \log_{10}\left[\frac{4.2 - p_t}{4.2 - 1.5}\right]$$

$$\beta_x = 0.40 + \frac{0.081 * (L_x + L_2)^{3.23}}{(SN+1)^{5.19} * L_2^{3.23}}$$

Sensors

Elastic Material
Quartz Element
Aluminum Housing

WIM System

Peizo Sensor

On-Site Data Acquisition System

Real Time Visual Information from WIM System

Summary Report   09/28/2007 11:11:38

SiteName   : ST000017
Date/Time : From 03/01/2007 00:00 To 06/01/2007 00:00
Period   :
Report   : Gross Weight Distribution by Classification
Unit   : Kip
Class   : 1, 2, 3, 4, 5, 6, 7, 8, 9, 10, 11, 12, 13
Directions : Toward Capital Away From Capital
Lanes   : 0, 2, 4, 6, 7, 5, 3, 1

			Classification						
Weight	1	2	3	4	5	6	7	8	9	10
Under 5	1165	157384	13162	0	0	0	0	0	0	0
5 to 10	20	35556	69962	0	0	0	0	0	0	0
10 to 15	1	24	1611	199	2318	8	0	2	3	0
15 to 20	0	0	53	430	1026	170	0	22	4	7
20 to 25	0	0	0	498	800	359	1	70	5	20
25 to 30	0	0	0	103	539	244	1	210	84	85
30 to 35	0	0	0	68	340	195	4	243	368	827
35 to 40	0	0	0	45	169	78	2	76	437	2205
40 to 45	0	0	0	20	31	70	0	11	240	666
45 to 50	0	0	0	16	7	73	1	11	110	111
50 to 55	0	0	0	4	0	86	7	4	68	18
55 to 60	0	0	0	2	1	87	9	6	63	31
60 to 65	0	0	0	0	1	30	5	5	38	45
65 to 70	0	0	0	0	0	16	11	4	37	83
70 to 75	0	0	0	1	0	0	23	1	71	84
75 to 80	0	0	0	0	0	0	14	0	116	107
80 to 85	0	0	0	0	0	0	4	0	101	207
85 to 90	0	0	0	1	0	0	0	0	49	519
90 to 95	0	0	0	0	0	0	0	1	15	932
95 to 100	0	0	0	0	0	0	0	0	2	1258
100 to 105	0	0	0	0	0	0	0	1	8	1869
105 to 110	0	0	0	0	0	0	0	0	2	1975
110 to 115	0	0	0	0	0	0	0	0	1	844
115 to 120	0	0	0	0	0	0	0	0	0	248
Above 120	0	0	0	0	0	0	0	0	0	69
Total	1186	192964	84788	1387	5232	1416	82	667	1822	12210
%	0.4	64	28.1	0.5	1.7	0.5	0	0.2	0.6	4

Example of Report on Traffic Classification

FIGURE 5.3 Commonly Used Piezoelectric Sensor, Installation on a Pavement, On-Site Data Acquisition System, Software Output, and an Example of Traffic Classification.

56   Pavement Engineering: Principles and Practice

TABLE 5.1
Vehicle Classification

Class	Description
0	Unclassified vehicles which do not fit into any other classification. Vehicles which do not activate the system sensors are also unclassified.
1	Motorcycles. All two- or three-wheeled motorized vehicles. This category includes motorcycles, motor scooters, mopeds, and all three-wheel motorcycles.
2	Passenger cars. All sedans, coupes, and station wagons manufactured primarily for the purpose of carrying passengers.
3	Other two-axle, four-tire single units. Included in this classification are pickups, vans, campers, and ambulances.
4	Buses. All vehicles manufactured as traditional passenger-carrying buses with two axles and six tires or three or more axles.
5	Two-axle, single-unit trucks. All vehicles on a single frame, including trucks and camping and recreation vehicles.
6	Three-axle, single-unit trucks. All vehicles on a single frame, including trucks and camping and recreational vehicles.
7	Four or more axle, single-unit trucks. All vehicles on a single frame with four or more axles.
8	Four or less axle, single-trailer trucks. All vehicles with four or less axles consisting of two units, one of which is a tractor or straight truck power unit.
9	Five-axle, single-trailer trucks. All five-axle vehicles consisting of two units, one of which is a tractor or straight truck power unit.
10	Six or more axle, single-trailer trucks. All vehicles with six or more axles consisting of two units, one of which is a tractor or straight truck power unit.
11	Five or less axle, multitrailer trucks. All vehicles with five or less axles consisting of three or more units, one of which is a tractor or straight truck power unit.
12	Six-axle, multitrailer trucks. All six-axle vehicles consisting of three or more units, one of which is a tractor or straight truck power unit.
13	Seven or more axle, multitrailer trucks. All vehicles with seven or more axles consisting of three or more units, one of which is a tractor or straight truck power unit.

where:

L_x = load on one single tire or one tandem-axle set (kips)
L_2 = axle code (1 for single axle, and 2 for tandem axle)
SN = structural number
p_t = terminal serviceability
β_{18} = value of β_x when L_x is equal to 18 and L_2 is equal to 1
w_{t_x} = total applied load from a given traffic
$w_{t_{18}}$ = total applied load from an 18-kip single axle

5.3.2 RIGID PAVEMENTS

$$\log_{10}\left[\frac{w_{t_x}}{w_{t_{18}}}\right] = 4.62 * \log_{10}(18+1) - 4.62 * \log_{10}(L_x + L_2) + 3.28 * \log_{10} L_2 + \frac{G_t}{\beta_x} - \frac{G_t}{\beta_{18}}$$

$$G_t = \log_{10}\left[\frac{4.5 - p_t}{4.5 - 1.5}\right]$$

$$\beta_x = 1.00 + \frac{3.63 * (L_x + L_2)^{5.20}}{(D+1)^{8.46} * L_2^{3.52}}$$

where D = slab thickness, inch.
The equivalency factor,

$$e_x = \frac{w_{t_{18}}}{w_{t_x}}$$

The equation for converting total applications (w_{t_x}) of a given axle load and configuration into an equivalent number of applications of the standard 18-kip single-axle load is as follows:

$$w_{t_{18}} = w_{t_x} * e_x$$

Examples of load equivalence factors are shown in Table 5.2.

For design of a pavement, it is important that the total ESALs expected over the design period of the pavement (that is being designed) are estimated. This is accomplished by determining the current ESALs and applying a growth factor (provided in *AASHTO Guide*, vol. 1, Table D-20, p. D-23; AASHTO, 1986 and p. D-24, AASHTO, 1993) for the expected growth rate (percentage) and

TABLE 5.2

Examples of Axle Load Equivalency Factors for Flexible Pavements, Single Axles and p_t = 2.5[a] (Table D.4, p. D-6, AASHTO, 1986)

Axle Load (kips)	Pavement Structural Number (SN)					
	1	2	3	4	5	6
2	0.0004	0.0004	0.0003	0.0002	0.0002	0.0002
10	0.078	.102	0.118	0.102	0.088	0.080
18	1.00	1.00	1.00	1.00	1.00	1.00
30	10.3	9.5	7.9	6.8	7.0	7.8

Examples of Axle Load Equivalency Factors for Rigid Pavements, Single Axles and p_t = 2.5 (Table D.13, p. D-15, AASHTO, 1986)

Axle Load (kips)	Slab Thickness, D (inches)								
	5	7	8	9	10	11	12	13	14
2	0.0002	0.0002	0.0002	0.0002	0.0002	0.0002	0.0002	0.0002	0.0002
10	0.097	0.089	0.084	0.082	0.081	0.080	0.080	0.080	0.080
18	1.00	1.00	1.00	1.00	1.00	1.00	1.00	1.00	1.00
30	8.16	7.67	7.79	8.28	8.79	9.14	9.35	9.46	9.52

[a] p_t = terminal serviceability.

TABLE 5.3
Growth Rate Factors

Analysis Period, Years (n)	No Growth	Annual Growth Rate, % (g)						
		2	4	5	6	7	8	10
10	10	10.95	12.01	12.58	13.18	13.82	14.49	15.94
15	15	17.29	20.02	21.58	23.28	25.13	27.15	31.77
20	20	24.30	29.78	33.06	36.79	41.00	45.76	57.28
25	25	32.03	41.65	47.73	54.86	63.25	73.11	98.35

design life (years). Examples of growth factors are shown in Table 5.3.

$$\text{Growth Rate Factor} = \frac{(1+g)^n - 1}{g}, \quad \text{for } g \neq 0$$

$$g = \frac{\text{rate}}{100}$$

Finally, the estimated accumulated ESAL should be split into two directions, and split further considering the multiple lanes. The recommended lane-split factors are shown in Table 5.4. An example calculation is shown in Table 5.5.

Quite often, since load station data are not available everywhere, many agencies determine an ESAL factor (also known as a *truck factor*) for each type of vehicle, considering the total number of axles for that vehicle. The vehicle types can be selected on the basis of the FHWA vehicle classification system (thirteen different types of vehicles, ranging from Class 1 to Class 13). Furthermore, an average truck factor can be determined by adding all of the products of number of axles times LEF, and dividing the product by the total number of vehicles of that specific type. These data from weighing stations can then be used directly to estimate the expected total ESALs on the pavement (that is, the design traffic, or design ESAL). AASHTO recommends the use of SN = 5/D = 9 inch and p_t = 2.5 for picking the LEFs from the tables and designing the thickness; and if designed thickness is different from the assumed thickness by more than 1 inch, repeat the design by assuming a different SN and hence picking a different LEF. An example calculation is shown in Tables 5.6 and 5.7.

TABLE 5.4
Lane Distribution Factors

Number of Lanes in Both Directions	% of 18-Kip ESAL Traffic in Design Lane
1	100
2	80–100
3	60–80
4 or more	50–75

Source: From AASHTO Guide for Design of Pavement Structures © 1993, by the American Association of State Highway and Transportation Officials, Washington, D.C. Used by permission.

TABLE 5.5

Example: Calculation of 18-Kip Equivalent Single-Axle Load (ESAL)

Vehicle Type	Current Traffic (A)	Growth Factor for 2% Growth (B)	Design Traffic (C)	ESAL Factor (Truck Load Factor) (D)	Design ESAL
	From Growth Factor Table	*A*B*365, for annual traffic*		*A*B*C*D*	
Passenger Cars	5,925	24.30	52,551,787	0.0008	42,041
Buses	35	24.30	310,433	0.6806	211,280
Panel and Pickup Trucks	1,135	24.30	10,066,882	0.0122	122,816
Other 2-Axle/4 Tire Trucks	3	24.30	26,609	0.0052	138
2-Axlw/6-Tire Trucks	372	24.30	3,299,454	0.1890	623,597
3 or more Axle Trucks	34	24.30	301,563	0.1303	39,294
All Single Unit Trucks					
3 Axle Tractor Semi-Trailers	19	24.30	168,521	0.8646	145,703
4 Axle Tractor Semi-Trailers	48	24.30	434,606	0.6560	285,101
5+Axle Tractor Semi-Trailers	1,880	24.30	16,674,660	2.3719	39,550,626
All Tractor Semi-Trailers					
5 Axle Double Trailers	103	24.30	913,359	2.3187	2,118,268
6 + Axle Double Trailers	0	24.30			
All Double Trailer Combos					
3 Axle Truck-Trailers	208	24.30	1,844,856	0.0152	28,042
4 Axle Truck-Trailers	305	24.30	2,705,198	0.0152	41,119
5+Axle Truck Trailers	125	24.30	1,108,688	0.5317	589,489
All Truck-Trailer Combos					
All vehicles	10,193		90,406,816		**Design ESAL = 43,772,314** (*summation of all the above numbers in this column*)

From W-4 Table   *From W-4 Table*

Design Lane Traffic Estimate, in 18-kip ESAL = 43,772,314*(0.5)*(0.9) = 19,697,541

For Directional Split   *From Lane Distribution Table*

Note: For analysis period of 20 years and assumed thickness, D = 9 inches, for a four-lane rural interstate highway with equal traffic on both directions.

TABLE 5.6

Grouping of Load Equivalency Factor (LEF) for Structural Number = 5, p_t = 2.5[a]

Single-Axle Load (lb)	LEF
Under 3000	0.0002
3000–6999	0.0050
7000–7999	0.0320
8000–11,999	0.0870
12,000–15,999	0.3600
26,000–29,999	5.3890

[a] p_t = terminal serviceability.

Source: From AASHTO Guide for Design of Pavement Structure © 1986, by the American Association of State Highway and Transportation Officials, Washington D.C. Used by permission.

TABLE 5.7

Example of Calculation of Load Equivalency Factor (LEF) for Structural Number = 5, p_t = 2.5[a]

Single-Axle Load (lb)	LEF	Number of Axles	18-Kip Equivalent-Axle Load (EAL)
Under 3000	0.0002	0	0
3000–6999	0.0050	1	0.005
7000–7999	0.0320	6	0.192
8000–11,999	0.0870	144	12.528
12,000–15,999	0.3600	16	5.760
26,000–29,999	5.3890	1	5.3890
Tandem-Axle Load (lb)	**LEF**	**Number of Axles**	**18-Kip EAL**
Under 6000	0.0100	0	0
6000–11,993	0.0100	14	0.140
12,000–17,999	0.0400	21	0.924
18,000–23,999	0.1480	44	6.512
24,000–29,999	0.4260	42	17.892
30,000–32,000	0.7530	44	33.132
32,001–32,500	0.8850	21	18.585
32,501–33,999	1.0020	101	101.202
34,000–35,999	1.2300	43	52.890
	18-kip EALs for all trucks		255.151

Therefore, for five-axle trucks, as weighed in this specific weigh station,

$$\text{Truck load factor} = \frac{18\,\text{kip EALs for all trucks}}{\text{Number of trucks}} = \frac{255.151}{165} = 1.5464$$

[a] p_t = terminal serviceability.

Note: The numbers in this table represent data obtained from a weigh station (W-4 form) for 165 trucks of five-axle, tractor semitrailer type.

Source: From AASHTO Guide for Design of Pavement Structure © 1986, by the American Association of State Highway and Transportation Officials, Washington D.C. Used by permission.

5.4 ALTERNATIVE LOAD EQUIVALENT FACTOR CONCEPT

The concept of LEF can be approached in a different way also (as mentioned in Yoder and Witczak, 1975). Note that the ratio of damage by a single pass of the axle in question to a standard axle is the LEF. If N_a passes of an axle cause failure (note that the "failure" has to be defined) of a pavement, as do N_s passes of a standard axle, then the damage due to one pass of the axle and the standard axle can be denoted by $1/N_a$ and $1/N_s$, respectively. According to the definition of LEF,

$$LEF = \frac{\frac{1}{N_a}}{\frac{1}{N_s}} = \frac{N_s}{N_a}$$

Tests with fatigue cracking and analysis of cracking data with respect to tensile strain data show that the number of repetitions to failure of a pavement by fatigue cracking, due to repeated tensile strain, can be expressed as follows:

$$N_f = k_1 \left(\frac{1}{\varepsilon_t} \right)^{k_2}$$

where N_f is the repetitions to failure, ε_t is the tensile strain in the asphalt mix layer, and k_1 and k_2 are constants obtained by plotting the experimental data (log ε_t versus log N_f). It follows that LEF can also be then expressed as follows:

$$LEF = \left(\frac{\varepsilon_{ta}}{\varepsilon_{ts}} \right)^{k_2}$$

where ε_{ta} and ε_{ts} correspond to the strains for the axle in question and the standard axle. The value of k_2 has been reported to be between 3 and 6, with most common values being 4–5.

The most simple approach (and proven to be a good approximation) is to use the fourth power law to convert any axle load to an LEF as follows:

$$LEF = \left(\frac{W_a}{W_s} \right)^4$$

For example,

$$LEF \text{ for a 40,000-lb load } = \left(\frac{40,000}{18,000} \right)^4 = 24.4$$

(In comparison, the AASHTO LEF for a 40-kip single-axle load for a flexible pavement with SN = 5 and p_t = 2.5 is 21.1, and that for SN = 6 and p_t = 2.5 is 23.0.)

5.5 EQUIVALENT SINGLE-WHEEL LOAD

For airport pavements, the concept of equivalent single-wheel load (ESWL, as opposed to ESAL for highways) has been used. For fixed levels of traffic (for example, x number of departures), charts relating pavement thickness to CBR (California Bearing Ratio, a test property used to estimate strength of soil, Chapter 7) of the subgrade have been developed for specific wheel loads. This development has been done on the basis of conversion of multiple-wheel gear loads to ESWL. A critical aircraft is considered, and the ESWLs for all aircrafts are converted to those for the critical aircraft. The ESWL is defined as the load on a single tire that will cause the same response (such as stress, strain, or deflection) at a specific point in a given pavement system as that resulting from multiple-wheel loads at the same point in the same pavement system.

In determining the ESWL, either the tire pressure or the tire contact area of the ESWL is considered to be similar to those of the tires in the multiple-wheel gear. The choice of the response in developing the ESWLs for different gears depends on the agency using it—specifically, on its experience in correlating the response of choice to field performance. Derivations of ESWL can be made according to various procedures, such as the Asphalt Institute (AI), U.S. Navy, and U.S. Army Corps of Engineers methods. Examples are shown in Table 5.8.

In the Federal Aviation Administration (FAA) airport pavement design process, there are two steps in determining the equivalent load. First, each of the aircrafts (single-gear, dual-gear, dual-tandem-gear, and wide-body aircrafts) is converted into an equivalent gear, and then the number of departures is converted into the equivalent annual departures of the design aircraft. The sum of all of the equivalent annual departures is the design traffic. Furthermore, the departures are converted

TABLE 5.8
Different Methods of Computing ESWL

Method	Equation	Considerations
U.S. Navy	For dual tires: $P_S = P_D / (1 + (S_D/100))$ For dual tandem: $P_S = \dfrac{P_{dt}}{\left[1 + \left(\frac{S_d}{100}\right)\right]\left[1 + \left(\frac{S_d}{100}\right)\right]}$	Single-layer, equivalent stress concept is utilized at a depth of 30 in. for two tires separated center to center distance of S_d, tandem spacing of S_t, with duals contributing P_d load and dual tandem applying P_{dt} load.
U.S. Army Corps of Engineers	$P_e = \dfrac{P_k \sum_{i=1}^{n} F_{i\,max}}{F_e}$ $F_e = \dfrac{1.5}{\left[1 + \left(\frac{z}{a}\right)^2\right]^{0.5}}$ F_e is determined from charts relating depth (in terms of radii) to deflection factor for various offsets (in terms of radii—see Figure 5.4.	Single-layer, equal interface deflection criteria, n number of tires, each tire of radius a applying P_k load, separated center, considering the maximum deflection under the multiple gear; determination involves selecting multiple planes and determining the point where the deflection is maximum.
Asphalt Institute	$L = \dfrac{P_d}{P_e}$ L is read off charts designed for specific subgrade CBRs, relating S_d/a to L, for different h/a ratios; h = thickness.	Two-layer interface deflection criteria for dual wheels, with asphalt mix modulus of 100,000 psi and subgrade soil modulus equal to 1500 times CBR (in psi).

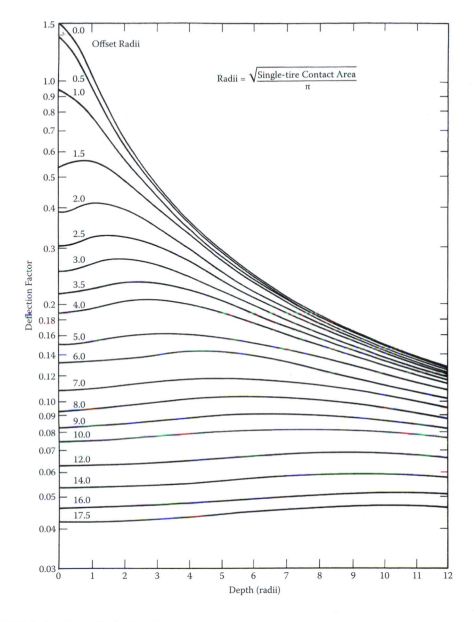

FIGURE 5.4 One-Layer Deflection Factor.
Source: From Yoder, E. J., and M. W. Witczak (1975) *Principles of Pavement Design*, 2nd edition © 1975. Reprinted with permission of John Wiley & Sons, Inc.

to "coverages" to take the lateral distribution of aircraft traffic into consideration. The following paragraphs explain the process.

5.5.1 Conversion to Equivalent Gear

The effect of a variety of aircrafts using an airport is considered by converting all aircrafts to the same landing gear type as the design aircraft, using conversion factors (same for flexible and rigid) that have been developed on the basis of relative fatigue effects. The conversion factors are as follows:

To Convert From	To	Multiply Departures By
Single wheel	Dual wheel	0.8
Single wheel	Dual tandem	0.5
Dual wheel	Dual tandem	0.6
Double dual tandem	Dual tandem	1.0
Dual tandem	Single wheel	2.0
Dual tandem	Dual wheel	1.7
Dual wheel	Single wheel	1.3
Double dual tandem	Dual wheel	1.7

5.5.2 CONVERSION TO EQUIVALENT ANNUAL DEPARTURE

The following formula is to be used:

$$\log R_1 = \log R_2 * \left(\frac{W_2}{W_1}\right)^{\frac{1}{2}}$$

where:

R_1 = equivalent annual departures by the design aircraft
R_2 = annual departures expressed in design aircraft landing gear
W_1 = wheel load of the design aircraft (95% of the gross weight is assumed to be carried by the main landing gear)
W_2 = wheel load of the aircraft in question

Each wide-body aircraft is treated as a 300,000-lb dual tandem aircraft for the determination of equivalent annual departures. An example is shown below.

Example

Assume an airport pavement is to be designed for the following forecast traffic.

Aircraft	Gear Type	Average Annual Departures	Maximum Take-Off Weight[a]	
			Lbs	Kgs
727-100	Dual	3760	160,000	(72,600)
727-200	Dual	9080	190,500	(86,500)
707-320B	Dual tandem	3050	327,000	(148,500)
DC-g-30	Dual	5800	108,000	(49,000)
cv-880	Dual tandem	400	184,500	(83,948)
737-200	Dual	2650	115,500	(52,440)
L-101 1-100	Dual tandem	1710	450,000	(204,120)
747-100	Double dual	85	700,000	(317,800)

[a.] Available from Federal Aviation Administration (1989)

Source: From Federal Aviation Administration, FAA Advisory Circular AC No. 150/5320-6D (1995), 25; see also Federal Aviation Administration, FAA Advisory Circular AC 150/5300-13 (1989)

Step 1. Determine design aircraft. Using charts provided in "Section 2. Flexible Pavement Design" in Federal Aviation Administration (1995), determine the thickness required for each aircraft. In this example, the thickest pavement is required for the 727-200 aircraft, which is, therefore, the design aircraft.

Step 2. Convert all aircrafts to landing gear of design aircraft.

Aircraft	Gear Type	Average Annual Departures	Multiplier	Equivalent Dual-Gear Departures
727-100	Dual	3760	—	3760
727-200	Dual	9080	—	9080
707-320B	Dual tandem	3050	1.7	5185
DC-g-30	Dual	5800	—	5800
cv-880	Dual tandem	400	1.7	680
737-200	Dual	2650	—	2650
L-101 l-100	Dual tandem	1710	1.7	2907
747-100	Double dual	85	1.7	145

Step 3. Convert equivalent dual-gear departures to equivalent annual departures for the design aircraft.

Aircraft	Equivalent Dual-Gear Departures	Maximum Take-Off Weight (lbs)	Wheel Load of Design Aircraft (190,500 * 0.95 * 0.5 * 0.5)	Equivalent Annual Departure of Design Aircraft
727-100	3760	160,000	45,240	1,891 ($\log R_1 = \log R_2 * (W_2/W_1)^{0.5}$) $R_2 = 3760$ $W_2 = 160,000$ $W_1 = 190,500$ $R_1 = 1891$
727-200	9080	190,500		9080
707-320B	5185	327,000		2764
DC-g-30	5800	108,000		682
cv-880	680	184,500		94
737-200	2650	115,500		463
L-101 l-100	2907	450,000		1184
747-100	145	700,000		83
Total design departures considering a 190,500-lb dual-wheel aircraft				16,241

5.6 TRUCK TIRE PRESSURE

The contact pressure at the interface of the tire and the pavement is important for determination of the structural response of the pavement. Quite often, the tire inflation pressure is assumed to be equal to the contact pressure. Either the tire inflation pressure is used, or, using an assumed elliptical or more commonly used circular contact area, the contact pressure is calculated from the load. The hot inflation pressure could be 10–15% more than the cold pressure, and can range from 90 to 120 psi, with loads ranging from 3600 lbs to 6600 lbs. Assumed tire contact areas are shown in Figure 5.5. The tire inflation pressure controls the tire contact stress at the center of the tire, whereas the load actually dictates the contact stress at the tire edges. An increasing trend is the replacement of duals

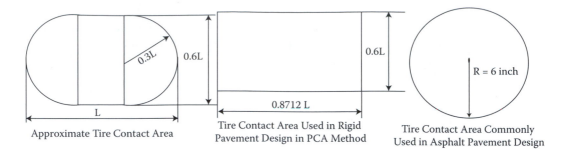

FIGURE 5.5 Assumed Tire Contact Areas.

with super-single tires, which have significantly higher pressures and hence more damaging effects on the pavement.

Tire contact pressure could be nonuniform across the contact area, depending on the condition of the tire. If the tire is underinflated, the pressure will be higher at the edges, and if it is overinflated, then the pressure is higher at the center. The existence of high shear stresses and/or tensile stresses due to anomalies in axle/loading/tire pressure can lead to top-down cracking in asphalt mix surface layers. Currently, this is not considered in the structural design of highway pavements.

5.7 TRUCK SPEED

For asphalt mix layers, the effect of loading time, which is dictated by the speed of the vehicle, is similar to the effect of temperature. In general, shorter loading times result in more elastic response and lower strains, whereas longer loading times result in more viscous response and larger strains. For structural design, the dynamic modulus can be determined at appropriate loading times to simulate the response under trucks moving at different speeds. For the same speed, any point at a greater depth from the surface will have a higher loading time compared to a point nearer to the surface. For an asphalt mix layer, one option is to assume that at the middle of the asphalt layer, the load is uniformly distributed over a circular area with a radius, r, where:

$$r = a + \frac{h}{2}$$

where a is the radius of the tire contact area and h is the thickness of the asphalt layer.

The loading time is expressed as t, where:

$$t = \frac{2a + h}{s}$$

where s = speed of the truck.

The other option is to consider that the moving load varies as a haversine function with time. The stress at any point can be considered to be zero or negligible when the load is at a distance greater than 6a from that point, where a is the radius of the tire contact area. The loading time can

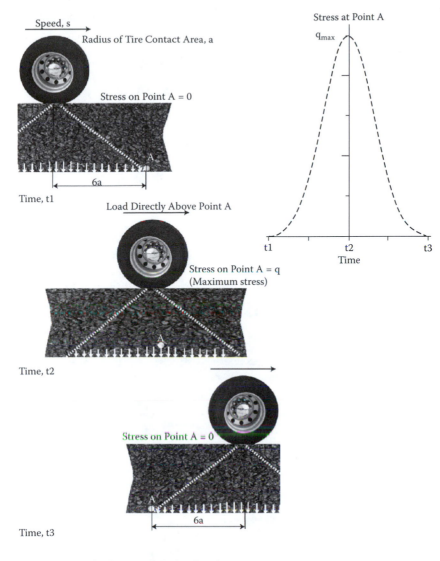

FIGURE 5.6 Stress at a Point Due to a Moving Load.

be approximately denoted as t:

$$t = \frac{12a}{s}$$

where s is the speed of the truck. The concept is illustrated in Figure 5.6.

5.8   AIRCRAFT LOADING, GEAR CONFIGURATION, AND TIRE PRESSURE

As the amount of load that an aircraft can carry has increased over the years, so have the number of gears and axles. At the same time, realizing that a higher number of gears means heavier weight and more fuel spent in carrying the aircraft itself, aircraft companies have also tried to reduce the number of gears, with a resulting increase in tire pressure and hence contact pressure on the

Aircraft	Gross Load on Main Landing Gear Group, lb	Gear Layout	Dimensions, in	Tire Pressure, psi
Airbus, A-300-B2	304,000	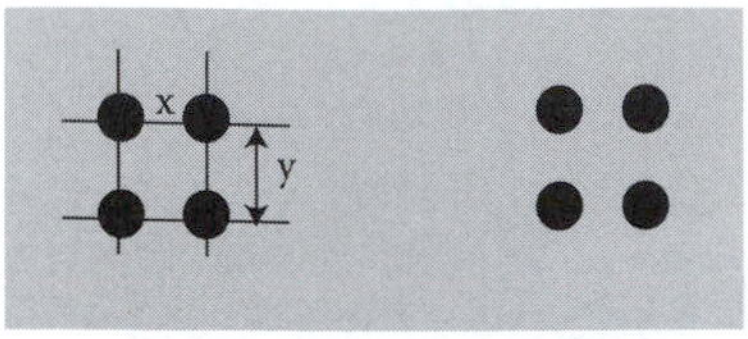	x = 35 y = 55	168
Airbus, A-330	460,000	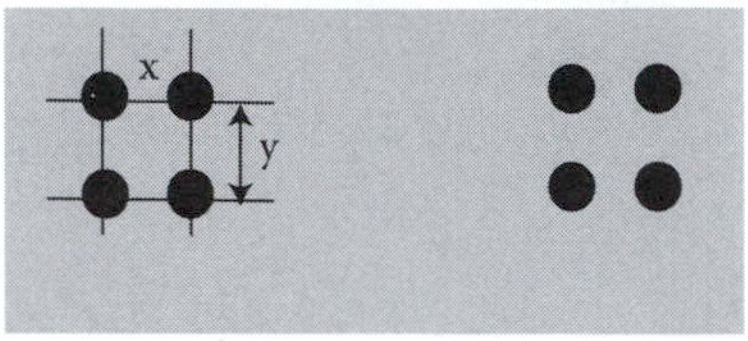	x = 55 y = 78	200
Airbus, A380	942, 700	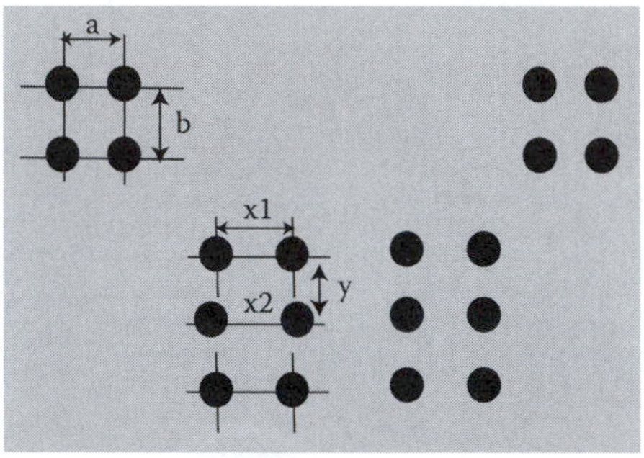	a = 53 b = 67 x1 = 60 x2 = 61 y = 67	194
Boeing, B-737-100	100,000	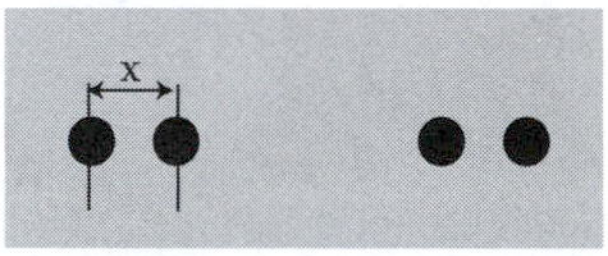	x = 55	148
Boeing, B-747-200	833,000	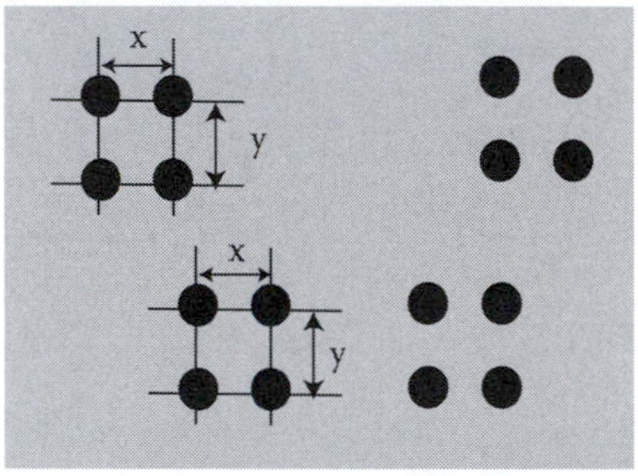	x = 44 y = 58	200
Boeing, B-777-200B	634,500	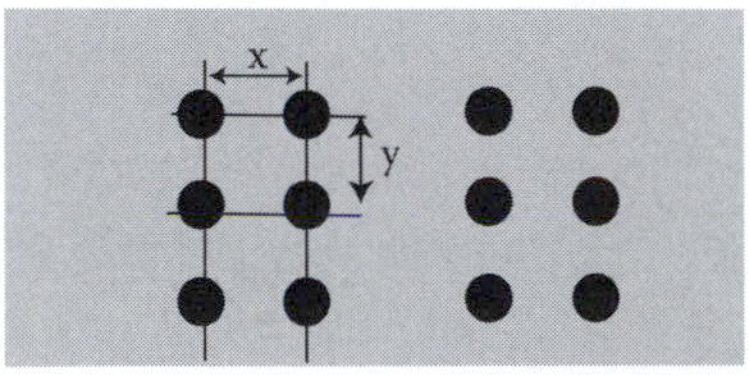	x = 55 y = 57	215

FIGURE 5.7 Gear Layout, Load, and Tire Pressure of a Few Commonly Used Civilian Aircrafts.

pavement surface. Figure 5.7 shows the load, gear configuration, and tire pressure of a few commonly used civilian aircrafts.

QUESTIONS

1. What is a weigh-in-motion (WIM) sensor, and how is it used?
2. Determine the load equivalency factor for the following cases, for both flexible and rigid pavements, for $p_t = 2.5$: pavement structural number (SN) = 5, D for rigid pavement = 10 in., axle loads = 2, 18, and 30 kips
3. Determine the total traffic for a 15-yr design period for the following case, for a four-lane new rigid pavement, with traffic growth of 2%. The truck load factors are provided for the different vehicles.

Vehicle Type	Current Traffic (A)	ESAL Factor (Truck Load Factor; D)
Passenger cars	5925	0.0008
Buses	35	0.6806
Panel and pickup trucks	1135	0.0122
Other two-axle/four-tire trucks	3	0.0052
Two-axle/six-tire trucks	372	0.1890
Three or more axle trucks	34	0.1303
All single-unit trucks		
Three-axle tractor semitrailers	19	0.8646
Four-axle tractor semitrailers	48	0.6560
Five-axle-plus tractor semitrailers	1880	2.3719
All tractor semitrailers		
Five-axle double trailers	103	2.3187
Six-axle-plus double trailers	0	
All double-trailer combos		
Three-axle truck-trailers	208	0.0152
Four-axle truck-trailers	305	0.0152
Five-axle-plus truck trailers	125	0.5317
All truck-trailer combos		
All vehicles	10,193	

4. Compute the load equivalency factor for loads of 9000, 18,000, 36,000, and 45,000 lbs by the approximate method.
5. Determine the ESWL for dual ties of a Boeing 737 (see Figure 5.7 for dimensions) using the U.S. Navy method.
6. Determine the time of loading for truck speeds of 10, 20, 40, 60, and 70 mph, and plot speed versus time of loading data.

6 Drainage

6.1 SOURCE AND EFFECT OF WATER

Precipitation, in the form of rain and snow, may result in accumulation of water on the surface, and find its way through joints and cracks inside a pavement structure. Water can also enter the subgrade soils through capillary action from a nearby water table. Water puddles or films on the surface, especially in ruts, can cause unsafe driving conditions through hydroplaning and skidding of vehicles, whereas water inside the pavement structure can result in deterioration of the mixtures and materials. If frost-susceptible soils are present, ice lenses can form due to freezing temperatures in the winter, causing heaves. And then thawing in the spring causes loss of support and consequent cracking on the surface. The sources and effects of water are summarized in Figure 6.1.

The damage due to the presence of water includes the following:

1. The aggregate/soil layers lose shear strength due to increase in moisture content or saturation, resulting in an increase in pore water pressure when subjected to traffic loads.
2. Repeated stress generation due to repeated freeze-thaw in asphalt as well as soil layers.
3. In the presence of water, vehicle loads cause hydrostatic pressures in soil layers, which cause movement of soil particles, especially near joints and cracks, leading to displacement of soil and hence loss of support. The end result is the failure of the pavement by cracking and deformation.

Of these three ways, the first two are taken care of in the drainage design, while the third one is considered in the materials selection and mixture design process.

Drainage refers to the system of providing ways to remove water from a pavement. In general, a drainage system consists of two separate subsystems—*surface* and *subsurface* drainage systems to avoid water coming from rain and/or snow, and to remove water coming in through surface voids and/or cracks and from groundwater, respectively.

6.2 ESTIMATING FLOW

It is necessary to understand some hydrologic concepts before starting the design of drainage structures. The amount of water that needs to be drained away comes as runoff over areas, and is dictated by the characteristics of the rainfall and the surface over which the water flows. This is illustrated with the rational method formula, which is the most commonly used equation to determine the runoff in drainage areas less than 80 hectares. The individual parameters in the formula are explained in the subsequent paragraphs.

The rational method equation is as follows:

$$Q = 0.00278 \, CIA$$

where:
Q = discharge, m³/s
C = runoff coefficient
I = rainfall intensity, mm/h
A = drainage area, hectares (ha) [1 ha = 10,000 m²]

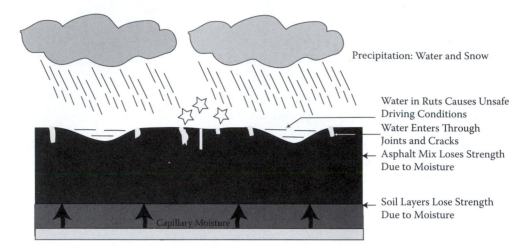

FIGURE 6.1 Sources and Effects of Water.

Runoff coefficient, C, is the ratio of runoff to rainfall. It is affected by many characteristics, the more important of which are the soil group, land use, and average land slope. If the drainage area consists of different types of surfaces, then a weighted C value is used as follows:

$$\text{Weighted C}, C_w = \frac{A_1 C_1 + A_2 C_2 + A_3 C_3 + \cdots}{A_1 + A_2 + A_3 + \cdots}$$

where, A_1, A_2, and A_3 are the subareas with different types of surfaces, and C_1, C_2, and C_3 are the corresponding runoff coefficients.

Typical values of runoff coefficients are shown in Figure 6.2.

Since the selected rainfall intensity depends on them, before discussing rainfall intensity, it is appropriate to discuss two other parameters, the *time of concentration* (T_c) and the *return period*. T_c is the time required for water to travel from the most hydraulically distant point of the drainage area to the drain system. It is used in determination of inlet spacing as well as pipe sizing. For inlet spacing, T_c is the time required for the water to flow from the hydraulically most distant point of the drainage area to the inlet (inlet time). This is the sum of the time required for water to flow across the pavement of overland back of the curb to the gutter plus the time required for flow to move through the length of the gutter to the inlet. If the total time of concentration to the upstream inlet is less than 7 minutes, then a minimum T_c of 7 minutes should be used. For pipe sizing, the T_c consists of the inlet time plus the time required for the water to flow through the storm drain to the point under consideration.

The time to flow overland can be determined from the chart shown in Figure 6.3 or by estimating the velocity from Figure 6.4, and dividing the length of flow by the velocity, whereas the time to flow within the storm drain can be determined simply by dividing the length of the pipe for runoff travel divided by the estimated normal velocity of water flow, which can be determined from Manning's equation, presented later in this chapter.

If there is more than one source of runoff to a given point in a drain system, the longest T_c is used. For municipal areas, a minimum T_c of 5 minutes is recommended.

6.2.1 RETURN PERIOD

It is not possible to design drainage structures for a maximum runoff that it could produce—say, in an interval of a long period of time. Therefore, for practical purposes, a realistic discharge, which

Recommended Coefficient of Runoff for Pervious Surfaces by Selected
Hydrologic Soil Groupings and Slope Ranges

Selected Hydrologic Soil Groupings And Slope Ranges

Slope	A	B	C	D
Flat (0 - 1%)	0.04-0.09	0.07-0.12	0.11-0.16	0.15-0.20
Average (2 - 6%)	0.09-0.14	0.12-0.17	0.16-0.21	0.20-0.25
Steep (Over 6%)	0.13-0.18	0.18-0.24	0.23-0.31	0.28-0.38

Source: Storm Drainage Design Manual, Erie and Niagara Counties Regional Planning Board.

Recommended Coefficient of Runoff Values for Various Selected Land Uses

Description of Area		Runoff Coefficients
Business:	Downtown areas	0.70-0.95
	Neighborhood areas	0.50-0.70
Residential:	Single-family areas	0.30-0.50
	Multi units, detached	0.40-0.60
	Multi units, attached	0.60-0.75
	Suburban	0.25-0.40
Residential (0.5 ha lots or more)		0.30-0.45
Apartment dwelling areas		0.50-0.70
Industrial:	Light areas	0.50-0.80
	Heavy areas	0.60-0.90
Parks, cemeteries		0.10-0.25
Playgrounds		0.20-0.40
Railroad yard areas		0.20-0.40
Unimproved areas		0.10-0.30

Source: Hydrology, Federal Highway Administration, HEC No. 19, 1984

Coefficients for Composite Runoff Analysis

Surface		Runoff Coefficients
Street:	Asphalt	0.70-0.95
	Concrete	0.80-0.95
Drives and walks		0.75-0.85
Roofs		0.75-0.95

Source: Hydrology, Federal Highway Administration, HEC No. 19, 1984

Characteristics	Soil Group			
	A	B	C	D
Runoff Potential	Low	Moderately Low	Moderately High	High
Primary type of soil	Deep well drained sands and gravel	Moderately deep to deep, moderately well to well drained soils with moderately fine to moderately coarse texture	Layer exists near the surface that impedes downward movement of water/soils with moderately fine to fine texture	Clays with high swelling potential, permanently high water tables, with claypan or clay layer at or near surface, shallow soils over nearly impervious parent material

FIGURE 6.2 Runoff Coefficients.
Source: From Model Drainage Manual, 2000 Metric Edition, © 2000, by the American Association of State Highway and Transportation Officials, Washington, D.C. Used by permission.

could result from the biggest rainstorm that is *probable* (as opposed to possible) within the design life of the drainage structure, is considered. Of course, the probability and hence the size of the biggest storm/flood would be different for different types of pavements—a lower probability and a bigger storm/flood would be considered for a more important highway than a less important one. For design

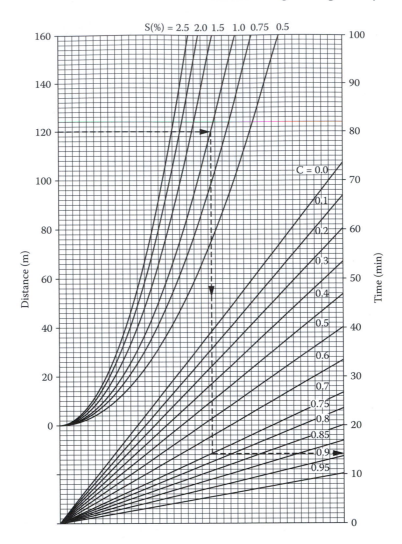

FIGURE 6.3 Direct Estimation of Overland Time of Flow.
Source: From Model Drainage Manual, 2000 Metric Edition, © 2000, by the American Association of State Highway and Transportation Officials, Washington, D.C. Used by permission.

purposes, all of the above considerations are made through two parameters, design frequency and return period (RP) or recurrence interval (RI). The frequency with which a given flood can be expected to occur is the reciprocal of the probability that the flood will be equaled or exceeded in a given year. For example, if a flood has a probability of 2% of being equaled or exceeded each year, over a long period of time, then the frequency is 1/(2/100) = 50, which means that the flood will be equaled or exceeded on an average of once in every 50 years. The RI or RP is also 50 years. Looking at this in another way, the probability of being exceeded = 100/RP. Suggested values of RP are provided in Table 6.1.

6.2.2 RAINFALL INTENSITY

Rainfall intensity, I, is the intensity of rainfall in mm/hour for a duration equal to the time of concentration. Intensity multiplied by the duration gives the amount of rain for that duration. The rainfall intensity-duration-frequency plots for any weather station could be generated from the weather

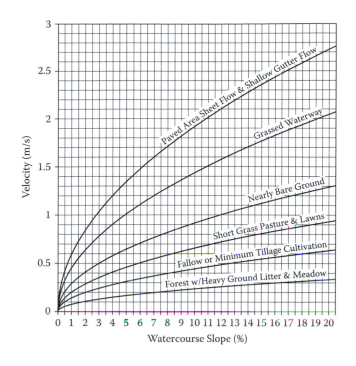

FIGURE 6.4 Estimation of Velocity of Flow to Compute Overland Time of Flow.
Source: From Model Drainage Manual, 2000 Metric Edition, © 2000, by the American Association of State Highway and Transportation Officials, Washington, D.C. Used by permission.

TABLE 6.1
Suggested Values of Return Period

Roadway Classification	Exceedance Probability	Return Period
Rural principal arterial system	(2%)	(50 years)
Rural minor arterial system	(4–2%)	(25–50 years)
Rural collector system, major	(4%)	(25 years)
Rural collector system, minor	(10%)	(10 years)
Rural local road system	(20–10%)	(5–10 years)
Urban principal arterial system	(4–2%)	(25–50 years)
Urban minor arterial street system	(4%)	(25 years)
Urban collector street system	(10%)	(10 years)
Urban local street system	(20–10%)	(5–10 years)

Note: Federal law requires interstate highways to be provided with protection from the 2% flood event, and facilities such as underpasses, depressed roadways, and the like, where no overflow relief is available, should be designed for the 2% event.

Source: From Model Drainage Manual, 2000 Metric Edition, © 2000, by the American Association of State Highway and Transportation Officials, Washington, D.C. Used by permission.

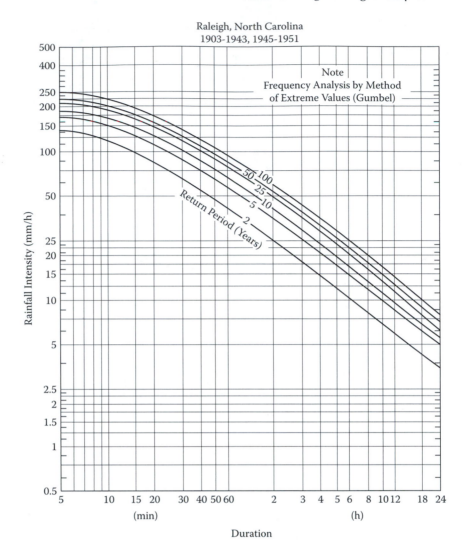

FIGURE 6.5 Example of Rainfall Intensity Curve.
Source: From Model Drainage Manual, 2000 Metric Edition, © 2000, by the American Association of State
Highway and Transportation Officials, Washington, D.C. Used by permission.

database at the NOAA Hydrometeorological Design Studies Center, Precipitation Frequency Data
Server (http://hdsc.nws.noaa.gov/hdsc/pfds/). An example is shown in Figure 6.5.

6.3 HYDROPLANING AND SURFACE DRAINAGE SYSTEM

If the water from rainfall exceeds the depth that is required for hydroplaning for a specific pavement width and rainfall intensity, hydroplaning can be expected. In a broad sense, the potential of
hydroplaning depends on vehicle speed, tire condition (pressure and tire tread), pavement micro
and macro texture, cross slope and grade, and pavement conditions such as rutting, depression,
and roughness. Apart from the driver's responsibility to control speed, the following steps could be
taken to reduce the potential of hydroplaning and/or prevent accidents:

1. Maximize transverse slope and roughness and use porous mixes (such as open graded friction course).
2. Provide gutter inlet spacing at sufficiently close spacing to minimize spread of water, and maximize interception of gutter flow above superelevation transitions.
3. Provide adequate slopes to reduce pond duration and depth in sag areas.
4. Limit depth and duration of overtopping flow.
5. Provide warning signs in sections identified as problem areas.

The basic principle of surface drainage is to design the pavement surface and the adjacent areas in such a way as to facilitate the quick flow of water falling in the form of rain and/or snow on the surface to the sides, and then drain it away to a nearest point of collection. This involves two major components—providing the adequate slope to the pavement surface and drainage channels with sufficient flow capacity along the pavement on either side.

The minimum grade for a gutter is 0.2% for curbed pavements. For sag vertical curves, a minimum of a 0.3% slope should be maintained within 15 m of the level point in the curve. In a very flat terrain, the slopes can be maintained by a rolling profile of the pavement. For one lane, the minimum pavement cross slope is 0.015 m/m, with an increase of 0.005 m/m for additional lanes. Slopes of up to 2% can be maintained without causing driver discomfort, whereas it may need to be increased to 2% for areas with intense rainfall.

In a multilane highway, if there are three or more lanes inclined at the same direction, it is recommended that each successive pair of lanes from the first two lanes from the crown line have an increased slope compared to the first two lanes (by about 0.5 to 1%), with a maximum pavement cross slope of 4%. Note that the allowable water depths on inside lanes are lower because of high-speed traffic on those lanes, and hence sloping of inside lanes toward the median should be done with caution. Pavements also have longitudinal slopes—with a minimum of 0.5% and a maximum of 3 to 6%, depending on the topography of the region.

A high level of macro texture on the pavement is desirable for allowing the rainwater to escape from the tire-pavement interface and reduce the potential of hydroplaning. Tining of PCC pavement, while it is still in plastic state, could be done to achieve this. For existing concrete pavements, macro texture can be improved by grooving and milling. Such grooving in both transverse and longitudinal directions is very effective in enhancing drainage. Use of porous asphalt mix layers such as open graded friction course allows rapid drainage of water. It is important that such layers are daylighted at the sides, and are constructed over layers that have been compacted adequately to prevent the ingress of water downward.

The coefficient of permeability of the surface mixture is dependent upon many factors, the more important of which are aggregate gradation and density. If the gradation consists of a relatively higher amount of coarse aggregate particles (compared to fine and filler materials), then the permeability is relatively high. On the other hand, permeability decreases with an increase in density.

Generally, the gradation is decided upon from other considerations, and the only available way for the pavement engineer to lower the permeability of the surface layer is through compaction and hence providing adequate density. Note that because of the effect of the gradation, the desirable density will be different for mixtures with different gradations. Also note that no matter how dense the surface layer is, some water will find its way through it, and hence there must be a way to get rid of this water—through the use of the subsurface drainage system (apart from the fact that the materials must be resistant to the action of water to a certain extent). The permeability of new rigid and asphalt mix pavement can be assumed to be 0.2 and 0.5 inches per hour, respectively.

There can be different types of drainage channels such as ditches, gutters, and culverts. The design of these channels means the design of the cross section. This design is accomplished with the help of Manning's formula (Daugherty and Ingersoll, 1954, equation):

$$Q = \frac{K}{n} S^{\frac{1}{2}} R^{\frac{2}{3}} A$$

where:

Q = pipe flow capacity, m³/s

S = slope of the pipe invert, m/m

n = pipe coefficient of roughness (0.012 for smooth pipe, and 0.024 for corrugated pipe; FHWA, 1992)

A = pipe cross-sectional area, m²

K = **1**

$$R = \frac{A}{P} = \frac{D}{4}$$

P = wetted perimeter of pipe, m

D = pipe diameter, m

The practical considerations for design include the elevation with respect to the subgrade (same or lower level than the subgrade), low construction and maintenance costs, and safeguard against slope failure (slope of 2:1 or less).

Curbs are concrete or asphalt mix structures provided along the side of the low-speed urban highways as well as bridge decks to facilitate collection of drained surface water from the pavement surface and protect pavement sides from erosion. Curbs are generally placed with gutters. In rural areas, roadside and median channels are provided instead of curbs and gutters.

For pavement drainage systems, two parameters are selected (Table 6.2), depending on the type of the pavement, design frequency, and spread (accumulated flow in and next to the roadway gutter, which can cause interruption to traffic flow, splash, or hydroplaning problems). The spread (T)

TABLE 6.2
Parameters for Drainage Systems

Road Classification		Design Frequency	Design Spread
High volume	< 72 km/h	10-year	Shoulder + 0.9 m
	> 72 km/h	10-year	Shoulder
	Sag point	50-year	Shoulder + 0.9 m
Collector	< 72 km/h	10-year	1/2 driving lane
	> 72 km/h	10-year	Shoulder
	Sag point	10-year	1/2 driving lane
Local streets	Low ADT	5-year	1/2 driving lane
	High ADT	10-year	1/2 driving lane
	Sag point	10-year	1/2 driving lane

Note: These criteria apply to shoulder widths of 1.8 m or greater. Where shoulder widths are less than 1.8 m, a minimum design spread of 1.8 m should be considered.

Source: From AASHTO (2000).

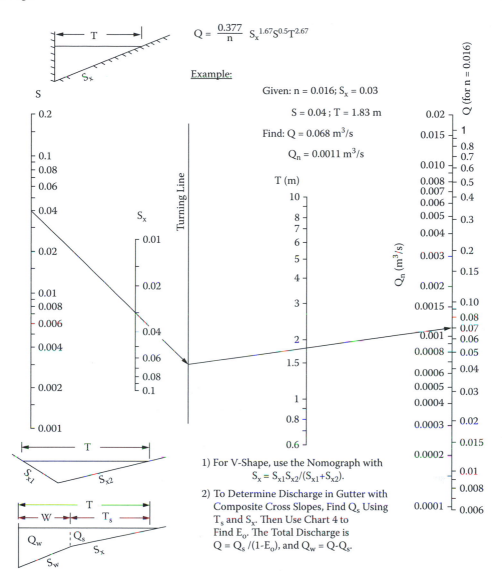

$$Q = \frac{0.377}{n} S_x^{1.67} S^{0.5} T^{2.67}$$

Example:

Given: $n = 0.016$; $S_x = 0.03$

$S = 0.04$; $T = 1.83$ m

Find: $Q = 0.068$ m³/s

$Q_n = 0.0011$ m³/s

1) For V-Shape, use the Nomograph with
$S_x = S_{x1}S_{x2}/(S_{x1}+S_{x2})$.

2) To Determine Discharge in Gutter with
Composite Cross Slopes, Find Q_s Using
T_s and S_x. Then Use Chart 4 to
Find E_o. The Total Discharge is
$Q = Q_s /(1-E_o)$, and $Q_w = Q-Q_s$.

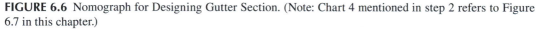

FIGURE 6.6 Nomograph for Designing Gutter Section. (Note: Chart 4 mentioned in step 2 refers to Figure 6.7 in this chapter.)
Source: From Model Drainage Manual, 2000 Metric Edition, © 2000, by the American Association of State Highway and Transportation Officials, Washington, D.C. Used by permission.

is constant for a specific design frequency—for higher magnitude storms, the spread can be allowed to utilize most of the pavement as an open channel.

The nomograph shown in Figure 6.6 can be used to design gutter sections. An example is shown on the plot. Manning's coefficients are provided in Table 6.3.

The next nomograph, Figure 6.7 (ratio of frontal to gutter flow), can be used to calculate the frontal flow for grate inlets and flow in a composite gutter section with width (W) less than the total spread (T), which can also be determined from Figure 6.8.

TABLE 6.3
Manning's Coefficients

Type of Gutter or Pavement	Manning's n
Concrete gutter, troweled finish	0.012
Asphalt pavement	
Smooth texture	0.013
Rough texture	0.016
Concrete gutter-asphalt pavement	
Smooth	0.013
Rough	0.015
Concrete pavement	
Float finish	0.014
Broom finish	0.016

Note: For gutters with a small slope, where sediment may accumulate, increase above *n* values by 0.002.

Sources: From Federal Highway Administration (1961); and AASHTO (2000).

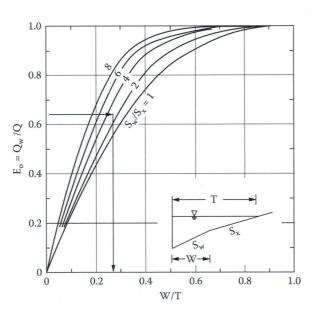

FIGURE 6.7 Ratio of Frontal Flow to Gutter Flow.
Source: From Model Drainage Manual, 2000 Metric Edition, © 2000, by the American Association of State Highway and Transportation Officials, Washington, D.C. Used by permission.

Example: Find Spread, Given Flow

Step 1. Given: $S = 0.01$, $S_x = 0.02$, $S_w = 0.06$, $W = 0.6$ m, $n = 0.016$, and $Q = 0.057$ m³/s trial value of $Q_s = 0.020$ m³/s.

Step 2. Calculate gutter flow: $Q_w = Q - Q_s = 0.057 - 0.020 = 0.037$.

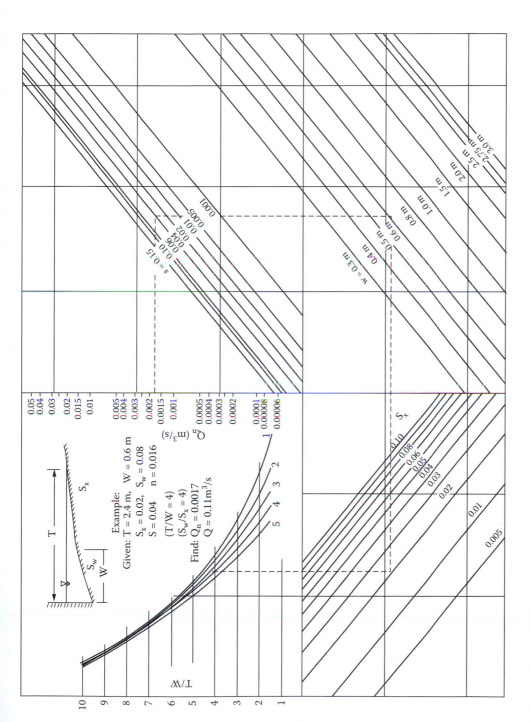

FIGURE 6.8 Flow in Composite Gutter Section.

Source: From Model Drainage Manual, 2000 Metric Edition, © 2000, by the American Association of State Highway and Transportation Officials, Washington, D.C. Used by permission.

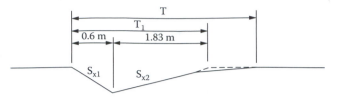

FIGURE 6.9 V-Type Gutter.
Source: From Model Drainage Manual, 2000 Metric Edition, © 2000, by the American Association of State Highway and Transportation Officials, Washington, D.C. Used by permission.

Step 3. Calculate ratios of $(E_o =) Q_w/Q = 0.037/0.057 = 0.65$, and $S_w/S_x = 0.06/0.02 = 3$. And, using Figure 6.7, find an appropriate value of W/T; W/T = 0.27.
Step 4. Calculate T: $T = W/(W/T) = 0.6/0.27 = 2.22$ m.
Step 5. Find spread above the depressed section: $T_s = 2.22 – 0.6 = 1.62$ m.
Step 6. Use Figure 6.6 to determine Q_s: $Q_s = 0.014$ m³/s.
Step 7. Compare Q_s (0.014) from Step 6 to the assumed value of Q_s (0.020); since they are not close, try another value of Q_s (for example, 0.023) and repeat until the calculated and assumed values are close.

The use of Figure 6.8 is illustrated with an example on the figure.

Figure 6.6 can be used to solve the flow for a V-gutter section (utilizing Figure 6.9), such as those in triangular channel sections adjacent to concrete median barriers, using the following steps:

1. Determine S, S_x, n, and Q.
2. Calculate $S_x = \frac{S_{x1}+S_{x2}}{S_{x1}+S_{x2}}$.
3. Solve for Q (flow) using Figure 6.6.

6.4 INLETS

Inlets are provided at regular intervals to collect surface water and convey them to the storm drains. The inlets could be of the grate or curb-opening type or a combination of both. Grate inlets are suitable for continuous grades and should be made bike safe where bike traffic is expected. Curb openings are more suitable for sag points, since they can let in large quantities of water, and could be used where grates are hazardous for bikes or pedestrians. Inlets are spaced at regular intervals, as explained below, and also at points such as sag points in the gutter grade, upstream of median breaks, entrance and exit ramps, crosswalks and intersections, immediately upstream and down-stream of bridges, side streets at intersections, the end of channels in cut sections, behind curbs, shoulders or sidewalks to drain low areas, and where necessary to collect snowmelt, and *not* in areas where pedestrian traffic is expected.

The inlet spacing should be calculated on the basis of collection of runoff. Inlets should be first located from the crest working downgrade to the sag point, by first calculating the distance of the first inlet from the crest and then computing the distances of the other successive inlets. The distance of the first inlet from the crest is calculated as follows.

$$L = \frac{10,000Q_t}{0.0028CiW},$$

where:

L = distance of the first inlet from the crest, m
Q_t = maximum allowable flow, m³/s
C = composite runoff coefficient for contributing drainage area
W = width of contributing drainage area, m
i = rainfall intensity for design frequency, mm/h

6.5 SUBSURFACE DRAINAGE SYSTEM

Subsurface drainage involves providing a subsurface system to remove subsurface water. This water can be from any of the following:

1. Water coming from nearby groundwater
2. Water infiltrating the surface layers through voids and/or cracks
3. Water seeping up from artesian aquifers
4. Water resulting from the thawing of ice lenses in soil layers

6.5.1 GROUNDWATER

Groundwater fluctuations can cause ingress of water, but generally the water table is low or lowered enough so as to avoid its effect. For cases where the groundwater table does need to be considered for seepage of water into the pavement, the following method can be used to estimate the inflow.

Refer to the general case illustrated in Figure 6.10.

Step 1. Determine the parameter, L_i, radius of influence: $L_i = 3.8 (H - H_0)$.
Step 2. Estimate q_1, inflow from above the bottom of the drainage layer:

$$q_1 = \frac{k(H - H_0)}{2L_i}$$

where k is the permeability of the soil in the cut slope.

Step 3. Determine the parameter, $\frac{L_i + 0.5W}{H_0}$.
Step 4. Determine the parameter, W/H_0.
Step 5. Use the chart in Figure 6.11 to estimate the parameter: $\frac{k(H-H_0)}{2q_2}$.

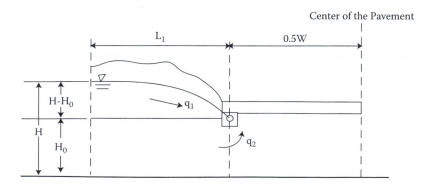

FIGURE 6.10 Flow from Groundwater.

84   Pavement Engineering: Principles and Practice

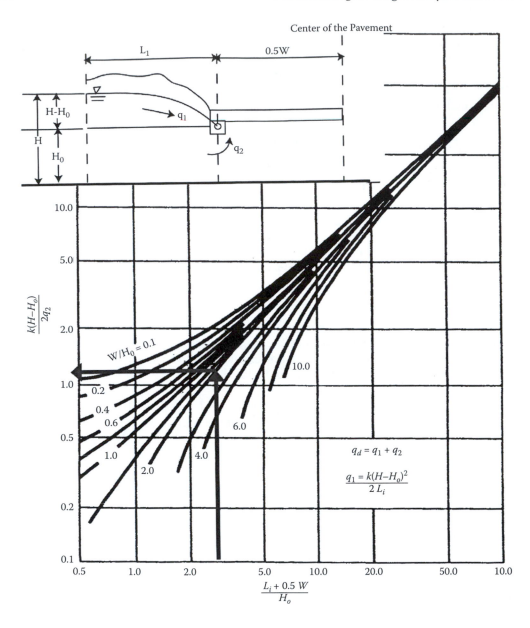

FIGURE 6.11 Estimation of K.
Source: From Garber/Hoel. *Traffic and Highway Engineering*, 3 E. © 2002 Nelson Education Ltd. Reproduced by permission. www.cengage.com/permissions.

Step 6. Knowing k, permeability of the soil in the subgrade; H and H_0, determine q_2, the seepage inflow from the bottom of the drainage layer.

Step 7. Determine the total lateral inflow into the drainage pipe: $q_d = q_1 + q_2$, for collector pipes on both sides; and $q_d = 2(q_1 + q_2)$, for a collector pipe on one side of the pavement only.

Step 8. Determine the groundwater inflow into the drainage layer per unit area: $q_g = \frac{2q_2}{W}$, for pavement sloped on both sides and collector pipes on both sides; and $q_g = \frac{q_1 + 2q_2}{W}$, for pavement sloped on one side with a collector pipe on one side.

6.5.2 Water Entering through Cracks

Water enters a pavement system through cracks and joints in the pavement surface or between the pavement and the shoulder, as well as in the shoulder and side ditches. The common method to determine the infiltration rate in an uncracked pavement is the use of the following equation:

$$q_i = I_c \left[\frac{N_c}{W} + \frac{W_c}{WC_s} \right] + k_p$$

where:

q_i = rate of pavement infiltration, m^3/day/m^2 (ft^3/day/ft^2)

I_c = crack infiltration rate, m^3/day/m (ft^3/day/ft); suggested value, (I_c) = 0.223 m^3/day/m, 2.4 ft^3/day/ft of crack (Ridgeway, 1976)

N_c = number of longitudinal cracks

W_c = length of contributing transverse joints or cracks, m (ft)

W = width of permeable base, m (ft)

C_s = spacing of contributing transverse joints or cracks, m (ft)

k_p = pavement permeability, m/day (ft/day); suggested permeability of uncracked specimens of asphalt pavement (AC) after being subjected to traffic and Portland Cement Concrete pavement (PCC) are on the order of 1×10^{-9} cm/s (15×10^{-5} ft/day) (Barber and Sawyer, 1952)

6.5.3 Artesian Aquifers

For seepage of water from artesian aquifers, the following equation is used to estimate the inflow:

$$q_a = K \frac{\Delta H}{H_0}$$

where:

q_a = inflow from artesian source, ft^3/day/ft^2 of drainage area

ΔH = excess hydraulic head (ft)

H_0 = thickness of the subgrade soil between the drainage layer and the artesian aquifer (ft)

K = coefficient of permeability (ft/day)

6.5.4 Melting Snow

Melting of snow and ice also contributes to water flow inside the pavement. Moulton's chart (Figure 6.12) can be used to estimate the inflow.

The required parameters are overburdened pressure that can be determined from the density of the materials in the layer above this layer, the permeability of the subgrade material, as well as the frost susceptibility of the soil. The average rate of heave can be determined experimentally or from Moulton's table (Table 6.4).

6.6 DESIGN OF SUBSURFACE DRAINAGE STRUCTURES

In general, the subsurface drainage system consists of a drainage layer in the subsurface part of the pavement and side drainage channels—the drainage layer of high-permeability material slopes away on both sides to intercept subsurface water and direct it sideways to drainage channels. It is important to provide filters in the drainage layer and the channels such that they remove water and water only, and do not let finer soil particles wash out with the water. The various components are shown in Figure 6.13.

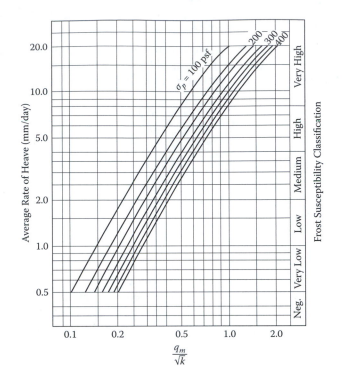

FIGURE 6.12 Chart for Estimating Inflow Due to Melting Snow.
Source: From Moulton (1980).

For rigid pavements, there is a separator layer between the subgrade (to prevent migration of fines) and the permeable base over which the PCC layer is placed. In asphalt pavements, the permeable base layer is placed under adequate thickness of the asphalt mix layer.

The resultant slope of the permeable base is given by the following:

$$S_R = \left(S^2 + S_x^2\right)^{0.5}$$

where:
 S_R = resultant slope, ft/ft
 S = longitudinal slope, ft/ft
 S_x = cross slope, ft/ft

The resultant length of the flow path is the following:

$$L_R = W[1 + (S/S_x)^2]^{0.5}$$

where:
 L_R = resultant length of flow path through permeable base, ft
 W = width of permeable base, ft

The steps in designing a drainage system for pavements consist of determining the amount of water, designing a permeable base and the separation layer, determining the flow to edge drains and spacing of outflows, and checking the outlet flow.

The coefficient of permeability, k, is an important factor and is primarily dictated by effective rain size, D_{10}, porosity, n, and percentage passing the 0.075 mm sieve. The addition of stabilizer

TABLE 6.4

Heave Rates for Various Soil Types

Unified Classification	Symbol	% < 0.02 mm	Heave Rate (mm/day)	Frost Susceptibility
Gravels and sand gravels	GP	0.4	3.0	Medium
Gravels and sand gravels	GW	0.7–1.0	0.3–1.0	Low
Gravels and sand gravels	GW	1.0–1.5	1.0–3.5	Low to medium
Gravels and sand gravels	GW	1.5–4.0	3.5–2.0	Medium
Silty and sandy gravels	GP—GM	2.0–3.0	1.0–3.0	Low to medium
Silty and sandy gravels	GW—GM and GM	3.0–7.0	3.0–4.5	Medium to high
Clayey and silty gravels	GW—GC	4.2	2.5	Medium
Clayey and silty gravels	GM—GC	15.0	5.0	High
Clayey and silty gravels	GC	15.0–30.0	2.5–5.0	Medium to high
Sands and gravelly sands	SP	1.0–2.0	0.8	Very low
Silty and gravelly sands	SW	2.0	3.0	Medium
Silty and gravelly sands	SP—SM	1.5–2.0	0.2–1.5	Low
Silty and gravelly sands	SW—SM	2.0–5.0	1.5–6.0	Low to high
Silty and gravelly sands	SM	5.0–9.0	6.0–9.0	High
Clayey and silty sands	SM—SC and SC	9.5–35.0	5.0–7.0	High
Silts and organic silts	ML—OL	23.0–33.0	1.1–14.0	Low to high
Silts and organic silts	ML	33.0–45.0	14.0–25.0	Very high
Clayey silts	ML—CL	60.0–75.0	13.0	Very high
Gravelly and sandy clays	CL	38.0–65.0	7.0–10.0	High
Lean clays	CL	65.0	5.0	High
Lean clays	CL—OL	30.0–70.0	4.0	High
Fat clays	CH	60.0	0.8	Very low

Source: Moulton (1980).

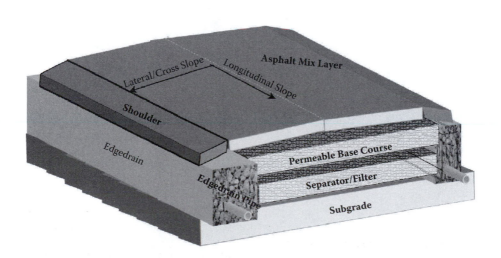

FIGURE 6.13 Various Components of Drainage System.

such as asphalt or Portland cement can counterbalance the effect of replacing the fine aggregate portion and hence the loss in stability.

Although k can be determined in the laboratory, quite commonly k is estimated from empirical equations such as the following:

$$k = C_k D_{10}^2$$

where:

    k = permeability, cm/s

    D_{10} = effective grain size corresponding to size passing 10%

    C_k = Hazen's coefficient, 0.8–1.2

Another equation proposed by Moulton (1980) is as follows:

$$k = \frac{(6.214 * 10^5) D_{10}^{1.478} n^{6.654}}{P_{200}^{0.597}}$$

where:

    n = porosity

    P_{200} = percentage passing No. 200 sieve

The void ratio or porosity has a significant effect on k and the amount of water that can stay within the soil, and this is important since all of the water within a soil cannot be removed, since some of the water would remain as thin film. The porosity that is effective in determining how much water can be removed is called the *effective porosity*.

The total porosity, or porosity, is defined as follows:

$$n = \frac{V_v}{V_T}$$

where:

    V_T = total volume

    V_v = volume of voids

    $V_v = V_T - V_S$

    V_S = volume of solid

$$V_v = V_T - \frac{\gamma_d V_T}{\gamma_w G_S}$$

    G_s = specific gravity of soil

    γ_d = dry unit weight of soil

    γ_w = unit weight of water

Effective porosity:

$$n_e = \frac{V_v - V_R}{V_T} = n - \frac{V_R}{V_T}$$

where

    V_R = volume of the water retained in the soil.

The volume of water retained in a soil:

$$V_R = \frac{\gamma_d w_c}{\gamma_w}$$

where w_c = water content of the soil after draining.
   If $V_T = 1$,

$$n_e = n - \frac{\gamma_d w_c}{\gamma_w}$$

If a test is conducted to determine the volume of water draining under gravity from a known volume of material, then the effective porosity can be computed as follows:

$$n_e = \frac{V_e}{V_T}$$

where V_e = volume of water draining under gravity.
   If W_L is the *water loss* percentage—that is, water drained from the sample—then the effective porosity can be expressed as follows (FHWA 1992):

$$n_e = \frac{n W_L}{100}$$

Reported values of W_L expressed as a percentage are as follows:

	< 2.5% Fines			5% Fines			> 5% Fines		
	Filler	Silt	Clay	Filler	Silt	Clay	Filler	Silt	Clay
Gravel	70	60	40	60	40	20	40	30	10
Sand	57	50	35	50	35	15	25	18	8

Source: FHWA (1992).

6.6.1 DESIGN OF PERMEABLE BASE

The permeable base can be designed according to one of the two available methods. The Moulton (1979) method is based on the idea that the thickness of the base should be equal to or greater than the depth of the flow, which means that the steady flow capacity of the base should be equal to or greater than the rate of inflow. The design equations are as follows:

   k = permeability
   S = slope
   L_R = length of drainage
   q_i = rate of uniform inflow
   H_1 = depth of water at the upper end of the flow path

$$\text{Case 1: } \left(S^2 - \frac{4q_i}{k} \right) < 0$$

$$H_1 = \sqrt{\frac{q_i}{k}} L_R \left[\left(\frac{S}{\sqrt{\frac{4q_i}{k-S^2}}} \right) \left(\tan^{-1} \frac{S}{\sqrt{\frac{4q_i}{k-S^2}}} - \frac{\pi}{2} \right) \right]$$

Case 2: $\left(S^2 - \dfrac{4q_i}{k} \right) = 0$

$$H_1 = \sqrt{\frac{q_i}{k}} L_R^{-1}$$

Case 3: $\left(S^2 - \dfrac{4q_i}{k} \right) > 0$

$$H_1 = \sqrt{\frac{q_i}{k}} L_R \left[\frac{S - \sqrt{S^2 - \frac{4q_i}{k}}}{S + \sqrt{S^2 - \frac{4q_i}{k}}} \right]^{\frac{S}{2\sqrt{S^2 - \frac{4q_i}{k}}}}$$

The equations can be solved with the use of a chart (Figure 6.14). The chart can be used to determine the maximum depth of flow, or the required k of the material, when the other parameters are known. The use of the chart is based on the assumption that H_1 equals H_{max}, which is the maximum depth of the flow.

The second approach is based on the concept of time to drain, specifically 50% drainage in 10 days, developed on the basis of a study with freeze-thaw-susceptible base courses by Casagrande and Shannon (1952).

For the following equations:

U = percentage drainage (expressed as a fraction, e.g., 1% = 0.01)
S_1 = slope factor = H/(LS)
H = thickness of granular layer
L = width of granular layer being drained
S = slope of granular layer
T = time factor = tkH/n_eL^2
 t = time for drainage, U, to be reached
 k = permeability of granular layer
n_e = effective porosity of granular material

If U > 0.5,

$$T = \left(1.2 - \frac{0.4}{S_1^{\frac{1}{3}}} \right) \left[S_1 - S_1^2 \ln\left(\frac{S_1 + 1}{S_1} \right) + S_1 \ln\left(\frac{2S_1 - 2US_1 + 1}{(2 - 2U)(S_1 + 1)} \right) \right]$$

If U ≤ 0.5,

$$T = \left(1.2 - \frac{0.4}{S_1^{\frac{1}{3}}} \right) \left[2US_1 - S_1^2 \ln\left(\frac{S_1 + 2U}{S_1} \right) \right]$$

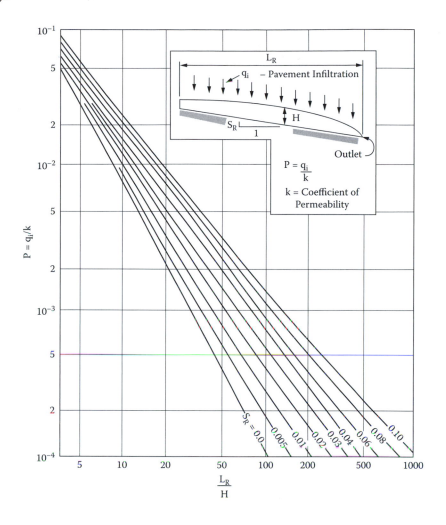

FIGURE 6.14 Nomograph for Solving Moulton's (1980) Equation.
Source: From Moulton (1980).

The equations can be solved by charts as shown in Figure 6.15.
Barber and Sawyer (1952) equations are as follows:

U = percentage drainage (expressed as a fraction, e.g., 1% = 0.01)
S_1 = slope factor = H/DS
H = thickness of granular layer
D = width of granular layer being drained
S = slope of granular layer
T = time factor = $(tkH)/(n_e L^2)$
t = time for drainage, U, to be reached
k = permeability of granular layer
n_e = effective porosity of granular material

For $0.5 \leq U \leq 1.0$,

$$T = 0.5S_1 - 0.48S_1^2 \log\left(1 + \frac{2.4}{S_1}\right) + 1.15S_1 \log\left[\frac{S_1 - US_1 + 1.2}{(1 - U)(S_1 + 2.4)}\right]$$

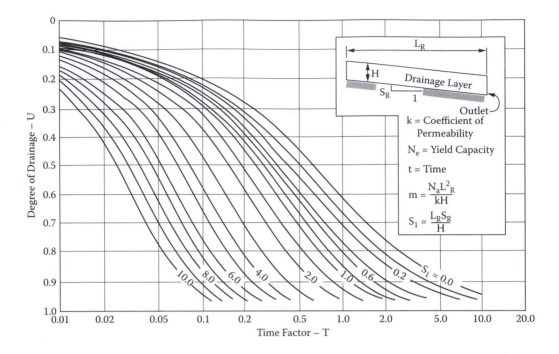

FIGURE 6.15 Nomograph for Solving Casagrande and Shannon's (1952) Equations.
Source: From Casagrande and Shannon, 1952, with permission from ASCE.

For $0 \leq U \leq 0.5$,

$$T = US - 0.48S_1^2 \log\left(1 + \frac{4.8U}{S_1}\right)$$

6.6.1.1 Materials for Permeable Base

The time to drain of a soil depends primarily on its coefficient of permeability, which is primarily dependent on the aggregate gradation. Gradations of typical materials (as suggested by FHWA, 1999) are given in Tables 6.5, 6.6, and 6.7. Generally materials with permeability exceeding 1000 ft per day are used for highways.

6.6.2 Design of Separator or Filter Layer

The separator layer can be made up of aggregates or a geotextile layer. The functions of this layer include preventing pumping of fines from the subgrade to the permeable base, providing a stable platform to facilitate the construction of the overlying layers, directing water infiltrating from above to the side drains or edge drains and preventing it from entering the subgrade, and distributing the loads over the subgrade without deflecting excessively. Separator layers with aggregates range in thickness from 4 to 12 inches and can provide the stable construction platform as well as distribute loads over the subgrade without deflecting. Generally, geotextiles are used over stabilized subgrades. The separator layer should be such that subgrade fines do not move up to the separator layer

TABLE 6.5

Typical Unstabilized Permeable Base Gradations

State	2 in.	1 1/2 in.	1 in.	3/4 in.	1/2 in.	3/8 in.	No. 4	No. 8	No. 16	No. 40	No. 50	No. 200
AASHTO #57		100	95–100		25–60		0–10	0–5				
AAHSTO #67			100	90–100		20–55	0–10	0–5				
Iowa			100					10–35			0–15	0–6
Minnesota			100	65–100		35–70	20–45			2–10		0–3
New Jersey		100	95–100		60–80		40–55	5–25	0–12		0–5	
Pennsylvania	100			52–100		33–65	8–40		0–12			0–5

Source: FHWA (1999).

TABLE 6.6

Typical Asphalt-Stabilized Permeable Base Gradations

State	1 in.	1/2 in.	3/8 in.	No. 4	No. 8	No. 200
California	100	90–100	20–45	0–10		0–2
Florida	100	90–100	20–45	0–10	0–5	0–2
Illinois	90–100	84–100	40–60	0–12		
Kansas	100	90–100	20–45	0–10	0–5	0–2
Ohio	95–100			25–60	0–10	
Texas	100	95–100	20–45	0–15	0–5	2–4
Wisconsin	95–100	80–95	25–50	35–60	20–45	3–10
Wyoming	90–100	20–50		20–50	10–30	0–4

Source: FHWA (1999).

TABLE 6.7

Typical Cement-Stabilized Permeable Base Gradations

State	1 1/2 in	1 in	3/4 in	1/2 in	3/8 in	No. 4	No. 8
California	100	88–100	X + 15		X + 15	0–16	0–6
Virginia		100		25–60		0–10	0–5
Wisconsin		100	90–100		20–55	0–10	0–5

Source: FHWA (1999).

and the fines from the separator layer do not move into the permeable base. The following requirements must be met:

D_{15} (separator layer) $\leq 5\ D_{85}$ (subgrade)
D_{50} (separator layer) $\leq 25\ D_{50}$ (subgrade)
D_{15} (base) $\leq 5\ D_{85}$ (separator layer)
D_{50} (base) $\leq 25\ D_{50}$ (separator layer)

The requirements of the aggregate separator layer are as follows:

1. Should have durable, crushed, angular aggregate.
2. Maximum Los Angeles abrasion loss of 50%.
3. Maximum soundness of 12 or 18% loss as determined by the sodium sulfate or magnesium sulfate tests (see Chapter 8).
4. Density should be at least 95% of the maximum density.
5. Maximum percentage of materials passing the No. 200 sieve is 12%.
6. Coefficient of uniformity is between 20 and 40.

Examples of typical gradations used by state highway agencies are provided in Table 6.8.

6.6.2.1 Geotextile Separator Layer

The requirements for geotextile separators have been set as follows (FHWA, 1998):

$$AOS\ or\ O_{95\ (geotextile)} \leq B\ D_{85\ (soil)}$$

where:
  AOS = apparent opening size, mm
   O_{95} = opening size in the geotextile for which 95% are smaller, mm
  AOS $\approx O_{95}$
   B = dimensionless coefficient
   D_{85} = soil particle size for which 85% are smaller, mm

For sands, gravelly sands, silty sands, and clayey sands (less than 50% passing 0.075 mm), B is a
function of C_u.

TABLE 6.8
**Typical Gradation Requirements
for Separator Layer**

Sieve Size	% Passing
$1\frac{1}{2}$ in.	100
$\frac{3}{4}$ in.	95–100
No. 4	50–80
No. 40	20–35
No. 200	5–12

Source: FHWA (2001).

For:

$$C_u \leq 2 \text{ or } \geq 8, \qquad B = 1$$
$$2 \leq C_u \leq 4, \qquad B = 0.5\, C_u$$
$$4 < C_u < 8, \qquad B = 8/C_u$$

where:

$$C_u = D_{60}/D_{10}$$

For silts and clays, B is a function of the type of geotextile:

For woven geotextiles, B = 1; $O_{95} \leq D_{85}$
For nonwoven geotextiles, B = 1.8; $O_{95} \leq 1.8\, D_{85}$
And for both, AOS or $O_{95} \leq 0.3$ mm

6.6.3 Design of Edge Drains

Guidelines from FHWA (1992) can be used for the design of edge drains. Generally, the edge drains are designed to handle the peak flow coming from the permeable base.

$$Q = Q_p \times L_o = (kS_xH)\, L_o$$

where:
   Q = pipe flow capacity, m^3/day (ft^3/day)
   Q_p = design pavement discharge rate, m^3/day/m (ft^3/day/ft)
   q_i = pavement infiltration, m^3/day/m^2 (ft^3/day/ft^2)
   W = width of the granular layer, m (ft)
   L_o = outlet spacing, m (ft)
   k = permeability of granular layer, m^3/day (ft^3/day)
   U = percentage drainage (as 1% = 0.01)
   S_x = transverse slope, m/m (ft/ft)
   H = thickness of granular layer, m (ft)
   t = time for drainage, U, to be reached, days
   n_e = effective porosity of granular material

For design based on the pavement infiltration flow rate, the design flow capacity of the edge drain is as follows:

$$Q = Q_p \times L_o = (q_i\, W)\, L_o$$

For design based on average flow to drain the permeable base, the design flow capacity is as follows:

$$Q = Q_p \times L_o = (WHn_e U(24/t))\, L_o$$

The pipe for the edge drain is designed according to the Manning equation (Daugherty and Ingersoll, 1954):

$$Q = \frac{K}{n} S^{\frac{1}{2}} R^{\frac{2}{3}} A$$

where:

Q = pipe flow capacity, m^3/s

S = slope of the pipe invert, m/m

n = pipe coefficient of roughness (0.012 for smooth pipe, and 0.024 for corrugated pipe; FHWA, 1992)

A = pipe cross-sectional area, m^2

$K = 1$

$R = \dfrac{A}{P} = \dfrac{D}{4}$, m

P = wetted perimeter of pipe, m

D = pipe diameter, m

$$Q = \frac{K}{n} S^{\frac{1}{2}} \left(\frac{D}{4}\right)^{\frac{2}{3}} \pi \left(\frac{D}{2}\right)^2$$

where:

Q is in m^3/sec

D is in m

K is in (m$^{1/3}$)/sec

In English units:

$$Q = \frac{53.01}{n} S^{\frac{1}{2}} D^{\frac{8}{3}}$$

where:

Q = ft^3/day

D = inches (3–4 in., generally)

The equations are generally used for determining the spacing of the pipes, by fixing a specific type of pipe and with a specific diameter. This is done by setting the pipe capacity equal to the discharge from the unit length of the pavement times the distance between the pipe outlets (spacing).

For infiltration flow:

$$L_o = Q/(q_i W)$$

For peak flow:

$$L_o = Q/kS_x H$$

For average flow:

$$L_o = (Q\,t)/(24 W H n_e U)$$

The spacing is generally used as 250 to 300 ft. Pipe diameters and spacing are usually determined also in consideration of maintenance requirements.

A filter layer in the form of geotextiles (or, less commonly, aggregates of different gradations) could be used wrapped around the pipe or as an envelope to the edge drains to prevent the inflow of adjacent soil into the pipes but allow the free flow of water into them; and if slotted pipes are used, the filter material must be such that it does not enter the pipes.

Koerner and Hwu (1991) recommend requirements of prefabricated edge drain filters as follows.

Requirement	Method	Value
Core strength	GRI GG4	$\geq 9{,}600$ lbf/in^2
Core flow rate	ASTM D4716	≥ 15 gal/min—ft
Geotextile permeability	ASTM D4491	≥ 0.001 cm/sec
Geotextile apparent opening size (AOS)	ASTM D4751	$\geq$ No. 100 sieve
Geotextile puncture	ASTM D3787	≥ 75 lb
Geotextile grab strength	ASTM D4632	≥ 180 lb
Geotextile tear strength	ASTM D4533	≥ 75 lb

Recommendations (NCHRP, 1994) are that geocomposite edge drains should be placed on the shoulder side, and the pavement side should be backfilled with suitable sand. The outlet drain pipe is selected so as to make sure that the capacity is equal to or greater than the capacity of the edge drain.

Recommendations (FHWA, 1989) regarding edge drains and outlets are as follows:

1. The preferable location of the edge drain is under the shoulder just adjacent to the pavement/shoulder joint, with the top of the pipe at the bottom of the layer to be drained, in a 12-inch trench.
2. The filter fabric could be avoided at the subbase–edge drain interface to prevent clogging of the filter by fines, and the trench backfill material should have adequate permeability.
3. Outlet spacing should be less than 500 ft, with additional outlets at the bottom of vertical sags, with rigid PVC outlets being desirable along with headwalls to protect it, prevent erosion, and help in locating the outlets.

6.7 CONSIDERATION OF DRAINAGE IN PAVEMENT DESIGN

Note that the effect of the quality of the drainage system provided in a pavement structure should and can be considered in its structural design. The basic idea is that if there is a good drainage system that removes water quickly from the pavement structure, then the total pavement structure will be thinner than in the case in which a good drainage system is not provided. For example, in the current AASHTO design procedure, "drainage coefficients" are recommended to be used according to the quality of drainage provided. Figure 6.16 explains the steps, which include determination of the quality of drainage based on the time required for removal of water and selection of an appropriate drainage coefficient, based on the quality of drainage and the percentage of time the pavement is exposed to moisture levels close to the saturation level.

6.8 PUMPING IN RIGID PAVEMENTS

Pumping is defined as the ejection of material from underneath the PCC slab as a result of water pressure. Water accumulated underneath a PCC slab will pressurize when the slab deflects under traffic load. The source of the ejected material can be from the base or subgrade. To initiate the pumping process, a small void space needs to be created beneath the slab. This could occur by compression of plastic soil or due to warping of the PCC slab at the joint area. A free-draining soil will not accumulate the necessary water. A fine soil-water suspension underneath the slab is formed and aids in the ejection of the soil. After repeated cycles of slab deflection and soil-water

Step 1
Determine Quality of Drainage:

Water Removal Time	Quality of Drainage
2 Hours	Excellent
1 Day	Good
1 Week	Fair
1 Month	Poor
Water will not Drain	Very Poor

Step 2
Determine Drainage Coefficients

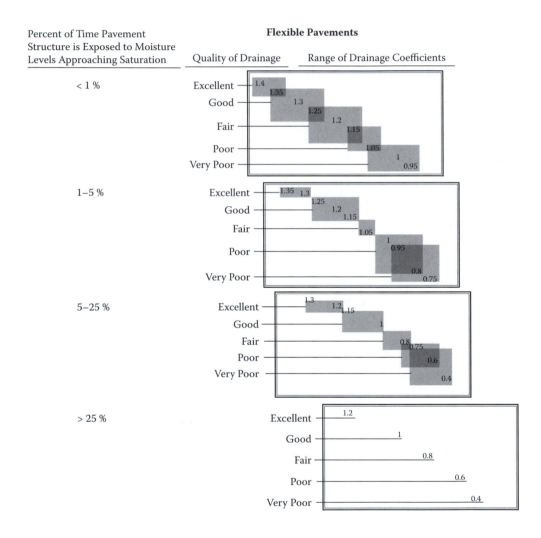

FIGURE 6.16a Drainage Coefficients for AASHTO Method.
Source: Prepared with data from AASHTO (1993).

suspension is ejected, a larger void space is created underneath the slab. At this stage faulting is observed, and eventually cracking and corner breaks could occur if significantly higher stresses are developed in the PCC slab. To minimize pumping, pump-susceptible soils such as clays and high-plasticity fine-grained soils should be avoided. Free-draining bases and erosion-resistant

Step 1
Determine Quality of Drainage:

Water Removal Time	Quality of Drainage
2 Hours	Excellent
1 Day	Good
1 Week	Fair
1 Month	Poor
Water will not drain	Very Poor

Step 2
Determine Drainage Coefficients

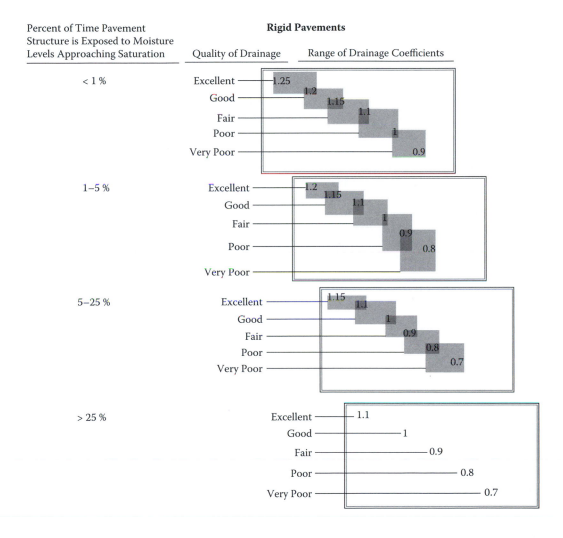

FIGURE 6.16b Drainage Coefficients for AASHTO Method (continued).
Source: Prepared with data from AASHTO (1993).

bases are highly recommended. A source of water is essential for pumping to occur. Water could be provided by the presence of a high water table or poorly functioning joint seals or cracks in the PCC slab. In any event, a well-draining base could greatly decrease the potential for pumping distress.

QUESTIONS

1. What are the primary sources of water in a pavement?
2. What are the components of surface and subsurface drainage systems in asphalt and concrete pavements?
3. Estimate the maximum rate of runoff for a 10-year return period using the rational method for the following case:

   Area: 25 ha, with two types of use: 60% unimproved and 40% residential single-family area; length of overland flow: 400 m; average overland slope: 1.5%
4. For a trapezoidal channel with Manning's coefficient of n = 0.012 with a longitudinal slope of 0.0015, side slope of 2:1, and depth of flow 2.5 ft (bottom width of 6 ft), what is the rate of flow?
5. For an open channel to carry 650 ft³/sec of water, determine the most efficient cross section for a rectangular channel. Assume n = 0.012 and S = 0.0012.
6. For the following conditions, determine the groundwater infiltration.

   New two-lane highway, lane width: 12 ft; shoulder width: 10 ft, length of cracks: 20 ft, rate of infiltration: 0.5 ft³/day/ft²; spacing of cracks: 25 ft; kp = 15 * 10⁻⁵ ft/day
7. Compute the location of the first gutter inlet from the crest, considering the following information:

   Maximum allowable flow = 0.25 m³/s
   Composite runoff coefficient = 0.45
   Intensity of rainfall = 120 mm/h
   Width of drainage area = 70 m

7 Soil

7.1 OVERVIEW

Soil can be defined as unconsolidated earth material composed of discrete particles with gas or liquids between, a relatively loose agglomeration of mineral and organic materials and sediments found above bedrock, or any earth material except embedded shale and rock.

The classification on the basis of particle size and gradation gives three groups of soils: coarse-grained or granular soils, consisting of sands and gravels; fine-grained or cohesive soils, consisting of clay; and silts, which lie in between.

The dominant factors for the coarse-grained soils are mass or gravitational forces between particles, and water has very little or no influence; whereas for fine-grained soils, electrical and chemical forces dominate particle interaction, and the amount and nature of pore fluid (such as water) affect the particle interaction significantly.

Particle size and gradation are important factors for coarse-grained soils. Uniform gradation means the soils consist of predominantly one size, and should have a high permeability. *Well graded* means the presence of different sizes in sufficient proportions, which would result in higher density and strength. *Gap graded* would mean the absence of certain sizes of particles in the soil.

Mineralogy also has significant influence on the properties of the soil, more so in the case of the clay particles than in the case of coarse-grained soil. Coarse-grained soils mostly consist of siliceous materials with quartz and feldspar particles.

The physical properties that characterize clay particles are that they are very small, platy, and negatively charged. There are primarily three clay minerals (Figure 7.1). Kaolinite ranges in size from 0.5 to 5 microns. The specific surface area (particle surface area/particle mass) ranges from 10 to 20 m^2/gram. In comparison, the specific surface of silt-size particles is less than 1 m^2/gram. Kaolinite has a stable structure with strong bonds, and is electrically almost neutral. The electrical activity is expressed by its cation exchange capacity, which for Kaolinite is 3–10 meq/100 gram. Illite ranges in size from 0.1 to 2 microns, with a specific surface of 80–100 m^2/gram. It has a moderately stable structure, with a cation exchange capacity of 10–40 meq/100 gram. Montmorillonite ranges in size from 0.1 to 1 micron, with a specific surface of approximately 800 m^2/gram. It has an unstable structure, with a cation exchange capacity of 80–120 meq/gram.

Water is bound to negatively charged clay particles (hydrated), and the absorbed cations (Li^+, Na^{++}, Mg^{++}, Ca^{++}) influence the nature of the bound water layer. The bound water influences particle interaction.

7.2 SOILS IN SUBGRADE

Tests performed for the characterization of subgrade soil consist of the following:

1. Grain size distribution (through sieve analysis)
2. Specific gravity
3. Atterberg limits
4. Organic content
5. Moisture retention and hydraulic conductivity
6. Compaction

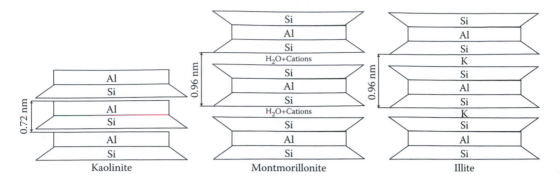

FIGURE 7.1 Different Clay Minerals.
Source: From Holtz, Robert D., Kovacs, William D., Introduction to Geotechnical Engineering, 1st edition © 1982, pg. 86. Reprinted by permission of Pearson Education, Inc., Upper Saddle River, NJ.

7. Frost susceptibility
8. Unfrozen moisture content
9. Resilient modulus or California Bearing Ratio (CBR)

Classification of the subgrade soil type and approximate judgment regarding the usability of the material can be made according to the American Association of Highway and Transportation Officials (AASHTO) and the American Society of Testing and Materials (ASTM). Subgrade soils (henceforth called *soils*) that are used for pavements can be primarily of two types—coarse grained (sands and gravel) and fine grained (silts and clays). Organic soils such as peat are generally not used in pavement subgrade. Tests are carried out on subgrade soils for characterization to select the proper materials, to develop specifications and hence quality control guidelines, and for estimation of properties such as modulus, which can be used in the structural design procedure for determination of responses under load. The starting point in characterizing the subgrade soil is the development of a boring/coring location plan based on existing soil maps. The soil samples are then tested in the laboratory for the different properties.

7.3 MASS-VOLUME RELATIONSHIPS

Any soil mass typically consists of three materials or "phases"—solid soil particles (or solid), water, and air. The relative proportions of these components greatly influence the other physical properties and load-carrying capacity of soils. Therefore, several parameters have been defined that help us determine one or two important parameters if a set of other parameters is known. Generally, a phase or block diagram is used to solve for such parameters. The parameters and different formulas are shown in Figure 7.2. It is important to note that these parameters help us to convert volume to mass and mass to volume, which is required for the estimation of materials during construction.

$$\text{Density: } \rho = \frac{M}{V} \text{ kg/m}^3; \quad \text{dry density: } \rho_d = \frac{M_s}{V} \text{ kg/m}^3$$

$$\text{Specific gravity of solids,: } G_s = \frac{M_s}{V_s \rho_w} \quad \text{void ratio: } e = \frac{V_v}{V_s} \quad \text{porosity: } n = \frac{V_v}{V} * 100\%$$

$$\text{Water content: } w = \frac{W_w}{W_s} * 100\%; \quad \text{degree of saturation: } S = \frac{V_w}{V_v} * 100\%$$

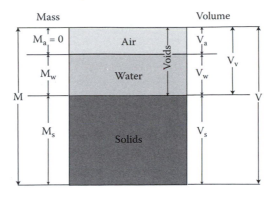

FIGURE 7.2 Soil Volumetric Properties.

7.4 GRAIN SIZE DISTRIBUTION: GRADATION

The grain size distribution or gradation of soil is one of its most important properties. A handful of soil from the ground would show us that the mass consists of particles of different sizes. The relative proportion of the particles of different sizes is very important in determining the soil's load-carrying capacity. The method to determine this proportion is called *particle size analysis*. In this procedure, particles of different sizes are mechanically separated by passing the soil through a stack of sieves, with progressively decreasing size openings from top to bottom. The dry weight of the soil in each sieve is then determined, and through a few steps the percentage passing each sieve is determined and plotted against the logarithm of sieve size opening. Generally the data are plotted (as a gradation curve) with the percentage passing in arithmetic scale in the y-axis and the sieve sizes in logarithmic scale in the x-axis. Note that the gradation of the soil is used in classification as well as indicators of different engineering properties, as mentioned later in this chapter.

For testing, either standard AASHTO 88 or ASTM D-422 can be used. The sieve sizes are selected on the basis of samples or specifications, and the particle size analysis of soil particles passing the 2.0 mm can be conducted using a hydrometer. Particles retained on a 0.075 mm sieve after hydrometer tests should be subjected to sieve analysis also, and then all of the results should be combined to produce one combined gradation plot for the soil. The data are analyzed as shown in Figure 7.3.

The hydrometer analysis is conducted on the principle of Stokes' law, which relates the terminal velocity of a particle falling freely in a liquid to its diameter, as follows:

$$v = \frac{D^2 \gamma_w (G_s - G_L)}{18\eta}$$

where:

$\quad$ v = velocity of settling soil particles
$\quad$ D = diameter of particle
$\quad$ γ_w = unit weight of water
$\quad$ G_s = specific gravity of solid particles
$\quad$ G_L = specific gravity of soil-water mixture
$\quad$ η = dynamic viscosity of soil-water mixture

The soil sample is suspended in water in a cylinder, using a dispersing agent such as sodium hexametaphosphate. The hydrometer, with a graduated scale on it, is then placed inside the liquid, and hydrometer readings are taken at the end of 120 s, and then at time intervals such as 5, 15, 30,

Particle Size Analysis Using Sieves

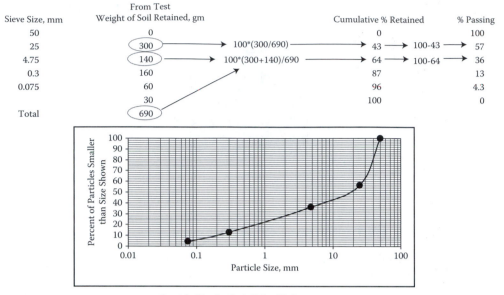

Sieve Size, mm	From Test Weight of Soil Retained, gm		Cumulative % Retained		% Passing
50	0		0		100
25	300	100*(300/690) →	43 →	100-43 →	57
4.75	140	100*(300+140)/690 →	64 →	100-64 →	36
0.3	160		87		13
0.075	60		96		4.3
	30		100		0
Total	690				

Particle Size Analysis Using Hydrometer

Time, min	Corrected Hydrometer Reading		Effective Depth, L, mm	Percent Passing	Particle Size, mm
0.25	47.15		77.62	90.528	0.0677
0.5	44.15		82.54	84.768	0.0494
1	43.15		84.18	82.848	0.0353
2	42.15		85.82	80.928	0.0252
4	41.15	Depends on Type of	87.46	79.008	0.0180
8	40.15	Hydrometer →	89.1	77.088	0.0128
15	39.15	and Actual Reading	90.74	75.168	0.0094
30	38.15		92.38	73.248	0.0067
60	36.15		95.66	69.408	0.0049
120	34.15		98.94	65.568	0.0035
240	30.15		105.5	57.888	0.0025
480	28.15		108.78	54.048	0.0018
1440	25.15		113.7	48.288	0.0011

38.15*0.96*100/50

Gs = 2.85, a = 0.96, w=50 Grams, K = 0.003842

$$0.003842 * \sqrt{\frac{92.38}{30}}$$

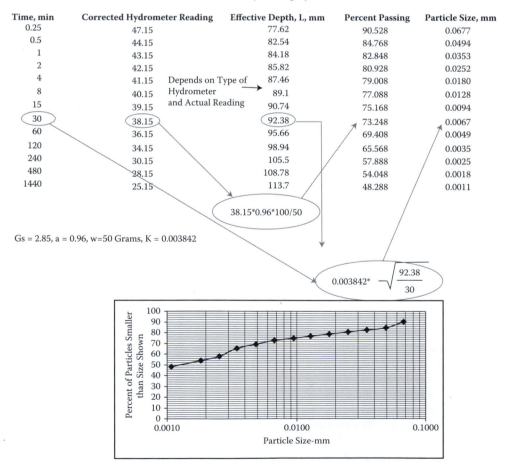

FIGURE 7.3 Particle Size Analysis.

60, 250, and 1440 minutes, as particles continues to settle (sedimentation) down in the liquid. The two equations used for the determination of percentage finer and diameters of the particle are as follows:

$$P = \frac{Ra}{w} * 100$$

where:

P = percentage of originally dispersed soil remaining in suspension
R = corrected hydrometer reading
a = constant depending on the density of the soil specimen
w = mass of soil, in grams

and

$$d = \sqrt{\frac{30nL}{980(G - G_1)T}}$$

where:

d = maximum grain diameter, mm
n = coefficient of viscosity of the suspending medium in Pa.s
L = distance from the surface of the suspension to the level at which the density of the suspension is being measured, mm, known as the effective depth
T = interval of time from beginning of sedimentation to the taking of the reading, min
G = specific gravity of soil particles
G_1 = specific gravity of the suspending medium

The above equation is simplified as follows, with values of K and L provided in standards:

$$d = K\sqrt{\frac{L}{T}}$$

An example is shown in Figure 7.3.

Depending on the gradation, the soil can be classified as poorly graded (short range of sizes), well graded (good range of sizes) or gap graded (certain sizes missing). This is done by characterizing the gradation curves with the help of certain "coefficients" as follows.

Coefficient of uniformity:

$$C_u = \frac{D_{60}}{D_{10}} \qquad \begin{array}{l} \text{poorly graded} \downarrow \\[6pt] \text{well graded} \uparrow \end{array}$$

Coefficient of curvature:

$$C_c = \frac{(D_{30})^2}{D_{60}D_{10}} \qquad \begin{array}{l} \text{well graded} \approx 1 - 3 \\[6pt] \text{gap graded} < \text{or} > 3 \end{array}$$

Note that D_x refers to grain size corresponding to x% passing.

Examples of different gradations are shown in Figure 7.4.

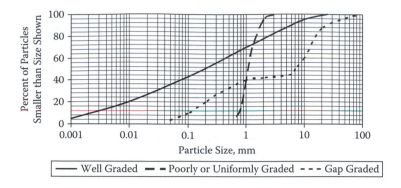

FIGURE 7.4 Different Types of Gradation.

7.5 EFFECT OF WATER

The presence of certain types of soil materials, such as clay, can cause the effect of water to be significant, and hence may make the soil not suitable for use in pavement subgrade. As the water content of the soil increases from very low to high, the soil properties change. The Atterberg Limit parameters—the liquid limit, plastic limit, and Plasticity Index of a soil—tell us how sensitive the soil is to the effect of water.

Shrinkage limit: Shrinkage limit is defined as the maximum calculated water content, at which a reduction in water content will not cause a decrease in the volume of the soil mass. AASHTO T-92 can be used for determination of shrinkage limit (SL).

Plastic limit: Plastic limit (PL) is the moisture content below which the soil is nonplastic.

Experimental definition: plastic limit is defined as the water content at which a soil thread just crumbles when it is rolled down to a diameter of 3 mm (approximately).

$$\text{Plastic Limit} = \left[\frac{(\text{mass of water})}{(\text{mass of oven dry soil})} \right] \times 100 \text{ Percent}$$

The AASHTO T-90 procedure is used for the determination of the plastic limit of the soils.

Liquid limit: Liquid limit (LL) is the moisture content below which the soil behaves as a plastic material. Liquid limit is defined as the water content at which a part of soil placed in a brass cup, cut with a standard groove, and then dropped from a height of 1 cm will undergo a groove closure of 12.7 mm when dropped 25 times. A nearly linear plot of the water content (%) versus log of number of blows (N) is prepared. The moisture content

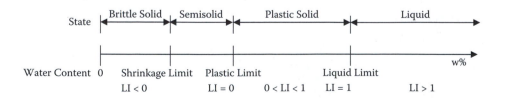

FIGURE 7.5 Atterberg Limits and State of the Soil. (Redrawn from Holtz and Kovacs, 1981.)

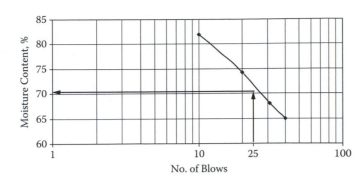

FIGURE 7.6 Plot of Number of Blows versus Moisture Content.

corresponding to 25 blows is the liquid limit. The AASHTO T-89 procedure is used for the determination of liquid limit.

$$\text{Plasticity Index (PI): PI} = \text{LL} - \text{PL}$$

$$\text{Liquidity Index (LI): LL} = \frac{(w\% - \text{PL})}{\text{PI}}$$

The different limits are shown in Figure 7.5.

Example

The following observations were obtained from a liquid limit test of a soil:

Number of blows: 10, 20, 31, 40.
Moisture content: 82.0%, 74.3%, 68.0%, 65.0%.
Two tests for plastic limit were done which gave values of 28.2 and 29.4, respectively.
Determine liquid limit and plasticity index of the soil.

Plot the moisture content versus number of blows data as shown in Figure 7.6, with number of blows on the x-axis in log scale and moisture content on the y-axis in arithmetic scale.

From Figure 7.6, LL = 71%
Average of two plastic limit tests, plastic limit = (28.2 + 29.4) / 2 = 28.8
Plasticity index = LL − PL = 71 − 28.8 = 42.2

7.6   SOIL CLASSIFICATION

7.6.1   AASHTO METHOD

Taking particle size analysis and Atterberg Limits into consideration, the AASHTO soil classification system has been developed, as shown in Table 7.1.

One can use this system to classify soil and make a conclusion regarding its suitability as sub-grade soil. The criteria for the AASHTO classification system are based on performance as highway construction material. There are two general groups—coarse grained and fine grained (or granular

TABLE 7.1

AASHTO M-145 Standard Specifications for Classification of Soils and Soil-Aggregate Mixtures for Highway Construction Purposes

| General Classification | Granular Materials (35% or Less Passing No. 200 Sieve) | | | | | | | Silt-Clay Materials (> 35% Passing No. 200 Sieve) | | | | Highly Organic |
| | A-1 | | A-3 | A-2 | | | | A-4 | A-5 | A-6 | A-7 (A-7-5 A-7-6) | A-8 |
	A-1-a	A-1-b	A-3	A-2-4	A-2-5	A-2-6	A-2-7	A-4	A-5	A-6	A-7-5 / A-7-6	A-8
Sieve Analysis, % Passing												
No. 10 sieve	≤ 50											
No. 40 sieve	≤ 30	≤ 50	≥ 51									
No. 200 sieve	≤ 15	≤ 25	≤ 10	≤ 35	≤ 35	≤ 35	≤ 35	≥ 36	≥ 36	≥ 36	≥ 36	
Characteristics of Fraction Passing No. 40 Sieve												
Liquid limit				≤ 40	≥ 41	≤ 40	≥ 41	≤ 40	≥ 41	≤ 40	≥ 41	
Plasticity Index	Max: 6		NP*	≤ 10	≤ 10	≥ 11	≥ 11	≤ 10	≤ 10	≥ 11	≥ 11	
Usual types of significant constituent materials	Stone fragments; gravel and sands		Fine sand	Silty or clayey gravel and sand				Silty soils		Clayey soils		Peat or muck
General rating as subgrade	Excellent to good							Fair to poor				Unsuitable

* *NP:* nonplastic; for A-7-5 soils, PI ≤ (LL-30); for A-7-6 soils, PI > (LL-30).

Source: From M 145–91 (2004) of *Standard Specifications for Transportation Materials and Methods of Sampling and Testing 27th Edition*, 2007, by the American Association of State Highway and Transportation Officials, Washington, D.C. Used by permission.

and cohesive). The distinction between coarse- and fine-grained soil is 35% passing the No. 200 sieve. There is also organic soil.

The groups range from A1 (best), which is the most coarse grained, to A8 (worst), which is the most fine grained. The parameter used for grouping is called the Group Index (GI), which rates soil as a pavement subgrade material, as shown below.

Group Index (GI)	Subgrade Rating
0	Excellent
0–1	Good
2–4	Fair
5–9	Poor
10–20	Very poor

The steps in classification consist of the following:

1. Use grain size distribution and Atterberg Limits data to assign a group classification system and a Group Index.

2. Compute the Group Index using the following equation:

$$GI = (F - 35)[0.2 + 0.005(LL - 40)] + 0.01(F - 15)(PI - 10)$$

where:
   F = portion passing No. 200 sieve
  LL = liquid limit
  PI = Plasticity Index
- Express the Group Index as a whole number.
- A Group Index value of less than zero should be reported as zero.
- For A-2-6 or A-2-7 soils, use only the second term for the calculation of the GI.
3. Express the AASHTO soil classification as the group classification followed by the Group Index in parentheses.

Example

The following data were obtained for a soil sample.

Sieve No.	% Passing
4	98
10	93
40	87
100	77
200	69

Plasticity tests: LL = 48%, PL = 26%.

Determine the classification of the soil, and state whether this material is suitable in its natural state for use as a subbase material.

Refer to Table 7.1. Since the percentage passing the No. 200 sieve is 69, the soil must be a silt-clay material. The LL is 48%; therefore, it should fall in either the A-5 or A-7 category. The PI = LL − PL = 22 > 10, so it must be an A-7 soil. In this case, LL − 30 = 18; PI > (LL − 30); therefore, the soil can be classified as an A-7-6 material.

$$GI = (69 - 35)[0.2 + 0.005(48 - 40)] + 0.01(69 - 15)(22 - 10) = 14.6 \text{ (very poor)}$$

Conclusion: unsuitable as a subbase material in its natural state.

7.6.2 Unified Soil Classification System (ASTM)

The first step in this system is to determine if the soil is highly organic (which has dark brown, gray, or black color; has an odor; is soft; and has fibrous materials) or not. If yes, it is classified as peat; and if not, the soil is subjected to particle size analysis and the gradation curve is plotted. Based on the percentage of weight passing the 3-in., No. 4, and No. 200 sieves, the percentages of gravel, sand, and fines are computed on the basis of Table 7.2. If less than 100% of the material passes through a 3-in. sieve, then base the classification on the basis of the percentage of the material that passes through the 3-in. sieve and note the percentage of cobbles and/or boulders. If 100% passes the 3-in. sieve, then check the percentage passing the No. 200 sieve—if 5% or more passes the No. 200 sieve, then determine liquid and plastic limits. Check whether more than 50% passes through the No. 200 sieve—if it does (fine-grained soil), then follow subsequent instructions for fine-grained soils; otherwise (coarse-grained soil), follow those for coarse-grained soils.

TABLE 7.2
ASTM Particle Size Classification (ASTM D-2487)

Sieve Size		Particle Diameter		Soil Classification	
Passes	Retained On	In.	Mm		
	12 in.	> 12	> 350	Boulder	Rock fragments
12 in.	3 in.	3–12	75.0–350	Cobble	
3 in.	¾ in.	0.75–3.0	19.0–75.0	Coarse gravel	
¾ in.	#4	0.19–0.75	4.75–19.0	Fine gravel	
# 4	#10	0.079–0.19	2.00–4.75	Coarse sand	
#10	#40	0.016–0.079	0.425–2.00	Medium sand	
#40	#200	0.0029–0.016	0.075–0.425	Fine sand	
#200		< 0.0029	< 0.075	Fine silt and clay	

Source: From Coduto, Donald P. Geotechnical Engineering: Principles and Practices, 1st edition © 1999, pg. 115. Reprinted by permission of Pearson Education, Inc., Upper Saddle River, NJ.

Fine-grained soils are primarily silt and/or clay, and Figure 7.7 is to be used for classification. The soils falling above the A line are classified as clays, and those falling below are classified as silts. The group symbols for fine-grained soils are as follows:

First Order	Second Order
M: predominantly silt	L: low plasticity
C: predominantly clay	H: high plasticity
O: organic	
CL ‡ lean clays	ML: silt
CH ‡ fat clays	MH: elastic silt

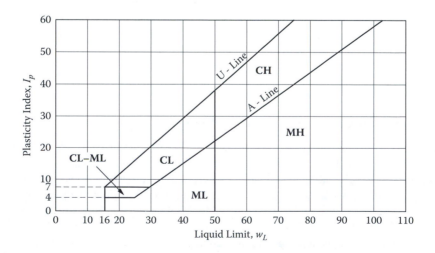

FIGURE 7.7 Plasticity Chart, ASTM D-2487.
Source: From Coduto, Donald P. Geotechnical Engineering: Principles and Practices, 1st edition © 1999, pg. 143. Reprinted by permission of Pearson Education, Inc., Upper Saddle River, NJ.

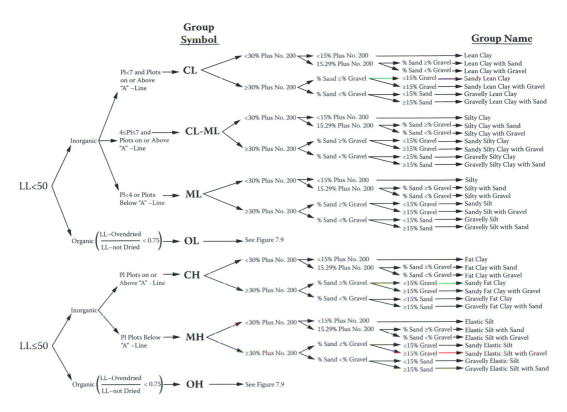

FIGURE 7.8 Chart for Fine-Grained Soils.
Source: Reprinted, with permission, from D2487-06 Standard Practice for Classification of Soils for Engineering Purposes (Unified Soil Classification System), copyright ASTM International, 100 Barr Harbor Drive, West Conshohocken, PA 19428.

The tests needed to determine whether a soil is organic or not are those testing the liquid limit on an unmodified and an oven-dried sample. The soil is considered to be organic if the liquid limit after oven drying is <75% of the original. For classification of inorganic soils, use Figure 7.7 and Figure 7.8, and for classification of organic soils, use Figure 7.7 and Figure 7.9.

Coarse soils are primarily sand and/or gravel. The symbols for coarse-grained soils are as follows:

First Letter	Second Letter
S: predominantly sand	P: poorly graded
G: predominantly gravel	W: well graded
	M: silty
	C: clayey

Use Figure 7.10 to classify coarse-grained soils, according to the following guidelines:

1. Use two-letter groups to describe soils with < 5% passing the No. 200 sieve.
2. Use four-letter groups to describe gradation and type of fines for soils with 5–12% passing the No. 200 sieve.
3. Use two-letter symbols to describe the fines for soils with > 12% passing the No. 200 sieve.

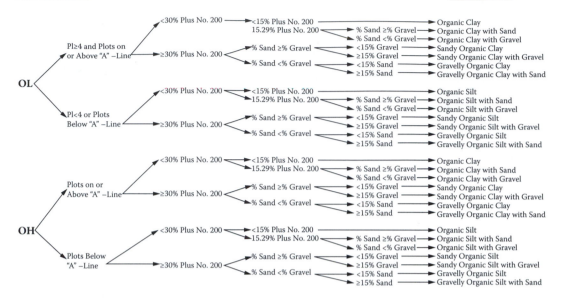

FIGURE 7.9 Chart for Organic Soil.
Source: Reprinted, with permission, from D2487-06 Standard Practice for Classification of Soils for Engineering Purposes (Unified Soil Classification System), copyright ASTM International, 100 Barr Harbor Drive, West Conshohocken, PA 19428.

7.7 DENSITY AND OPTIMUM MOISTURE CONTENT

Compacting loose soil is the simplest way to improve its load-bearing capacity. Water is added to the soil, which is compacted in lifts, to lubricate the soil particles and air in the compaction process. As water is added to the soil, the density of the soil increases due to compaction. However, beyond certain water content, even though the soil becomes more workable, the unit weight of the soil would decrease. This is shown in Figure 7.11.

This can be explained by the fact that beyond the "optimum" water content, the water cannot force its way into trapped air pockets, and hence takes up the space originally taken by the soil solids—that is, it separates the soil particles and hence causes a corresponding decrease in density.

Generally, the density of the soil solids (generally referred to as *dry density*) is taken as an indicator of the load-carrying capacity of the soil and as an indicator of the degree of compaction of the soil—a higher density relates to higher stiffness and/or strength. Moreover, a high density also ensures low permeability, and minimizes settlement and volume change due to frost action, swelling, and shrinkage. A high density of the solids can be achieved in the field only by applying compactive effort (through rollers) and moisture. To achieve this, it is important to determine in the laboratory the "optimum" moisture content of the soil—the word *optimum* here refers to that moisture content which produces the highest dry density, for a given compactive effort. In soils with 10% or more passing the No. 200 sieve, the change in dry density versus change in moisture content can be expressed in the form of the curve shown in Figure 7.11. The moisture content corresponding to the maximum dry unit weight is called the *optimum moisture content*.

Note that the moisture content versus dry density curve is different for different compactive efforts for the same soil. The test that is carried out to determine the optimum moisture content is commonly referred to as the *Proctor compaction test*. The compactive effort to be applied for compaction of soils during the Proctor test is dependent on the level of compactive effort expected in the field—the higher the expected effort, the higher is the effort in the laboratory.

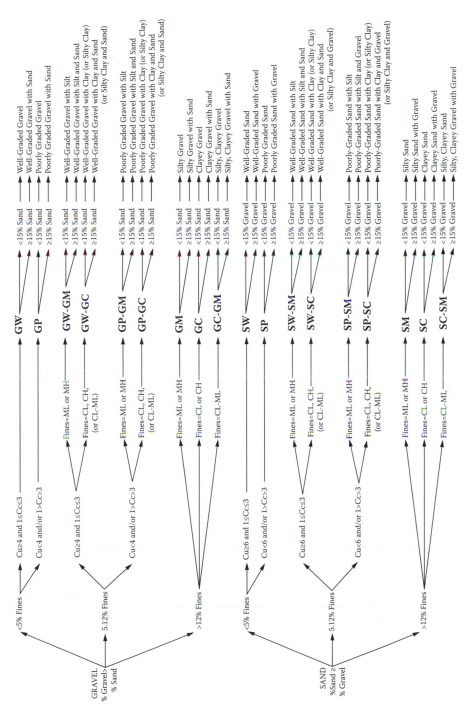

FIGURE 7.10 Chart for Coarse-Grained Soils.

Source: Reprinted, with permission, from D2487-06 Standard Practice for Classification of Soils for Engineering Purposes (Unified Soil Classification System), copyright ASTM International, 100 Barr Harbor Drive, West Conshohocken, PA 19428.

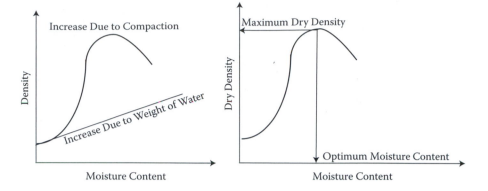

FIGURE 7.11 Moisture Content versus Density.

Accordingly, there are two standards—AASHTO T-99 (Moisture-Density Relations of Soils Using a 2.5 kg [5.5 lb] Rammer and a 305 mm [12 in.] Drop; energy level of 12,400 ft lb per ft^3 of soil) and AASHTO T-180 (Moisture-Density Relations of Soils Using a 4.54 kg [10 lb] Rammer and a 457 mm [18 in.] Drop; energy level of 56,200 ft lb of energy per ft^3 of soil).

Once the optimum moisture content and the maximum dry density are determined from the laboratory, these values are used for specifying moisture levels (generally, a little dry of the optimum) and density (generally, a percentage of the maximum). One practice is to use a fill in the site to determine the soil lift thickness and number of passes of compaction equipment that is required for the available equipment (or, if necessary, with other equipment) to achieve the required compaction level, and then prepare the method specification accordingly.

An example of determination of optimum moisture content is shown in Figure 7.12.

7.8 HYDRAULIC CONDUCTIVITY

Based on results of experiments conducted by D'Arcy (1856), the flow of water through soils is characterized by the following expression:

$$v = ki$$

where:

 v = the velocity of flow
 i = the hydraulic gradient (head loss per unit length of flow)
 k = a coefficient

Generally k is referred to as a coefficient of permeability or permeability or hydraulic conductivity. The permeability of different types of soils is shown in Table 7.3.

Permeability for soils can be measured in the laboratory by using either a constant head (ASTM D-2434, AASHTO T-215) or a falling head method (ASTM D-4630). In the constant head test, a constant head of water is maintained in a soil sample, and the flow rate (volume divided by time) is used along with the sample cross-sectional area to determine the permeability. In a falling head test, the level of water in a standpipe is noted at regular time intervals as the water is made to flow through a soil sample, and the permeability is calculated on the basis of the total length and area of the soil sample, the area of the standpipe, and the initial and final levels of water in the standpipe in the specified time.

In the absence of test data, correlations with gradation properties can be used to estimate the permeability of different types of soils. For example:

$$k = C_k D_{10}^2$$

Sample No.	Volume, ft^3	Weight, lb	Bulk density, lb/ft^3	Moisture Content, %	Dry density, lb/ft^3
1	0.03333	4.16	124.8	4	120.0
2		4.39	131.7	6.1	124.1
3		4.6	138	7.8	128.0
4		4.68	140.4	10.1	127.5
5		4.57	137.1	12	122.4
6		4.47	134.1	14	117.6

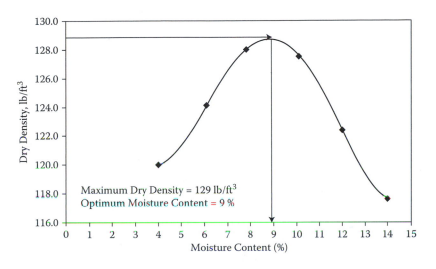

Example Calculation: Bulk Density = (4.16)/(0.03333) = 124.8 lb/ft^3;
Dry Density = Bulk Density/(1 + Moisture Content) = (124.8)/(1 + 0.04) = 120.0 lb/ft^3

FIGURE 7.12 Example Problem.

where:

k = permeability, cm/s

D_{10} = effective grain size (cm) corresponding to size passing 10%

C_k = Hazen's coefficient, 0.8 – 1.2 (applicable for 0.1 mm < D_{10} < 3 mm, C_u < 5)

TABLE 7.3
Hydraulic Conductivity of Soils

Soil	Hydraulic Conductivity, k	
	cm/s	ft/s
Clean gravel	1–100	$3*10^{-2}$–3
Sand-gravel mixtures	10^{-2}–10	$3*10^{-4}$–0.3
Clean coarse sand	10^{-2}–1	$3*10^{-4}$–$3*10^{-2}$
Fine sand	10^{-3}–10^{-1}	$3*10^{-5}$–$3*10^{-3}$
Silty sand	10^{-3}–10^{-2}	$3*10^{-5}$–$3*10^{-4}$
Clayey sand	10^{-4}–10^{-2}	$3*10^{-6}$–$3*10^{-4}$
Silt	10^{-8}–10^{-3}	$3*10^{-10}$–$3*10^{-5}$
Clay	10^{-10}–10^{-6}	$3*10^{-12}$–$3*10^{-8}$

Source: From Coduto, Donald P. Geotechnical Engineering: Principles and Practices, 1st edition © 1999, pg. 222. Reprinted by permission of Pearson Education, Inc., Upper Saddle River, NJ.

In general, coarse-grained soils have higher permeability compared to fine-grained soils. As such, coarse-grained soils have good drainage capabilities and are preferred for pavement applications compared to fine-grained soils.

7.9 FROST SUSCEPTIBILITY

In certain areas, where freezing cold weather is common during the winter months, frost heaves are common sights. In spring these heaves break apart under traffic, leading to potholes. The main causes of this distress are the environmental conditions and the existing soil in the lower pavement layers.

Soils that have low permeability and retain water, and change in volume in freezing temperatures, are known as being frost susceptible. The degree of heaving in such soils cannot be explained only on the basis of the expanding of water into ice.

Portions of a pavement can experience a rise or "heave" (frost heaves) due to the formation of ice lenses in these frost-susceptible materials, which typically contain excessive amounts of fine materials (i.e., those less than 0.02 mm). Formation of ice lenses leads to pressure exerted by the ice and soil. There is differential movement in the pavement structure, and as a result there is roughness and cracking. Most of the water that forms the ice lenses is drawn up by capillary action from water tables. When the soil thaws in warm weather in spring (from the surface downward), this new water cannot drain quickly, leading to a significant increase in the moisture content of the subgrade, and hence a significant reduction in shear strength. This is illustrated in Figure 7.13. The effects are relatively more significant in pavements with untreated layers.

The process starts with the soil and a water table close to the soil. A soil such as silt, which has relatively high ability to draw water through small pores (high capillarity, 0.9 to 9 m rise compared to 0.15 m rise in coarse sand) as well as ability to move water in the pores relatively rapidly (moderate permeability, 0.9 m/day compared to 0.0009 m/day for clay), is considered to be frost susceptible. Such soil would draw water from a relatively close water table, and this water would freeze in the winter time as the freezing front moves downward. The expansion of water upon freezing into ice lenses exerts pressure on the layers above them, resulting in surface heaves. With the coming of spring, as the temperature rises, the ice lenses start thawing from top downward. This water in the subgrade and the void created (by the difference in volume of ice and water) result in a weaker saturated subgrade and a weaker pavement structure as a whole.

Subsequently, when traffic moves over the heaved surface, which has no or little support underneath, it breaks down into smaller pieces, forming potholes. Frost-susceptibility tests can be conducted by subjecting soils to freeze-thaw cycles, noting the heave rate during the first 8 hours and then determining the California Bearing Ratio, a strength-indicating test property, on the thawed soil after allowing time for drainage of the water from the soil. The criteria shown in Table 7.4 can be used for classification of the soil.

Plots of frost susceptibility and heaves, shown in Figure 7.14, can be used for the frost susceptibility classification of a soil. The use of dielectric constant devices, both in place and the laboratory, has been developed to evaluate and monitor freezing and thawing processes in the soil.

The frost susceptibility of the soil is affected primarily by the size and distribution of voids (where ice lenses can form), and is determined through empirical correlations from the soil classification (Unified Soil Classification System) and the percentage finer than 0.02 mm. Table 7.5 relates these parameters to the frost codes.

Frost heaves are nonuniform and tend to make pavements rougher, and expressions have been developed to relate the variability of soil frost susceptibility and the winter roughness. Affected pavement surface profiles can be characterized in terms of dominant wavelength, which has been found to range from 10 to 80 m.

Seasonal frosts can cause nonuniform heave and loss of soil strength during melting, resulting in possible loss of density, development of pavement roughness, restriction of drainage, and cracking and deterioration of the pavement surface. Whether such effects are possible or not depend on

Temperature Below Freezing Point

Resulting Heave at the Surface

Asphalt Mix Layer

Frost Susceptible Soil
Soil is Pushed Up
Due to Expansion

Ice Lenses Form
From Water Under
Freezing Temperatures

Capillary Action Feeds
Water Into Soil

Water Table

Heaves Caused by Freezing in Winter

Temperature Above Freezing Point

Loss of Support

New Water from Thawed Soil
Cannot Drain Quickly and Reduce
Strength of Soil

Water Table

Loss of Support Due to Spring Thaw

FIGURE 7.13 Freeze-Thaw in Soils.

TABLE 7.4
Frost Susceptibility Classification

Frost Susceptibility Classification	8-Hour Heave Rate (mm/day)	Thaw CBR (%)
Negligible	< 1	> 20
Very low	1–2	20–15
Low	2–4	15–10
Medium	4–8	10–5
High	8–16	5–2
Very high	> 16	< 2

Source: Chamberlain (1987).

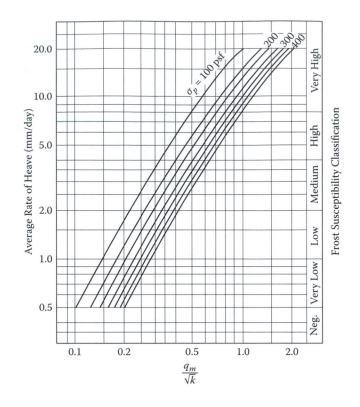

FIGURE 7.14 Frost Susceptibility Classification Chart.
Source: From Moulton (1980).

TABLE 7.5
Frost Codes and Soil Classification by the Federal Aviation Administration

Frost Group	Type of Soil	Percentage Finer than 0.02 mm by Weight	Soil Classification
FG-1	Gravelly soils	3 to 10	GW, GP, GW-GM, GP-GM
FG-2	Gravelly soils sands	10 to 3 to 15	GM, GW-GM, GP-GM, SW, SP, SM, SW-SM SP-SM
FG-3	Gravelly soils	Over 20	GM, GC
	Sands, except very fine silty sands	Over 15	SM, SC
	Clays, PI above 12		CL, CH
FG-4	Very fine silty sands	Over 15	SM
	All silts		ML, MH
	Clays, PI = 12 or less		CL, CL-ML
	Varied clays and other fine-grained banded sediments		CL, CH, ML, SM

Note: The higher the frost group number, the more susceptible the soil, i.e., soils in frost group 4 (FG-4) are more frost susceptible than soils in frost groups 1, 2, or 3.

Source: From FAA Advisory Circular No. 150/5320-6D (FAA, 1995).

TABLE 7.6
Frost Penetration Depths

Frost Penetration Depth, In Degree Days (°F)	Soil Unit Weight (pcf)			
	100	115	125	150
200	20.5	21.5	23.8	25.5
400	27.5	30.5	35	38.5
600	34	38	44.5	49
800	40	44.5	54	59
1000	45	51	62	69
2000	69.5	79	102	113
3000	92	105	140	156
4000	115	130	177	205
4500	125	145	197	225

Source: FAA Advisory Circular No. 150/5320-6D (FAA, 1995).

whether the subgrade soil is frost susceptible, there is free moisture to form ice lenses, and the freezing front is deep enough to penetrate the subgrade soil.

During the design of pavements, the effect of the frost susceptibility needs to be checked by determining the depth of frost penetration—measures should be taken if the depth is found to penetrate into the frost-susceptible soil layer. The frost penetration depth can be estimated from Table 7.6. The required inputs are the Air Freezing Index (degree days °F) and the unit weight of the soil (in pcf). The frost penetration depth is calculated from the interpolation of the data in Table 7.6, developed on the assumption of a 12-in. (300-mm) thick rigid pavement or a 20-in. (510-mm) thick flexible pavement. Note that the Air Freezing Index is a measure of the combined duration and magnitude of below-freezing temperatures occurring during any given freezing season. The average daily temperature is used in the calculation of the Freezing Index. For example, if the average daily temperature is 12 degrees below freezing for 10 days, the Freezing Index would = 12 degrees * 10 days = 120 degree days. It is recommended that the design Air Freezing Index be based on the average of the 3 coldest winters in a 30-year period, if available, or the coldest winter observed in a 10-year period. Air freezing indices are available at http://www4.ncdc.noaa.gov/ol/9712/AFI-seasonal.pdf.

The steps that could be taken to avoid or reduce the potential of the freeze-thaw problem are replacing the frost-susceptible soil, lowering the water table, increasing the depth of the pavement with nonfrost-susceptible soil over the frost-susceptible soil, and providing means of draining away excess water, such as that from thawing.

Frost protection can be provided by completely or partially removing the soil to the depth of frost penetration, or by assuming a reduced modulus of the soil during design (so as to result in a thicker pavement structure, which in turn would provide a cover of nonfrost-susceptible materials on the frost-susceptible soil).

7.10   SWELL POTENTIAL

Soils consisting of minerals such as Montmorillonite would expand significantly in contact with water. The presence of such soils in the subgrade would lead to differential movement, surface roughness, and cracking in pavements. Dominant wavelengths of affected pavements have been reported to range from 3.0 to 10.0 m.

The swell potential of a soil is primarily dictated by its percentage of clay-size particles and the Plasticity Index (PI), as illustrated below by one of the many equations developed (Gisi and Bandy, 1980):

$$S(\%) = a \, A^b C^{3.44}$$

where S (%) is the swell potential.

$$A = \text{activity} = \frac{PI}{C-5}$$

where:

PI = Plasticity Index
C = percentage of clay size particles (finer than 0.002 mm)
$a = 3.28 * 10^{-5}$, $b = 2.259$, for PI ≤ 20
$a = 2.40 * 10^{-5}$, $b = 2.573$, for $21 \leq$ PI ≤ 30
$a = 1.14 * 10^{-5}$, $b = 2.559$, for $31 \leq$ PI ≤ 40
$a = 0.72 * 10^{-5}$, $b = 2.669$, for PI > 40

The swelling potential can be determined as the percentage change in volume due to absorption of water when tested for the California Bearing Ratio (CBR) or in a Proctor mold. In the latter test, a soil is placed in a standard Proctor mold at its optimum moisture content, with a surcharge of 1 psi, and the mold with a perforated base in a pan full of water. The test is actually conducted on two specimens at OMC +3% and OMC –3% moisture contents, and the results are used to determine the percentage swell at the OMC.

The use of total suction characteristic curves has also been developed for the identification of swelling potential of soils.

The presence of swelling soil can affect the performance of the pavement and deteriorate its serviceability. The loss in serviceability with time due to the effect of swelling soils can be modeled as an exponential equation. Generally, soils exhibiting a swell percentage of greater than 3 (as determined with the CBR test, ASTM D-1883) are required to be treated. The Federal Aviation Administration (in AC 150/5320-6D; FAA, 1995) recommends modified compaction efforts and careful control of compaction moisture, as well as removal of such soils and replacement by stabilized materials, as shown in Table 7.7.

7.11 STIFFNESS AND STRENGTH OF SOILS

The most important characteristic feature that needs to be determined for a soil is its response, such as strain and deformation to load (or, more specifically, stress) from the traffic. In this respect, both "stiffness" and "strength" are important parameters that need to be measured in the laboratory. However, note that some design methods may use stiffness only, and hence do not require the use of strength tests, while others may require it.

For a strength test, a triaxial testing system is used in general, to simulate the triaxial state of stress in the field. Shear failure in soils is caused by slippage of soil particles past each other, and the resistance against such failure (shear strength) is provided by two components, friction and cohesion.

Frictional strength, commonly denoted as Normal Force * Coefficient of Friction, is expressed in soils in terms of the effective friction angle (or effective angle of internal friction), φ', which is defined as follows:

$$\varphi' = \tan^{-1} \mu$$

where μ is the coefficient of friction.

Shear strength, S, due to friction is defined as $S = \sigma' \tan \varphi'$, where σ' is the effective normal stress.

TABLE 7.7
Federal Aviation Administration Recommendations for Treating Soils with Swelling Potential

Swell Potential (Based on Experience)	% Swell Measured (ASTM D-1883)	Potential for Moisture Fluctuation	Treatment
Low	3–5	Low	Compact soil on wet side of optimum (+2% to +3%) to not greater than 90% of appropriate maximum density.[a]
		High	Stabilize soil to a depth of at least 6 in.
Medium	6–10	Low	Stabilize soil to a depth of at least 12 in.
		High	Stabilize soil to a depth of at least 12 in.
High	Over 10	Low	Stabilize soil to a depth of at least 12 in.
		High	For uniform soils (i.e., redeposited clays), stabilize soil to a depth of at least 36 in. below pavement section, or remove and replace with nonswelling soil. For variable soil deposits, depth of treatment should be increased to 60 in.

[a] When control of swelling is attempted by compacting on the wet side of optimum and reduced density, the design subgrade strength should be based on the higher moisture content and reduced density.

Note: Potential for moisture fluctuation is a judgmental determination and should consider proximity of water table, likelihood of variations in water table, other sources of moisture, and thickness of the swelling soil layer.

Source: From FAA AC No. 150/5320-6D (FAA, 1995).

The frictional strength is dictated by many factors, including density, mineralogy, shape of the soil participles, gradation, void ratio, and the presence or absence of organic materials. The other component of soil shear resistance, cohesion (C), present in some soils (cohesive), is defined as the strength that exists in the soil even if the effective stress (σ') appears to be zero. C is a result of different factors such as cementation, electrostatic attraction, adhesion, and negative pore water pressures, and is significantly affected by factors such as density, moisture, and drainage conditions.

Therefore, the combined shear resistance of the soil can be expressed as follows:

$$S = C' + \sigma'\tan\varphi'$$

Shear stress-strain curves for sand and gravel soils generally show ductile curves, whereas those for clays show brittle behaviors. A difference in test condition would result in a difference in stress-strain behavior, and as such, tests for the determination of shear strength are conducted under standard conditions or expected field conditions.

For evaluation of shear strength of soils, tests are conducted at different normal stress levels, and the failure shear stress corresponding to each normal stress level is determined. A series of such data points are connected to obtain the Mohr-Coulomb failure envelope and determine the cohesion and frictional components of the shear strength equation. This is explained in Figure 7.15.

Laboratory shear strength measurements can be conducted using the Direct Shear Test (ASTM D-3080), the Unconfined Compression Test (ASTM D-2166), or the Triaxial Test (ASTM D-2850 and ASTM D-4767).

In the direct shear test, a cylindrical soil sample is subjected to a vertical and a shear force, and the shear stress and shear displacements data are obtained. The test is repeated with different samples for different values of the normal force, and the peak shear stress in each case is obtained. A plot of peak shear stress versus the normal stress provides the information for the components of the shear strength equation. Although simple and inexpensive, and suitable for sandy soils, the drawbacks include nonuniform strain, failure along a specific plane, and inability to control drainage conditions (especially important for clay soils).

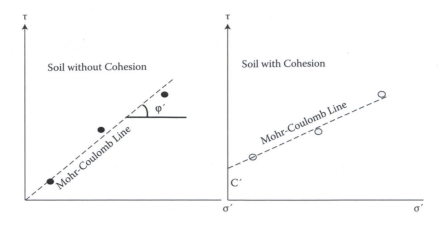

FIGURE 7.15 Mohr-Coulomb Failure Envelope.
Note: The τ – *intercept* of this line is the effective cohesion, C'. The *slope* of this line is the effective frictional angle, φ'. The Mohr-Coulomb line is often referred to as the *failure envelope*—points in the soil that have (σ',τ) values that fall *below* this line *will not* fail in shear, whereas the points with (σ',τ) values that fall *above* this line will fail in shear.

The unconfined compression test uses an unconfined cylindrical soil sample on which normal stress is applied and the failure stress is noted. The shear strength is obtained from the unconfined compression strength. It is a simple, inexpensive, and quick test but lacks the ability to apply confining stress, which is usually present in the field, and hence underestimates the shear strength (since confinement increases the shear strength).

The triaxial compression test is the most sophisticated of the three tests; in it, a soil sample is subjected to confining stress (by pressurized water) and loaded to failure. This test enables the determination of pore water pressure, if any, and hence the calculation of the effective stress (effective stress = total stress – pore water pressure). Depending on the soil and the specific project, triaxial testing can be conducted under unconsolidated undrained, consolidated drained, or consolidated undrained conditions.

7.11.1 California Bearing Ratio (CBR) Test (AASHTO T-193)

This procedure was originally developed by the California Division of Highways in the 1920s, and was later modified by the U.S. Army Corps of Engineers, and finally adopted by ASTM and AASHTO.

The CBR test is actually a penetration test that uses a standard piston (3 sq in.) which penetrates the soil at a standard rate of 0.05 in. per minute. A unit load is recorded at several penetration depths, typically 0.1 and 0.2 in. The CBR value is computed by dividing the recorded unit load by a standard unit load that is required to penetration for a high-quality crushed stone material (1000 and 1500 psi for 0.1 and 0.2 in., respectively). The CBR test is conducted on a soaked sample of soil—the soaking in water (for 96 hours) is conducted to simulate the worst (saturated) condition under which the pavement would perform. Figure 7.16 shows a mold with compacted soil (soaking) in a water bath.

7.11.2 Resilient Modulus Test (AASHTO T-307)

The main disadvantage of all of the strength tests presented above is that the loading is static as opposed to dynamic, which is the case due to traffic loading in the field, and that the soil sample is subjected to failure, which is not quite the case in the field, where stress levels at the subgrade level are in most cases too small to cause failure. These two deviations from actual field conditions make the strength tests relatively less useful for predicting deformation and strains under real-world traffic applications.

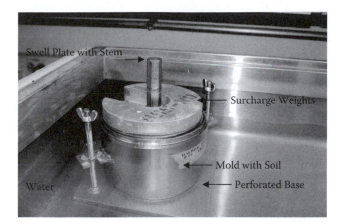

FIGURE 7.16 California Bearing Ratio (CBR) Mold with Soil in Water Bath.

The more appropriate test to characterize subgrade soil is the stiffness test—one which applies dynamic loads on a soil in a way in which the traffic loads are applied in the field, and also at levels that are more representative of those coming from the traffic. The resilient modulus test is one such widely recognized and increasingly used test procedure. Resilient modulus is the elastic modulus to be used with the elastic theory. The elastic modulus based on the recoverable strain under repeated loads is called the *resilient modulus*, or M_R. It is defined as the ratio of the amplitude of the repeated axial stress to the amplitude of the resultant recoverable axial strain.

$$M_R = \frac{\sigma_d}{\varepsilon_r}$$

where σ_d = deviator stress, which is the axial stress in an unconfined compression test or the axial stress in excess of the confining pressure in the triaxial compression test, and ε_r is the recoverable strain (Figure 7.17 and Figure 7.18).

The basic principle of this test is to apply a series of stresses (with rest periods in between) under triaxial conditions and measure the resulting strain. A haversine or a triangular stress pulse is applied in order to simulate the traffic loading on pavements.

As the magnitude of the applied load is very small, the resilient modulus test is a nondestructive test, and the same sample can be used for many tests under different loading and environmental conditions. This modulus is input in the mechanistic part of the mechanistic-empirical design

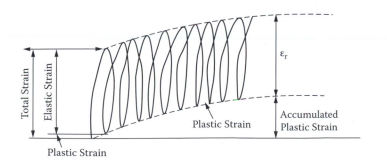

FIGURE 7.17 Plastic and Elastic Strain during Resilient Modulus Testing.

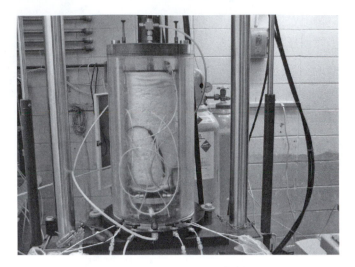

FIGURE 7.18 Soil inside Cell during Resilient Modulus Testing.
Courtesy: Mingjiang Tao, Worcester Polytechnic Institute.

procedure to estimate the response (such as strain) due to a traffic load. Generally, pulse loading and rest periods of 100 ms and 900 ms are used, respectively.

The resilient modulus of the soil is determined at different stress levels (to simulate different levels of traffic load). These differences happen in terms of the deviator stress as well as the confining stress. Generally, designers try to model the modulus (y) with respect to one or more relevant stresses (such as deviator stress, as x, and/or confining stress, as x). Note that fine-grained soils generally show stress-softening behavior, which means modulus decreases with an increase in deviator stress, whereas coarse-grained soils show stress-hardening behavior, which means an increase in modulus with an increase in the sum of the principal stresses. These important observations indicate the significant difference that can be expected in the response from the same soil materials at different depths (and hence difference stress levels) of a pavement. In the resilient modulus test, the stress levels are kept low enough to ensure an elastic response of the soil. However, researchers have used experiments to correlate elastic strain with permanent strain of soils.

Confining stress is applied in this stress using a membrane and water pressure, whereas axial stress is applied through an actuator mounted on a reaction-loading frame. In the standard resilient modulus test for soils, a soil specimen is subjected to multiple (such as 16) stress stages, each consisting of the application of a specific confining stress and a fixed count of pulsed axial deviator stress. The software, which controls the test, begins testing the soil specimen for the next stress state as one state ends, until all of the stress states are completed. The first stress state is a conditioning state. Using sensors and software for the vertical and confining stress, the vertical and lateral specimen displacements are measured, and for each stress state the mean and standard deviation of the different parameters for the last five pulses are reported. An ASCII file is created with the data, which can then be imported to a spreadsheet for further analysis.

Any outlier, such as negative deviator stress values, very high or low values of resilient modulus, and deviator stress values, should be discarded before statistical analysis of the data is conducted. Next, the resilient modulus and the various stress data are evaluated to develop models relating the resilient modulus to one or more of the stress variables, according to the following equation:

$$M_r = K_1[f(\sigma)]^{K_2}$$

where:

M_r = resilient modulus, kPa.

K_1, K_2 = regression constants which represent the nonlinear elastic coefficients and exponents. They should be determined for each specimen to ensure that the coefficient of correlation exceeds 0.90.

σ = stress, kPa.

$f(\sigma)$ = a stress parameter.

Three values of stress parameter can be chosen: first stress invariant or bulk stress, J_1; second stress invariant/octahedral stress, J_2/τ_{oct}; and octahedral stress, τ_{oct}.

$$J_1 = \sigma_1 + \sigma_2 + \sigma_3$$

$$= 3\sigma_c + \sigma_d$$

$$J_2 = \sigma_1\sigma_2 + \sigma_2\sigma_3 + \sigma_3\sigma_1$$

$$= \sigma_c^2 + 2(\sigma_c(\sigma_c + \sigma_d))$$

$$\tau_{oct} = 0.5\sqrt{\left((\sigma_1 - \sigma_2)^2 + (\sigma_2 - \sigma_3)^2 + (\sigma_1 - \sigma_3)^2\right)}$$

$$= 0.5\sqrt{\left(2(\sigma_d)^2\right)}$$

where, σ_1, σ_2, and σ_3 are the principal stresses; σ_c is the confining stress; and σ_d is the deviator stress.

The equation can be simplified by applying log on both sides as follows:

$$\log M_r = \log K_1 + K_2 \log f(\sigma)$$

An example of raw data and analysis is shown in Figure 7.19.

The resilient modulus also needs to be conducted at different moisture and density levels to determine their effect, such that the response of a soil layer under varying environmental/construction conditions can be determined during design. One way of considering the effect of differing moisture- and stress-level conditions on the resilient modulus is to determine an effective roadbed soil resilient modulus, by considering the reduction of the resilient modulus of the soils in the different months.

The AASHTO (1993) standard guide for design of pavement provides relationships between resilient modulus and layer coefficients, whereas the relationship between CBR and resilient modulus is provided by the UK Transportation Research Laboratories (Powell et al., 1984). The relationships are as follows:

$$M_r = 2555 \, (CBR)^{0.64}$$

where M_r, resilient modulus, is in psi.

$$M_r = 30{,}000(a_i/0.14)$$

where M_r is in psi, and a_i is the AASHTO layer coefficient.

Sequence Number	Pulse Group	Conf Stress (kPa)	Dev Stress (kPa)	Mr (MPa)	Mr (kPa)	Bulk Stress (kPa)	
0	1	44.8	24.0	34.5	34500.0	158.4	
0	2	44.8	24.8	35.8	35800.0	159.2	
0	3	44.8	24.7	37.0	37000.0	159.1	
0	4	44.8	24.7	38.7	38700.0	159.1	
1	1	44.8	11.0	37.4	37400.0	145.4	
1	2	44.8	11.8	36.5	36500.0	146.2	
1	3	44.8	11.6	37.6	37600.0	146.0	
1	4	44.8	11.9	37.0	37000.0	146.3	
2	1	44.8	24.1	38.8	38800.0	158.5	
2	2	44.6	24.8	38.9	38900.0	158.6	
2	4	44.8	24.8	38.7	38700.0	159.2	
3	1	45.1	35.9	39.9	39900.0	171.2	
3	2	44.8	36.6	39.4	39400.0	171.0	
3	3	44.3	36.6	39.3	39300.0	169.5	
3	4	44.8	36.5	39.1	39100.0	170.9	
4	1	44.8	48.8	40.0	40000.0	183.2	
4	2	44.8	70.0	49.4	49400.0	204.4	
4	3	44.6	49.6	40.0	40000.0	183.4	
4	4	44.8	49.7	40.6	40600.0	184.1	
5	2	44.8	61.6	41.3	41300.0	196.0	
5	3	44.8	61.6	41.8	41800.0	196.0	
5	4	44.8	61.6	41.7	41700.0	196.0	
6	1	31.9	10.8	28.2	28200.0	106.5	
6	2	31.9	11.7	27.8	27800.0	107.4	
6	3	31.9	11.8	28.5	28500.0	107.5	
6	4	32.2	11.6	28.6	28600.0	108.2	
7	1	32.2	23.8	28.3	28300.0	120.4	
7	2	32.2	25.1	28.8	28800.0	121.7	
7	3	31.7	24.8	28.9	28900.0	119.9	
8	1	32.2	35.7	30.1	30100.0	132.3	
8	2	31.9	36.6	30.2	30200.0	132.3	
8	3	31.9	36.9	30.4	30400.0	132.6	
8	4	32.2	36.9	30.3	30300.0	133.5	
9	1	32.2	48.7	32.6	32600.0	145.3	
9	2	31.9	49.7	32.7	32700.0	145.4	
9	3	32.2	176.7	95.6	95600.0	273.3	
9	4	31.9	49.7	33.2	33200.0	145.4	
10	1	32.2	60.4	34.8	34800.0	157.0	
10	2	31.9	61.5	34.7	34700.0	157.2	
10	4	32.2	61.6	35.0	35000.0	158.2	
11	2	18.8	11.9	20.0	20000.0	68.3	
11	3	18.5	11.9	20.6	20600.0	67.4	
11	4	18.8	12.0	20.5	20500.0	68.4	
12	1	19.1	153.3	106.6	106600.0	210.6	← Outlier
12	2	18.8	24.9	20.4	20400.0	81.3	
12	3	18.2	24.6	20.4	20400.0	79.2	
12	4	18.8	24.7	20.7	20700.0	81.1	
13	1	18.5	35.4	22.2	22200.0	90.9	
13	2	18.8	36.4	22.5	22500.0	92.8	
13	3	18.8	36.8	22.9	22900.0	93.2	
13	4	18.5	36.7	22.9	22900.0	92.2	
14	1	18.8	48.4	25.0	25000.0	104.8	
14	2	18.8	49.8	25.2	25200.0	106.2	
14	3	18.8	49.5	25.2	25200.0	105.9	
14	4	18.8	49.6	22.2	22200.0	106.0	← Outlier
15	1	19.1	60.1	26.6	26600.0	117.4	
15	2	18.8	61.3	26.0	26000.0	117.7	
15	3	18.8	61.4	25.9	25900.0	117.8	
15	4	18.8	61.4	25.9	25900.0	117.8	

FIGURE 7.19 Example of Raw Data and Analysis of Resilient Modulus Testing of Soil.

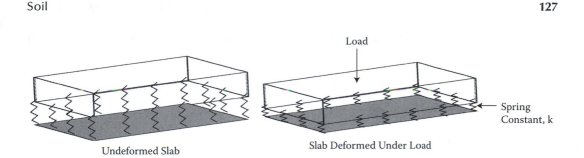

FIGURE 7.20 Slab on Springs and Spring Constant.

7.12 SUBGRADE SOIL TESTS FOR RIGID PAVEMENTS

When considering the magnitude of stresses induced in a slab under loading, the influence of the subgrade is defined by its modulus of subgrade reaction (k). The modulus of subgrade reaction is defined as the ratio of the pressure applied to the subgrade using a loaded area divided by the displacement experienced by that loaded area. (Figure 7.20 shows a slab with springs deflecting under load.)

7.12.1 PLATE LOAD TEST

The plate load test (Figure 7.21) applies a load to a steel plate bearing on the subgrade surface. The plate load is usually applied using a hydraulic jack mounted against a vehicle used for the reaction frame. The resulting surface deflection is read from dial micrometers near the plate edge but away from the loading point. The value of k is found by dividing the pressure exerted on the plate by the resulting vertical deflection, and is expressed in units of N/mm^3, MN/m^3, kg/cm^3, or $lbs/in.^3$ (pci). k is established by plate-bearing tests with a load plate diameter of 750 mm (30 in.). A modification is needed if a different plate diameter is used: for a 300-mm diameter plate, k is obtained by dividing

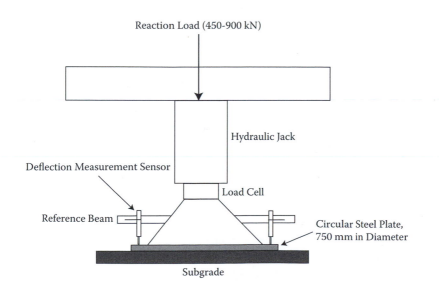

FIGURE 7.21 Schematic of Plate Load Test.

the result by 2–3, and for a 160-mm diameter plate, it is divided by 3.8. The modulus of subgrade reaction is determined by the following equation:

$$k = \frac{P}{\Delta}$$

where:

k = modulus of subgrade reaction
P = applied pressure (load divided by the area of the 762-mm [30-in.] diameter plate) (or, stress)
Δ = measured deflection of the 762-mm (30-in.) diameter plate

---

Example

A plate load test was conducted on two types of soils, a subgrade and a 10-in. base over the subgrade. Results from the test are shown in Figure 7.22. Determine (k) for the subgrade and (k) for the subgrade-base combination.

SOLUTION

The modulus of subgrade reaction is determined from the slope of the stress versus deformation curve.

(k) = stress/deformation

As can be seen from Figure 7.22, the subgrade curve has two slopes: a steep initial slope of 350 pci, and a more shallow slope of 133 pci. At this stage, the designer must decide on the level of deformation the subgrade will be subjected to. Otherwise, a more conservative value of 133 should be selected. For the base + subgrade behavior, a relatively constant relationship is observed and the slope is approximately 277 pci.

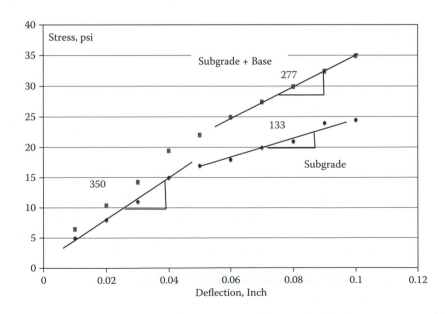

FIGURE 7.22 Example of Results from a Plate Load Test.

7.13   SUBBASE AND UNSTABILIZED BASE

Subbase and unstabilized base materials are generally characterized in the same ways. The characterization tests are the same as those mentioned for subgrades, except that organic content is not generally determined.

7.14   SOIL STABILIZATION CONCEPTS AND METHODS—CHEMICAL AND MECHANICAL

Soil stabilization refers to improvement of soil so that it performs a required function. Soil stabilization techniques can be classified into two groups, mechanical and admixture. Mechanical stabilization can be through densification or compaction, addition of granular material and compaction, and the use of reinforcement, such as geotextile. Admixture stabilization (which also includes compaction as part of the process) can be done with the use of additives, such as lime, Portland cement and asphalt.

The stabilization process or mechanism results in the modification of the properties of the soil. Compaction results in increase in strength or stiffness, decrease in permeability and compressibility. The use of additives (admixtures or modifiers) results in change in basic properties of the soil. For example, the addition of lime and cement and granular materials reduce plasticity. Cementing agents such as asphalt help develop bonds between soil particles, and hence increases cohesive strength, and decreases the absorption of water. Water retaining agents such as calcium chloride ($CaCl_2$) reduce evaporation rates and helps in reducing dust in construction sites as well as unpaved roads.

7.14.1   MECHANICAL STABILIZATION BY DENSIFICATION OR COMPACTION

Compaction is the artificial process of increasing the density or unit weight and decreasing the void ratio of a soil mass. The energy consumed during compaction is called compactive effort.

The mechanism of compaction consists of reorientation of soil particles, fracture of soil particles and breaking of bonds between them, and bending or distortion of soil particles. For coarse grained soil the primary mechanism is reorientation with some fracture, and the compactive effort must be high enough so as to overcome the friction between particles to make them reoriented. Water helps in the compaction by lubrication of the contacts. Effective compaction requires reduction in frictional forces between the particles. Vibration is effective in moving particles relative to each other and reducing the normal force and hence the friction.

In fine grained soil the primary mechanism of compaction is reorientation and distortion. The compactive effort must be high enough to overcome the cohesive/interparticle forces. Increased water content can decrease cohesion. Effective compaction requires high pressure and shearing or kneading action, to overcome cohesion. Sheepfoot roller (with small contact area), tamping foot roller and rubber tire roller are effective.

7.14.1.1   Effect of Compaction on Soil Properties

For cohesionless soil (granular soils) the effect of compaction are in terms of increase in strength and stiffness, and decrease in compressibility and permeability. The properties are not affected by moisture content. The density that could be achieved with a specific compactive effort influences the properties. For cohesive soils there can be difference in properties, even at the same dry density (due to difference in the soil structure, which can be either flocculated or dispersed). Generally higher the density, higher is the strength. Note that soils compacted dry of OMC will have higher permeability. The swelling potential is higher when density is high and degree of saturation is low. The potential for shrinkage is higher for low density and high degree of saturation.

7.14.1.2   Field Compaction

It is important to know the desired soil properties and select a laboratory method for controlling density, and hence develop specifications on percent compaction or percent maximum density. The water content should range about the OMC.

The equipment and procedure for compaction should be selected such as to allow the level of compaction at the desired moisture content to be achieved economically.

The factors that influence field compaction are:

1. Type of roller (vibratory rollers may cause fluffing, so the last few passes are made with vibration turned off).
2. Size of rollers—weight and contact area/pressure.
3. Lift (Layer) thickness—thicker the lift less effective is the compaction.
4. Number of passes.
5. Strength and stiffness of the underlying material. (Whether the compaction effort densifies the soil or actually shears it depends on this factor. One can start with lower compactive effort initially, and then gradually go to higher compactive effort.)

7.14.1.3   Field Control

Compaction in the field needs to be controlled to ensure adequate density. This is done with the help of specifications and measuring devices. Specifications are generally written to control density and moisture content. There is a target density and a target water content range about the OMC. The average may be close to the target but there may be some variability. Standard deviation values, from historical data, are used to set tolerances. Generally a one sided acceptance limit is set for density, and the specification might read, 100% pay if relative compaction $\geq$ 93%.

7.14.1.4   Measuring Devices

For density one can use sand cone or balloon density measurement. It involves direct measurement of mass and volume to calculate density. The disadvantage is that it is time consuming. Better method, which can be used for both density and moisture content, is using a nuclear gage, for which the key is calibration. These measurements are done instantaneously.

The nuclear gage can be used either in direct or backscatter mode. In direct method, a hole is drilled and the probe is inserted in it, and the signal coming out of the probe tip (hydrogen atom) is picked up, measuring average condition in between the depth of the probe and the surface. In backscatter mode the device is kept on the surface and a definite depth of measurement is associated with this method.

7.14.2   Use of Geosynthetics

Geosynthetics (planar products manufactured from polymeric materials) are also used for stabilization of subgrades. Of the different types of geosynthetic, geotextiles and geogrids are primary used, either for separation of the subgrade from the aggregate base (to avoid contamination of the base with the subgrade materials) and/or for strengthening of the subgrade (particularly for very poor and/or organic soils and unpaved roads). For strengthening of the subgrade, sufficient stiffness and good interlock with the surrounding materials is necessary. Sometimes nonwoven/grids are more effective than woven geotextiles. Hence the selection of the geosynthetic should be done after careful consideration of the existing soil conditions and objective of stabilization. The construction process consists of rolling out the geotextile on the cleaned and leveled subgrade and pulling out to avoid wrinkles. Adjacent rolls are overlapped (by say 30 inch) and the fabrics are kept in place by pins or staples or base materials. A woven or nonwoven geotetxile could be used for strengthening the subgrade, or a geogrid could be used in conjunction with a separator geotextile. The relevant test properties include tensile strength (ASTM D4632 and D4595), tear strength (ASTM D4533), puncture strength (ASTM D 4833), permittivity (ASTM D4491), apparent opening size (ASTM D4751) and resistance to ultraviolet light (ASTM D4355). Good information is available from the International Geosynthetic Society (http://www.geosyntheticssociety.org/indexigs.htm).

7.14.3 Lime Treatment of Soils

Lime is often used for soil stabilization. (Detailed information can be obtained from the National Lime Association, www.lime.org.) Limestone, $CaCO_3$ or Dolomite, $CaCO_3 + MgCO3$, is heated in a rotary kiln to produce quicklime, $CaO + CO_2$ or $CaO + MgO + CO_2$. Water is added to it to make it react, $+ H_2O \rightarrow$ producing hydrated lime, $Ca(OH)_2 + Mg(OH)_2$. In a pound of quicklime there is more Ca ++ ions than in hydrated lime, and hence is more effective. There is not much difference between dolomite or limestone, as far as stabilization is concerned.

The mechanisms of soil-lime reaction consist of the following.

1. The modification process happens through cation exchange. Ca++ is substituted for absorbed cations in the clay soil-water system. This process is most effective when the absorbed cations are sodium, Na+. They are replaced by Ca++. Li+ are most easily replaced although they are not that common. It is replaced by Na+, which is the most common, and which is replaced by K+, which can be replaced by Mg++ and Ca++. Two valence materials have stronger replacing power.
   Replacement power:

$$Li + <Na + <K + <<Mg ++ <<Ca ++ \text{ (Holtz and Kovacs, 1981)}$$

2. Flocculation and agglomeration process which allows the end to side contact, and soil particles behave as larger particles.
3. Cementing of soil particles due to carbonation of lime. In the presence of CO_2, $CaCO_3$ is produced from CaO. Also, there is Pozzolanic reaction, in which lime reacts with alumina and silica from the clay minerals to form cementitious compounds, similar to the process during hydration of Portland cement. This reaction occurs only when there is excess lime after the completion of the cation exchange process.

The modification process (cation exchange) is a faster reaction, compared to the other mechanisms. The lime content required generally ranges from 3 to 5% (of hydrated lime by weight of the soil). As a result of this mechanism, the plasticity index (PI) is decreased, as the liquid limit decreases and the plastic limit increases. The effective particle size increases, and as the moisture content remains the same, it appears that the soil is drier, and more workable. Volume change and swell potential also decreases.

The cementing mechanism can occur only after Ca++ requirements for modification have been satisfied. The requirement generally ranges from 5–10% hydrated lime. Adding more than 10% lime is generally not economically viable. This mechanism results in increase of strength, as can be measured with unconfined compressive strength. It is also common to add pozzolona (fly ash) to the soil to provide silica for the cementing action. This process is called lime-fly ash stabilization.

Lime treatment is used in subgrade soils, and subbase and base materials. For subgrade soils (clay), lime can be used to reduce the potential of change in volume, whereas, for marginal base and subbase materials it is used for improving their quality. The main benefit from lime is the reduction in plasticity, and hence the reduction in the effect of water. This leads to improvement in both constructability and performance of the pavement. Soils already containing Ca++ are not affected by lime treatment, and the secondary effect of strength gain is, in some cases (such as sand-gravel mix) less than what could be achieved with the addition of Portland cement.

The amount of lime needed depends on whether it is to be used as a modifier alone or also for increasing the strength. The modification process results in reduction of soil plasticity, which is a measurement of the effect of water on the soil, and the applicable tests are the Atterberg limit tests, LL, PL, and PI. Occasionally swell tests are run to note either pressure or volume change. The cation exchange occurs rapidly and hence these tests can be performed quickly. The lime content is expressed as a percentage of hydrated lime by dry weight of the soil. For the same lime, generally 33% more hydrate lime (by weight) is required compared to quick lime. There may or may not be consideration of the strength and stiffness increase in design, and how permanent the strength increase is, is unknown.

The cementing action results in an increase in strength, and this can be evaluated by running unconfined compressive strength or indirect tensile strength tests. The addition of lime may result in increase in strength in a layer, such as in subbase, which could lead to a reduction of the designed layer. Generally, for subgrade materials, the lime is added purely as a modifier.

The construction process of lime stabilization involves the following steps.

1. Scarification and pulverization, done after cutting down the material to the required slope and grade. The scarification process loosens up the soil and makes mixing easy.
2. Spreading lime, dry or as a slurry, depending on whether the area is windy, as well as costs. Considerations include cost, environmental, production rate and natural water content of the soil.
3. Preliminary mixing and watering, which can be eliminated for easily workable subgrades and subbase materials. Heavy clays require two mixings, and rotary mixers are the most efficient. Water is required for compaction, to get lime distributed uniformly through the soil, and is needed for the pozzolanic reaction.
4. Preliminary curing, to give time to react throughout the soil. (The above two steps are mostly for high plasticity subgrade, to make them friable and workable.)
5. Second lime application.
6. Final watering and mixing.
7. Compaction.
8. Finishing; subgrade trimmers are used for proper grading.

Curing, which may involve spraying asphalt binder, is required to prevent moisture loss, erosion and damage from traffic, as well as construction traffic. Construction can also be done with a central pugmill mixing, for subbase and base course stabilization. In this process the mixing of soil, water lime and fly ash, if needed are added in proportion and mixed in the pugmill. Spreading is done by trucks and the rest of the processes are similar to in-place application of lime.

Climatic consideration in the stabilization process is important. If temperature is too low, hydration or modification reaction with lime will not occur. The rule of thumb is that lime stabilization should not be started unless the temperature is 40-45°F and rising.

Safety consideration is also important. Quicklime is dangerous for the skin and dusting in urban areas could be a problem.

Lime slurry pressure injection (LSPI) is also used for stabilization. In this process lime slurry is injected into the soil under pressure, ranging from 20–200 psi, at spacing of 5 ft, for example, and to a depth of 7 to 12 ft, depending on the application, but generally deep enough to go below the depth of seasonal moisture fluctuations.

The volume of slurry is generally 10 gallons per foot of depth, and 2.5 to 3 lbs of lime per gallon of water. Because the permeability of clays (soils for which lime treatment is necessary), fractures, seams and fissures in the soil are necessary for effective application of this process. This process is most efficient in the dry season when cracks and fissures are the widest.

This process can be used as pretreatment or a post treatment, for stabilization of building foundations and subgrades for railroads, streets and airport pavements (to prevent swelling, for example), and also in levees and slopes to reduce shrinkage potential.

Examples of modern lime spreading and stabilization equipment are shown in Figure 7.23.

7.14.4 Cement Treatment of Soil

In cement treatment, sufficient Portland cement is added to the soil to produce a hardened mixture, resulting from hydration reaction. (Detailed information can be obtained from the Portland Cement Association, www.cement.org, as well as the American Concrete Pavement Association, www.pavement.com.) Portland cement, which is produced from a mixture of limestone, clay, shales, and iron ore, contain Ca, Si, and Al, which reacts with water, to produce hydrated calcium silicate or aluminate. Unlike lime, cement-soil reaction does not require external source of Al, and Si, and hence is more effective in this reaction than lime.

Cement stabilized soils can be categorized into a few groups.

1. Soil cement, which consists of a mix of natural soil of low or marginal quality (such as natural clay or clayey sand) and cement. The objective of this statibilization is to make such soils usable as subgrade soils. The feasibility of using cement becomes an issue in such cases where the demand for the cement can exceed 20%.
2. Cement treated base (CTB) is a mix of granular materials of reasonably high quality (often they meet base course specifications) and cement, used as high quality stiff base course in pavements with high loads and traffic volume. The mix can contain wither crushed or uncrushed base course aggregate, or with a blend of both materials. The cement content is generally kept lower than 4% to prevent excessive reflective cracking in pavements.
3. Econocrete, which is a mix of low quality aggregates (natural or reclaimed aggregates, which do not meet the standard specifications) and cement, used principally as a subbase (such as in airport pavements).

The use of CTB under a relatively thin (say 3 in.) of HMA surface layer would result in reflection cracks. While breaking the bond between the CTB and the HMA with a crushed stone layer is one option, it results in a structure with a weaker layer above a stiffer layer. Since the option of using a very thick HMA is not economical, it is often advised not to use cement stabilized materials in flexible pavements (with HMA surface layer), or follow guideline regarding the minimum thickness of HMA with CTB. Since the spacing of the cracks is a function of the strength of the CTB, a lower percentage of cement would result in a larger spacing of the cracks. Therefore, the maximum allowable cement content could be determined from the minimum thickness of HMA layer that is needed for controlling the cracks.

The modification with cement results in three things—reduction in plasticity, decrease in volume change capacity and increase in strength and stiffness. Because cementation may not be permanent, the reduction in plasticity is perhaps the most important process. The increase in strength, due to cementation, may be lost due to weathering, such as through repeated freeze-thaw cycles. The hydration reaction of Portland cement and water that results in cementitious compounds that cement soil particles together is a time dependent process. In the initial cementation process, there is rapid strength increase; then there is a slow long term strength gain. There is release of Ca++ ions as a result of Pozzolanic reaction, and the slow process is the carbonation reaction where Ca reverts back to $CaCO_3$ compound.

7.14.4.1 Mixture Design Process

The mix design process for cement modified soils (meant for reduction of plasticity and permanent or temporary strength gain) is simple, whereas for cement stabilized soils (meant for increasing the strength of the soil; change in strength is permanent; durability is important) is more complex.

7.14.4.1.1 Mix Design for Cement Modified Soils

Atterberg Limit tests are run on raw and then on modified soils usually at 1, 24 and 48 hours after mixing to establish plasticity modification. Note that most of the modification happens within an hour. Moisture density tests are conducted to establish compaction requirements. Strength increase is not considered but compressive early strength test is run to check the requirements of minimum strength.

7.14.4.1.2 Mix Design for Cement Stabilized Soils

The purposes of the mix design are to determine the amount of Portland cement to adequately harden soil (adequacy is determined by the development and maintenance of strength during weathering), determine the amount of water needed for compaction (OMC is usually sufficient for hydration), and determine how density affects strength, stiffness, and durability of the mix.

Step 1: Selection of preliminary cement content from design guidelines-poorer the soil, the higher is the amount of cement required; the amount needed goes up significantly if there is organic material

Step 2: Conducting moisture-density tests—different for that of raw soil—this is given in ASTM D558 and AASHTO T134; this step helps in establishing OMC and maximum dry density for compaction durability, strength and construction compaction control.

Step 3: Conducting durability tests—wet dry tests—ASTM D559, AASHTO T135; freeze-thaw tests—ASTM D560, AASHTO T136. The purpose is to select cement content so that soil cement or CTB will stay hard and maintain strength when subjected to alternate wetting/drying or freezing/thawing cycles. Strength may be considered as secondary requirement but it has been determined that cement content established by durability insure adequate long term strength. Test procedure calls for 3 cement contents, and one freeze thaw and one wet dry specimen at each cement content. Specimens are 4 in. in diameter and 1/30 ft^3 in volume. The specimens are compacted to OMC and maximum dry density (from step 2) moisture density tests performed on material with median cement content. The specimens are moist cured for 7 days and then subjected to wet dry or freeze/thaw tests, as follows. For wet dry tests, samples are soaked for 5 hours at room temperature (77°F), and then oven dried for 42 hours at 160°F. Wire brush is used to apply 2 strokes per area. This is repeated for 12 cycles. After 12th cycle, it is dried at 230°F and weighed. The weight loss is calculated as % of the original dry weight of the sample. Note to find the dry weight one needs the moisture content of the sample.

The moisture density curve to get max dry density is plotted for the median cement content. The same compactive effort is needed when preparing wet/dry, or freeze/thaw samples.

For freeze thaw test the following steps are used. Freeze at −10°F for 24 hours; thaw in moist room at 70°F and 100% relative humidity for 23 hours. Brush after each thaw cycle; repeat for 12 cycles; compute weight loss.

Selection of final cement content is based on the weight loss, as follows (PCA method) Weight loss for different soil types (AASHTO group):

$$A - 1, A - 2 - 4, A - 2 - 5, A - 3 \le 14\%$$

$$A - 2 - 6, A - 2 - 7, A - 4, A - 5 \le 10\%$$

$$A - 6, A - 7 \le 7\%$$

For compressive strength requirement, supplementary tests to durability tests are conducted at cement content selected on the basis of durability tests

For this test 4–2 in. diameter specimens are used. The samples are moist cured for 2, 7, and 28 days and tested at various times. Strength criterion is based on minimum strength; additional requirement could be that strength increases with age.

The construction process for soil cement is very similar to that with lime.

The objective include pulverization of the soil, mixing with the proper amount of cement, compacting to maximum optimum water content to get maximum dry density or a percentage thereof, and maintaining moisture to allow hydration of cement. The steps are as follows:

1. Shape to proper line and grade (should be done prior to adding cement since once it starts to hydrate it is difficult to line and grade)
2. Scarify, pulverize and pre-wet—important in dry windy condition, since if the interface is dry and windy a significant amount of cement is lost
3. Spread Portland cement—done with tanker, with spreader bar attached; small jobs may use bags of cement, each bag weighs 94 lbs (approximately 1 ft^3).
4. Add water and mix—twin shaft roller may be used just as is used in lime
5. Compact—initial, usually done with sheep foot or vibratory roller, and finish, done with rubber tired or steel wheel roller
6. Final shaping and finishing—to reestablish the line and grade—must be accomplished quickly before hydration. Typical finishing operations include shaping with a motor grader, rolling with a steel wheel roller and then with a rubber tired roller to seal the surface. Curing is generally done for 7 days, either with moist cloth or burlap of by spraying with an emulsion (say 0.15 to 0.3 gallon per square yard). The surface is then sanded to protect surface, if traffic is expected.

The mixing process can also be conducted in a plant. In this process, the subgrade is compacted and shaped to line and grade soil, and the cement and water are mixed in a central plant A screw conveyor, or compressed air is used to get cement from silo to pugmill in the plant. The mix is then transported to the job site in dump trucks. It must be spread by mechanical spreaders for uniform thickness. It is advisable to breakup the construction area into smaller elements and do mixing and compaction. Such proper sequencing of construction operation is required to prevent damage to unfinished sections in the case of rain.

Temperature is an important consideration—soil cement work is not started if it is not 40°F and rising (April to October is suitable for most part of the U.S.) There should not be freezing for 7 days; otherwise preliminary strength gain will be destroyed. The surface must be protected from heavy traffic and erosion from rain. Application of emulsion provides abrasion resistance to protect from constructing equipment. Rain is actually good as it prevents moisture loss; however a prime coat is needed to resist erosion.

For quality control, checks are required for compaction (dry density), water content and cement content. Cement content can be checked by putting a 1 sq yd burlap on the pavement, and weighing it after the cement has been spread. Figure 7.23 shows a modern cement stabilization equipment.

7.14.5 ASPHALT (BITUMINOUS) TREATED SOIL

Asphalt viscosity has to be reduced to be mixed with soil. In the hot process, heated asphalt cement is mixed with heated aggregate mixture to produce high quality Hot Mix Asphalt (HMA). In the cold process, a mix of asphalt emulsion and soil is produced giving a low quality material. Clean sand and natural gravels are used. (Note that foamed asphalt is also increasing being used for asphalt stabilization—for details see the section on Full Depth Reclamation in Chapter 18.)

7.14.5.1 Stabilization Mechanism with Asphalt Treatment

It improves the properties of the soil by the following mechanisms.

1. Waterproofing: The soil particles are coated with asphalt, and water does not get in contact with the soil particles and hence makes the soil less sensitive to water and lowers the absorption of water.
2. Cementation: The asphalt increases the cohesion of the mix.

Wirtgen SW 16TA Lime Spreader
Tank Capacity: 16.5 m³, Spreading Width: 2500 mm
Spreading Quantity: 2-50 l/m², Manual Operation
Spreading Auger

Wirtgen SW 16MC Lime and Cement Spreader
Tank Capacity: 16.5 m³, Spreading Width: 2460 mm
Spreading Quantity: 2-50 l/m², Cellular Wheel
Automatic Operation

Lime Stabilization with Wirtgen WR 2500
Working Width: 8 ft, Working Depth: 0-20 in
Engine Output: 690 HP, Weight: 63,000 lb

Lime Stabilization with Wirtgen WR 2000
Working Width: 6 ft 7 in, Working Depth: 0-20 in
Engine Output: 400 HP, Weight: 49,060 lb

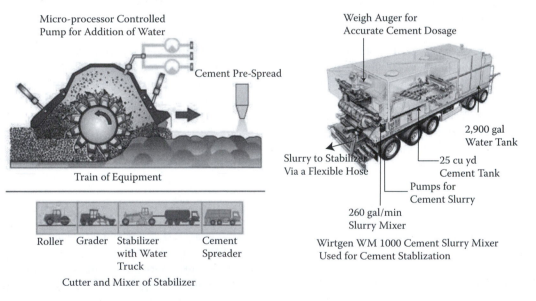

FIGURE 7.23 Examples of Equipment Used in Soil Stabilization.
Courtesy: Mike Marshall, Wirtgen GmbH.

Asphalt emulsion is used with fine grained soils primarily to make them waterproof. It may be used in conjunction with lime treatment. For fine grained soils with maximum LL-40, PI-18, 4–8% asphalt is used.

Clean sand is also used, to produce sand asphalt, for pavement layers as well as impervious liners or slope protection. Generally 4–10% asphalt is used, when the primary requirement is cementation. General rules include limiting the PI at 12%, and a also limit on maximum passing the #200 sieve, such as 12 or 25%.

Gravel and sand gravel mixtures are used to improve substandard materials having unacceptably high fine content (–#200 material). The primary function of such mixes is waterproofing and less strength change and volume change potential when exposed to water. General rules include a limit of 12% for the PI, and maximum percent passing the #200 sieve is 15%. Generally 2–6% asphalt is used.

7.14.5.2 Mix Design Procedure

While the Marshall mix design procedure can be used for HMA (discussed in Chapter 14), the asphalt soil mixes could be designed using the Hubbard field Mix Design procedure (ASTM D915, ASTM D1138—note that both standards have been withdrawn from ASTM). The basic procedure for the Hubbard field mix design consists of compaction using different asphalt contents and checking the strength with extrusion test. The extrusion requirements are as follows:

Fine grained soils
Room temperature—extrusion value before soaking 1,000 min
Extrusion value after soaking 7 days—400 min
Expansion, 5% max, absorbed water—7% maximum
Sand and gravel
Sands: Extrusion value minimum, 1200 (tested while submerged at 140°F—there is a an air curing period prior to testing).

The properties considered in Marshall Mix Design are: Stability, Flow, air voids and voids in mineral aggregate.

Construction considerations for asphalt treated soils: Asphalt treated soils are mixed in-place or central mix operation, depending on the location of the material to be stabilized.

The steps in the in-place mixing operation consist of establishing line and grade, pulverizing the material, adding emulsion, mixing and compaction. The application of emulsion and the mixing can be done with a distributor truck with a spray bar, and a mixer, or both steps can be accomplished with one single pass of a specialized equiupment. Adequate moisture in the soil is necessary for proper mixing, and the timing of compaction is critical to avoid rutting and shoving.

A travel plant or a traveling pug mill can also be used for mixing and is used mostly for cold mix. Plant Mix (Central Mix) operation is seldom used for fine grained soils. Emulsion and unheated aggregates are metered and blended in a pugmill. Generally a continuous plant is used, where aggregates and emulsion are fed and mixed continuously. Hauling, spreading and compaction follow, and timing of the above operations are critical. The timing should be made relative to the time of the break of the emulsion. The key is the consideration of the type of emulsion used—rapid, medium or slow setting (determines how fast the breaking occurs).

Rapid setting emulsion is used in surface treatments or chip seals; it involves spraying a layer of asphalt by a distributor truck and putting aggregates with another truck behind it. In many cases this happens in a trafficked road, so rapid setting emulsion is desirable. Medium setting emulsion is used for short haul jobs, and slow setting emulsions are used for long haul jobs. By varying the type and amount of the emulsifying agent one controls the setting time (RS, MS or SS) of the emulsion.

For construction control, checking the density and asphalt content (with nuclear gage, for example) are required.

Examples of modern equipment used for soil stabilization are shown in Figure 7.23.

7.15 DUST CONTROL

Dust control on unpaved roads is necessary because of the following:

1. Dust decreases safety in roads by reducing visibility.
2. Dust in roads causes environmental hazards.
3. Loss of road materials causes increases in cost of maintenance of roads, as well as maintenance of vehicles.

Results of several dust control studies in the U.S. [The Forest Service, United States Department of Agriculture (USDA), see http://www.fs.fed.us/eng/pubs/html/99771207/99771207.html] as well as other countries such as South Africa (Jones, 1999) and Australia (Foley et al., 1996) are available for guidance. The amount of dust that is generated depends on a number of factors such as silt content of the soil, speed, weight and number of vehicles as well as rainfall. Dust palliatives can be grouped into different types, such as water, water absorbing chemicals (salts), organic petroleum products (example, asphalt, and resins), organic non-petroleum product (such as lignosulfate), electrochemical products (example, enzymes), synthetic polymer products (such as polyvinyl acetate) and clay (for example, bentonite).

Water is readily available, and is easy to apply but it evaporates quickly and does not last long. Asphalt emulsions are also relatively easy to apply, but are relatively expensive and could run off to side ditches. Hygroscopic materials, such as salts ($NaCl$, $CaCl_2$, and $MgCl_2$) draw water from the air and keep the ground moist and hence reduce the potential of formation of dust. $NaCl$ is not very effective since it requires relative humidity > 75% to work. $CaCl_2$ and $MgCl_2$ are effective in relative humidity of as low as 30-40%. They are capable of drawing moisture from atmosphere in low relative humidity and keep the soil more moist than it would actually be without the salts. Generally $CaCl_2$ or $MgCl_2$ are in put down in solution by distributor truck and mixed in with the soil with motor grader. The advantages of $CaCl_2$, $MgCl_2$ are that they are relatively inexpensive and easy to apply. The disadvantages are that they are corrosive and cause rusting of equipment/vehicles, and salts in ground water and surface water are not desirable, particularly in agricultural areas. Also, salts may dissolve in rain and get washed away, and hence not provide a permanent solution to the dust problem. Sometimes the salts could make a road surface more slippery.

Lignosulfates are processed pulp liquor from paper mills, typically consisting of 50% solids. When sprayed down it acts as a binder and holds the particles together. Advantages are that it is relatively inexpensive, effective, noncorrosive, environmentally safe, and gives some hardening of the surface and makes it more trafficable. The disadvantages are that it may leach away, and the treated material may not be easily regradable. Emulsified petroleum resins are also used as a dust palliative. Containing about 60% resin, this product is generally diluted with water and applied at the rate of ½ to ¾ gallon per square yard. These commercial products are relatively expensive, but provide the advantages of the ability to control the depth of penetration by varying the amount of concentrate, and the ability to store products for long term (such as through winter). The buildup of resin residue in the soil allows decreasing the percentage of concentrate over succeeding applications. (For example, see: http://www.goldenbearoil.com/prod-coherex.htm.)

QUESTIONS

1. What are the different soil-forming minerals?
2. Classify the following inorganic soil according to the AASHTO procedure:

Sieve	% Passing	Atterberg Limits
No. 40	96	LL = 35
No. 200	55	PL = 18

3. Compute void ratio, degree of saturation, and dry density for the following soil:
   Volume = 0.37 ft³
   Mass = 50 lb
   Mass after drying = 46 lb
   Specific gravity = 2.700

4. For a soil with a specific gravity of 2.650, with the following results from a laboratory compaction test, determine the optimum moisture content.

Moisture Content (%)	Total Density (lb/ft³)
10	95
15	120
20	130
25	125
30	108

5. What are the different types of soil compaction rollers?
6. Describe some soil stabilization techniques.

8 Aggregates for Asphalt and Concrete Mixes

8.1 DEFINITION, PARENT ROCK, AND TYPES

Aggregate is defined as processed soil. Natural aggregates are obtained from quarries or river-beds. The process of obtaining aggregates consists of blasting or dredging. Large-size particles are crushed to obtain usable sizes that range from 50 mm to less than 0.075 mm for most pavement mixes. Since aggregates are derived from parent rocks, their characteristics depend on the properties of the parent rocks. The parent rocks can be classified in terms of their geologic origins, which dictate their chemical composition and hence many of the other key properties that affect their behavior and performance. Artificial aggregates are obtained as by-products of other industrial processes (such as slag from the production of steel).

Elements combine to form minerals, and minerals form different types of rocks. There are five groups of minerals that make up most rocks: silicates (containing primarily silicon and oxygen), carbonates (with carbon, oxygen, and other elements), oxides (oxygen and various metals), sulfates (sulfur, oxygen, and metals), and sulfides (sulfur and metals). Examples of minerals are quartz, SiO_2; mica, $K(Mg,Fe)_3 (AlSi_3O_{10})(OH)_2(Mg,Fe)_3(OH)_6$ (Chlorite); and feldspar, $KAlSi_3O_8$.

Geologically, parent rocks can be divided into three types: igneous, sedimentary, and tertiary. Extrusive igneous rocks are formed from magma—a viscous liquid composed of silicates, which is erupted onto the earth's surface as ash, lava flows, or solid chunks. These rocks are classified according to their texture and mineralogy, and are generally finer grained than the intrusive igneous rocks (which are not visible on the earth's surface, but remain below it). Extrusive rocks include andesite, basalt, and rhyolite. In the case of intrusive igneous rocks, magma is forced into other rocks as crosscutting or parallel (to layers of other rocks), and may consist of large areas consisting of thousands of square miles. Such large masses tend to contain a greater amount of silica (50–60%), compared to smaller dark-colored intrusions, which are sometimes referred to as *trap rock*. Intrusive rocks include granite, diorite, gabbro, and peridotite. Generally, good-quality aggregates can be produced from different types of igneous rocks, which may contain a wide variety of minerals.

Sedimentary rocks are those that are formed as a result of the consolidation of sedimentary materials formed by the reduction (clastic processes) of rock particle size through weathering and/or abrasion, or through the consolidation of chemical precipitates such as marine plant and animal deposits (carbonate rocks). In the clastic process, the weathering and abrasion action can be caused by wind, water, ice, or gravity, whereas the cementation of the materials is caused by deposition of silica or carbonate materials carried by groundwater, and compressed by the weight of the overlying materials. Examples of sedimentary materials formed by the clastic process are shale (from clay), siltstone (from silt), sandstone (from sand), and conglomerate (from gravel).

Examples of carbonate rocks are limestone and dolomites, formed by the consolidation and cementation of shells of marine plants and animals, and also from fine carbonate mud from marine water. Both silica-rich materials, such as sand and silt, as well as clastic sedimentary rocks can be found within or interlayered with carbonate rocks.

Sedimentary rocks may also form under specific chemical conditions, forming materials such as iron and gypsum.

Metamorphic rocks are formed by the alteration (recrystallization) of igneous or sedimentary rocks through high pressure and temperature. Parallel orientation and platy appearance for such rocks are common, as a result of shearing action. Sometimes, metamorphic rocks are formed by the action of heat resulting from their vicinity to large intrusive magma or other sources of heat. An example is the formation of marble from carbonate rock. A metamorphic equivalent of igneous rocks such as granite is gneiss, whereas metamorphic equivalents of sedimentary rock such as shale are slate and schist.

Natural sand and gravel are unconsolidated sedimentary materials, which are formed from the breakdown of rocks through the action of ice, wind, or water, and generally consist of smooth and rounded particles. The nature of the parent rock dictates the quality of sand and gravel that result from its weathering. Hence sands produced from igneous and metamorphic rocks or those produced from the ice- or water-weathering action of stronger rocks at higher elevations tend to be stronger than those produced from shales or siltstones, or those derived from weaker rocks in low-lying areas. Climate, under which weathering takes place, also has an important effect on the quality of sands and gravels. For example, those resulting from prolonged and deep chemical weathering under humid conditions are of better quality than those formed by the weathering of rocks by a continental ice sheet.

Weathering can be physical, chemical, or combined. Physical weathering takes place as a result of changes in temperature, humidity, and cycles of freezing and thawing or wetting and drying. Chemical weathering always results in the release of salts and colloids, which may be deposited and influence the quality of some rocks.

8.2 SUITABILITY FOR APPLICATION

For evaluation of suitability of aggregates from different parent rocks for different applications, it is important to understand that within the similar type of rocks (such as igneous), there could be wide-ranging variations, such as in composition and texture. For example, igneous rocks can be porous or dense, and the hardness and abrasion resistance depend on the size of the individual crystals that make up these rocks (they are greatest when crystals are less than 2 mm in size). Compositional factors are important; for example, certain high-silica glassy materials are unsuitable for use in Portland cement applications because of their tendency to react with the alkali in the cement.

Rocks of the same group with similar materials but different texture produce different-quality aggregates. For example, although all of them belong to the igneous group, quartz have tightly inter-locked mineral grains (giving good-quality aggregates), whereas micas are without interlocking and are present as weak layer deposits, and olivines consist of structures with rounded grains. Variations during the crystallization process of low-viscosity and low-silica igneous material can also result in segregation of minerals (magmatic differentiations), resulting in variation in rocks. Variations in metamorphic rocks are also common. For example, slates, with platy structure, are formed as a result of shearing action at low temperature, whereas schists formed at higher temperature may have more sound structure. For sedimentary rocks, variations can be in the form of amount of clay (shale has high clay content) as well as in cementation, often within formations, the extent of cementation being in many cases dictated by the age of the sedimentary rock formation.

In most cases, pavement engineers have to use locally available aggregates. There are certain characteristics of different types of aggregates that must be considered during the mix and structural design of pavements. Petrographic examination (ASTM C-295) of aggregates can be conducted to identify different parent rock materials and minerals in an aggregate sample. Such examination can provide valuable information regarding the suitability of aggregates for asphalt and concrete mixtures, particularly with respect to their durability, such as resistance against breakdown under freeze-thaw or moisture actions. Petrographic examination may include tests ranging from detailed visual examination to investigation using a polarizing microscope, differential thermal analysis, and X-ray diffraction. Figure 8.1 shows an example of results of petrographic examination of an aggregate sample.

Percent of Rock Type Retained in Each Sieve Size

Lithology	#4	#8	#16	#30	#50	#100	#200
Igneous							
granitic	17.6	23.8	21.9	22.3	21.3	20.2	23.0
felsite	2.6	0.7	0.6	1.3	0.0	0.0	0.0
gabbro	0.0	0.0	1.2	2.0	0.0	0.0	0.0
basalt	0.7	0.0	1.2	0.0	0.0	0.0	0.0
Metamorphic							
meta-granite	26.0	27.2	15.0	10.5	6.3	2.6	0.0
quartzite	8.5	7.9	9.2	8.5	5.7	3.9	0.0
meta-sediment	5.2	4.6	4.0	3.3	2.5	0.0	0.0
phyllite	3.3	4.0	2.3	1.3	1.9	0.0	1.3
phyllite, friable	0.7	0.0	0.0	0.0	0.0	0.0	0.0
schist, hard	5.9	1.3	0.6	3.3	1.3	0.7	0.0
schist, friable	1.3	0.0	0.0	0.0	0.0	0.0	0.0
mylonite	0.7	1.3	0.6	0.0	0.0	0.0	0.0
gneiss	0.7	0.7	0.0	0.0	0.0	0.0	0.0
Sedimentary							
siltstone	3.9	4.6	13.2	12.4	12.6	13.1	12.5
siltstone, friable	0.0	1.3	0.0	0.0	0.0	0.0	0.0
siltstone, iron oxide cemented	0.0	1.3	0.6	0.0	0.0	0.0	0.0
sandstone	13.1	11.3	9.2	5.2	3.8	6.5	2.6
sandstone, iron oxide cemented	0.0	1.3	0.6	2.0	1.3	0.0	0.0
greywacke	6.5	4.0	2.3	7.8	3.1	0.0	0.0
shale	2.6	2.6	6.4	3.3	5.0	1.3	3.9
chert, hard	0.7	0.7	1.2	0.6	0.6	0.7	0.0
iron oxide	0.0	0.7	0.6	0.6	0.0	0.0	0.0
Mineral							
feldspar	0.0	0.7	5.2	7.2	16.4	19.6	17.8
quartz, undulatory extinction	0.0	0.0	3.5	7.2	13.2	14.4	17.8
quartz, unit extinction	0.0	0.0	0.0	0.0	1.9	3.9	9.9
mica	0.0	0.0	0.6	0.6	3.1	12.4	7.2
heavy mineral	0.0	0.0	0.0	0.6	0.0	0.7	1.3
amphibole	0.0	0.0	0.0	0.0	0.0	0.0	1.3
pyroxene	0.0	0.0	0.0	0.0	0.0	0.0	0.7
magnetite	0.0	0.0	0.0	0.0	0.0	0.0	0.7
Total	100.0	100.0	100.0	100.0	100.0	100.0	100.0

FIGURE 8.1 Example of Results of Petrographic Examination of a Sand.
Note: The table shows classification by parent rock and minerals, and the figure shows a thin section of material less than 0.075 mm in diameter under plane polarized light—majority particles with quartz, feldspar, and mica.

8.3 PRODUCTION

Natural aggregates are either obtained from rock quarries or dredged from river beds. The production of aggregates is a complex process consisting of many steps. The process starts with the identification of a viable deposit by a geologist. Many factors are considered before setting up a quarry. These factors include development of the facility, hauling, processing, regulatory expenses and administrative costs, as well as expected sales and profits. A thorough step-by-step exploration is generally performed to identify the location of the quarry.

Standard topographical quadrangle maps and aerial photos are used to identify sand and gravel deposits, and such deposits with coarser aggregate sizes are preferred over those with finer sizes. Maps of geological bedrocks are studied for the identification of different strata before deciding on a quarry location for crushed aggregates (stones). Information from aerial photos and geophysical methods, and information from existing quarries in the area, may also be used. Once a potential site is located, it is important to study and understand outcrops as well as to investigate the uniformity of the different strata (outcropping and not outcropping), and geologic mapping may be conducted, especially if it is of a complex nature. To determine the reserves in a potential quarry, sampling can be taken from outcrops or pits, or by taking a core after drilling.

One important step is the determination of the total amount of overburden, which is the material above the mineable aggregate resource that should be removed. Such determination can be made with the help of exploration with drilling and/or with geophysical methods such as seismic or radar technology.

The amount of aggregate reserve (volume) in a proposed quarry can be calculated by simply multiplying the thickness of the layers by the area or, as in the case of layers of variable thickness, calculated from the sum of reserve for separate cross sections, drawn at regular intervals through the quarry. The tonnage of crushed aggregate is calculated by multiplying the volume by weight of usable rock per unit volume. Typical specific gravities of common raw rock are shown in Table 8.1.

$$\text{Yield(ft}^3\text{/ton)} = \frac{2,000}{62.4 * \text{Specific Gravity}}$$

$$\text{Reserve tonnage} = \frac{\text{Volume of reserve(ft}^3\text{)}}{\text{Yield(ft}^3\text{/ton)}}$$

TABLE 8.1
Specific Gravities of Different Rock Type

Rock Type	Specific Gravity
Andesite	2.4–2.8
Basalt	2.7–3.2
Traprock	2.8–3.1
Dolomite	2.7–2.8
Gabbro	2.9–3.1
Granite	2.6–2.7
Limestone	2.7–2.8
Marble	2.6–2.9
Sandstone	2.0–3.2

Source: Barksdale (1991).

The usable tonnage is the in-place reserve less the unrecoverable material from buffer zones, pit slopes, waste, haul roads, and ramps.

The production of aggregates from rocks in a quarry consists of two primary steps: removal of rock from the quarry by blasting, and sizing and separating different-sized particles with the help of a series of crushers and screens. The blasted particles, for example 8 inches in diameter, are fed into a jaw crusher, after screening the minus-1-inch material that results from the breaking of the weathered rocks around the edges of the sound rock. The crushed material from the jaw crusher is screened and then separated into different size stockpiles, whereas the larger fraction (for example, plus-1-inch material) is fed into secondary cone crushers for further crushing. Although the type of crushing equipment used is dictated by the rock and the properties of the end product that is desired, all crushers primarily use four types of mechanisms for reducing the size of particles—impact, attrition, shearing, and compression.

Generally, once production is started from a quarry, the aggregates are tested for their different properties (called *source properties*); and depending on the results of the tests, the aggregates are approved for use in specific applications, such as by the state department of transportation.

Excellent information on all types of aggregates used in the construction industry can be obtained from the National Stone, Sand & Gravel Association (n.d.; http://www.nssga.org).

8.4   OVERVIEW OF DESIRABLE PROPERTIES

Specific properties that are critical for good performance in different applications have been identified, and tests have been developed to measure them. Such tests have also been standardized by agencies, such as the ASTM. Although there are common desirable features, specific applications require specific properties of aggregates. For pavements, such applications include use as structural layers (in roads as well as railroads), drainage layers, asphalt mixtures, and cement concrete.

8.4.1   PROPERTIES CRITICAL FOR STRUCTURAL LAYERS

Aggregates could be used as base layers to decrease the stress coming from traffic (above) to the subbase and subgrade layers (below). For proper functioning, aggregates must be of sufficient strength, and if used as unsurfaced roads, they must possess good wearing resistance as well. When used as a ballast layer, directly below the ties in a railroad, to provide support to the ties and reduce the stress to the layers underneath, aggregates must be resistant to degradation action of traffic and maintenance equipment, cycles of wet-dry and freeze-thaw, accumulation of fines or binding of the aggregate layers, and the potential of swelling or frost heaves.

8.4.2   PROPERTIES CRITICAL FOR DRAINAGE LAYERS

When used as drainage layers, the proportion of different-sized particles in the aggregate (gradation) must be such as to allow sufficient drainage of water and work as a filter to prevent erosion and clogging due to finer materials.

8.4.3   PROPERTIES CRITICAL FOR ASPHALT MIX LAYERS

For use in asphalt mixes, the general requirements include cleanliness, proper shape, and resistance against the effects of traffic- and environment-related abrasion and weathering.

8.4.4   PROPERTIES CRITICAL FOR CEMENT CONCRETE LAYERS

In the blends used for cement concrete mixes, limitations are generally applied on the amount of fine aggregates to prevent the presence of clay particles. The aggregate must be resistant to the effects of abrasion and weathering due to the traffic as well as the environment, and must be such as to make the mix sufficiently workable and provide adequate strength and durability.

The different properties that are relevant to the different applications and tests are presented below.

8.5   GRADATION FOR ASPHALT PAVEMENTS

Gradation refers to the relative proportion of different-sized particles in an aggregate blend. Gradation of a blend is determined from sieve analysis. Conversely, a blend can be separated into different sizes by sieving. Generally, in a quarry the gradation of each stockpile aggregate material is determined first, and then the gradations are analyzed to determine in what proportion the aggregates from the different stockpiles can be combined to produce a **specific** gradation. This gradation is generally specified (in specifications) by pavement project owners (such as state departments of transportation). The specifications are developed on the basis of experience and results of laboratory and field studies. Figure 8.2 shows a variety of gradations for dense graded asphalt mixes with the same nominal maximum aggregate size (NMAS), which is defined as one sieve size larger than the first sieve to retain more than 10%. The maximum aggregate size is defined as one sieve size larger than the NMAS. The variation in gradation for the different NMASs is due to the fact that gradations are specified to remain close to the maximum density gradation.

The maximum density line can be obtained from the FHWA 0.45 power gradation chart by joining the origin to the maximum aggregate size by a straight line. Figure 8.2 shows the maximum density lines for the different NMASs in the FHWA 0.45 power plot. The maximum density lines are based on the concept proposed by Fuller and Thompson (1907, also known as *Fuller's curve equation*), which is as follows:

$$P = 100 * \left(\frac{d}{D} \right)^n$$

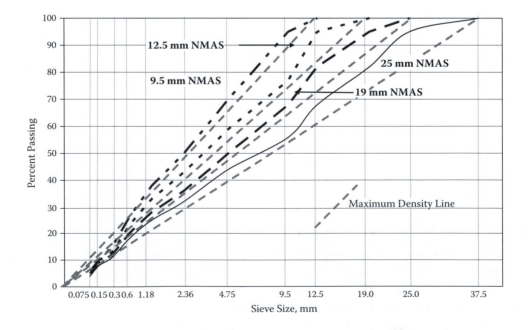

FIGURE 8.2 Different Gradations and Maximum Density Lines. (Note: Sieve Size Raised to the Power 0.45.)

where:
   P = percentage passing
   d = diameter of sieve
   D = maximum size of the aggregate
   n = a coefficient, first proposed as 0.5 and then modified to 0.45

   Note that gradations are specified so that they remain approximately parallel to this maximum density line but remain a few percentage points above or below—to remain close to the maximum gradation line so as to provide a stable structure but also retain enough space for accommodating adequate asphalt binder for durability. More sophisticated methods of developing gradations for providing aggregate interlock and adequate space for asphalt cement binder (to provide durability) have been developed. These methods, such as the voids in coarse aggregate (VCA) method as well as the Bailey method, are based on volumetric calculations with proper consideration of unit weight of individual aggregates.

   Gradations for different types of mixes are shown in Figure 8.3. Note that the gradations are different for different types of asphalt mixes, which are prepared for different applications. For common paving applications, dense graded mixes are used. To improve drainage of water and friction, open graded friction course (OGFC) is used, whereas stone matrix asphalt (SMA) is used as a high-rut-resistant mix. The gap gradation (certain fractions missing) helps in achieving a high permeability in the OGFC mixes. In the case of SMA, the percentage passing the 4.75 mm sieve is important; it is kept at a relatively low number to allow the interlock of coarse aggregate particles and hence develop strong resistance against rutting. Note, however, that the percentage passing the 0.075 mm sieve is relatively high, to provide a sufficient amount of filler material to stiffen the relatively large amount of asphalt binder used to provide a stable fine matrix, and prevent draindown of asphalt binder.

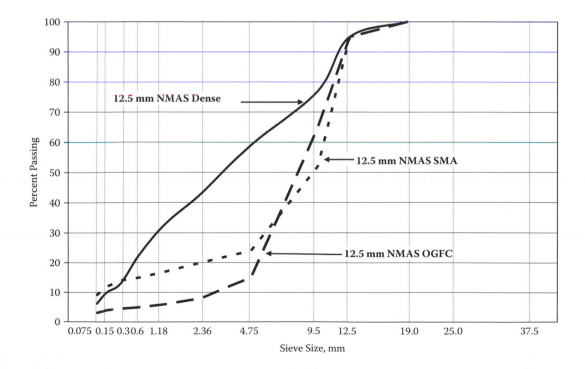

FIGURE 8.3 Gradations for Different Types of Asphalt Mixes.

FIGURE 8.4 Stack of Sieves in Sieve Shaker.

8.5.1 AGGREGATE TESTS

In the ASTM C136: Standard Method for Sieve Analysis of Fine and Coarse Aggregates test, a weighed sample of dry aggregates is separated by running them through a set of sieves. The sieves are stacked in such a way so that the openings decrease from top to bottom. After separating the particles of the different sizes, the percentage of particle passing each sieve is calculated to determine the gradation, or particle size distribution. The weight of the test sample depends on the size of the aggregates used. Generally mechanical sieve shakers are used, which could impart vertical and lateral motion to the sieves, during the sieving process. Guidelines for determining the adequacy of the sieving process are provided in the standard. Generally, 8-inch-diameter round sieves are used. Dry sieve analysis provides an estimate of particle size distribution, whereas washed sieve analysis provides an accurate indication. A stack of sieves in a sieve shaker is shown in Figure 8.4.

The steps required to combine aggregates from different stockpiles to produce a specific gradation are presented in Figure 8.5, with an example.

Note that in an asphalt mix, the surface of the aggregates is covered with asphalt binder. The total surface area of the aggregates dictates the amount of asphalt that is needed for covering the aggregates adequately, that is, as stated commonly, with an adequate film thickness. Although the concept is purely theoretical, as there is no uniform or one film thickness for the entire mix, it is a logical one, and can be used to ensure an optimum "coating" of aggregates. Fine aggregates have larger surface area, and hence blends with a higher amount of fine aggregates will have a higher surface area compared to those with a low amount of fines. Relatively thin films of asphalt can negatively affect the durability of the mix, whereas thick films can adversely affect its resistance against rutting. An optimum film thickness in the range of 8–12 microns has been mentioned in the literature.

Blending of aggregates in a mix design consists of the following steps.
Step 1: Consider all the aggregates make up 100 % of a batch
Step 2: Determine the gradation of the individual stockpile aggregates, for example,

Sieve Sizes	Percent Passing for Different Stockpiles				
	3/8	1/2	Sand	Stone Sand	Dust
19.00	100.0	100.0	100.0	100.0	100.0
12.50	100.0	81.0	100.0	100.0	100.0
9.50	93.0	15.0	100.0	100.0	100.0
4.75	24.0	1.7	98.7	95.3	100.0
2.36	7.5	1.5	79.0	77.4	84.5
1.18	5.8	1.4	57.4	50.0	57.4
0.60	5.1	1.3	39.7	31.0	40.5
0.30	4.5	1.2	16.9	15.7	29.2
0.150	3.5	1.0	6.9	5.1	21.1
0.075	1.6	0.7	2.8	1.6	14.2

Step 3: Based on approved designs and/or experience, select percentages of a batch from the different stock piles, for example:

Batch Size (lbs):		100.00
	Percentage of Batch	Cumulative
3/8	15	15.00
1/2	20	35.00
Sand	30	65.00
Stone Sand	25	90.00
Dust	10	100.00

Step 4: Compute the contribution of each stockpile to each aggregate size by multiplying the percentage passing each aggregate of a stockpile with the percentage of the stockpile in the total aggregate batch, for example:

	3/8 inch stockpile		
Sieve Size	Percent Passing A	Percent in Batch B	Percent of Total Batch (= A*B%)
19.00	100.0	15	15.0
12.50	100.0	15	15.0
9.50	93.0	15	14.0
4.75	24.0	15	3.6
2.36	7.5	15	1.1
1.18	5.8	15	0.9
0.60	5.1	15	0.8
0.30	4.5	15	0.7
0.150	3.5	15	0.5
0.075	1.6	15	0.2

FIGURE 8.5 Steps to Combine Different Stockpile Aggregates to Produce a Specific Blend.

The surface area of aggregates could be determined from surface area factors, as presented in Table 8.2. The factors are multiplied by the percentages, and then added up. An example of calculation is also shown.

Aggregates have an important role in proportioning concrete. Aggregates affect the workability of fresh concrete tremendously. The aggregate particle size and gradation, shape, and surface

1/2 inch stockpile			
Sieve Size	Percent Passing A	Percent in Batch B	Percent of Total Batch (= A*B%)
19.00	100.0	20	20.0
12.50	81.0	20	16.2
9.50	15.0	20	3.0
4.75	1.7	20	0.3
2.36	1.5	20	0.3
1.18	1.4	20	0.3
0.60	1.3	20	0.3
0.30	1.2	20	0.2
0.150	1.0	20	0.2
0.075	0.7	20	0.1

Sand stockpile			
Sieve Size	Percent Passing A	Percent in Batch B	Percent of Total Batch (= A*B%)
19.00	100.0	30	30.0
12.50	100.0	30	30.0
9.50	100.0	30	30.0
4.75	98.7	30	29.6
2.36	79.0	30	23.7
1.18	57.4	30	17.2
0.60	39.7	30	11.9
0.30	16.9	30	5.1
0.150	6.9	30	2.1
0.075	2.8	30	0.8

Stone Sand stock pile			
Sieve Size	Percent Passing A	Percent in Batch B	Percent of Total Batch (= A*B%)
19.00	100.0	25	25.0
12.50	100.0	25	25.0
9.50	100.0	25	25.0
4.75	95.3	25	23.8
2.36	77.4	25	19.4
1.18	50.0	25	12.5
0.60	31.0	25	7.8
0.30	15.7	25	3.9
0.150	5.1	25	1.3
0.075	1.6	25	0.4

Dust stockpile			
Sieve Size	Percent Passing A	Percent in Batch B	Percent of Total Batch (= A*B%)
19.00	100.0	10	10.0
12.50	100.0	10	10.0
9.50	100.0	10	10.0
4.75	100.0	10	10.0
2.36	84.5	10	8.5
1.18	57.4	10	5.7
0.60	40.5	10	4.1
0.30	29.2	10	2.9
0.150	21.1	10	2.1
0.075	14.2	10	1.4

FIGURE 8.5 (Continued).

Step 5: Compute the final blend by adding up the contribution to each aggregate size from each of the different stockpiles, in a horizontal layout table, and compare with the limits in the specification, for example:

Stockpile	Percent	Combined Gradation, % Passing Sieves, mm									
	batched	19.0	12.5	9.5	4.75	2.36	1.18	0.60	0.30	0.15	0.075
3/8	15	15.0	15.0	14.0	3.6	1.1	0.9	0.8	0.7	0.5	0.2
½	20	20	16.2	3	0.34	0.3	0.28	0.26	0.24	0.2	0.14
Sand	30	30	30	30	29.61	23.7	17.22	11.91	5.07	2.07	0.84
Stone Sand	25	25.0	25.0	25.0	23.8	19.4	12.5	7.8	3.9	1.3	0.4
Dust	10	10.0	10.0	10.0	10.0	8.5	5.7	4.1	2.9	2.1	1.4
Total	100.0	100	96	82	67	53	37	25	13	6	3.0
Upper Limit		100.0	100.0	90.0		58.0					10
Lower Limit		100.0	90.0			28.0					2

Step 6: If any of the percentages are outside the acceptable limits then the blend percentages can be readjusted to produce a more acceptable gradation. For example, in the data shown here, the blend can be modified as follows:

Batch Size (lbs):		100.00
	Percentage of Batch	Cumulative
3/8	15	15.00
1/2	20	35.00
Sand	35 (from 30)	65.00
Stone Sand	20 (from 25)	90.00
Dust	10	100.00

Stockpile	Percent	Combined Gradation, % Passing Sieves, mm									
	batched	19.0	12.5	9.5	4.75	2.36	1.18	0.60	0.30	0.15	0.075
3/8	15	15.0	15.0	14.0	3.6	1.1	0.9	0.8	0.7	0.5	0.2
½	20	20	16.2	3	0.34	0.3	0.28	0.26	0.24	0.2	0.14
Sand	35	35.0	35.0	35.0	34.5	27.7	20.1	13.9	5.9	2.4	1.0
Stone Sand	20	20.0	20.0	20.0	19.1	15.5	10.0	6.2	3.1	1.0	0.3
Dust	10	10.0	10.0	10.0	10.0	8.5	5.7	4.1	2.9	2.1	1.4
Total	100.0	100	96	82	68	53	37	25	13	6	3.1
Upper Limit		100	100	90		58					10
Lower Limit		100	90			28					2

FIGURE 8.5 (Continued).

texture will influence the amount of concrete that is produced with a given amount of paste (cement plus water). The selection of the maximum size aggregate is governed by the thickness of the slab and by the closeness of the reinforcing steel. The maximum aggregate size should not be obstructed to flow easily during placement and consolidation.

The maximum size of coarse aggregate should not exceed one-fifth of the narrowest dimension between sides of forms or three-fourths of the clear space between individual reinforcing bars or the clear spacing between reinforcing bars and form walls. For unreinforced slabs on grade, the maximum aggregate size should not exceed one-third of the slab thickness.

8.6   SPECIFIC GRAVITIES AND ABSORPTION

Specific gravity of a material is defined as the ratio of the weight of a unit volume of the material to the weight of an equal volume of water at 23°C. This parameter is needed to calculate volumetric properties in a mix, and any calculation that involves determination of mass from volume or

TABLE 8.2
Surface Area Factors and Calculations

Sieve Size	Surface Area Factors (square feet per pound)
% passing maximum sieve	2
% passing No. 4	2
% passing No. 8	4
% passing No. 16	8
% passing No. 30	14
% passing No. 50	30
% passing No. 100	60
% passing No. 200	160

Example Calculation

Sieve Size	% Passing	Calculation	Surface Area (square feet per pound)
1/2	100	1 * 2	2
3/8	90		
No. 4	75	0.75 * 2	1.5
8	62	0.62 * 4	2.5
16	50	0.5 * 8	4.0
30	40	0.4 * 14	5.6
50	30	0.3 * 30	9
100	10	0.1 * 60	6
200	4	0.04 * 160	6.4

$\Sigma = 37.0$ sq ft/lb

Source: Reprinted from Hot Mix Asphalt Materials, Mixture Design, and Construction by permission of National Asphalt Pavement Association Research and Education Foundation.

vice versa. There are three different specific gravities that could be used for determination of volumetric properties in asphalt mixes. The different specific gravities are because of the effect of aggregate pores and absorption. The bulk and apparent specific gravities are determined from the same test, whereas the effective specific gravity is calculated. The applicable ASTM standards are C-127 and C-182. Bulk and apparent specific gravities are explained in Figure 8.6.

Bulk-specific gravity is defined as the ratio of the oven dry weight in air of a unit volume of a permeable material (including both permeable and impermeable voids normal for the material) at a stated temperature to the weight of an equal volume of gas-free distilled water at the stated temperature.

$$\text{Bulk specific gravity, } G_{sb} = \frac{\text{Weight of dry aggregate}}{\text{volume of dry aggregate } + \text{ external voids}}$$

$$= \frac{\text{Weight in air (dry)}}{\text{weight in SSD condition } - \text{ weight in submerged condition}}$$

Apparent specific gravity is the ratio of the oven dry weight in air of a unit volume of an impermeable material at a stated temperature to the weight of an equal volume of gas-free distilled water

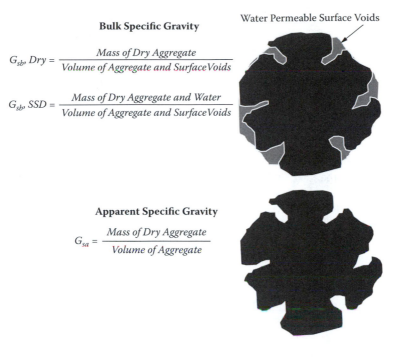

Bulk Specific Gravity

$$G_{sb}, Dry = \frac{Mass\ of\ Dry\ Aggregate}{Volume\ of\ Aggregate\ and\ Surface Voids}$$

$$G_{sb}, SSD = \frac{Mass\ of\ Dry\ Aggregate\ and\ Water}{Volume\ of\ Aggregate\ and\ Surface Voids}$$

Apparent Specific Gravity

$$G_{sa} = \frac{Mass\ of\ Dry\ Aggregate}{Volume\ of\ Aggregate}$$

Water Permeable Surface Voids

FIGURE 8.6 Bulk and Apparent Specific Gravities.

at the stated temperature.

$$\text{Apparent specific gravity, } G_{sa} = \frac{\text{Weight of dry aggregate}}{\text{volume of dry aggregate}}$$

$$= \frac{\text{Weight in air (dry)}}{\text{weight in air } - \text{ weight in submerged condition}}$$

Effective specific gravity is the ratio of the oven dry weight in air of a unit volume of a permeable material (excluding voids permeable to asphalt) at a stated temperature to the weight of an equal volume of gas-free distilled water at the stated temperature.

$$\text{Effective specific gravity, } G_{se} =$$

$$\frac{\text{Weight of dry aggregate}}{\text{volume of dry aggregate } + \text{ volume of voids between asphalt and aggregate}}$$

Effective specific gravity is calculated from the theoretical maximum specific gravity of asphalt mix.

The *absorption* of an aggregate is defined as the water content of the aggregate in its saturated surface dry condition.

$$\text{Absorption, } \% = \frac{\text{Saturated surface dry weight of aggregate, g} - \text{oven dried weight of aggregate, g}}{\text{oven dried weight of aggregate, g}}$$

This parameter is needed to determine how much of the total asphalt added is absorbed in the aggregates, and hence how much remains on the surface of the aggregate as effective asphalt. Generally, the water absorption of aggregate is determined and used as an indicator of the asphalt absorption.

The AASHTO T-84 (ASTM C-128) Specific Gravity and Absorption of Fine Aggregates test procedure consists of the following steps. About 1000 gram of oven-dried fine aggregate is immersed in water in a flask (pycnometer) with known weight. After 24 hours of immersion, the aggregate is dried on a flat surface, using warm air, until saturated surface dry (SSD) condition is reached. The SSD condition is detected at that moisture content when a lightly compacted mass of fine aggregate in a cone first slumps when the cone is removed. About 500 gram of the SSD aggregate is then placed in the flask and weighed. The flask is then filled with water, and its weight is recorded. Finally, the fine aggregate is removed from the flask, oven dried, and weighed.

In the AASHTO T-85 (ASTM C-127) Specific Gravity and Absorption of Coarse Aggregates test procedure, about 5 kg of washed coarse aggregate (retained on a 4.75 mm sieve) is oven dried and weighed, and then immersed in water for 24 hours. The aggregates are then removed from water, drained and surface dried (saturated surface dry) until there is no visible film of water on the surface, and weighed. The SSD aggregates are next submerged underwater in a wire-mesh basket and weighed.

The absorption of the aggregate blend (combined coarse and fine aggregates) is determined on the basis of a percentage of the two fractions. Maximum limits on absorption are specified, which may range from 0.5 to 5%.

8.7   CLEANLINESS AND DELETERIOUS MATERIALS

Deleterious or undesirable materials can be of different types, such as organic materials, vegetation (roots, sticks), clay lumps, clay coating on aggregates, and metal oxides. Such impurities are undesirable since they negatively affect the performance of asphalt mixtures. For example, organic materials can absorb water and lead to deterioration of the mix, clay lumps can form during production and subsequently disintegrate in the asphalt mix, clay dust can coat aggregates and hence negatively affect the bond between asphalt and aggregate, and metal oxides can react with water, causing swelling, popouts, and staining in pavements. Petrographic examination can identify impurities in many cases, and washing the aggregates reduces the impurities to a great extent. However, tests must still be conducted to ensure that the aggregates are clean to a minimum acceptable level, and/or the amount of impurities is below a maximum acceptable level.

In the AASHTO T-176 (ASTM D-2419) Plastic Fines in Graded Aggregates and Soils by Use of the Sand Equivalent test, aggregates passing the 4.75 mm sieve are placed and stirred in a graduated, transparent cylinder filled with water and a flocculating agent (a mixture of calcium chloride, glycerin, and formaldehyde). After settling, the sand in the fine aggregate separates from the clay, if any; the heights of the sand and clay columns in the flask are read off; and the ratio of the former to the latter is determined as the sand equivalent. Depending on the layer, minimum sand equivalents are specified, which can range from 25 to 60.

The AASHTO T-90 (ASTM D-4318) Determining the Plastic Limit and Plasticity Index of Soils test involves the determination of liquid limit and plastic limit, and the Plasticity Index (PI), which is the difference between liquid limit and plastic limit, for materials passing the 0.42 mm sieve. Maximum limits on PI (such as 4) are specified on either the fraction passing the 0.42 mm sieve or that passing the 0.075 mm sieve.

In the AASHTO T-112 (ASTM C-142) Clay Lumps and Friable Particles in Aggregates test, washed aggregates are dried and sieved to remove the fraction passing the 0.075 mm sieve. Samples are then soaked in water for 24 hours, after which particles are tested for removal by wet sieving and breaking with fingers. If a particle can be removed by wet sieving or breaking, it is classified as a clay lump or friable particle, and the total percentage of such particles on the basis of the weight of the sample is determined. Maximum limits, such as 1%, are generally specified.

The AASHTO T-113 Light Weight Pieces in Aggregate test involves the determination of the percentage of lightweight pieces in coarse and fine aggregates by separating (floating) them in a liquid of suitable specific gravity. Liquid such as zinc chloride solution (specific gravity = 2.0) can be used to separate particles such as those of coal.

The steps in the AASHTO T-11 Material Finer than 75μm (No. 200 Sieve) in Mineral Aggregate by Washing test are washing and wet sieving aggregates to determine the percentage of material passing the 75 μm sieve. Allowable ranges of material finer than 75 μm are generally specified, and depending on the layer and type of mix, they can range from 2 to 10%.

In the AASHTO T-P57 Methylene Blue Value test, a sample of material passing the 75 μm sieve is dispersed in distilled water using a magnetic stirrer, and the methylene blue solution is titrated into the dispersed mix until the sample cannot absorb any more reagent. This end point is indicated by a blue ring when a drop of the mix is placed on a filter paper. The methylene blue value (amount, MBV) is proportional to the amount of organic material or clay in the aggregate, and a maximum limit is generally specified.

8.8   TOUGHNESS OR RESISTANCE AGAINST ABRASION LOSS

For both asphalt and concrete mixes, the resistance against degradation (breaking down of larger particles into smaller sizes) by abrasion, or toughness, can be measured by the Los Angeles (LA) Abrasion (also sometimes referred to as the LA Degradation) test.

In the AASHTO T96 (ASTM C131) Resistance to Degradation of Small Size Coarse Aggregate by Abrasion and Impact in the Los Angeles Machine test, a 5,000g blend sample of coarse aggregate is placed in a steel drum, containing 6–12 steel balls (each weighing 420 gram). The drum is rotated for 500 revolutions, at a speed of 30–33 rpm, and during this time, the steel balls and aggregates are lifted and dropped about 69 cm from a shelf within the drum. The tumbling action leads to the shattering of brittle particles by impact, and degradation of smaller particles by surface wear and abrasion. At the end of tumbling, the aggregates are sieved dry over a 1.77 mm sieve (through which none of the aggregates passed before the sieve), and the percentage of the material passing this sieve is reported as the Los Angeles degradation value. A high value would indicate the potential of generation of dust, and breakdown during construction and in the field, although it may fail to identify such potential in lightweight aggregates such as slag. Maximum limits are specified, and may range from 30 to 50%. Values may range from 10 (for aggregates from hard igneous rock) to 60 (for aggregates from soft limestone).

The ASTM D-6928, AASHTO T-P58 MicroDeval Abrasion Loss test consists of placing aggregates in a jar with water and 3/8-inch diameter steel balls, and rotating the jar (Figure 8.7) at

FIGURE 8.7 MicroDeval Equipment.

100 rpm for 2 hours. The breakdown of the aggregate is then evaluated. The aggregate sample is soaked in water for 24 hours before tumbling in the jar. At the end of the tumbling, the sample is washed and dried, and the amount of material passing the No. 16 sieve is determined. This value is then used as a percentage of the original sample weight to determine the percentage of abrasion loss. A maximum limit, such as 18%, is specified.

8.9 PARTICLE SHAPE AND SURFACE TEXTURE

Angular particles tend to provide greater interlock and internal friction (and hence shear strength), and particles with rough texture (as found mostly in crushed aggregates) are desirable because of increased strength and voids for accommodating more asphalt (and hence more durability). Also, the occurrence of flat and/or elongated particles is undesirable, since they affect compaction and tend to increase voids, break down during construction and under traffic, and produce particles with uncoated (with asphalt) surfaces and negatively alter the volumetric properties.

For coarse aggregates, the ASTM D-3398 Index of Aggregate Particles Shape and Texture can be used. The volume of voids between packed, uniform-size particles is an indicator of the combined effect of shape, angularity, and surface texture of the aggregates. Clean, washed, and oven-dried aggregates are separated into different sizes, and aggregates of each size are compacted in a separate mold with a tamping rod, using first 10 blows per layer of three layers and then 50 blows per layer of three layers. The voids are determined from the weight of the mold with the aggregate and the bulk-specific gravity of the aggregates. The Particle Index is calculated for each fraction, and then a combined index is calculated on the basis of the relative proportion (percentage of the blend by weight) of the different sizes.

$$I_a = 1.25V_{10} - 0.25V_{50} - 32.0$$

where:

I_a = Particle Index value
V_{10} = percentage voids in the aggregate compacted with 10 blows per layer
V_{50} = percentage voids in the aggregate compacted with 50 blows per layer

Particle Index values of blends may range from 6 (for rounded aggregates with a smooth surface) to 20 (for angular crushed aggregates with a rough texture). For the same air voids, mixes with aggregates with a higher Particle Index value tend to show higher strength.

The ASTM D-4791 Flat or Elongated Particles in Coarse Aggregates test is also used. Flat or elongated particles are defined as those aggregates by particles which have a ratio of width to thickness or length to width greater than a specified value, such as 3:1, 4:1, or 5:1 (most common). Tests are conducted on particles of each size, using a proportional caliper, on a representative sample, and then the total value is calculated as a percentage, in terms of number or mass. Maximum limits on the percentages of flat and elongated particles are specified, which may range from 5 to 20, for example, for a ratio of 5:1.

The ASTM D-5821 Determining the Percentage of Fractured Particles in Coarse Aggregates test involves the determination of percentage of coarse aggregates (that are retained on the 4.75 mm sieve) with one or more fractured face(s), which is defined as a face that exposes the interior surface of a particle. A particle is said to have a fractured face if, for example, at least 25% of the area (ASTM standard) of the face is fractured. During the test, from a specific sample, the aggregate particles identified as those with fractured faces are separated from those that are not, and the percentage of particles with one (or two) fractured faces is determined. Minimum limits on the percentage of fractured particles in an aggregate blend are specified, and may range from 30 to 90, for both one and two or more fractured faces, dependent on the layer, with surface layers having requirements for higher percentages.

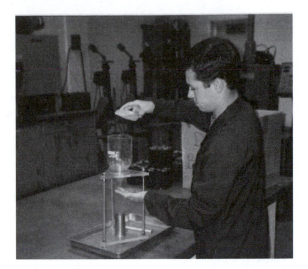

FIGURE 8.8 Testing for Uncompacted Voids for Fine Aggregates.

In the AASHTO T-P56 Uncompacted Voids in Coarse Aggregates test, coarse aggregate is dropped through a hopper into a cylinder with known weight and volume. The excess aggregate on top of the cylinder is struck off, and the weight of the filled cylinder, along with the known bulk-specific gravity of the aggregates, is used for determination of the uncompacted void content.

For fine aggregates, the ASTM C-1252, AASHTO TP-33, Uncompacted Void Content test is used (Figure 8.8). The test consists of pouring the fine aggregate blend of specified gradation through a funnel into a cylinder with a volume of 100 cm^3. The cylinder is weighed after striking off the excess particles. Using this weight, the bulk-specific gravity of the aggregate and the known volume and weight of the cylinder, the uncompacted void content is determined. A high uncompacted void content of, say, 45% indicates a fine aggregate with angular shape and rough texture, and tends to produce asphalt mixes with relatively high resistance against the potential of rutting.

8.10   DURABILITY/SOUNDNESS

Aggregates could be subjected to repeated cycles of weathering action due to wetting and drying or freezing and thawing of trapped water (due to expansion upon freezing). Tests are conducted to determine the potential of breakdown when subjected to such stresses.

In the AASHTO T-104 (ASTM C-88) Soundness of Aggregate by Use of Sodium Sulfate or Magnesium Sulfate test, the sample is immersed in a solution of sodium or magnesium sulfate of specified strength for a period of 16 to 18 hours at a temperature of $21 \pm 1°C$. The sample is removed from the solution and drained for 15 ± 5 minutes, and then dried at $110 \pm 5°C$ to a constant weight. During immersion, the sulfate salt solution penetrates into the permeable pores of the aggregate, and drying dehydrates the sulfate salts in the pores. Upon further immersion, the sulfate salts rehydrate, and the resulting expansion and stress simulate the freezing of water and associated stress. The same sample is then subjected to four more cycles (for a total of five cycles) of immersion and drying. The sample is then sieved, and the weight loss for each fraction and then the weighted average loss of the sample are determined. A higher loss indicates a potential of loss of durability through weathering action of the environment. Maximum limits (such as 12% loss after five cycles) are specified.

There are three different versions of the AASHTO T-103 Soundness of Aggregate by Freezing and Thawing test. In procedure A, samples are immersed in water for 24 hours and then subjected to freezing and thawing, while in procedure B the samples are vacuum (25.4 mm of mercury) saturated, using a 0.5% (by mass) solution of ethyl alcohol and water, before subjecting them to freezing and thawing (in the same solution). Procedure C is similar to B, except that water is used instead of the alcohol-water solution. Samples are dried and sieved, at the end of the final cycle (cycles may range from 16 to 50 cycles, depending the procedure), and the weight loss for each size is determined.

In the AASHTO T-210 (ASTM D-3744) Aggregate Durability Index test, a washed and dried sample of coarse aggregate is agitated in water for a period of 10 minutes. The washed water and fraction passing the 75 μm sieve are mixed with calcium chloride solution and placed in a plastic sand equivalent cylinder for 20 minutes. Next, the level of the sediment column is read, which is used for calculating the durability index.

8.11   EXPANSIVE CHARACTERISTICS

Expansion of aggregates (such as steel slag) as a result of hydration can lead to swelling in asphalt mixes and resulting disintegration of asphalt pavements. Tests can be conducted to identify the potential of expansion.

The ASTM D-4792 Potential Expansion of Aggregates from Hydration Reaction test method involves compaction of the aggregate to its maximum density in a CBR mold, and submerging it in water at $71 \pm 3°C$ for a period of at least 7 days. A perforated plate is kept on top of the compacted aggregate. Percentage expansion of the sample is determined from dial readings of the height of the sample submerged in water, and the initial height of the sample. Maximum permissible expansion, such as 0.5%, is specified.

8.12   POLISHING AND FRICTIONAL CHARACTERISTICS

Polishing is defined as the loss of microtexture and the gradual smoothing and rounding of exposed aggregates. High potential of polishing in aggregates by traffic is undesirable as it leads to lowering of friction and hence decrease in skid or frictional resistance, which is a measure of how quickly a vehicle can be stopped, say, under adverse/wet conditions.

In the AASHTO T-278 (ASTM E-303) Measuring Surface Frictional Properties Using the British Pendulum Tester test, aggregate particles are placed in a small mold along with a bonding agent (such as epoxy). A thin film of water is applied over the sample, and then a pendulum with a rubber shoe is released such that it touches and goes over the sample. The pendulum goes past the sample to a certain height, depending on the friction offered by the aggregate sample. This height, or the length of the arc, is read off. This zero-polishing reading is then used, along with readings taken after polishing the sample surface (for, say, 9 hours) with a British polish wheel (AASHTO T-279, ASTM D-3319) or other polishing device, to determine the polishing resistance.

In the ASTM D-3042 Insoluble Residue in Carbonate Aggregate test, the percentage of noncarbonated (acid-insoluble) materials in carbonate aggregates is determined. Five hundred grams of a sample (retained on a 4.75 mm sieve) is placed in a glass beaker with 1 liter of hydrochloric acid solution, and the mix is agitated until effervescence stops. This is repeated with an additional 300 ml of acid. Then the beaker is heated to 110°C, with new acid added in increments until effervescence stops. Next, the aggregate residue is washed over a 75 μm sieve, the plus-75-μm material is weighed, and the weight is expressed as insoluble residue (as a percentage of the original sample weight). The higher the amount of insoluble residue, the higher the amount of siliceous material, and hence the higher the resistance against polishing.

8.13　AGGREGATE TESTS SPECIFICALLY FOR CONCRETE

8.13.1　Fineness Modulus (FM; ASTM C125)

This property assesses the coarseness of the fine aggregate, and it is used in combination with the maximum nominal size coarse aggregate to estimate the proportions of both fine and coarse aggregates used in a concrete mix, to satisfy workability requirements. The fineness modulus is calculated by summing the cumulative percentage retained in the following sieves and dividing the sum by 100: 9.5 mm, 4.75 mm, 2.36 mm, 1.18 mm, 0.6 mm, 0.3 mm, and 0.15 mm. It varies between 2.5 and 3.1 usually. The higher the FM, the harsher and rougher the concrete mix is. However, it is more economical to have a mix with a higher FM, because the mix will not require as much water, and therefore, a relatively lower amount of cement could be used for a specific water-cement ratio.

8.13.2　Gradation

The gradation requirements for aggregates used in concrete are provided in ASTM C-33 and AASHTO M-80 for coarse aggregates, and ASTM C-33 and AASHTO M-6 for fine aggregates. These gradations assist the designer in providing a workable concrete in which the coarse aggregates are coated and suspended in the mortar (sand, cement, and water). The gradations differ on the basis of nominal maximum size of the coarse aggregate. The maximum aggregate size is defined as the smallest sieve through which 100% of the aggregate passes. The nominal maximum aggregate seize is defined as the smallest sieve that can retain up to 15% of the aggregates. The larger the maximum size, the greater the percentage of coarse aggregates in a mix. The limits are imposed by the thickness of the slab and the gap between the reinforcements, if any. Figure 8.9 shows examples of gradations for coarse and fine aggregates, along with ASTM limits.

8.13.3　Bulk Density and Voids in Aggregates Test

The Bulk Density and Voids in Aggregates (AASHTO T-19; also known as the *dry-rodded unit weight*) test is conducted for the volumetric mix design of PCC. The steps in this are filling a container with a known volume (0.5 ft³) with dry coarse aggregate using a specified procedure, and then measuring the weight of the filled container and determining the bulk density by dividing the weight of the aggregates needed to fill the container by the volume of the container.

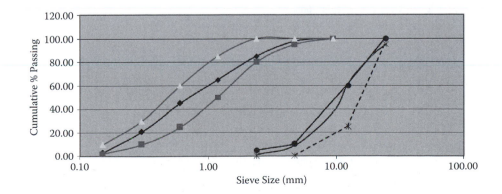

FIGURE 8.9 Examples of Gradations for Coarse and Fine Aggregates for PCC Pavement.

8.14   ARTIFICIAL AGGREGATES

Artificial aggregates are those that are not naturally occurring, but rather are formed from waste products or as by-products of industrial processes. The feasibility of producing lightweight aggregates (synthetic lightweight aggregates, or SLA, weighing less than 70 lb per cubic foot) suitable for use in pavements, from waste fly ash and mixed waste plastics, has been proven. Such products are generally formed through intense heating of several materials, such as heating shale/clay/slate in excess of 1000°C to produce ceramic SLA. Slag, a by-product of the iron and steel industry, is used in pavement mixtures and is noted for its good skid resistance (and relatively high absorption). Blast furnace (or iron) slags, most widely used in the pavement industry, are formed as a by-product in blast furnaces, whereas steel slags are formed in basic oxygen and electric arc furnaces. While blast furnace slag is lighter than natural aggregates, steel slags are heavier. The light weight of the blast furnace slag provides the advantage of fewer truckloads for the same volume of materials. Steel slags, on the other hand, have a high angle of internal friction, and are very suitable for rut-resistant pavement layers. The National Slag Association (n.d.; http://www.nationalslag.org) provides excellent information on this topic.

QUESTIONS

1. What are the primary types of rocks? How are they formed?
2. For the given information from sieve analysis, prepare a 0.45 power plot.

Sieve Size, Inches	% Passing
2	100
1 1/2	100
1	100
3/4	100
1/2	97
3/8	75
No. 4	20
No. 8	15
No. 16	10
No. 30	8
No. 50	7
No. 100	5
No. 200	3

3. Four different aggregates are to be combined to produce a blend. The percentages of the different aggregates in the blend are 20, 50, 10, and 20, and the corresponding bulk-specific gravities are 2.650, 2.720, 2.600, and 2.690. Compute the bulk-specific gravity of the combined aggregates in the blend.
4. Determine the bulk-specific gravity of an aggregate whose apparent specific gravity is 2.690 and absorption is 1.5%.
5. Combine the following aggregates in the right proportion to produce a blend with the given target gradation:

Sieve Size, mm	CA7 % Passing	Washed Sand % Passing	Dry Sand % Passing	% Passing Target
19.00	100	100	100	100.0
12.50	98.8	100	100	100.0
9.50	62.1	100	100	93.0
4.75	2.1	98.1	100	62.0
2.36	0	81	99	45.0
1.18	0	69.9	80.7	35.0
0.60	0	50.6	59.9	25.0
0.30	0	20.7	40	15.0
0.150	0	5.1	24.2	9.0
0.075	0	1.9	17.2	4.0

6. What is the importance of the fineness modulus for PCC?
7. Determine the fineness modulus for the following aggregate, and comment on the effect on fresh concrete.

Sieve Size, mm	Cumulative % Passing
9.50	100.00
4.75	98.00
2.36	85.00
1.18	65.00
0.60	45.00
0.30	21.00
0.15	3.00

9 Asphalt and Emulsions

9.1 ASPHALT BINDER

In the United States *asphalt* refers to the binder, also known as *bitumen*. Asphalt can refer to a mixture of bitumen and aggregates (as in Europe), which is usually called *asphalt mix* in the United States. In this book, the U.S. definition will be followed—asphalt will mean the binder.

Asphalt is a material that is made up of predominantly hydrocarbons; small amounts of sulfur, nitrogen, and oxygen; and trace amounts of metals such as vanadium, nickel, iron, magnesium, and calcium. Analysis of a typical asphalt would indicate the following: carbon, 82–88%; hydrogen, 8–11%; sulfur, 0–6%; oxygen, 0–1.5%; and nitrogen, 0–1%.

Traditionally "penetration," "viscosity," and "shear modulus" tests have been used to characterize asphalts. While discussing different types of asphalt, results from one of these tests are provided to make a relative comparison of the "softness/hardness" or "stiffness" of the material. In layman's language, the higher the penetration or the lower the viscosity/modulus, the "softer" the asphalt binder is, and vice versa.

9.2 NATURALLY OCCURRING ASPHALTS

Asphalt can be obtained in different forms, although refined asphalt from fractional distillation of crude oils is the most commonly used. The naturally occurring asphalts are lake asphalts, rock asphalt, and Gilsonite.

9.2.1 LAKE ASPHALT

The primary source of lake asphalt is a deposit in a lake in the southern part of Trinidad. This asphalt has been deposited as a result of surface seepage in the bed of the lake. Asphalt is excavated from the lake bed and transported to various places. This asphalt is a mixture of different materials, including pure asphalt binder. This asphalt is also known as Trinidad lake asphalt (or TLA). The refinement of the excavated material is done by heating it to 160°C and vaporizing the water. The material is then sieved to remove large-size/other particles and organic matter. This refined product has about 54% binder, 36% mineral matter, and 10% organic material. TLA has a penetration of about 2 mm, with a softening point of 95°C. This material was once used alone or in combination with softer asphalt for regular use in asphalt pavement construction. These days, its use is confined to mostly high-stiffness asphalt mixes for pavements and roofing applications.

9.2.2 ROCK ASPHALT

Rock asphalts are obtained from rocks that are impregnated with asphalts. They are formed by the entrapment of asphalt in impervious rock formations. These asphalts are obtained from mines or quarries. Their deposits can be found in Switzerland, France, and Italy, as well as in Utah and Kentucky in the United States. The concentration of asphalt in such deposits can range from 7 to 13%. Although used in some of the earliest roads in Europe and the United States, it is used very little at present, mostly in waterproofing and high-stiffness asphalt mix applications.

9.2.3 GILSONITE

Gilsonite is found in Utah in the United States. It has a penetration of zero with a softening point of 115–190°C. It is used in combination with regular asphalt to alter its characteristics to reduce the penetration and increase its softening point. Its use is mostly limited to bridge and waterproofing applications, and high-stiffness asphalt mixes.

9.3   REFINED ASPHALT FROM CRUDE OIL

Refined, regularly used asphalt for paving applications is produced from crude oil. Crude oil is formed as a result of the action of heat from within the earth's crust and weight of many layers of deposits of remains of marine organisms and vegetation over millions of years. Natural gas and oil reservoirs are detected by surveys and recovered by drilling through impermeable rocks. The oil is produced in mainly four regions of the world—the Middle East, Russia, the United States, and Central America. The characteristics of the crude oil differ significantly from source to source. Of the 1500 different sources of crude oils in the world, only a few are actually found to be suitable for obtaining asphalt.

Asphalt is manufactured from crude oil starting with a process called *fractional distillation*. This process separates the different hydrocarbons with different boiling points that make up crude oil. Fractional distillation is carried out in tall steel towers. The crude oil is first heated in a furnace to a temperature around 350°C, and the mix of liquid and vapor is then directed to the lower part of the tower, which has a pressure greater than the atmospheric pressure. The vapors rise up in the tower through holes in the horizontal layers that divide the tower vertically into segments. The vapors lose heat as they rise through the holes in the trays. When the temperature of the vapor falls below its boiling point, it condenses and the liquid that accumulates in one of the trays is then drawn off. The lightest fractions of the crude oil which remain as vapor are taken off from the top of the tower, and include propane and butane. Further down the tower, kerosene and then gas oil are recovered. The heaviest fraction (long residue) which comes from the bottom of the tower is a complex mixture of hydrocarbons, and is distilled again at a reduced pressure (10–15 mm Hg and 35–45°C) in a vacuum distillation tower. One of the products of this step is the short residue, which is used as the feedstock for producing different grades of asphalt. As the viscosity of the short residue is dependent on the source of the crude and the temperature and pressure during the process, the conditions are adjusted on the basis of the source of the crude oil to produce an asphalt with a penetration in the range of 35 to 300 dmm (1 dmm is 0.1 mm).

Air blowing or oxidation is used to modify the proportion of the short residue at a temperature of 240 to 320°C. The result of such oxidation is a decrease in penetration, increase in softening point, and decrease in temperature susceptibility. Such modification, which can be in continuous, semi, or full (oxidation) mode, results in the formation of asphaltenes and an increase in size of the existing asphaltenes.

The three processes that take place are oxidation, dehydrogenation, and polymerization. In the continuous blowing process, the penetration and softening point of the asphalt depend on the viscosity of feedstock, temperature in the blowing column, residence time in the blowing column, origin of the crude oil, and air-to-feed ratio. Fully blown or oxidized asphalts are produced by vigorous air blowing or by blending with relatively soft flux.

The properties in this case are dependent on the viscosity and chemical nature of the feed. The specific grade of asphalt is produced by selecting a suitable feedstock, controlling viscosity of the feed, and the conditions in the tower.

9.4   SAFE DELIVERY, STORAGE, AND HANDLING OF ASPHALTS

Considerations must be made regarding delivery, storage, and handling of asphalts to make sure that adequate safety is maintained, and that asphalt properties are not adversely affected. The safety factors are discussed first, followed by factors that could affect asphalt properties adversely.

9.4.1 Causes of Hazards and Precautions

The main hazard from asphalt results from the fact that it is usually applied at high temperatures which facilitate the emission of fumes of hydrocarbons and H_2S (the amount doubles for every 10–12°C increase in temperature). Exposure to even small amounts of H_2S is fatal. Although relatively very high temperatures are required for spontaneous burning of asphalt (400°C), small amounts of H_2S from asphalts can react with rust on tank walls to form iron oxide. This material reacts readily with O_2 and can self-ignite and also ignite carbonaceous deposits on the roof and walls of the tank. Also, if water comes in contact with hot asphalt, it is converted into steam, and foam is produced in the asphalt with a volume increase of approximately 1400 times. In order to prevent these three hazardous conditions, the following steps must be taken.

1. Test areas where H_2S may be present, and check if they are gas-free before entering.
2. Openings in asphalt tanks such as areas to roofs should be restricted, and the manhole should be kept closed to prevent entry of O_2.
3. Prevent water from entering asphalt tanks.

9.4.2 Health Hazards

Some benzene compounds are found in polycyclic aromatic compounds (PACs), which are present in crude oils—however, the concentration in asphalt that is refined from crude oil is very low. Intimate and prolonged skin contact of asphalt as well as asphalt emulsion should be avoided to prevent risk of irritation of skin and eyes.

Specific instructions on how to treat skin for first-, second-, and third-degree burns; eye burns; and inhalation should be clearly displayed and available in areas where asphalt is handled. Under normal working conditions, exposure to particulate asphalt, hydrocarbons, and H_2S that are emitted during heating of asphalt is well below allowable maximum limits. However, inhalation can cause usually temporary irritation to the eyes, nose, and respiratory tract; headache; and nausea. Such exposure should be minimized, and if it occurs, the affected person should be removed to fresh air and, if necessary, given medical attention as soon as possible.

9.4.3 Precautions and Good Practices

Proper equipment should be used for protecting against burns from hot asphalt. Standard gear includes helmet and neck aprons, heat-resistant gloves, safety boots, and overalls. Soiled garments should be replaced/dry cleaned as soon as possible. Persons handling asphalts can use barrier creams to help subsequent cleaning, for which only approved skin cleaners and warm water should be used.

Personnel should be properly equipped and trained to handle fires should one break out. Direct water jets should not be used since they may cause foaming and spread the hot asphalt. Dry chemical powder, foam, vaporizing liquid, or inert gases can be used to extinguish small fires. Portable fire extinguishers must be placed at strategic, permanent, and conspicuous areas. The local fire station should be consulted to formulate a plan if initial efforts fail to extinguish a fire.

Spills and splashes of hot asphalt should be avoided during sampling. Sampling areas should be well lit and have safe in and out areas, and adequate protective clothing must be worn. Sampling by dipping a weighted can should be done for only small samples and be avoided for cutback tanks because of the presence of flammable atmosphere in tank vapor spaces. The best way is to use properly designed sample valves from pipelines and tanks. The proper sample valve remains hot by the product so that there is no blockage in closed position, and also such that the plunger of the valve extends into fresh asphalt and allows obtaining fresh liquid asphalt when opened.

Regarding leaching of components from asphalt, it has been shown that prolonged contact with water leads to leaching of PACs, but the levels reach equilibrium rapidly and the numbers are one order of magnitude below the limits of potable water.

Users of asphalt (laboratories, plants, and job sites) should review materials safety data sheets and follow all Occupational Safety and Health Administration (OSHA) standards such as the OSHA Hazard Communication Standards (e.g., 29 CFR 1910.1200; see http://www.ilpi.com/msds/ osha/1910_1200.html; OSHA, n.d.).

9.5 ASPHALT BINDER PROPERTIES

Since asphalt binders are by-products of the petroleum distillation process, by nature, if heated to a very high temperature, they can give off vapors which can get ignited and pose safety hazards in plants. They can also become stiffer at low temperatures and relatively soft at high temperatures. Furthermore, asphalts lose components through volatilization as well as oxidation, when exposed to air/oxygen and/or high temperature, and as such, they may undergo a drastic change in properties. In fact, asphalts tend to become "harder" with time—what is commonly known as *age hardening*. Therefore, tests have been developed to characterize asphalts in terms of their relevant properties, such as those described above—safety, temperature susceptibility, and age hardening. These tests are presented below.

9.5.1 Specific Gravity: ASTM D-70

Specific gravity of a material is defined as the ratio of the mass of the material at a given temperature to the mass of an equal volume of water at the same temperature. For asphalt binder, the ASTM D-70 procedure, Density of Semi-Solid Bituminous Materials (Pycnometer Method), is used for the determination of specific gravity and density. In this procedure, a calibrated pycnometer with a sample of known weight is first weighed, and then filled with water. The setup is then brought to the test temperature and weighed. The density (Specific Gravity * Density of Water) is then calculated from the mass of the sample and the water displaced by the sample in the pycnometer, and the specific gravity is calculated. Note that the density of water varies between 999.1 to 997 kg/m^3 between a temperature of 15 and 25°C. If the density of water is approximately taken as 1 gm/cm^3, then the density is numerically equal to specific gravity. The specific gravity of the asphalt is expressed along with the temperature at which both asphalt and water were used in the test. Specific gravity is used in converting weight to volume and vice versa, in paying asphalt suppliers/contractors, as well as in determining the effective specific gravity of aggregate.

9.5.2 The Cleveland Open Cup Method (Flash Point): ASTM D-92

This test involves heating the asphalt in a brass cup, and passing a small flame over the cup periodically, until a quick flash occurs. The temperature at which the flash occurs is called the *flashpoint*. It is conducted to determine the temperature range which can be used during the production of HMA without causing safety hazards.

9.5.3 Solubility Test: ASTM D-2042

In this test, a sample of asphalt is mixed with a solvent and then poured through a glass fiber pad filter. The retained impurities are washed, dried, and weighed. This test is run to minimize the presence of such impurities as dust and organic materials in the asphalt.

9.5.4 Spot Test: AASHTO T-102

In this test, visual evaluation of a drop of asphalt and solvent mix on a filter paper is made. Results are reported as acceptable or not acceptable, based on the absence or presence of a dark area in the spot on the filter paper. This test was developed to identify asphalts which have been damaged due to overheating (cracking—molecules thermally broken apart) during the asphalt production process.

9.5.5 Penetration: ASTM D-5

In this test, a container of asphalt maintained at 77°F is placed under a needle. The needle, under a weight of 100 gram, is allowed to penetrate the asphalt in the container for 5 seconds. The depth of penetration, expressed in dmm, is reported as the penetration. The purpose of this test is to classify asphalt, check consistency, and evaluate the overheating of asphalt. Generally five different grades are available: 40–50, 60–70, 85–100, 120–150, and 200–300. A 60–70 grade means an asphalt binder that has a penetration value ranging from 60 to 70 dmm, and is harder than the 85–100 grade and softer than the 40–50 grade. A binder with a lower penetration grade such as 40–50 would be used in a project in a warmer location, whereas those with higher grades (such as 120–150) would be used in a project in a colder region.

9.5.6 Viscosity Tests

Viscosity is defined as the ratio of applied shear stress to applied shear strain. It can be measured by direct tests (such as a sliding plate viscometer), but is generally measured by indirect procedures. This test helps in measuring the resistance of the asphalt to flow. Two types of viscosity tests are generally run: absolute viscosity (expressed in poise) (flow in vacuum) (ASTM D-2171, AASHTO T-202) at 140°F, and kinematic viscosity (expressed in centistokes) (by gravity flow) (ASTM D-2170, AASHTO T-201) at 275°F.

In the absolute viscosity test, a sample of asphalt is introduced through one side of a capillary tube, maintained in a bath at 140°F. After reaching 140°F, the asphalt is made to flow by applying a vacuum on the other side of the tube, and the time to flow through two marked points in the tube is measured. The time is then used, along with the calibration constant of the tube, to determine the viscosity.

The kinematic viscosity test is similar to the absolute viscosity test, but runs at a higher temperature, at which the asphalt is sufficiently fluid to flow. A different type of tube is used for this test, and after reaching the test temperature, only a small vacuum is applied on the other side of the tube to initiate the flow of asphalt due to gravity. The kinematic viscosity is then determined by multiplying the time taken by the asphalt to flow between two timing marks and the calibration factor of the tube.

Viscosity tests are used for classification of asphalts (absolute viscosity), as well as on conditioned asphalts to evaluate overheating/overaging of asphalt. ASTM D-3381 presents the specifications for the different types of viscosity-graded asphalt. Viscosity grades are generally expressed as an *AC*, followed by a number expressing its viscosity divided by 100. For example, an AC 20 asphalt binder means a binder with absolute viscosity of 2000 poise. Generally higher viscosity asphalts are used in warmer climatic regions, whereas lower viscosity materials are used in colder regions.

Kinematic viscosity is used to determine suitable mixing and compaction temperature for HMA. An equiviscous temperature (temperature at which viscosities are similar) is used for mixing and compacting different types of mixes, such that the effect of the asphalt can be made uniform. Generally the viscosity for mixing and compaction is specified as 170 ± 20 centistokes and 280 ± 30 centistokes (for mixing and compaction, respectively; Asphalt Institute 1984). Generally the absolute viscosity is determined at 60°C, and the kinematic viscosity is measured at 135°C, and then a plot of temperature (in Rankine) versus kinematic viscosity (log-log) is plotted with two data points, one with the measured value at 135°C and another one calculated from the absolute viscosity at 60°C as follows:

$$\text{kinematic viscosity at 60°C} = \frac{\text{absolute viscosity at 60°C} * 100}{0.98 * G_b}$$

where kinematic viscosity is in centistokes, absolute viscosity is in poise, and G_b is the specific gravity of the asphalt binder at 15°C.

9.5.7 SOFTENING POINT (RING AND BALL) TEST: ASTM D-36

This test is conducted to determine the temperature at which asphalt starts flowing. This test involves the placement of a steel ball over a ring of asphalt in a bath, and heating the assembly at 5°C/minute. The temperature at which the asphalt cannot support the ball anymore and sinks is reported as the softening point. This is not used for paving asphalts that much, but is used for roofing asphalts.

9.5.8 FRAASS BREAKING POINT TEST: BS EN 12593, BS 2000-8

This test is conducted to determine the very low temperature stiffness of asphalt binder. In this test a steel plate coated with a thin layer of asphalt binder is slowly bent and released repeatedly, while the temperature is reduced constantly at the rate of 1°C per minute, until the asphalt binder cracks.

9.5.9 DUCTILITY: ASTM D113

In this test, a sample of asphalt is stretched until it breaks or until the testing equipment limit is reached. The asphalt sample is stretched at a rate of 5 cm/minute at 77°F, and the ductility is measured in cm. If run at a lower temperature, the rate of stretching is reduced. This test is run to ensure the use of nonbrittle asphalts.

9.5.10 THIN FILM OVEN TEST (TFOT): ASTM D1754

This is a conditioning procedure, in which a 50 gram asphalt sample is placed in a flat-bottom pan on a shelf in a ventilated oven maintained at 325°F. The shelf is rotated at 5–6 revolutions per minute for 5 hours. The loss in penetration or increase in viscosity as well as change in weight of the sample are usually measured after the heating. The use of this test is to evaluate the effect of heat and air on asphalts, and identify those that would harden excessively due to volatilization and/or oxidation.

9.5.11 ROLLING THIN FILM OVEN TEST (RTFOT): ASTM D-2872, BS EN 12591, AASHTO T-240

In this test, eight bottles with asphalt are placed in a rack inside an oven maintained at 325°F (Figure 9.1), and the rack is rotated around a heated air-blowing orifice for 75 minutes. As the rack rotates, asphalt binder inside each bottle flows inside, exposing the fresh surface to the air. The mass

FIGURE 9.1 Rolling Thin Film Oven Test (RTFOT) Equipment.

loss, as well as viscosity, after this process is checked. This test is run to classify asphalts (aged residue, or AR, process) and to ensure minimization of mass loss and change in properties.

9.6 ASPHALT BINDER PROPERTIES AND PAVEMENT DISTRESS AND PERFORMANCE

For asphalt pavements, rutting at high temperature and cracking at low temperature are the two primary topics of interest regarding asphalt. During the production of HMA and during its life on the pavement, the asphalt could become too brittle. It is important to select an asphalt that would not become too brittle because of the asphalt itself, because of the mix components, or because of construction. Hence, there are several factors that must be checked to ensure this, such as the properties of asphalt, using the TFOT, using the correct amount of miner filler (aggregates less than 0.075 mm in diameter) in the mix, and compacting the HMA to a minimum density, such that the voids are below a maximum allowable level.

9.6.1 AGING OF ASPHALT BINDER

The composition of asphalt binder is affected by air (oxygen), ultraviolet radiation, and fluctuations in temperature. The primary effect is the hardening or stiffening of the asphalt binder, both during the production process as well as during the service life of the pavement. This hardening needs to be "predicted" and taken into consideration during the selection of the asphalt binder, since excessive hardening of the binder increases the potential of cracking of asphalt mixes. Hardening of the binder takes place on the surface of the pavement as well as in the mix inside the pavement.

Different mechanisms, caused by the different environmental factors, are responsible for the hardening of asphalt binder. Although 15 different factors have been identified (as listed below), the first four are considered to be the more important ones:

1. Oxidation in the dark
2. Volatilization
3. Steric or physical factors
4. Exudation of oils
5. Photo-oxidation by direct light
6. Photo-oxidation in reflected light
7. Photochemical reaction by direct light
8. Photochemical reaction by reflected light
9. Polymerization
10. Changes by nuclear energy
11. Action of water
12. Absorption by solid
13. Absorption of components at a solid surface
14. Chemical reactions
15. Microbiological deterioration

The four important factors are explained below.

1. *Oxidation*: Asphalt binder components oxidize and form heavier and more complex molecules, increasing its stiffness and decreasing its flexibility. The rate of oxidation is affected by the temperature and the asphalt binder film thickness, with higher temperature and thinner films facilitating oxidation. For temperatures above 100°C, the rate of oxidation doubles for every 10°C increase in temperature.
2. *Volatilization*: Depending on the temperature and the surface area exposed, the asphalt binder loses lighter components by volatilization.

3. *Steric or physical hardening*: Reorientation of molecules and slow crystallization of waxes are responsible for causing steric hardening of asphalt binder at ambient temperatures.
4. *Exudation of oils*: Depending on its chemical nature as well as the porosity of the aggregate, oils from the asphalt binder are exuded into the aggregates in an asphalt mix to a different extent.

The hardening of asphalt occurs at different extents and because of different factors during storing, during mixing (during production), and in the pavement. When stored inside a tank in bulk, and not exposed to oxygen, very little hardening occurs, even at high temperatures. Proper layout of storage tanks should be used to ensure avoiding exposure of a larger surface area of asphalt binder to oxygen during circulation of the binder in the tank. During the mixing process, asphalt binder is subjected to relatively high temperature, and is spread over aggregates in films with thickness generally ranging from 5 to 15 microns. The time of mixing can be different for different types of mixing plants (shorter for drum plants, and hence shorter exposure to oxygen and less hardening effect on the asphalt binder), and for specific asphalt binder selected, the temperature and the amount of asphalt should be controlled properly to make sure that the asphalt film is of sufficient thickness to minimize hardening. Hardening increases with a decrease in film thickness. If the asphalt mix is stored in a silo, some hardening will take place as a result of the introduction of air along with the mix. The silo should, therefore, be airtight, and be as full as possible, to minimize the presence of air. Air within the silo may be removed by the introduction of inert gases, and any oxygen inside the silo could also react and form carbon dioxide and prevent further oxidation. If no fresh air is allowed inside the silo during storage, very little or no oxidation could occur between this time and transportation and laydown.

It is important to note that any gain in lowering the viscosity of the asphalt binder by overheating the binder during mixing will be lost by the subsequent increased rate of oxidation and resulting high viscosity, and the increased viscosity will be detrimental to the performance of the asphalt mix.

The hardening of the asphalt binder during storing, mixing, transportation, and laydown is grouped under *short-term aging*, whereas the hardening of the asphalt binder in the pavement, during its service life, is called *long-term aging*. This hardening is primarily affected by the amount of air voids in the compacted asphalt mix in the pavement. Since a higher amount of air voids would introduce a higher amount of air, oxidation and hence hardening can be minimized by providing adequate compaction and lowering voids, preferably below 5%. In practice, it may not be possible to compact to such an extent to lower the voids below 5% for all types of asphalt binder, and in general the average air voids of dense graded mixes are kept below 8%. Mixes with similar air voids may not have the same rate of hardening if the distribution of air voids in the mixes is different—mixes with interconnected voids would have higher air permeability and a faster rate of hardening.

Note that the hardening of the asphalt binder in the asphalt mix on the surface of the pavement occurs at a faster rate than that of the asphalt binder in the mix inside the pavement, because of the relatively higher amount of oxygen available, higher temperatures, and photo-oxidation by ultraviolet radiation. This top oxidized layer of asphalt could be eroded by water and, while present, could prevent further oxidation. However, adequate thickness of the asphalt film should be ensured through the provision of an adequate amount of asphalt binder. Generally, thicker films (that is, higher asphalt content) are provided for compacted asphalt mixes with relatively high amounts of air voids. Optimum film thickness can range from 9 microns to above 12 microns for dense and open graded asphalt mixes, respectively.

Note that while excessive hardening of the surface mix is always detrimental to the performance of the pavement, the hardening of binder, resulting in the increase in stiffness of the asphalt mix *base* course ("curing"), is also considered to be a beneficial effect with respect to the structural design of the pavement.

9.6.1.1 Hardening

There are parameters, based on penetration and viscosity tests, that could be used to evaluate the "hardening" of asphalt due to volatilization as well as oxidation.

$$\text{Retained penetration, } \% = \frac{\text{Penetration after TFOT}}{\text{Penetration before TFOT}}$$

$$\text{Viscosity ratio} = \frac{\text{Viscosity after TFOT}}{\text{Viscosity before TFOT}}$$

In order to ensure good performance of an asphalt pavement, the temperature susceptibility of asphalt must be controlled, since it is always desirable to have an asphalt that would possess sufficient stiffness at high temperature to prevent rutting and possess sufficient ductility at low temperature to prevent cracking.

There are three parameters that could be used to measure temperature susceptibility of asphalts: the Penetration Index (PI), the pen-vis number (PVN), and viscosity-temperature susceptibility (VTS), as explained below.

$$\text{Penetration Index (PI)} = \frac{20 - 500\text{A}}{1 + 50\text{A}}$$

$$\text{A} = \frac{\text{Log(penetration at } T_1) - \text{Log(penetration at } T_2)}{T_1 - T_2}$$

The PI of asphalt generally ranges between −1 and +1, and a higher PI indicates lower temperature susceptibility.

$$\text{Pen-vis Number (PVN)} = \frac{\text{L} - \text{X}}{\text{L} - \text{m}}(-1.5)$$

where:
L = log of viscosity at 275°F for PVN = 0.0
m = log of viscosity at 275°F for PVN = −1.5
X = log of viscosity at 275°F

Typical values of the PVN range between −2 and 0.5, and a higher PVN indicates a lower temperature susceptibility.

$$\text{Viscosity-temperature susceptibility (VTS)} = \frac{\text{Log(log viscosity at } T_2) - \text{Log(log viscosity at } T_1)}{\text{Log } T_1 - \text{Log } T_2}$$

Typical values of VTS range from 3.36 to 3.98. The lower the VTS, the lower the temperature susceptibility.

9.7 STIFFNESS

A significant amount of research has been conducted on defining, measuring, and evaluating stiffness of asphalt. Some of these are indirect, while the rest are direct measurement techniques, although the approach varies significantly. One method is to determine the stiffness of the asphalt from the stiffness of the mix from the nomograph shown in Figure 9.2.

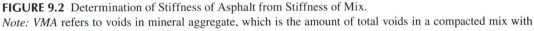

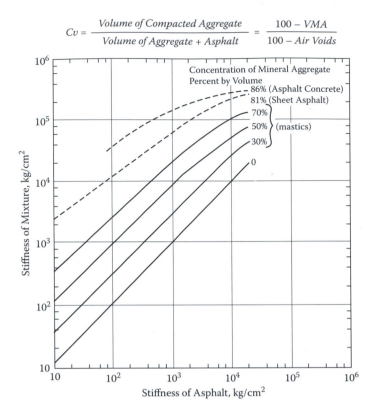

$$Cv = \frac{Volume\ of\ Compacted\ Aggregate}{Volume\ of\ Aggregate + Asphalt} = \frac{100 - VMA}{100 - Air\ Voids}$$

FIGURE 9.2 Determination of Stiffness of Asphalt from Stiffness of Mix.

Note: VMA refers to voids in mineral aggregate, which is the amount of total voids in a compacted mix with respect to the total volume of the mix.

Source: From Van der Poel, 1954, reprinted from Hot Mix Asphalt Materials, Mixture Design, and Construction by permission of National Asphalt Pavement Association Research and Education Foundation.

Another of the earliest developed methods is the use of the softening point (ring and ball test) and the Penetration Index, as shown in Figure 9.3.

Van der Poel, who developed this procedure, showed that for the same time of loading, two asphalt binders with the same PI have the same stiffness at temperatures that are equidistant from their respective softening points. A modification of this method is to use the corrected PI and the corrected softening point, which is done as follows. The penetration is measured at three temperatures, and the corrected PI (pen/pen) is determined from the plot of temperature versus penetration on the bitumen test data chart (BTDC). The BTDC is a chart which can be used to present penetration, softening point, Fraass breaking point, and viscosity data on the same chart. The chart (Figure 9.4), with two horizontal scales for the temperature and Fraass breaking point, and two vertical scales for penetration and viscosity, is prepared in such a way that data from asphalt binders with "normal" temperature susceptibilities plot as straight lines on the chart.

The corrected softening point is determined from the point at which the extended penetration straight line and the 12,000 poise ordinate on the BTDC intersect. A further modification of the process is the use of the PVN instead of the PI, as shown in Figures 9.5 and 9.6.

9.7.1 Viscosity for Stiffness

For asphalts, the viscosity parameter is often used as a measure of stiffness, using the definition of absolute viscosity—the ratio of shear stress to the rate of shear. A large number of viscometers have

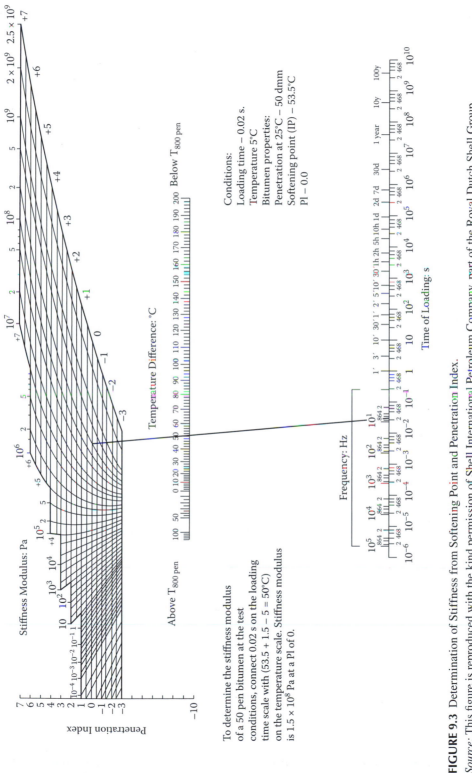

Stiffness Modulus: Pa

Penetration Index

To determine the stiffness modulus
of a 50 pen bitumen at the test
conditions, connect 0.02 s on the loading
time scale with (53.5 + 1.5 − 5 = 50°C)
on the temperature scale. Stiffness modulus
is 1.5×10^8 Pa at a PI of 0.

Conditions:
Loading time − 0.02 s.
Temperature 5°C
Bitumen properties:
Penetration at 25°C − 50 dmm
Softening point (IP) − 53.5°C
PI − 0.0

Temperature Difference: °C

Frequency: Hz

Time of Loading: s

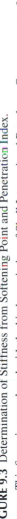

FIGURE 9.3 Determination of Stiffness from Softening Point and Penetration Index.

Source: This figure is reproduced with the kind permission of Shell International Petroleum Company, part of the Royal Dutch Shell Group.

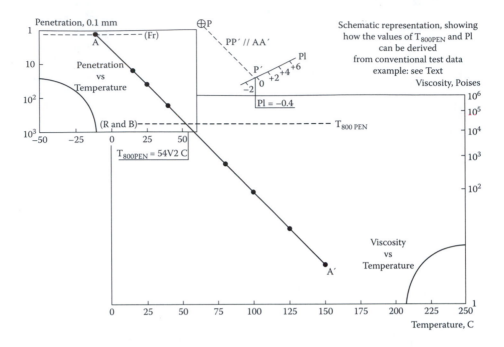

FIGURE 9.4 Bitumen Test Data Chart.
Source: From Heukelom, 1973, reprinted with kind permission from the Association of Asphalt Paving Technologists.

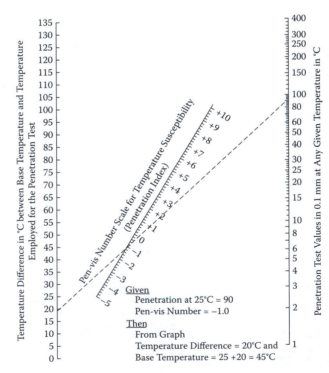

FIGURE 9.5 Relationship between Temperature, Penetration, and Pen-Vis Number.
Source: From McLeod, 1976, and Roberts et al., 1996, reprinted from Hot Mix Asphalt Materials, Mixture Design, and Construction by permission of National Asphalt Pavement Association Research and Education Foundation.

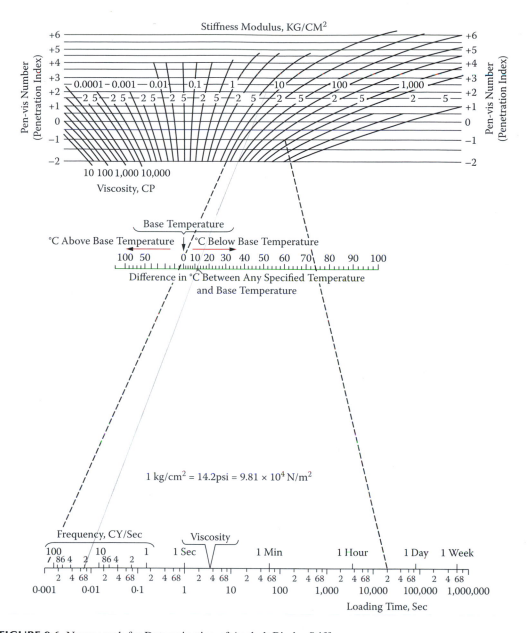

FIGURE 9.6 Nomograph for Determination of Asphalt Binder Stiffness.
Source: From McLeod, 1976, and Roberts et al., 1996, reprinted from Hot Mix Asphalt Materials, Mixture Design, and Construction by permission of National Asphalt Pavement Association Research and Education Foundation.

been developed over the years, which may be of rotational or capillary tube or sliding plate type. While high-temperature (> 140°F) viscosity is suitably measured with a capillary tube–type viscometer, as mentioned earlier, for low temperatures cone and plate viscometers and sliding plate viscometers are most widely used.

In the sliding plate viscometer, a sample of asphalt is sandwiched between two plates, one of which is clamped to a frame. A load applicator needle is placed onto the asphalt sample, using an adapter. At the start of the test, the support beneath the asphalt sample and the unclamped plate is removed and the asphalt sample is subjected to shearing load. The shearing stress and the shear displacement, as measured by a transducer, are used to determine the viscosity of the asphalt.

9.8 VISCOELASTIC NATURE OF ASPHALT AND DIRECT MEASUREMENT OF STIFFNESS

The viscoelastic nature of an asphalt mix comes primarily from the viscoelastic behavior of the asphalt binder. Depending on the temperature and rate of loading, an asphalt binder could behave as an elastic material or as a viscous material, or in a way that reflects both its elastic and viscous nature. At low temperatures and short periods of loading the response is elastic, whereas at high temperature and long periods of loading the response is viscous. At intermediate temperatures and periods of loading, asphalt binders show viscoelastic behavior. The lag between the shear stress and shear strain (Figure 9.7) is called the *phase angle*, δ, and can range from 0, for a perfectly elastic material, to 90° (one-quarter of a cycle) for a perfectly viscous material. Several models that have been developed by researchers capture the dependence of asphalt binder stiffness on temperature and periods of loading.

Conceptually, this can be expressed as follows:

$$\text{Stiffness, } S_{t,T} = \frac{\sigma}{\varepsilon_{t,T}}$$

where $S_{t,T}$ is the stiffness at loading time t at temperature T, σ is the stress, and $\varepsilon_{t,T}$ is the strain at loading time t at temperature T. The stiffness relevant for asphalt binder is related to shear failure. To determine stiffness or modulus related to shear failure, one needs to measure shear stress and shear strain. The shear stress during a test can be applied in different ways, to be consistent with the specific mode of failure in the pavement. For example, it can be applied as a sinusoidal wave (dynamic loading to simulate the effect of traffic), with varying stress, using a constant amplitude and frequency. Or, the load can be applied as a constant load for a long period of time (as in a creep test to simulate the effect of the environment). The shear modulus can be determined as the ratio of shear stress to strain at a specific frequency. Note that the frequency is related to the period or time of loading. Under dynamic conditions of loading, the stiffness (or modulus) is equal to three times the shear modulus.

The dynamic tests could be conducted for shorter loading periods, and creep tests for longer loading periods, and the combined test data could be used for obtaining stiffness versus loading-time information, over the range of loading times expected in a pavement (Figure 9.8). Note that stiffness versus loading time plots for a specific asphalt binder obtained at different temperatures would coincide if shifted by different factors on the same plot. This property of asphalt binder is often referred to when asphalt binders are said to be "thermorheologically simple."

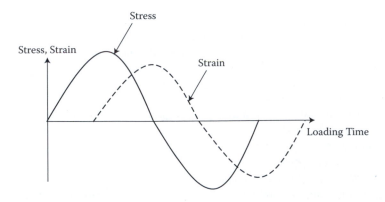

FIGURE 9.7 Concept of Lag between Stress and Strain.

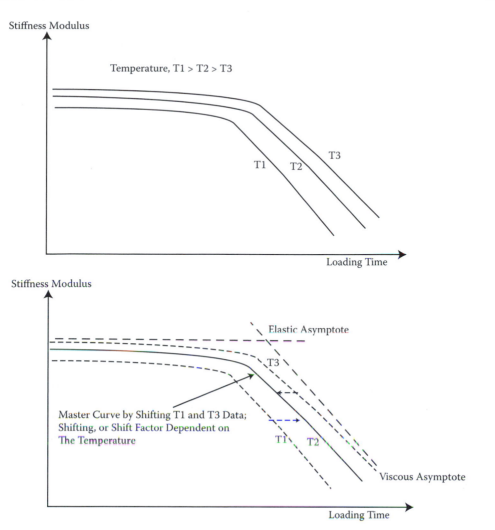

FIGURE 9.8 Effect of Loading Time and Temperature on Stiffness.

9.9 TENSILE BEHAVIOR

The behavior of asphalt binder in tension is usually expressed in terms of two parameters—breaking stress and strain at failure. At higher stiffness, the strain at failure is a function of the elastic modulus. High stress due to low temperatures occurs when the asphalt binder has relatively high stiffness, and the strain at failure is very low. The strain at failure at a specific temperature can be predicted from the PI, the softening, and a specific time of loading.

Toughness and tenacity: in this test (Figure 9.9), which is similar to the tensile strength test, the stress to cause the elongation as well as the area under the stress/strain curve are determined. Generally this test is useful for polymer-modified asphalt binders. The test results show two distinct areas under the stress/strain curve—the relatively small one under the beginning to the yield point of the asphalt, and the second larger area under a milder peak and prolonged elongation. The area of total curve (first and second combined) is called *toughness*, whereas the area under the second (large) part, which comes mainly from the contribution of the polymer modification, is known as the *tenacity*.

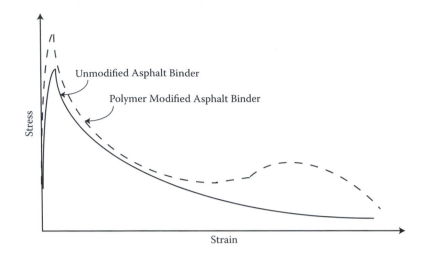

FIGURE 9.9 Concept of Toughness and Tenacity Test.
Source: Redrawn from Shell (2003).

9.10 SUPERPAVE (SUPERIOR PERFORMING ASPHALT PAVEMENTS)

Superpave is the HMA mix design system that resulted from the $150 million, 5-year-long Strategic Highway Research Program (SHRP) in the United States between 1987 and 1993. The major objectives behind the development of Superpave were to identify and define properties of asphalt binders, aggregate, and hot mix asphalt that influence pavement performance, and develop test methods for performance-based specifications. Even though research continues to improve the Superpave system, at present, the Superpave system is being used by most of the state departments of transportation in the United States.

The major elements of Superpave are (1) Superpave performance-graded (PG) asphalt binder specifications, (2) aggregate properties, and (3) the Superpave mix design and analysis system. The asphalt binder specifications and tests are described in the following paragraphs.

The two major outcomes in the area of asphalt binder specifications are the introduction of performance-related tests and the shift from the traditional approach of testing asphalt at the same temperature for different specified values of test results to testing asphalt at different temperatures for the same specified values of test results. The new tests relate to three asphalt binder characteristics—temperature susceptibility, viscoelastic behavior, and age hardening. The condition of the asphalt binders during the tests simulates the condition of the asphalt binder at supply, during production of HMA, and in service in the pavement. The modes of failure for which the asphalt binder characteristics are specified include rutting and cracking (temperature and fatigue related). Several relatively new asphalt tests are used routinely now, or are being implemented, as a result of the introduction of the Superpave system.

9.10.1 HIGH-TEMPERATURE VISCOSITY

High-temperature viscosity measurements, such as those required for determination of suitable mixing and compaction temperature and to check for suitability of pumping during mixing, can be conducted with a rotational viscometer. The test consists of rotating a spindle with a sample of asphalt binder maintained at the test temperature, measuring the torque required to maintain a constant rotation speed, and calculating the viscosity from this measurement. For plotting the kinematic viscosity versus temperature data, this test could be used to determine the viscosities at 135°C and 165°C (and avoid the need to convert absolute viscosity to kinematic viscosity, as discussed earlier). Kinematic viscosity is measured in centipoises and reported in the Pa-s (SI) unit, and the viscosities used for determining the equiviscous temperatures are 0.17±0.02 Pa-s for mixing and 0.28±0.03 Pa-s for compaction. The different parameters used for the calculation of kinematic viscosities are as follows.

Viscosity at 60°C, absolute viscosity, expressed in poises (P, cgs unit) or centipoises (CP), where 100 cP = 1 P. The SI unit is Pa-s,

$$1 \text{ Pa-s} = \frac{1 \text{ kg}}{\text{ms}} (1\text{P} = 0.1 \text{ Pa-s})$$

$$\text{Kinematic viscosity} = \frac{\text{Absolute Viscosity}}{\text{Density}} = \frac{\text{Pa} - \text{s}}{\frac{\text{kg}}{\text{m}^3}} = \frac{\frac{\text{kg}}{\text{ms}}}{\frac{\text{kg}}{\text{m}^3}} = \frac{\text{m}^2}{\text{s}} = \text{m}^2\text{s}^{-1}$$

In rotational viscometer, the absolute viscosity is measured as the ratio of shear stress to rate of shear, and the kinematic viscosity is expressed in the same unit as absolute viscosity (Pa-s), as follows:

$$\frac{\text{Absolute Viscosity}}{\text{Density}} = \frac{\text{Pa} - \text{s}}{\text{CF} * \text{G}_b * \frac{1{,}000 \text{kg}}{\text{m}^3}} = \frac{0.001 \text{Pa} - \text{s}}{\text{CF} * \text{G}_b * \frac{1\text{kg}}{\text{m}^3}} = \frac{0.001 \text{ Pa} - \text{s}}{\approx 1 * \frac{1\text{kg}}{\text{m}^3}}$$

where:
   CF = a correction factor for the test temperature, CF = $1.0135 - 0.0006 * (T)_{\text{test}}$.
   G_b = the specific gravity of asphalt binder at 15°C (approximately 1.03)

The product of CF and G_b for a temperature range of 135–165°C is approximately equal to 1.

Note that the field compaction temperature should be based on the above guidelines but adjusted according to a number of factors such as air and base temperature, wind speed, haul distance, roller type, and lift thickness. One excellent tool for simulating different conditions and determining appropriate compaction temperature is the PaveCool software developed by the Minnesota Department of Transportation (2000–2006; http://www.mrr.dot.state.mn.us/research/MnROAD_Project/restools/cooltool.asp).

9.10.2 COMPLEX MODULUS AND PHASE ANGLE

The dynamic shear rheometer (DSR; test method: AASHTO T-315; see Figure 9.10) is used to characterize the viscoelastic behavior of asphalt binder, and evaluate its rutting and cracking potential. The basic principle used for DSR testing is that asphalt behaves like elastic solids at low temperatures and like a viscous fluid at high temperatures. These behaviors can be captured by measuring the complex modulus (G*) and phase angle (δ) of an asphalt binder under a specific temperature and frequency of loading. The G* and δ parameters (Figure 9.10) are measured by the DSR by applying a torque on the asphalt binder between a fixed and an oscillating plate, and measuring the resultant strain. G* is defined as the ratio of shear stress to shear strain, and δ is the time lag between the stress and the strain in the asphalt binder. These properties are determined at high and intermediate pavement service temperatures. The high temperature is determined from the average of maximum HMA pavement temperature over a 7-day period through summer, which is obtained from data from weather stations. The intermediate testing temperature is determined from an average of 7-day average maximum and minimum design temperature. Figure 9.11 shows results of DSR tests conducted on different asphalts at different temperatures. Dividing G* by the angular velocity (10 radians per second used in DSR testing), one can obtain complex dynamic shear viscosity from DSR tests.

9.10.3 AGING TESTS

Short-term aging is done by subjecting the asphalt binder to the rolling thin film oven (RTFO) test, explained earlier. Long-term aging is simulated with the use of the pressure aging vessel (PAV; ASTM-D6521, AASHTO R-28; Figure 9.12). The PAV test consists of subjecting the RTFO residue

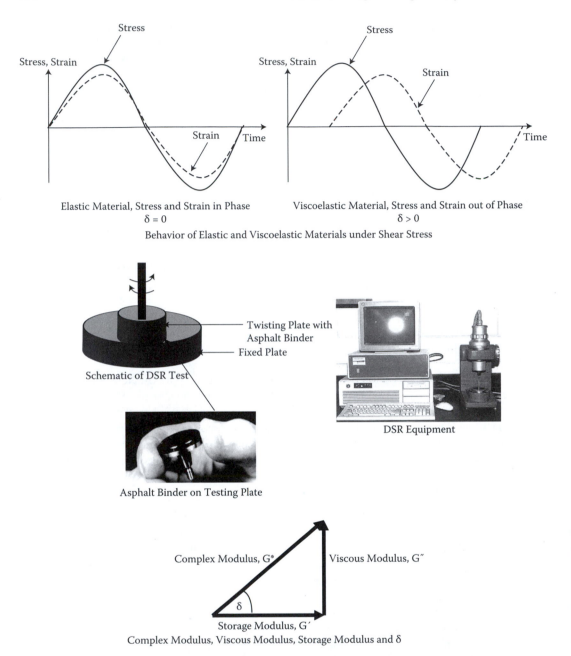

FIGURE 9.10 Dynamic Shear Rheometer (DSR) Test for Asphalts.

to high pressure and temperature inside a closed vessel. The PAV is placed inside an oven for maintaining the specified high temperature, and a cylinder of clean, dry compressed air is used to supply and regulate air pressure.

9.10.4 DSR Tests Conducted on Aged Asphalt

The asphalt binder in its unaged condition, short-term aged condition (RTFO conditioned), and long-term aged condition (PAV conditioned) is tested with the DSR for evaluation of its rutting and fatigue characteristics. The rutting potential is evaluated by the rutting factor, defined as $G^*/\sin \delta$,

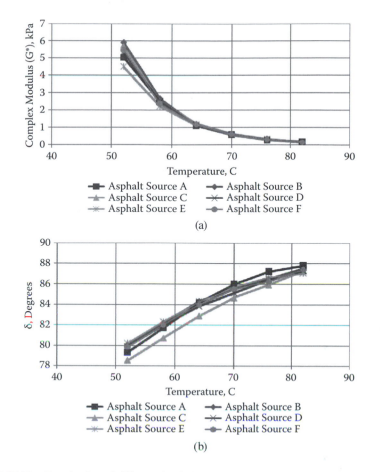

FIGURE 9.11 DSR Test Results from Different Asphalts.

FIGURE 9.12 Pressure Aging Vessel (PAV) Equipment with Air Supply Cylinder.

of which G* and δ can be obtained from DSR testing. Since higher G* means greater resistance to rutting, and lower δ means a more elastic asphalt binder, higher G* and lower δ values are desirable to make the HMA less susceptible to rutting. Considering rutting as a stress-controlled, cyclic loading phenomenon, it can be shown that rutting potential is inversely proportional to G*/Sin δ. Hence, a minimum limit is set on the value of G*/Sin δ. Testing asphalt for the rutting factor prior to long-term aging is critical, since with aging the stiffness of the asphalt binder increases and resistance to rutting increases (Bahia and Anderson, 1995). Hence, the DSR test for rutting potential is conducted on RTFO residue. It is also conducted on unaged asphalt binder, since in some cases RTFO may not indicate the actual aging of the binder during the production and placement of HMA.

The asphalt binder is also tested with the DSR to determine the fatigue cracking potential of HMA. The fatigue potential is measured by the fatigue factor, defined as G* Sin δ. This is done because it can be shown that under a strain-controlled process, the work done for fatigue is proportional to G* Sin δ, and hence a lower value of G* Sin δ means a lower potential of fatigue cracking. This can be explained by the fact that as G* decreases, the asphalt binder becomes less stiff and is able to deform without building up large stresses, and as δ value decreases, the asphalt binder becomes more elastic and hence can regain its original condition without dissipating "energy." An upper limit is set on the G* Sin δ value in the Superpave specification, in order to minimize the potential of fatigue cracking (Bahia and Anderson, 1995). Since the asphalt binder becomes stiffer due to aging during its service life and becomes more susceptible to cracking, the DSR test for fatigue factor determination is conducted on PAV-aged samples.

9.10.5 LOW-TEMPERATURE STIFFNESS (ASTM D-6648, AASHTO T-313)

The bending beam rheometer (BBR) is used to evaluate the stiffness of asphalt binder, and its cracking potential at low temperature. The BBR (Figure 9.13) consists of a loading mechanism in which an asphalt binder beam is subjected to a transient creep load at a constant low temperature. The creep stiffness and the rate at which the stiffness changes with loading time give an indication of its cracking potential at the test temperature. The higher the stiffness (s), the lower the rate (m), and the greater the potential for cracking. A lower slope or rate of change in stiffness means a reduced ability of the asphalt binder to relieve thermal stress by flowing. Hence, in the Superpave specification, a maximum limit is set on the value of stiffness, and a minimum limit is set on the value of slope. Since the asphalt binder becomes stiffer with aging and more susceptible to thermal cracking, the BBR test is conducted on PAV-aged samples.

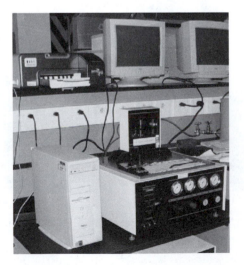

FIGURE 9.13 Bending Beam Rheometer.

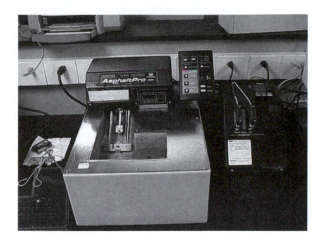

FIGURE 9.14 Direct Tension Testing Equipment.
Courtesy: Denis Boisvert and Alan Rawson, New Hampshire Department of Transportation.

9.10.6 Direct Tension Test (ASTM D-6723, AASHTO T-314)

In some cases, such as modified binders, a measure of stiffness is not adequate to characterize the cracking potential. This is because these binders may have a higher stiffness, but they may not be susceptible to cracking, as they are more ductile and can stretch without breaking. The direct tension test (DTT; Figure 9.14) is used to characterize such binders. It is used for testing any binder which has a stiffness exceeding the specified stiffness at low temperatures. The DTT consists of a tensile-testing machine, which measures the strain at failure of a dog bone–shaped asphalt binder specimen, which is subjected to tensile load at a constant rate. A minimum limit is set on the strain at failure.

9.10.7 Superpave Requirements

In the Superpave system of specifications, the requirements for results from tests on asphalt binders are the same. However, the temperature at which the requirements are met is used to classify the asphalt binder. For example, a binder classified as PG 64-22 means that the binder meets high-temperature physical property requirements up to a temperature of 64°C, and low-temperature physical property requirements down to –22°C. There is also a provision of using higher grade asphalt (PG 76-22, for example) for mixes to be used in slow transient or standing load areas.

A specific asphalt is classified (according to the PG classification system) through the following steps.

1. Check flashpoint (minimum = 230°C).
2. Check viscosity (ASTM 4402) at 135°C; maximum: 3 Pas.
3. Check at which highest temperature, for the neat asphalt, $G*/Sin\delta \geq 1.00$ kPa, when tested with the DSR at 10 rad/s; note temperature (high pavement temperature: 46/52/58/64).
4. Run RTFOT (AASHTO T240) or TFO (AASHTO T-179).
5. Check mass loss; maximum: 1.00%.
6. Check at which highest temperature, for the RTFOT/TFO aged asphalt, $G*/Sin\delta \geq 2.2$ kPa, when tested with the DSR at 10 rad/s; note temperature (high pavement temperature: 46/52/58/64)
7. Run pressure aging vessel test with the RTFOT/TFO-aged binder at 90°C for PG 46/52 (as identified above), 100°C (58/64/70/76/82), and 110°C (desert climate, 70/76/82).

8. Check at which temperature, for the RTFOT/TFO-PAV-aged asphalt, G* Sin δ ≤ 5000 kPa, when tested with the DSR at 10 rad/s; note temperature (intermediate pavement service temperature: 10/7/4 for PG 46, 25/22/19/16/13/10 for PG 52, 25/22/19/16/13 for PG 58, 31/28/25/22/19/16 for PG 64, 34/31/2825/22/19/16/13 for PG 70, 37/34/31/2825/22 for PG 76, and 40/37/34/31/28 for PG 82).

9. Check lowest temperature at which the S value is ≤ 300 MPa and m value is ≥ 0.300 in the test BBR run for 60 seconds (low pavement temperatures: –36 to 0, in 6-degree decrements).

10. If stiffness exceeds 300 MPa in step 9, run direct tension test and check minimum temperature at which failure strain is ≥ 1%, when run at 1.0 mm/min (low pavement temperatures: –36 to 0, in 6-degree decrements).

To illustrate the concept, Figure 9.15 is presented. It shows the complex modulus and phase angles at different temperatures for different asphalt grades. Note that the lower the high-temperature grade, the lower the complex modulus, and the higher the phase angle at a specific temperature. Also note that the PG 82-22 binder is a polymer-modified binder, and hence shows significantly different properties than the rest. Generally, asphalt binders with high range (*range* means the difference between the high and low temperatures) are polymer modified.

The asphalt binder to be used in a specific project is selected on the basis of the geographical location of the project, the pavement temperature and air temperature, and the specific reliability that the user wants to have in selection of the binder. The process is illustrated below.

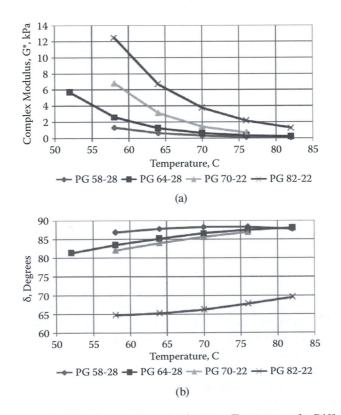

FIGURE 9.15 Plots of Complex Modulus and Phase Angle versus Temperature for Different Grades.

Example

Consider a project located in Worcester, MA, and another one in Auburn, AL. The selection of the binder could be done with the help of freely available software, LTPPBIND, which is based on the original SHRPBIND software. The LTPPBIND (version 3.1 beta, accessed 11/20/07) software can be obtained from http://www.fhwa.dot.gov/pavement/ltpp/ltppbind.cfm (FHWA, n.d.).

- *Project No. 1*: Surface layer; needs reliability of 98% for both high and low temperatures
- Location: Worcester, MA; latitude: 42.27, longitude: 71.87
- Weather station: Worcester Municipal Airport
- High air temperature: mean: 29.9°C, standard deviation: 1.4, minimum: 27.4, maximum: 33 (from 34 years of data)
- Low air temperature: mean: –21.7°C, standard deviation: 2.4, minimum: –25, maximum: –16 (from 35 years of data)
- Low air temperature drop: mean: 24.2°C, standard deviation: 2.4, minimum: 20.5, maximum: 31 (from 35 years of data)
- Degree days over 10°C: mean: 2119, standard deviation: 105, minimum: 1893, maximum: 2345 (from 34 years of data)
- Pavement temperature: for high temperature, reliability level at 48.6°C is 50%, whereas reliability level at 52°C is 98%. Choose 52°C. For low temperature, reliability at –16°C is 56%, whereas reliability at –22°C is 98%. Choose –22°C. Appropriate binder will be PG 52-22, or any binder that has a high-temperature level above 52°C and low-temperature level below –28°C (for example, a PG 64-28).
- *Project No. 2*: Surface layer: needs reliability of 98% for high temperature and low temperature
- Location: Auburn, AL; select nearest weather station at Opelika, AL; latitude: 32.63; longitude: 85.38
- High air temperature: mean: 35.1°C, standard deviation: 1.4, minimum: 32.7, maximum: 38.5 (from 31 years of data)
- Low air temperature: mean: –12.1°C, standard deviation: 3.5, minimum: –21.5, maximum: –6 (from 33 years of data)
- Low air temperature drop: mean: 28.6°C, standard deviation: 2.6, minimum: 25, maximum: 36 (from 33 years of data)
- Degree days over 10°C: mean: 3699, standard deviation: 141, minimum: 3414, maximum: 3957 (from 31 years of data)
- Pavement temperature: for high temperature, reliability level at 61.9°C is 50%, whereas reliability level at 64°C is 98%. Choose 64°C. For low temperature, reliability at –10°C is 90%, whereas reliability at –16°C is 98%. Choose –16°C. Appropriate binder will be PG 64-16, or any binder that has a high-temperature level above 64°C and low-temperature level below –16°C (for example, a PG 70-22).

9.10.7.1 Explanation

The steps in the software are as follows. A high-temperature algorithm is used for the determination of the high-temperature PG. An equation, developed from a database of pavement temperatures and rutting damage, is used to determine a PG damage (PGd, or a "base PG" with 50% reliability at the surface layer), then an estimate of variability of pavement temperature (CVPG) is obtained using an equation, and the base PG is adjusted for any other desired level of reliability. Finally, the PG is adjusted for any other depth of the pavement, using another equation. Note that the determined PG

TABLE 9.1
Average Grade Adjustments "Grade Bumps"

Speed	Base Grade	Traffic Loading Equivalent Single-Axle Loads, Millions			
		<3	3–10	10–30	>30
Fast	52	0	10.3	16.8	19.3
	58	0	8.7	14.5	16.8
	64	0	7.4	12.7	14.9
	70	0	6.1	10.8	12.9
Slow	52	3.1	13	19.2	21.6
	58	2.9	11.2	16.8	19
	64	2.7	9.8	14.9	17
	70	2.5	8.4	12.9	14.9

grade is valid for fast-moving traffic with a design volume of 3 million equivalent single-axle loads (ESAL), and that adjustments (known as *grade bumping*) could also be made to accommodate other traffic volumes and speed. The adjustment factor (a calculated average for different PG, traffic volume, and speeds), shown in Table 9.1, needs to be added to the base PG to determine the design PG.

$$PG_d = 48.2 + 14\, DD - 0.96\, DD^2 - 2\, RD$$

$$CVPG = 0.000034\, (Lat-20)^2\, RD^2$$

$$PG_{rel} = PGd + Z\, PGd\, CVPG / 100$$

where:
PG_{rel} = PG at a reliability, °C
PG_d = damage-based performance grade, °C
CVPG = yearly PG coefficient of variation, %
DD = average yearly degree-days air temperature over 10°C, × 1000
RD = target rut depth, mm
Lat = latitude of site, degrees
Z = from standard probability table, 2.055 for 98% reliability

For adjusting PG with depth:
$$T_{pav} = 54.32 + 0.78\, T_{air} - 0.0025\, Lat^2 - 15.14 \log_{10}(H + 25) + z\,(9 + 0.61\, S_{air}^2)^{0.5}$$

where:
T_{pav} = high AC pavement temperature below surface, °C
T_{air} = high air temperature, °C
Lat = latitude of the section, degrees
H = depth to surface, mm
S_{air} = standard deviation of the high 7-day mean air temperature, °C
z = standard normal distribution table, z = 2.055 for 98% reliability

The low pavement temperature for different reliabilities is determined using the following equation.

$$T_{pav} = -1.56 + 0.72\, T_{air} - 0.004\, Lat^2 + 6.26 \log_{10}(H + 25) - z\left(4.4 + 0.52\, S_{air}^2\right)^{0.5}$$

where:
T_{pav} = low AC pavement temperature below surface, °C
T_{air} = low air temperature, °C
Lat = latitude of the section, degrees

TABLE 9.2
Typical Grades Available in the United States

High-Temperature Grade	Low-Temperature Grade						
	−10	−16	−22	−28	−34	−40	−46
46							
52							
58							
64							
70							
76							
82							

H = depth to surface, mm
S_{air} = standard deviation of the mean low air temperature, °C
z = standard normal distribution table, z = 2.055 for 98% reliability

Table 9.2 shows the different grades of asphalts generally available in the United States. Note that higher range asphalt binders are mostly modified binders (such as with polymer), and grades wider than the typical ranges presented in the table can be produced for specialty applications.

The steps in the selection of an asphalt binder for a specific project can thus be summarized as follows.

1. Determine the type of traffic loading that is expected on the constructed pavement (standing/slow, 50 km/hour or less/fast, 100 km/hour or more).
2. Determine the high pavement temperature.
3. Determine the low pavement temperature.
4. Determine the design volume of traffic.
5. Select the asphalt grade—slow or standing traffic, and higher traffic volumes require a higher grade.

9.11   RECOVERY OF ASPHALT BINDER FROM ASPHALT MIX

The asphalt binder from an asphalt mix could be recovered for testing, with the help of the solvent extraction method, using a rotary evaporator (Figure 9.16). The asphalt mix is mixed with a solvent (such as methylene chloride) to remove the asphalt binder from the aggregates to the solution. In the

FIGURE 9.16 Rotary Evaporator Used for Extraction of Asphalt Binder.

next step the mineral filler and very fine aggregates are separated from the solution by filtration and centrifuging. In the
final step, the asphalt binder-solvent mix is heated in a flask
maintained at a specific temperature in an oil bath, such that
the solvent is evaporated and condensed elsewhere, while
the pure asphalt binder remains in the flask. If the asphalt
binder is not required for testing, then the amount of asphalt
binder present in a mixture of asphalt and aggregate can be
determined by the use of the ignition test (AASHTO T-308,
Determining the Asphalt Binder Content of Hot Mix Asphalt
[HMA] by the Ignition Method), in which the mix is heated
to a very high temperature (500–600°C) to burn the asphalt
off, and the difference between the initial and final weight of
the mix is determined (Figure 9.17).

9.12 ADHESION PROPERTIES

The primary function of the asphalt binder is to bind the
aggregate particles together, or work as an adhesive. The

FIGURE 9.17 Ignition Test Equipment.

aggregate-asphalt binder adhesion is affected by a number of factors, and if not considered carefully, water can destroy it and adversely affect pavement performance. These factors are related to
the properties of the aggregate, the asphalt binder, and the asphalt mix. External factors, such as
rainfall and temperature, are also important.

The surface energy concept has been used and researched to explain the adhesive bond between
aggregate and asphalt binder, and its failure. The use of liquid with a polarity opposite to that of an
aggregate with unbalanced surface charge, and energy, leads to the formation of an adhesive bond.
Water, if present in addition to asphalt binder, could meet the surface energy requirement of an
aggregate, and hence displace asphalt binder (stripping) as the preferred coating.

The nature of the aggregate is of primary importance. Generally, siliceous quartz and granite,
which are also classified as "acidic type," have less affinity for asphalt binder (and hence make it
more difficult to provide a bond between the aggregate and the binder) than basic rocks such as
basalts and limestone. A significant increase in the adsorption of asphalt can be caused by the
presence of a fine microstructure of pores, voids, and micro-cracks on the aggregate surface. The
surface texture of the aggregate, viscosity of asphalt during mixing, and polarity of asphalt affect
the adhesion significantly.

9.13 ASPHALT EMULSIONS

Emulsions are produced by dispersing asphalt binder as droplets of small diameter with the help of
an emulsifying agent. The emulsifying agent(s) and other additives, if needed, provide the electrical
charges to the droplets of asphalt binder so that they repel each other and remain dispersed in water.
Common anionic emulsifiers, which impart negative charges, are made from the saponification of
fatty acid–type materials, such as rosins and lignins (wood by-products) with sodium or potassium
hydroxide. Cationic emulsifiers are generally made from fatty amines (such as diamines) and hydrochloric acids, as well as from fatty quaternary ammonium salts.

The successful use of a specific emulsion depends on the type of the aggregate with which
it is used. Depending on the charge, there are primarily two types of emulsions that are used in
pavement mixes—anionic, containing negatively charged droplets of asphalt binder, and cationic,
containing positively charged droplets. The emulsifier keeps the emulsion in equilibrium (stabilizes
the emulsion) and dictates the breaking (coalescing of droplets) rate and adhesion of the asphalt
binder to the aggregate.

Emulsions are produced in a colloid mill, containing a high-speed (1000 to 6000 revolutions per minute) rotor, through which hot asphalt binder and emulsifier solutions are passed at specific temperatures. The shearing action of the rotor forms asphalt binder droplets, which get coated with emulsifier and as a result get charged, which prevents them from coalescing. High-pressure mills are used to facilitate the production of emulsion from higher viscosity grade asphalt binders, and static mixers with baffle walls, through which the materials can be pumped at high speed, can also be used instead of a colloid mill. Generally an emulsion may contain about 28% droplets smaller than 1 μm in size, 57% droplets 1–5 μm in size, and the rest 5–10 μm in size. In general, emulsions contain 50–75% of asphalt.

9.13.1 PROPERTIES

The desirable properties of emulsion include adequate stability, proper viscosity and breaking characteristics, and strong adhesion. The emulsion must be stable during storage, transportation, and applications, and must be sufficiently fluid (but not too fluid, to be optimum) to facilitate application. The emulsion should break in a timely manner, in contact with the aggregates, and form strong bonds.

Stability of an emulsion refers to the ability to keep the asphalt droplets well dispersed throughout the emulsion. Because the density of asphalt is slightly higher than the aqueous phase of the emulsion at ambient temperature, the droplets would tend to settle down over time, form clumps by agglomeration, and ultimately form larger globules by coalescing. Allowable ranges of storage temperatures vary from 20–60°C (minimum) to 50–85°C. Slow- and medium-setting emulsions as well as high floats have a higher range of storage temperature.

The viscosity of emulsion can be increased by increasing the amount of asphalt binder; increasing the viscosity of the continuous phase, such as by increasing the emulsifier content; and decreasing the droplet sizes, such as by increasing the flow rate through the colloid mill or decreasing the viscosity of the asphalt binder used for emulsion. Emulsion viscosity is measured with an orifice viscometer, in which viscosity is determined from the time required for a specific amount of emulsion to flow through a standard orifice.

The breaking characteristics of an emulsion can be affected by the asphalt content, composition of the aqueous phase, droplet size distribution, environmental conditions, type of aggregates, and use of breaking agents. The rate of breaking can be increased with an increase in asphalt and emulsifier content, and by reducing the size of asphalt droplets. Note that the breaking process is accelerated by the evaporation of water from the emulsion, as well as by the absorption of the emulsifier into the aggregates. High temperatures and low humidity are favorable to breaking, while the presence of dust coating on the aggregate surface could accelerate the breaking process without causing any adhesion of the asphalt binder to the aggregate surface. The presence of calcium and magnesium ions on the aggregate surface would also accelerate the breaking process. If the emulsifier is absorbed from the solution into the aggregates at the desired rate (such as through a rough-textured porous surface), the lowering of the charge on the droplets causes rapid coalescence and good adhesion of the asphalt binder to the aggregate surface. The breaking can also be accelerated with the use of breaking agents, sprayed along with the emulsion or just after the application of the emulsion, as well as with the help of mechanical compaction (using a roller).

The adhesive nature of emulsion is made possible by good "wetting" of the aggregate surface (high surface tension on the aggregate surface) and the creation of a thermodynamically stable system of minimum surface energy. There are many factors that affect this process, including the aggregate type (such as acidic or basic), type and amount of emulsifier (such as cationic or anionic), grade of the asphalt, and size of the droplets.

It is important to note that although the viscosity of the asphalt binder has no effect on the viscosity of an emulsion formed with it, there are several properties of the asphalt binder which do affect the emulsion properties significantly. These include its electrolyte content, which can cause an increase as well as a decrease in viscosity depending on its amount, density (a higher density causes lower stability), and acid content (depending on the type of the emulsion, it may improve or deteriorate stability).

Emulsions are also made with polymer-modified binders (polymer is generally added to reduce temperature susceptibility), generally requiring higher temperature for manufacturing. However, one beneficial effect of polymer is the ability to use a higher asphalt content in the emulsion without affecting its viscosity (and hence ease of use) adversely.

9.13.2 Tests for Asphalt Emulsions

Particle charge test. In this test, a current is sent through an anode and cathode immersed in an emulsion. If the cathode shows an appreciable layer of asphalt binder at the end of the test, then the emulsion is characterized as cationic.

Viscosity test. The Saybolt Furol viscosity test is conducted to measure viscosity, either at 25°C or 50°C.

Demulsibility test. This test is conducted to check the rate of breaking of rapid-setting emulsions when applied on aggregates or soils. The test consists of mixing an emulsion with a solution of water and either calcium chloride or dioctyl sodium sulfosuccinate (for anionic and cationic emulsions, respectively), and sieving the mix. The concentration of the solution and the minimum amount of asphalt retained on the sieve (as a result of coalescence) are generally specified. A similar test is conducted for the SS grades, using finely ground Portland cement, and washing over a 1.40 mm sieve. Generally, limits are specified on the amount of material that is retained on the sieve, to ensure the ability of the emulsion to mix with the soil/aggregate with high surface area before breaking.

Identification test for CRS (ASTM D244). In this test, sand washed with hydrochloric acid and alcohol is mixed with emulsion for 2 minutes, and then the uncoated area is compared to the coated area.

Identification test for CSS (ASTM D244). In this test, washed and dried silica sand is mixed with the emulsion until they are completely coated, and the mix is cured for 24 hours and then placed for 10 minutes in a beaker with boiling water. Then the sample is spread on a level surface and the part of the mix coated, as percentage of the mix is determined.

Settlement and storage stability tests. In this test, a specified volume of emulsion is made to stand in a graduated cylinder for either 24 hours (for the storage test) or 5 days (for the settlement test), and then samples are taken from the top and bottom of the cylinder, weighed, and heated to evaporate the water. The difference in weight of the residue from the top and the bottom parts, if any, is determined.

Sieve test. In this test, an emulsion is poured over a 850 μm sieve, and the sieve is rinsed with mild sodium oleate solution and distilled water for anionic emulsions and distilled water only for cationic emulsions. Then the sieve and retained asphalt binder are dried and weighed, and the amount of asphalt binder that is retained on the sieve is determined.

Coating ability and water resistance test. This test is conducted for MS emulsions to check their ability to coat aggregates, remain as a film during mixing, and resist washing action of water after mixing. The test consists of two steps. In the first step, the aggregate is coated with calcium carbonate dust and then mixed with emulsion. Half of the mix is placed on an absorbent paper for visual inspection of coating. The rest is sprayed with water and rinsed until the water runs clear, and then placed on the absorbent paper for inspection of coating. In the second step, the first step is repeated but the emulsion is added to the aggregate only after coating the aggregates with water.

Field coating test. In this test, measured amounts of aggregate and emulsion are mixed with the hands, and the coating is observed during the 5-minute mixing process. Then the coated aggregates are placed in a container, which is filled with water, and the water is drained off five times (after refilling). Next the aggregates are checked for coating. A good coating indicates a full coating of aggregate, a fair coating indicates an excess of coated over uncoated area, and a poor coating indicates an excess of uncoated over coated area.

Unit weight test. The weight of an asphalt emulsion in a standard measure of a known volume is found out to determine its unit weight.

Residue and oil distillate by distillation. The asphalt, oil, and water can be separated in an emulsion for subsequent testing by distillation, which consists of heating the emulsion in an aluminum alloy still at a temperature of 260°C for 15 minutes.

Residue by evaporation. Residue from the emulsion can also be obtained by evaporating it in an oven at 163°C. This test should not be run if the residue needs to be tested for the float test. The residue can be tested for specific gravity (ASTM D-70, AASHTO T-228).

Solubility in trichroloethylene (ASTM D-2042, AASHTO T-44). In this test, the residue is dissolved in a solvent, and the insoluble (contaminants) part is separated by filtration.

Penetration test (ASTM D-5, AASHTO T-49), ductility test (ASTM D-113, AASHTO T-51), and ring and ball softening point test (ASTM D-36, AASHTO T-53). Similar to that described for asphalt binders.

Float test (ASTM D-139, AASHTO T-50). This test, conducted to determine resistance to flow at high temperatures, consists of forming a test plug of residue asphalt (from emulsion) in a brass collar and screwing the collar at the bottom of an aluminum float placed in a 60°C water bath. The time required by the water to break through the plug is determined, and a higher time indicates a higher resistance. Asphalt emulsions showing high float test numbers (called *high floats*) possess a gel-like structure and are resistant to flow from high temperatures during the summer.

Breaking index. In this test, which is run for rapid-setting (RS) emulsion, silica sand is added and mixed with the emulsion under controlled conditions of rate and temperature. *Breaking index* is defined as the weight of the sand needed in grams to break the emulsion.

Vialit test. This test simulates the formation of chip seal—aggregate embedded by rolling onto an emulsion-coated surface—in a small metal plate. After curing, the aggregates are treated with water and dried. Finally, a 500 gram steel ball is dropped from a specified height three times on the reverse side of the plate, and the weight loss of the aggregate is measured.

Zeta potential. In this test, the speed of movement of asphalt binder droplets in an aqueous medium is measured to determine the intensity of charge.

Elastic recovery after ductility. This test follows the ductility test, and involves determining the percentage recovery of the thread that has been used in the ductility test, when cut into half.

Force ductility. The test is similar to the ductility test, but the force required to cause the elongation of the asphalt thread is measured.

Tensile strength test. In this test, the force (stress) required to pull an asphalt thread in the vertical direction to a specified elongation (strain) is determined.

Torsional recovery. In this test, a shaft and arm assembly is immersed in a heated sample of the residue asphalt, and is twisted 180° inside the sample at a lower temperature. The percentage recovery after a specified time is then determined.

9.13.3 CLASSIFICATION OF EMULSIONS AND SELECTION

Emulsions are classified according to their charge (cationic [C] and anionic [no symbol]), as mentioned earlier. They are also classified according to their rate of breaking or setting—rapid (rapid setting, or RS), quick (QS), medium (MS), or slow (SS). A number is used for denoting the relative viscosity of the emulsion. Hence, *CRS-1* means a cationic rapid-setting emulsion with less viscosity than an MS-2 emulsion, which is anionic and medium setting. A *h* or a *s* following the emulsion name would indicate the use of hard and soft asphalt binder, respectively, whereas *HF* indicates "high float" properties as measured in the float test.

The separation of water from the emulsion is called *breaking*, the coalescing of asphalt binder droplets is called *setting*, whereas the formation of a continuous adhesive layer of asphalt binder (through breaking) on the surface of the aggregates is called *curing*. Breaking is generally accompanied by the formation of black and sticky material from the more fluid and brownish emulsion.

Emulsions are graded according to their rate of breaking (SS, MS, QS, and RS), the nature of their charge (cationic [C] and anionic [no symbol]), the presence of hard and soft asphalt (H or S), and the results of the float test (HF if high results). The following are the ASTM and AASHTO specified grades of emulsion.

ASTM D-977, AASHTO M-140: RS-1, RS-2, HFRS-2, MS-1, MS-2, MS-2H, HFMS-1, HFMS-2, HFMS-2H, HFMS-2S, SS-1, SS-1H

ASTM D-2397, AASHTO M-208: CRS-1, CRS-2, CMS-2, CMS-2h, CSS-1, CSS-1h

Emulsions can be selected for specific applications according to the following guidelines.

Cape seal: RS-2, CRS-2
Dust palliative, mulch treatment, and crack filler: SS-1, SS-1h, CSS-1, CSS-1h
Fog seal: MS-1, HFMS-1, SS-1, SS-1h, CSS-1, CSS-1h
Immediate-use maintenance mix: HFMS-2s, CRS-2, CMS-2
Microsurfacing: CSS-1h
Mixed in-place open-graded aggregate mix: MS-2, HFMS-2, MS-2h, HFMS-2h, CMS-2, CMS-2h
Mixed in-place well-graded aggregate mix, sand mix, and sandy soil mix: HFMS-2s, SS-1, SS-1h, CSS-1, CSS-1h
Plant mix hot or warm mix asphalt: MS-2h, HFMS-2h
Plant mix cold open-graded aggregate mix: MS-2, HFMS-2, MS-2h, HFMS-2h, CMS-2, CMS-2h
Plant mix cold dense-graded aggregate mix and sand mix: HFMS-2s, SS-1, SS-1h, CSS-1, CSS-1h
Prime coat: MS-2, HFMS-2, SS-1, SS-1h, CSS-1, CSS-1h
Sand seal: RS-1, RS-2, HFRS-2, MS-1, HFMS-1, RS-1, CRS-2
Sandwich seal: RS-2, HFRS-2, CRS-2
Single- and multiple-surface treatments: RS-1, RS-2, HFRS-2, RS-1, CRS-2
Slurry seal: HFMS-2s, SS-1, SS-1h, CSS-1, CSS-1h
Stockpile maintenance mix: HFMS-2s
Tack coat: MS-1, HFMS-1, SS-1, SS-1h, CSS-1, CSS-1h

QUESTIONS

1. What are the different sources of asphalt binder?
2. Of the three different asphalts for which DSR results are shown below, pick the one that is suitable for a project in a climate where the maximum pavement temperature is 76°C, on the basis of results from unaged asphalt only:

Asphalt	Temperature (°C)	Complex Modulus, kPa	Phase Angle, degree
A	76	1.20	87
B	76	0.90	86
C	76	1.08	88

3. For a bending beam rheometer test, the stiffness is calculated as follows:

$$S = \frac{PL^3}{4bh^3\delta(t)}$$

where:
  S = creep stiffness, MPa
  P = applied load, N
  L = distance between supports of the beam, mm
  b = width of the beam, mm
  h = thickness of the beam, mm
  $\delta(t)$ = deflection of the beam, mm

If $P = 0.981$ N, $L = 102$ mm, $b = 12.5$ mm, $h = 6.25$ mm, and $\delta = 0.3$ mm, determine S.

4. Download the LTPPBIND software, and determine the most suitable asphalt binder grade for your town/city.

10 Concrete Fundamentals for Rigid Pavements

10.1 CONCRETE

Portland cement concrete (PCC) is the most commonly used construction material in the world. Approximately half of all PCC used in the United States is consumed in the construction of PCC pavements.

The constituents of Portland cement concrete include coarse aggregates (natural stone or crushed rock), fine aggregate (natural and manufactured sand), Portland cement, water, and mineral or chemical admixtures. The paste phase is made up of the Portland cement, water, and air (entrapped and entrained). The paste phase is the glue or bonding agent that holds the aggregates together. The mortar phase is the combination of paste and fine aggregate. Aggregates occupy approximately 60–75% of the total volume of the concrete. The aggregates should be well graded to efficiently utilize the paste. It is desirable that each aggregate particle is coated with paste. The quality of the concrete is therefore highly dependent on the quality of the paste, the aggregates, and the interface or bond between the two.

The quality of the hardened concrete is strongly influenced by the quality of the paste, which in turn is highly dependent on the amount of water used in mixing the concrete. Therefore, the water-to-cement ratio by mass (W/C) is an important parameter that relates to many concrete qualities. Excessive amounts of mix water (a higher W/C ratio) will increase the porosity, lower the strength (compressive, tensile, and flexural), increase the permeability and lower watertightness and durability to aggressive chemicals, decrease bond strength between concrete and reinforcement, and increase volume changes (more cracking) due to moisture changes (wetting and drying) and temperature changes. Therefore, it is important to use the least amount of water for mixing the concrete, provided that it can be placed and consolidated properly. Stiffer mixes with less water are more difficult to work with; however, mechanical vibration is available for proper consolidation.

To produce a concrete that is strong, resists abrasion, and is watertight and durable, the following sequence should be followed: proper proportioning, batching, mixing, placing, consolidating, finishing, and curing.

The ease of placing, consolidating, and finishing concrete without segregation of the coarse aggregates from the mortar (paste and sand) phase is called *workability*. The degree of workability is highly dependent on the concrete type and method of placement and consolidation. Mechanical placement and consolidation can handle much stiffer mixes. Other properties related to workability include consistency, segregation, pumpability, bleeding, and finishability. The slump test is used to measure consistency, which is closely related to workability. A low-slump concrete has a stiff consistency and may be difficult to place and consolidate properly. However, if the mix is too wet, segregation or separation of large-size aggregates and honeycombing could occur. The consistency should be the driest that is most practical for placement using the available equipment (Powers, 1932).

Concrete mixtures with the larger aggregate size and a continuous gradation will have less space to fill with paste and less aggregate surface area to coat with paste; therefore, less cement and water are needed. Stiffer mixtures usually result in improved quality and economy, granted the concrete is properly consolidated.

10.2 AGGREGATES

Aggregates should be of sufficient quality with adequate strength, soundness, and environmental durability so as not to negatively impact the performance of the hardened concrete. The larger portions of crushed rock, stone, and sand are called *coarse aggregate* with sizes larger than the No. 4 sieve (4.75 mm). *Fine aggregate* is natural sand or crushed rock that has a particle size smaller than the No. 4 sieve.

For concrete pavements subject to cycles of freezing and thawing, aggregate and cement paste durability is critical to the pavement performance. For aggregates, the frost resistance is related to the porosity, absorption, permeability, and pore structure. If an aggregate particle absorbs enough water to reach critical saturation, then the pore structure is not able to accommodate the hydraulic pressures associated with the freezing of water. If enough inferior aggregates are present in the concrete, concrete distress and disintegration may occur. If the problem aggregate particles are present at the surface, popouts can occur. Popouts appear as conical breaks that are removed from the concrete surface.

Cracking of concrete pavements caused by freeze-thaw deterioration of the aggregate within the concrete is called D-cracking (see Figure 11.2). The visual distress of D-cracking resembles the letter *D*, where closely spaced cracks emerge parallel to the transverse and longitudinal joints and progress outward toward the center of the slab. Assessing aggregates for freeze-thaw durability can be done by past performance or by laboratory testing of concrete. The Freeze-Thaw Testing of concrete (ASTM C666; AASHTO T161) involves conditioning concrete beams through cycles of freeze-thaw (freeze in air and thaw in water, or freeze and thaw in water), and testing for dynamic modulus of the beam before and after conditioning. If the conditioned modulus falls below 60% of the unconditioned modulus, then the mix is not accepted. Other tests include resistance to weathering using immersion cycles in sodium or magnesium sulfate solutions. Salt crystal growth in the aggregate pores creates a pressure similar to the hydraulic pressure from freezing water. A rapid pressure-release test developed by Janssen and Snyder (1994) has also been correlated to D-cracking potential. In this test, aggregates are placed in a high-pressure chamber. The pressure is released rapidly, and unsound aggregates with ineffective pore structure fracture.

10.3 CEMENT

Portland cements are hydraulic cements that chemically react with water to produce calcium silicate hydrates, which are considered to be the glue that gives Portland cement concrete its strength. Portland cement is produced by combining raw material (or feed) containing calcium, silica, alumina, and iron compounds. Appropriate amounts of this raw feed are combined in pulverized or slurry form and sintered or burned in an inclined rotary kiln (Figure 10.1 shows a schematic of the cement production process).

The raw feed is introduced into the kiln, and as the kiln rotates, the materials will slide down the kiln at a slow rate. The kiln temperature ranges from approximately 200°C at the entry to

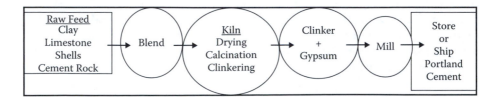

FIGURE 10.1 Schematic of Cement Production Process.

TABLE 10.1
Clinkering Process for Cement Production

Up to 700°C	Raw materials are free flowing; water is lost, and clay compounds recrystallize.
700–900°C	Calcination continues; CO_2 is liberated; reactive silica combines with CaO to start formation of C_2S (belite).
1150–1200°C	Solid solution; particles interact; small belite crystals form.
1200–1350°C	Solid solution; interaction of belite and CaO to form C_2S (alite).
1350–1450°C	Agglomeration and layering continue; belite crystals decrease in amount and increase in size; alite increases in size and amount.
Cooling	Upon cooling, C_3A and C_4AF crystallize in liquid phase; belite develops lamellar structure.

1450°C at the exit and burner location. Table 10.1 shows the clinkering (solid solution) reactions in the kiln. At the end of the clinkering process, clinker nodules measuring 1–25 mm are produced. To produce Portland cement, the clinker and approximately 5% gypsum are combined and ground to a very fine powder averaging 15 μm in diameter (and a surface area of about 300–600 m²/kg).

Since the chemical equations for cement chemistry are long and complex, a shorthand (abbreviated) notation was developed and is in use today by cement and concrete technologists, as shown in Table 10.2.

A clinker chemical analysis is typically given in oxide form, as shown in Table 10.3. From the chemical analysis, the quantity of each of the four main minerals (C_3A, C_4AF, C_3S, and C_2S) may be calculated using the "Bogue" equations (Taylor, 1997).

Clinker contains four main compounds which constitute over 90% of Portland cement. Table 10.4 shows the compound name, the approximate chemical formula (note: cement chemistry stoichiometry is not exact), the shorthand or abbreviated name, and a range for the compound composition for typical ASTM C-150 cements.

10.3.1 Types of Portland Cement

There exist many different types of Portland cements in the world to meet the needs for specific purposes. Manufactured Portland cements usually meet the specifications of ASTM C-150 (Standard Specifications for Portland Cement), AASHTO M85 (Specification for Portland Cement), or ASTM 1157 (Performance Specification for Hydraulic Cements). ASTM C-150 provides standards for eight different Portland cements, as shown in Table 10.5.

TABLE 10.2
Shorthand (abbreviation) for Cement Chemistry

Symbol	For Compound	Symbol	Compound
C	CaO_3	M	MgO
S	SiO_2	T	TiO_2
A	Al_2O_3	H	H_2O
F	Fe_2O_3	$\bar{S}$	SO_3
K	K_2O	$\bar{C}$	CO_2
N	Na_2O		

TABLE 10.3
Example of a Typical Clinker
Analysis (oxide weight %)

SiO_2	21.6	K_2O	0.6
Al_2O_3	5.3	Na_2O	0.2
Fe_2O_3	2.9	SO_3	1.0
CaO	66.6	LOI	1.4
MgO	1.0	IR	0.5
Total	98.9		

Type I. General-purpose cement suitable for all uses where special properties or other cement types are not required.

Type II. Used where precaution against moderate sulfate attack is warranted. Type II cements should be specified where water-soluble sulfate (SO_4) in soil is between 0.1 and 0.2% or the concentration of SO_4 in water is between 150 and 1500 ppm. The moderate sulfate resistance of Type II cement is due to its limited C_3A content (< 8%). Sulfate degradation of hydrated Portland cement is due to the formation of additional ettringite when external sulfate is introduced. Additional formation of ettringite and other calcium aluminate products in the paste may cause cracking and degradation of the concrete.

Type III. Provides higher strength at an early age (1 week or less) compared to Types I and II. The strength is comparable to that of Type I in the long term (months and years). Type III cement is chemically similar to Type I; however, it is ground much finer. The increase in surface area allows for more reaction contact sites between the cement and water and, therefore, results in a faster hydration product formation.

Type IV. Used where the rate and amount of heat generated from hydration must be minimized. Strength rate development is much lower than in other types. This type of cement is specially ordered and is not readily available.

Type V. Used where precaution against severe sulfate exposure is warranted. Type V cements should be specified where the water-soluble sulfate (SO_4) in soil is greater than 0.2% or the concentration of (SO_4) in water is greater than 1500 ppm. The high sulfate resistance of Type V cement is due to its limited C_3A content (< 5%). The sulfate degradation mechanism is similar to that for Type II cement. ASTM C-150 and AASHTO M85 allow both chemical and physical standards to assure the sulfate resistance of Type V cement. However, only one approach may be specified but not both. ASTM C-452 is used to assess the potential expansion of mortar specimens exposed to sulfate.

TABLE 10.4
Compound Name, Approximate Chemical Formula, Abbreviated Name, and Range for the Compound Composition for Typical ASTM 150 Cements

Compound Name	Approximate Chemical Formula	Abbreviation	Potential Compound Composition Range (%)
Alite	$3CaO.SiO_2$	C_3S	45–55
Belite	$2CaO.SiO_2$	C_2S	15–30
Tricalcium aluminate	$3CaO.Al_2O_3$	C_3A	4–10
Tetracalcium aluminoferrite	$4CaO.Al_2O_3.Fe_2O_3. Fe_2O_3$	C_4AF	8–15

Note: Composition balance is made up of alkali sulfates and minor impurities. Free lime = 1.0% CaO; the balance is due to small amounts of oxides of titanium, manganese, phosphorus, and chromium.

TABLE 10.5
Types of Cements

Type I	Normal
Type IA	Normal, air entraining
Type II	Moderate sulfate resistance
Type IIA	Moderate sulfate resistance—air entraining
Type III	High early strength
Type IIIA	High early strength—air entraining
Type IV	Low heat of hydration
Type V	High sulfate resistance

10.4   WATER

Potable water without any pronounced taste or odor is suitable for making PCC. If the water is not potable, it can also be used if the 7-day compressive strength of mortar cubes with this water is at least 90% of the 7-day compressive strength of mortar cubes made with distilled water (ASTM C-109 and AASHTO T106). The suitability of the water can also be checked using tests for setting time of the cement according to ASTM C-191 and AASHTO T131.

10.5   HYDRATION

Cement reacts with water to form a hydration product that is the basis for the "glue" that gives hydrated Portland cement its bonding characteristics. This is a chemical reaction that involves four main cement compounds: tricalcium silicate, dicalcium silicate, tricalcium aluminate, and tetracalcium aluminate ferrite.

When water is added to cement and mixed together, the reaction is exothermic and heat is generated. Monitoring the rate of heat generation produced is a good indicator of the rate of reaction of Portland cement compounds. An illustrative example of the heat evolution curve produced by cement using a calorimeter (ASTM C-186) is shown in Figure 10.2. It shows the stages of heat evolution during Portland cement hydration as determined by a calorimeter.

As water is added to cement, some of the clinker sulfates and gypsum dissolve, producing an alkaline and sulfate-rich solution. The C_3A phase reacts vigorously with water to form an aluminate-rich gel (Stage I in Figure 10.2). The gel reacts with the sulfates in solution to form small rod-like crystals of ettringite (see Figure 10.3). The hydration of C_3A is a strongly exothermic reaction

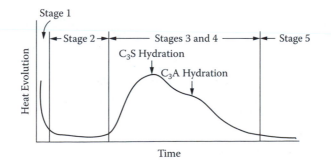

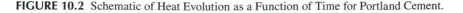

FIGURE 10.2 Schematic of Heat Evolution as a Function of Time for Portland Cement.

TABLE 10.6
Portland Cement Compound Hydration Reactions (oxide form)

$2\,C_3S + 11\,H = C_3\,S_2\,H_8 + 3\,CH$
Tricalcium silicate + water = calcium silicate hydrate (CSH) + calcium hydroxide

$2C_2S + 9\,H = C_3\,S_2\,H_8 + CH$
Dicalcium silicate + water = calcium silicate hydrate (CSH) + calcium hydroxide

$C_3A + 3C\,\bar{S}_3\,H_2 + 26\,H = C_6A\bar{S}_3\,H_{32}$
Tricalcium aluminate + gypsum + water = Ettringite

$2C_3A + C_6A\bar{S}_3\,H_{32} + 4\,H = 3C_4A\bar{S}_3\,H_{12}$
Tricalcium aluminate + Ettringite + water = Calcium monosulfoaluminate

$C_3A + CH + 12\,H = C_4A\,H_{13}$
Tricalcium aluminate + calcium hydroxide + water = Tetracalcium aluminate hydrate

$C_4AF + 10\,H + 2CH = C_6AF\,H_{12}$

but only lasts a few minutes (approximately 7 minutes) and is followed by a period of a few hours of relatively low heat evolution. This is called the dormant, or induction, period (Stage II). The first part of the dormant period corresponds to when concrete can be placed, and it lasts a couple of hours. As the dormant period progresses, the paste becomes stiff and unworkable. This period is related to initial set in concrete.

At the end of the dormant period, the alite (C_3S) and belite (C_2S) compounds start to hydrate, with the formation of calcium silicate hydrate (CSH) and calcium hydroxide (CH). This corresponds to the main period of cement hydration (Stage III), rate of strength increase, and final set for concrete. The cement grains react from the surface inwards as water diffuses through layers of formed hydration product, and the anhydrous particles decrease in size. In stage IV, C_3A hydration continues as well, as water diffuses through cement particles and newly exposed crystals become accessible to water. The rate of early strength gain is influenced in this period. In addition, conversion of ettringite crystals to monosulfate aluminate crystals is also occurring. This peak occurs between 12 and 90 hours. In stage V, the slow steady formation of hydration products establishes the rate of later strength. The ferrite compounds continue their slow hydration at this phase, although their contribution to strength gain is limited. The period of maximum heat evolution for Portland cement mixes occurs approximately between 12 and 20 hours after mixing, and then dissipates. For blended cements and the use of pozzolans such as fly ash, silica fume, and slag, the heat evolution periods may be different. Table 10.6 shows the hydration reactions.

Figure 10.3 shows the cement microstructure during hydration.

FIGURE 10.3 Scanning Electron Micrograph Showing Cement Hydration Products at Various Ages; Phases Include CSH, $Ca(OH)_2$, and Ettringite.

10.6 STEEL IN CONCRETE

Steel reinforcing bars (commonly known as *rebar*) are produced by pouring molten steel into casters and then running it through a series of stands in the mill. Most bars are deformed, which means that a raised pattern is rolled along the perimeter. The deformed ribs improve the bond between the steel and the concrete, and help transfer the loads between the two materials. Plain bars are also fabricated. Some are used for dowels, since at least one-half of a dowel bar should be smooth and lubricated to allow for free expansion and contraction. The specifications for concrete rebar are given in the following ASTM standards:

- A615/A615M-05a: Standard Specification for Deformed and Plain Billet-Steel Bars for Concrete Reinforcement (covers grades 40 and 60/soft metric grades 420 and 520)
- A616: Standard Specification for Rail-Steel Deformed Bars for Concrete Reinforcement (covers grades 50 and 60)
- A617: Standard Specification for Axle-Steel Deformed Bars for Concrete Reinforcement (covers grades 40 and 60)
- A706/A706M-96b: Standard Specification for Low-Alloy-Steel Deformed and Plain Bars for Concrete Reinforcement (grade 60 only)

The commonly used steel grades are given in Table 10.7. The allowable stress is usually given as two-thirds of the yield strength.

The size designations (Table 10.8) are the number of eighths of an inch in the diameter of a plain round bar having the same weight per foot as the deformed bar. So, for example, a No. 6 bar would have the same mass per foot as a plain bar 6/8 in. in diameter. The metric size is the same dimension expressed to the nearest millimeter.

Table 10.9 shows steel type, yield strength (40–75 ksi, 300–520 MPa), and ASTM standard for the different types of steels used for reinforcement.

Welded wire mesh is steel wire welded together to form a flat sheet with a square grid pattern. A common grid size is 150mm × 150mm (6" × 6"), and a common steel wire thickness is 4 mm (1/8"). Wire fabric or bar mats are used in PCC slabs to control temperature cracking. The stresses developed in the concrete slab during these volume changes are mainly due to friction that develops. However, these reinforcements do not increase the structural capacity of the slab. The wire fabric aids in holding the cracked concrete pavement together and allows for load transfer through aggregate interlock. Additionally, the reinforcement allows for greater spacing between joints.

TABLE 10.7
Commonly Used Grades of Steel

English Grade	International System of Units (SI) Grade	Minimum Yield Strength	
		psi	MPa
Grade 40	Grade 280	40,000	280
Grade 50	Grade 350	50,000	350
Grade 60	Grade 420	60,000	420
Grade 75	Grade 520	75,000	520

TABLE 10.8
Sizes and Dimensions

Bar Designation Number	Nominal Diameter (in.; does not include the deformations)	Metric Designation Number	Weight (lb/ft)
3	0.375	10	0.376
4	0.500	13	0.668
5	0.625	16	1.043
6	0.750	19	1.502
7	0.875	22	2.044
8	1.000	25	2.670
9	1.128	29	3.400
10	1.270	32	4.303
11	1.410	36	5.313
14	1.693	43	7.650
18	2.257	57	13.60

TABLE 10.9
Steel Type, Strength, and Applicable Standard

Mark	Meaning	Applicable ASTM Standard by Grade					
		40 and 50	60	75	300 and 350	420	520
S	Billet	A615	A615	A615	A615M	A615M	A615M
I	Rail	A616	A616	—	A996M	A996M	—
IR	Rail meeting supplementary requirements S1	A616	A616	—	—	—	—
A	Axle	A617	A617	—	A996M	A996M	—
W	Low alloy	—	A706	—	—	A706M	—

QUESTIONS

1. What is the importance of cement chemistry with respect to strength, durability, and setting time of concrete?
2. Why is gypsum added to cement clinker?
3. What are the different applications for the different types of cement?
4. Describe the hydration process.
5. Why is steel used in PCC pavements?

11 Distress and Performance

A brand-new pavement at the start of its design life is expected to be one without any "distress" or undesirable features. Such features, which include rutting, cracking, patching, or roughness, are "undesirable" from the point of view of performance of the pavement—the more distress, the shorter the pavement's life—and at some point, the distresses are so great in intensity (for example, deep ruts) as well as extent (for example, 75% of the wheelpath area in a project area has cracks) that the pavement is considered to be "failed" or at the end of its design life.

The following sections explain the possible distresses and their relationship to the performance of a pavement.

11.1 DISTRESSES IN ASPHALT PAVEMENTS

There are different forms of distresses in asphalt pavements, each tied to a specific reason (such as poor mix design, construction, or environmental conditions) or a combination thereof, and most happening as a result of traffic. Figure 11.1 lists the common distresses in asphalt pavements. The different distresses are described in the following paragraphs, in alphabetical order.

11.1.1 BLEEDING

Bleeding is the appearance of asphalt binder on the surface of the pavement. This is a surface defect caused by excessive asphalt binder in the surface asphalt mix layer. It is measured in square meters. The different conditions of bleeding include discoloration, covering of aggregate with a thin reflective surface, and loss of texture.

11.1.2 BLOCK CRACKING

Block cracking refers to a pattern of cracks that divide the surface into approximately rectangular pieces (> 0.1 m²). Such cracks occur due to shrinkage of asphalt mix because of volume changes in the base or subgrade. It is measured in square meters, and its severity can be described as low (crack width ≤ 6 mm, or sealed cracks whose width cannot be measured, with sealant in good condition), moderate (crack width > 6 mm but ≤ 19 mm, or any crack with mean width ≤ 19 mm and adjacent low-severity random cracking), and high (crack width > 19 mm or any crack with mean width ≤ 19 mm and adjacent moderate- to high-severity random cracking).

11.1.3 CORRUGATIONS

Corrugations are ripples formed laterally across an asphalt pavement surface. These occur as a result of lack of stability of the HMA at a location where traffic starts and stops or on hills where vehicles brake downgrade. The causes of the lack of stability are too much or too soft asphalt, a high sand content, and an excessive presence of smooth and rounded aggregate in the mix.

11.1.4 DELAMINATION

Delamination is the separation of the top wearing layer from the layer underneath. It is caused by poor bond or by failure of the bond between the two layers. The poor bond can be due to improper surface preparation or tack coat before the application of the wearing layer and/or a relatively thin

FIGURE 11.1 Common Distresses in Asphalt Pavements.
Courtesy: Ed Kearney.

wearing layer. The loss of bond can be caused by environmental factors such as ingress of water and repeated freeze-thaw cycles.

11.1.5 EDGE CRACKS

Edge cracks are found in pavements with unpaved shoulders. They are crescent shaped or fairly continuous cracks which intersect the pavement edge. The cracks are located within 0.6 m of the pavement edge adjacent to the shoulder. Longitudinal cracks outside of the wheelpath and within 0.6 m of the pavement edge are also classified as edge cracks.

The cracking is due to the lack of lateral (shoulder) support, base or subgrade weakness caused by frost action, and inadequate drainage. It is measured in meters, and severity is reported as low (cracks without breakup or loss of material), moderate (loss of material and breakup in up to 10% of the cracked area), or high (loss of material or breakup in more than 10% of the cracked area).

11.1.6 FATIGUE CRACKS AND EDGE FATIGUE CRACKS

Fatigue cracks generally begin at the wheelpath. Edge fatigue cracks are formed due to poor under-lying support at the edge of the pavements with paved shoulders. This distress, found on wheelpaths, begins as a series of interconnected cracks and develops into a chicken wire/alligator pattern. Inter-connected cracks give rise to many-sided, sharp-angle pieces, usually with a maximum length of 0.3 m. Repeated tensile stress/strain at the bottom of the asphalt mix layer, caused by traffic, leads to fatigue cracking. It starts from the bottom and moves upward. The cracked area must be quan-tifiable in order to be counted as a fatigue crack. The cracked area should be measured in square meters, for each of the three severity levels: low, medium, and severe. If, within an area, there are different severity levels and they cannot be distinguished, the entire area should be rated as for the highest severity rating. A *low*-severity area is one in which there are no or few connecting cracks, the cracks are not spalled or sealed, and there is no evidence of pumping. *Moderate*-severity areas are those in which the cracks are interconnected, forming a complete pattern, and cracks may be slightly spalled or sealed, with no evidence of pumping. In a *high*-severity area, the cracks are interconnected with moderate or severe spalls, forming a complete pattern; pieces may move under traffic; cracks may be sealed; and pumping may be evident.

11.1.7 LONGITUDINAL JOINT CRACKS

Longitudinal joint cracks develop at construction joints because of poorly paved joints or improper construction techniques. They are measured in meters, and their severity is expressed in the same manner as in block cracking.

11.1.8 POLISHED AGGREGATE

Polished aggregate is the exposure of coarse aggregate due to wearing away of the asphalt binder and fine aggregates from the surface asphalt mix. It is measured in square meters.

11.1.9 POTHOLES

Bowl-shaped holes, with a minimum plan dimension of 150 mm in the pavement surface, are called *potholes*. There are four main causes of potholes: (1) insufficient thickness of the pavement to sup-port traffic through winter-spring freeze-thaw cycles; (2) poor drainage, leading to accumulation of excess water; (3) failures at utility trenches and castings; and (4) paving defects and unsealed cracks. For a distress survey, the number of potholes and the square meters of affected area are recorded. Severity levels can be reported as low (< 25 mm deep), moderate (25 mm < depth < 50 mm), or high (> 50 mm deep).

11.1.10 Raveling

Raveling refers to wearing away of the pavement surface by loss of asphalt binder and displacement of aggregates. The process starts as loss of fines and can continue to a situation with loss of coarse aggregate and a very rough and pitted surface. Raveling is caused by the action of water that finds its way through the surface of the pavement because of poor compaction and hence low density and relatively high voids, and it can initiate the potholing process. It is measured in square meters.

11.1.11 Reflective Cracking

Reflective cracks are those in asphalt overlays caused by discontinuities in the pavement structure underneath. This can be due to old cracked asphalt pavement or joints in concrete pavement underneath. The severity levels are recorded in the same manner as in block cracking.

11.1.12 Rutting

A *rut* is defined as a longitudinal depression in the wheelpath, with or without transverse displacement. It can be measured with a straight edge or a profiler at regular intervals. A rut is a physical distortion of the surface, and it also prevents the cross drainage of water during rains, leading to accumulation of water in the ruts and increasing the potential of hydroplaning-related accidents. Generally a rut depth of 0.5 in. is considered a rutting failure. Rutting is the result of repeated loading, which causes accumulation, and increase of permanent deformations. Rutting can be (1) low- to moderate-severity rutting—one-dimensional densification or vertical compression near the center of the wheelpath, caused by densification of mixes with excessive air voids in the in-place mix under traffic; (2) moderate- to high-severity rutting—a depression in the wheelpath along with humps on either side of the depression, caused by lateral flow due to plastic deformation, resulting from shear failure of the mix under traffic, and generally associated with very low air voids in the mix; and (3) rutting accompanied by cracks on the surface of the pavement, caused by rutting in underlying layers, such as the subgrade or subbase.

11.1.13 Slippage Crack

Slippage cracks are typically crescent- or half-moon-shaped cracks produced when vehicles brake or turn, which causes the pavement surface to slide or push. This is caused by a low-strength HMA or a lack of bond between the surface and lower courses.

11.1.14 Thermal Cracks

Thermal cracking occurs in the form of transverse cracking, which is defined as cracks that are predominantly perpendicular to the pavement centerline. These cracks occur at regular intervals. Thermal cracks can be caused by the fracture of asphalt mix due to a severe drop in temperature or by thermal fatigue caused by repeated low- and high-temperature cycles. The severity of thermal cracking is measured by the width of the crack. A low-severity transverse crack is one with a mean width of ≤ 6 mm, or a sealed crack with sealant material in good condition and a width that cannot be determined. A medium-severity transverse crack is one with a mean width > 6 mm and ≤ 19 mm, or any crack with a mean width ≤ 19 mm and adjacent low-severity random cracking, while a high-severity crack has a mean width > 19 mm, or ≤ 19 mm but with adjacent moderate- to high-severity random cracking. Only cracks that are > 0.3 m in length are counted. The number and length of transverse cracks at each severity level are recorded. The entire transverse crack is to be rated at the highest severity level that is present for at least 10% of the total length of the crack. The length, in m, is the total length of the crack. If the cracks are sealed, then the length of cracks with sealant in good condition (for at least 90% of the crack) should be measured at each severity level. Part of the thermal transverse crack extending into a load-induced fatigue crack area is not counted.

11.2 DISTRESSES IN CONCRETE PAVEMENTS

The different types of distresses found in PCC pavements are described below. Figure 11.2 summarizes the different distresses.

11.2.1 CORNER BREAKS

A corner portion of the slab is separated by a crack. The crack intersects the adjacent transverse and longitudinal joints, and is approximately at a 45° angle with the direction of traffic. The length of the sides is from 0.3 m to one-half the width of the slab on each side of the corner. Corner breaks are measured by counting the number that is encountered per unit length. For a low severity level, the crack is not spalled more than 10% of the length of the crack, and there is no faulting. For a moderate severity level, the crack is spalled > 10% of its total length, or the faulting of the crack or joint

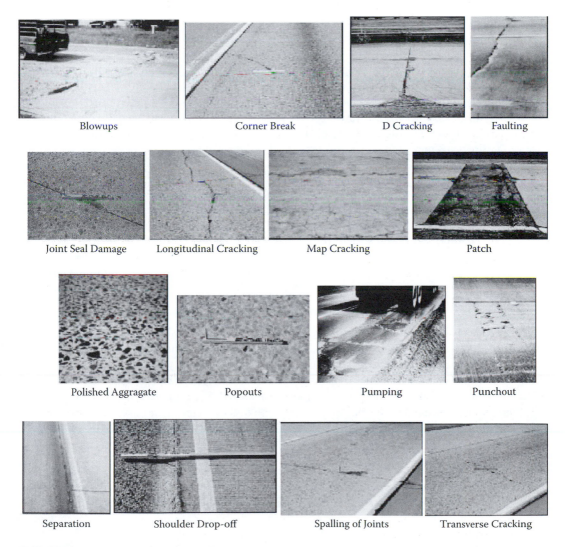

FIGURE 11.2 Distress in PCC Pavements.
Source: From Federal Highway Administration, *Distress Identification Manual for the Long Term Pavement Performance Program, FHWA-RD-03-031*, June (Washington, DC: U.S. Department of Transportation, 2003).

is < 13 mm and the corner piece is not broken into two or more pieces. For a high severity level, the crack is spalled at moderate to high severity > 10% of its total length, or the faulting of the crack or joint is ≥ 13 mm, or the corner piece is broken into two or more pieces or contains patch material.

11.2.2 DURABILITY CRACKING (OR "D" CRACKING)

In this type of distress, the cracking is due to freezing and thawing and usually initiates at construction joints. These are usually closely spaced crescent-shaped hairline cracks that appear like the letter *D*. D-cracks are measured by recording the number of slabs with "D" cracking and square meters of area affected at each severity level. The slab and affected area severity rating is based on the highest severity level present for at least 10% of the area affected. For low-severity distress, D-cracks are close together, with no loose or missing concrete pieces, and no patches in the affected area. For moderate severity, D-cracks are well defined, and some small concrete pieces are loose or have been displaced. For high severity, D cracking has a well-developed pattern, with a significant amount of loose or missing material, and may have been patched.

11.2.3 LONGITUDINAL CRACKING

These are stress cracks that are predominantly parallel to the pavement centerline and are measured in meters. For low-severity distress, crack widths are < 3 mm in width, with no spalling and no measurable faulting, or are well sealed and with a width that cannot be determined. For moderate-severity distress, crack widths range between 3 mm and 13 mm, or with spalling < 75 mm, or faulting up to 13 mm. For high-severity distress, crack widths are < 13 mm; or with spalling < 75 mm or faulting < 13 mm.

11.2.4 TRANSVERSE CRACKING

These are stress cracks that are predominantly perpendicular to the pavement centerline and are measured in meters. For low-severity distress, crack widths are < 3 mm, with no spalling and no measurable faulting, or are well sealed similar to longitudinal cracking. For moderate-severity distress, crack widths are > 3 mm and < 6 mm, or with spalling < 75 mm, or faulting up to 6 mm. For high-severity distress, crack widths are > 6 mm, or with spalling >75 mm, or faulting > 6 mm.

11.2.5 SPALLING OF TRANSVERSE JOINTS

This distress consists of cracking, breaking, chipping, or fraying of slab edges within 0.3 m from the face of the transverse joint, and is measured in frequency of occurrence and meters. For low-severity distress, spalls are < 75 mm wide, measured to the face of the joint, with loss of material, or spalls with no loss of material and no patching. For medium-severity distress, spalls are 75–150 mm wide, measured to the face of the joint, with loss of material. For high-severity distress, spalls are > 150 mm wide, measured to the face of the joint, with loss of material, or broken into two or more pieces, or contain patch material.

11.2.6 MAP CRACKING AND SCALING

Map cracking: A series of interconnected cracks that extend into the upper surface of the slab. Usually, larger cracks are oriented in the longitudinal direction of the slab and are interconnected by finer transverse or random cracks. These are measured in frequency of occurrence and in square meters.

Scaling: Scaling is the deterioration and flaking of the upper concrete slab surface, normally in areas of 3 to 13 mm, and may occur anywhere over the pavement. This is measured by the number of occurrences and the square meters of affected area.

11.2.7 Polished Aggregate

Exposed and polished coarse aggregate occur due to surface mortar and paste loss which results in a significant reduction of surface friction. Distress is measured in square meters of affected surface area. The level of severity is not applicable for this distress. Diamond grinding also removes surface texturing, but this condition should not be recorded as polished aggregate.

11.2.8 Popouts

Popouts are small pieces of pavement broken loose from the surface, ranging in diameter from 25 to 100 mm, and ranging in depth from 13 to 50 mm.

11.2.9 Blowups

Blowups are slab length changes resulting in localized upward movement of the pavement surface at transverse joints or face cracks, and are usually accompanied by shattering of the concrete in that area. Distress is measured by recording the number of blowups. Severity levels are not applicable. However, severity levels can be defined by the relative effect of a blowup on ride quality and safety.

11.2.10 Faulting of Transverse Joints and Cracks

This occurs as a result of elevation difference across a joint or crack. Distress is measured in millimeters, to the nearest millimeter at a location 0.3 m and 0.75 m from the outside slab edge (this is approximately the location of the outer wheelpath). For a widened lane, the wheelpath location will be 0.75 m from the outside lane edge stripe. Faulting is recorded as positive (+) if the "approach" slab is higher than the "departure" slab; if the approach slab is lower, faulting is recorded as negative (–). Faulting on PCC pavements can be measured using a FHWA-modified Georgia Faultmeter.

11.2.11 Lane-to-Shoulder Dropoff

This is a difference in elevation between the edge of the slab and the outside shoulder, and usually occurs when the outside shoulder settles. Distress is measured at the longitudinal construction joint between the lane edge and shoulder. Distress is recorded to the nearest millimeter at 15–25-m intervals. The recorded value is negative (–) if the traveled surface is lower than the shoulder.

11.2.12 Lane-to-Shoulder Separation

This is a widening of the joint between the edge of the slab and the shoulder. Distress is measured to the nearest millimeter at intervals of 15 to 25 m along the lane-to-shoulder joint. It should be documented if the joint is well sealed or not at each shoulder location. Severity levels are not applicable.

11.2.13 Patch/ Patch Deterioration

A patch is a portion greater than 0.1 m^2, or all of the original concrete slab that has been removed and replaced, or additional material applied to the pavement after original construction. For a low-severity-level distress, the patch displays low severity distress of any type, there is no measurable faulting or settlement, and pumping is not evident. For a moderate severity level, the patch has moderate severity distress of any type or faulting or settlement up to 6 mm, and pumping is not evident. For a high severity level, the patch has a high-severity distress of any type, or faulting or settlement > 6 mm, and pumping may be evident.

11.2.14 Water Bleeding and Pumping

This is the ejection or seepage of water from beneath the pavement through cracks. The residue of very fine materials deposited on the surface which were eroded (or pumped) from the support layers, causing staining, aids in recognizing pumping phenomenon. This is measured by recording the number of occurrences of water bleeding and pumping and the length in meters of affected pavement with a minimum length of 1 m. The combined length of water bleeding and pumping cannot exceed the length of the test section.

11.2.15 PUNCHOUTS

Punchouts are broken areas enclosed by two closely spaced (commonly < 0.6 m) transverse cracks, a short longitudinal crack, and the edge of the pavement or a longitudinal joint. They also include "Y" cracks that exhibit spalling, breakup, or faulting. The number of punchouts should be recorded at each severity level. It should be noted that the cracks which outline the punchout are also recorded under "longitudinal cracking" and "transverse cracking." Punchouts that have been completely repaired by removing all broken pieces and replacing them with patching material (rigid or flexible) are rated as patches. For low-severity distress, transverse cracks are tight and may have spalling < 75 mm or faulting < 6 mm, with no loss of material and no patching. For moderate distress, spalling > 75 mm and < 150 mm or faulting > 6 mm and < 13 mm exists. For a high severity rating, spalling is > 150 mm, or concrete within the punchout is punched down by > 13 mm, is loose and moves under traffic, is broken into two or more pieces, or contains patch material.

11.2.16 JOINT SEAL DAMAGE

Joint seal damage is any condition that enables incompressible materials or a significant amount of water to infiltrate the joint from the surface. Common types of joint seal damage are extrusion, hardening, adhesive failure (bonding), cohesive failure (splitting), complete loss of sealant, and the intrusion of foreign material in the joint such as weed growth in the joint. This distress is measured by recording the number of longitudinal joints that are sealed and the length of sealed longitudinal joints with joint seal. The severity level is not applicable.

11.3 CONSIDERATION OF PERFORMANCE

A pavement is constructed for the safe and smooth passage of traffic. There is a need to quantify the extent to which the pavement is serving its purpose, or its "performance." Such quantification can be done with respect to distress conditions exhibited by the pavement at any time after its construction.

The AASHO Road Test introduced the concept of serviceability to measure performance. It was necessary to characterize the pavement sections in terms of their condition, in order to develop relationships between the performance and the factors affecting the performance. First, the concept of present serviceability rating (PSR) was introduced. PSR (Figure 11.3) is defined as the judgment of an observer as to the current ability of a pavement to serve the traffic it is meant to serve.

The PSR ranges from 1 to 5, starting from a perfect 5 and decreasing with the passage of traffic. To characterize pavements in terms of the serviceability rating, the Pavement Serviceability Index (PSI) was then introduced on the basis of observed distress conditions.

$$PSI = 5.03 - 1.91\log(1 + \overline{SV}) - 1.38RD^2 - 0.01\sqrt{C + P} \text{ (flexible pavement)}$$

$$PSI = 5.41 - 1.80\log(1 + \overline{SV}) - 0.09\sqrt{C + P} \text{ (rigid pavement)}$$

where:
   SV = mean of the slope variance in the two wheelpaths (measured with the CHLOE profilometer or Bureau of Public Roads [BPR] roughometer).
   C, P = measures of cracking and patching in the pavement surface.
      C = total linear feet of Class 3 and Class 4 cracks per 1000 ft² of pavement area. A Class 3 crack is defined as opened or spalled (at the surface) to a width of 0.25 in. or more over a

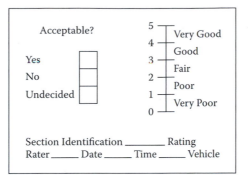

FIGURE 11.3 Concept of Present Serviceability Rating.

distance equal to at least one-half the crack length. A Class 4 crack is defined as any crack which has been sealed.

RD = average rut depth, in.

P = expressed in terms of ft² per 1000 ft² of pavement surfacing.

The basic idea was that just like PSR (numerically the same value as PSI), PSI would drop over time (that is, with the passage of traffic), starting from an initial p_0 to a terminal p_t. The curve can be expressed with an equation as follows (Figure 11.4):

$$p_o - p = (p_0 - p_t)\left(\frac{W}{\rho}\right)^{\beta}$$

where:

β and ρ depend on pavement structure (thickness and stiffness) and loading

β determines the shape of the graph

ρ is the number of loads at which $p = 1.5$

W = cumulative load

Note that performance at any time can be expressed as the area under the serviceability curve from the beginning to the point under consideration. The serviceability of the pavement at any time can be "raised" by rehabilitation, such as an asphalt mix overlay.

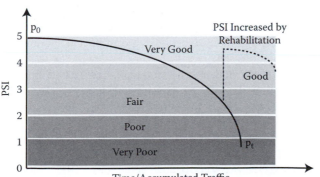

FIGURE 11.4 Plot of Pavement Serviceability Index (PSI) versus Time.

11.4 DAMAGE

In the AASHTO design process, the loss of serviceability (ΔPSI) is referred to as *damage*. This damage is considered to be caused by traffic, environmental conditions, and age. In mechanistic-empirical pavement design processes, the performance of the pavement is expressed in terms of distresses. The distresses are predicted on the basis of models (commonly referred to as *transfer functions*) relating mechanistic response (stress/strain) to observed distress. Note that such models need to be calibrated for specific regions, climate, materials, and construction conditions. In more sophisticated methods, a damage index is calculated from a mechanistic response, and the damage index is used to compute accumulation of distress with time. Damage is defined as the deterioration of the pavement due to the action of traffic over different environmental conditions. Such deterioration happens as a result of change in the engineering properties of the pavement layer materials.

This damage does not happen altogether at the same time, but rather progressively—or, more precisely, in increments—with the passage of every vehicle or, if expressed in time, every hour during its service. Note that the damage at every hour is not the same—it can be higher, for example, due to a heavier vehicle moving over it at that hour, or due to a lowering of the modulus of the asphalt mix layer due to a high temperature that hour. Therefore, the most rational approach is to consider and compute the damage in each and every increment by considering the relevant traffic (class of vehicle) and the pavement material properties (with respect to environmental conditions such as temperature) for that increment period (say, 1 hour). The damage at any increment can be expressed as follows:

$$D = \frac{n}{N}$$

where n is the calculated load applications and N is the allowable number of load applications. Note that the allowable number of loads is dependent on the condition of the pavement layer at any increment period—hence, N is different for different increment periods.

The total damage at any point is computed by summing up all of the damages over time, up to that time, as follows (commonly referred to as *Miner's hypothesis*; Miner, 1945):

$$\text{Total Damage} = \sum_{i=1}^{n} \sum_{j=1}^{m} \sum_{k=1}^{o} \frac{n_{ijk}}{N_{ijk}}$$

where i, j, and k are the different categories over which the summation of the damage is made. Such categories can include different time increments, traffic wander, and truck class. The basic idea is that for each and every time increment (that is, for a unique set of material properties tied to a unique set of environmental conditions), the damage is determined for each and every type and movement of the vehicle. The more categories used in the calculation of increments, the more precise is the damage calculation.

This total damage over time can then be related to distress, and then finally the distress can be expressed as a function of time. For any specific type of distress, when total damage (sometimes referred to as the *cumulative damage factor*, or CDF) reaches 1, the pavement is said to have failed (say, by cracking).

In a simpler approach, instead of considering the damage in each and every increment individually, effective or weighted values of each variable could be used to determine the total damage. For example, the concept of effective soil (roadbed/subgrade) resilient modulus is recommended in the AASHTO design procedure. The method consists of the laboratory or in-place determination (or prediction through the correlation of other properties) of resilient modulus of the soil for different moisture conditions (wet/dry) that are expected in the different moisture seasons. Then an effective resilient modulus value is calculated which has the same (equivalent) effect as the combined effect of the different moduli under different moisture conditions. This effective modulus is calculated on the basis of relative damage due to variation of the soil modulus in different seasons. The procedure is illustrated with an example in Figure 11.5.

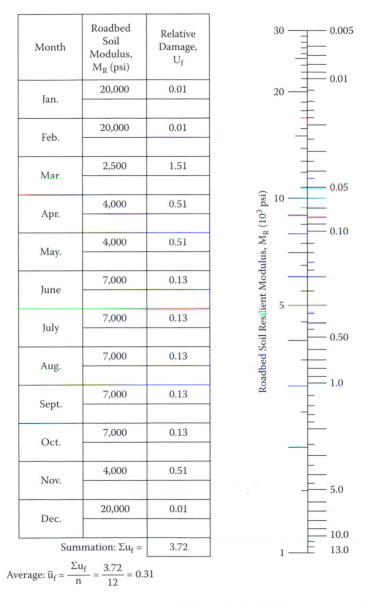

Month	Roadbed Soil Modulus, M_R (psi)	Relative Damage, U_f
Jan.	20,000	0.01
Feb.	20,000	0.01
Mar.	2,500	1.51
Apr.	4,000	0.51
May.	4,000	0.51
June	7,000	0.13
July	7,000	0.13
Aug.	7,000	0.13
Sept.	7,000	0.13
Oct.	7,000	0.13
Nov.	4,000	0.51
Dec.	20,000	0.01
Summation: $\Sigma u_f =$		3.72

Average: $\bar{u}_f = \dfrac{\Sigma u_f}{n} = \dfrac{3.72}{12} = 0.31$

Effective Roadbed Soil Resilient Modulus, M_R (psi) = <u>5,000</u> (corresponds to $\bar{u}_f$)

FIGURE 11.5 Calculation of Effective Soil Modulus Based on Relative Damage.
Source: From AASHTO Guide for Design of Pavement Structures © 1993, by the American Association of State Highway and Transportation Officials, Washington, D.C. Used by permission.

QUESTIONS

1. List the distresses for asphalt and concrete pavements, along with their causes.
2. From a review of information from your local newspapers, county, and/or municipality, find out the most common pavement distress in your locality. Can you suggest some corrective actions?
3. Determine the effective resilient modulus of a soil if the moduli at different months are as follows:

Month	Modulus (psi)
January	18,000
February	18,000
March	3000
April	5000
May	5000
June	8000
July	8000
August	8000
September	8000
October	8000
November	8000
December	10,000

12 Consideration of Major Distress Mechanisms and Material Characterization for Asphalt Pavements

The different types of distresses have been presented in Chapter 11. The major distresses are considered in the mix and structural design of asphalt mixes. This consideration consists of three basic parts: explaining and formulating the problem, conducting appropriate tests to determine the relevant material properties (characterization), and using models to relate the amount/extent of distress to the relevant material properties, and hence predict the distress. All of these three parts are presented in the following sections.

12.1 FATIGUE CRACKING

The mechanism of fatigue cracking can be divided into two parts: (1) the occurrence of tensile stress/tensile strain in the asphalt mix layer, and (2) the repetitive occurrence of such tensile stress/strain under traffic repetitions. The occurrence of tensile stress/strain results from the elastic behavior of the pavement structure under traffic loading, whereas the initiation and formation of cracks result from "fatigue" behavior of the layer. The high tensile stresses/strains responsible for causing fatigue cracks can occur because of a relatively thin or weak asphalt mix layer, very high load or tire pressure, or a relatively weak subgrade and/or base and/or subbase, due to moisture or low density due to inadequate compaction.

If a beam is subjected to loading from above then directly under the load, the beam would tend to assume a convex downward shape, with compressive stresses/strains on the top part and tensile stress/strain in the bottom part. Due to the elastic nature of the pavement, in such a situation, it springs back to its original position, without the stresses/strains, when the load is removed. However, under the next load, it undergoes the same cycle. Under the heaviest possible load, if the pavement structure is such that the tensile stress/strain (which is responsible for causing cracks) is very small, then the pavement can sustain many such cycles/repetitions without any damage. If the tensile stress/strain is moderate, then it can sustain a certain number of repetitions without damage, and if the tensile stress/strain is very high, then it can sustain a relatively small number of repetitions before permanent damage.

There are two analysis steps that are necessary to predict how many repetitions the pavement can sustain before it gets damaged. Assuming a standard (high load), one can determine the tensile stress/strain at the bottom of the asphalt mix layer of an assumed pavement structure (if there is tensile stress/strain in the layer, then it is the maximum at the bottom), using layered elastic analysis, as explained in Chapter 2. Then, conducting an experiment to determine how many repetitions the pavement can sustain before there is damage, one can determine how many repetitions it can sustain for this specific tensile stress/strain. Now let us get into the details of these steps.

If one assumes a pavement structure, the elastic moduli, thickness, and Poisson's ratio of the different layers are needed to determine the tensile stress/strain under a specific load. These properties can be used from experience. The load can be considered as the highest allowable load—say,

9000 lb—by considering a maximum allowable load of 18,000 lb per axle. The Poisson's ratio of the generally used pavement materials does not differ greatly, and a value of 0.35–0.4 can be assumed.

Now if the pavement structure is such that the asphalt mix layer on the top is very thin, one may determine that there is no tensile stress/strain in the layer, and all through its depth there is only compressive stress/strain. In that case, there is no potential of forming fatigue cracks, as explained here. There can be situations between two extremes: (1) the asphalt mix layer is such that it shows a very high level of tensile stress/strain at the bottom, and (2) the asphalt mix layer and the pavement structure are such that there is a very small amount of tensile stress/strain at the bottom. Note that whether or not there will be a high tensile stress/strain is dictated by the thickness and stiffness of the asphalt mix layer as well as the stiffness of the layers underneath (and supporting) the asphalt mix layer.

Next comes the experimentation part to determine how many repetitions of the tensile stress/strain the pavement can sustain before there is damage. This needs the development of a model, based on statistical analysis of experimental data, of the following form:

$$N_f = f(\sigma \text{ or } \varepsilon)$$

where N_f = number of repetitions to failure, and $f(\sigma \text{ or } \varepsilon)$ indicates "function" of initial stress or strain. From this relationship, one can determine an N_f for a specific σ or ε as determined with the layered elastic analysis. Before proceeding any further, note that N_f is a function of the tensile (σ or ε) for a specific material or, more precisely, for a material with a specific modulus (or stiffness). A simple analogy to illustrate this concept is that a stiffer paper clip can be bent fewer times before it snaps, compared to a more flexible paper clip, provided they are both "bent" (or strained) to the same extent. Indeed, a more complete way to express N_f would be to say,

$$N_f = f(\sigma \text{ or } \varepsilon \text{ and } E)$$

where E is the dynamic modulus.

Note that the same asphalt mix has different E at different temperatures. So a test run with the same mix, but at different temperatures, would yield the required data for constructing the N_f model. How does one arrive at the above model?

Suppose one conducts an experiment with an asphalt mix beam over four supports and loaded at two points in the middle, commonly referred to as the *third-point loading test*. This loading configuration results in a uniform bending moment over the middle portion of the beam. The loading is applied repetitively, to simulate the repeated loading of traffic. Note that the intent of this testing is to cause cracking failure in the beam through initiation and propagation of the cracks by loading. The test can be conducted in two ways. Either a load is applied such that the stress in the beam remains constant from the beginning, or the load is varied such that the strain in the beam remains constant. If a constant stress is used, as the beam gets damaged, the strain will keep on increasing, while in the constant strain model, as the beam gets damaged, the load will need to be decreased to keep the strain at the same level.

Whether to use constant stress or constant strain depends on the thickness of the asphalt mix layer that is being considered for design. For thicker layers, strains would quickly increase with a decrease in the stiffness of the asphalt mix layer, and the constant stress mode is applicable; whereas for thinner layers, where strains would remain constant and are primarily dictated by the stiffness of the layers underneath the asphalt layer, constant strain is applicable. In a constant stress test in the laboratory, the strain will increase and the beam would fail relatively quickly after the initiation of the crack. In a controlled strain test, the time for crack propagation is long, and failure would take a relatively long time. To end the test, sometimes "failure" is defined as the point where the modulus of the mix has decreased to one-half of its original modulus. For asphalt pavement layers less than 6 inches in thickness, typically constant strain is generally used.

Center-point loading is used for tests with a rubber pad underneath the beam to simulate an elastic foundation, while third-point loading (most commonly, AASHTO T321, Method for Determining the Fatigue Life of Compacted Hot Mix Asphalt [HMA], Subjected to Repeated Flexural Bending) is generally used in tests without an elastic foundation. Note that other tests include the indirect tension test, where a compressive load is used on the cylindrical specimen to develop tensile stresses along the diametral axis; tests with either rectangular or trapezoidal section cantilever beams; and the direct tension test of rectangular beams.

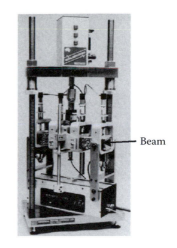

Beam

Typically, 12–15-inch long and 2–3-inch wide and deep asphalt mix beam samples are used in this test, where the load is applied at the third points in the form of a haversine wave, with 0.1 s duration and then 0.4 s rest (Figure 12.1). Tests are run to span a range of stress such that the beams fail within a range of a small (1,000) to large (say, 1,000,000) number of repetitions.

FIGURE 12.1 Fatigue Tests of Asphalt Mix Beam.

Each generates one data point, consisting of a strain and the number of repetitions to failure.

The common form of equation used to express N_f as a function of the initial stress or strain and modulus is

$$N_f = K_1(\varepsilon_t)K_2(E)K_3$$

where K_1, K_2, and K_3 are regression constants.

Maximum tensile stress and strain are calculated as follows:

$$\sigma_t = (0.357\ P)/(bh^2)$$

$$\varepsilon_t = (12\delta h)/(3L^2 - 4a^2)$$

where,

σ_t = maximum tensile stress (Pa)
P = load applied (N)
b = width of beam (m)
h = thickness of beam (m)
ε_t = maximum tensile strain (m/m)
δ = maximum deflection at center of beam (m)
a = space between inside clamps in the beam (0.119 m)
L = length of beam between outside clamps (0.357 m)

12.1.1 MATERIAL CHARACTERIZATION TESTS

Before moving on to the various fatigue cracking models, it is appropriate to discuss the three important tests that are run to characterize HMA.

12.1.1.1 Indirect Tensile Strength: Test Method

In the indirect tensile test (ASTM D-4123, SHRP Protocol P07), a load is applied diametrically to a cylindrical sample of HMA until it fails. The load is applied at a deformation rate of 2 inches per minute (Figure 12.2).

Indirect tensile strength (or ITS) = $\dfrac{2P}{\pi dt}$, P = load (lb – f), d = diameter (in.), t = thickness (in.)

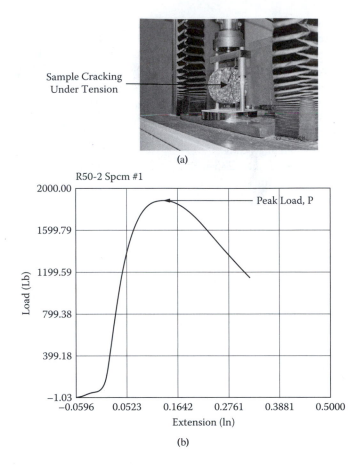

(a)

(b)

FIGURE 12.2 Indirect Tensile Strength Test and Results.

For the example shown in Figure 12.2, the peak load is 1880 lb for a 4-inch diameter sample with a thickness of 2 inches.

$$\text{ITS} = \frac{2 * 1,880}{\pi * 4 * 2} = 149.6 \text{ psi}$$

12.1.1.2   Resilient Modulus

In resilient modulus testing (ASTM D-4123, SHRP Protocol P07), a haversine loading is applied to a sample to determine the resulting deformation, and hence calculate resilient modulus, which is defined as the ratio of stress and resilient strain (as opposed to viscous strain) in an asphalt mix sample. Generally, for asphalt mix samples, the test is carried out in indirect tensile mode, as shown in Figure 12.3. The haversine load utilized in the protocol has a period of 0.1 s, followed by an appropriate rest period, generally 0.9 s. The tensile strength of each replicated set is determined prior to testing by performing an indirect tensile test on a companion specimen. The magnitude of the applied load causes tensile stress levels within the specimen equivalent to 30, 15, and 5% of the tensile strength at 25°C, at 5°C, 25°C, and 40°C, respectively; and the seating load is 3, 1.5, and 0.5% (10% of the applied load) of the specimen tensile strength measured at 25°C, at each of the three test temperatures, respectively. The sample is placed under a diametrical repeated loading equipment with linear variable displacement transducers (LVDTs) placed on both sides of the horizontal diameter to measure horizontal deformations. The loading is applied on top through a

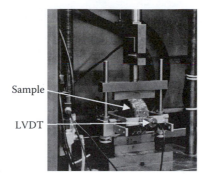

Sample

LVDT

FIGURE 12.3 Resilient Modulus Testing of Asphalt Mix Sample.

horizontal metal bar, and the sample is rested at the bottom on a similar bar. The horizontal bar provides the distributed load along the edge of the sample. An example is shown in Figure 12.4.

12.1.1.3   Dynamic Modulus (AASHTO TP62-03)

Dynamic modulus (E*) is defined as the absolute value of the complex modulus calculated by dividing the maximum (peak-to-peak) stress by the recoverable (peak-to-peak) axial strain for a material subjected to sinusoidal (repeated vertical) loading (Figure 12.5).

The dynamic modulus is the absolute value of the complex modulus

$$|E*| = \sqrt{\left(\frac{\sigma_0 \cos\phi}{\varepsilon_0}\right)^2 + \left(\frac{\sigma_0 \sin\varphi}{\varepsilon_0}\right)^2} = \frac{\sigma_0}{\varepsilon_0}$$

where σ_o and ε_o are the stress and strain amplitudes, respectively. An example is shown in Figure 12.6.

The dynamic modulus of asphalt mixes can also be determined from an empirical equation that relates it to other properties of the mix as well as temperature and loading times. The following is the widely used equation, which has been modified a number of times, and was originally developed by Fonseca and Witczak (1996).

$$\log E* = 3.750063 + 0.02932 * (p_{200}) - 0.001767 * (p_{200})^2 - 0.002841 * (p_4) - 0.058097 * (V_a)$$

$$- 0.802208 \left(\frac{V_{beff}}{V_{beff} + V_a}\right)$$

$$+ \frac{3.871977 - 0.0021 * (p_4) + 0.003958 * (p_{38}) - 0.000017 * (p_{38})^2 + 0.005470 * p_{34})}{1 + e^{(-0.603313 - 0.313351 \log(f) - 0.393532 \log(\eta))}}$$

where:

    $E*$ = dynamic modulus
  p_{200} = percentage passing the No. 200 sieve
    p_4 = percentage retained on the No. 4 sieve
    V_a = percentage of air voids
 V_{beff} = effective binder content by volume
  p_{38} = percentage retained on 3/8-inch sieve
  p_{34} = percentage retained on 3/4-inch sieve
    f = loading frequency
    η = binder viscosity

Source of equation: p. 2.2.11, NCHRP, 2004.

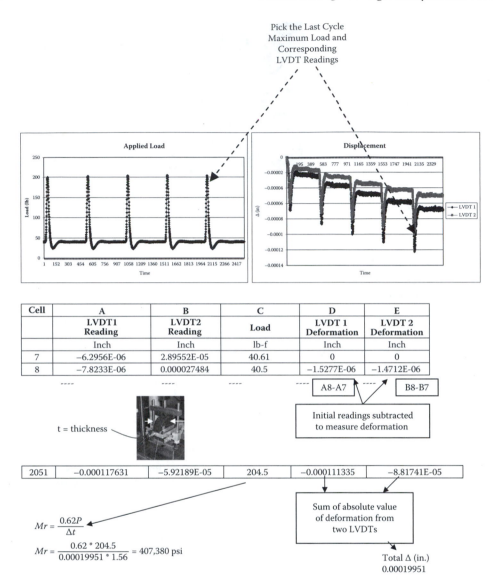

Cell	A	B	C	D	E
	LVDT1 Reading	LVDT2 Reading	Load	LVDT 1 Deformation	LVDT 2 Deformation
	Inch	Inch	lb-f	Inch	Inch
7	−6.2956E-06	2.89552E-05	40.61	0	0
8	−7.8233E-06	0.000027484	40.5	−1.5277E-06	−1.4712E-06
----	----	----	----	----	----

A8-A7   B8-B7

Initial readings subtracted to measure deformation

t = thickness

| 2051 | −0.000117631 | −5.92189E-05 | 204.5 | −0.000111335 | −8.81741E-05 |

Sum of absolute value of deformation from two LVDTs

$$Mr = \frac{0.62P}{\Delta t}$$

$$Mr = \frac{0.62 * 204.5}{0.00019951 * 1.56} = 407,380 \text{ psi}$$

Total Δ (in.)
0.00019951

FIGURE 12.4 Example of Resilient Modulus Calculation.

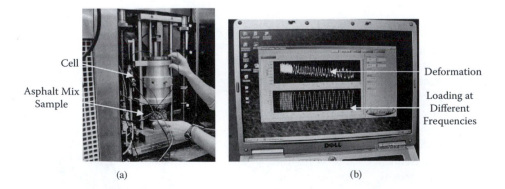

(a)   (b)

FIGURE 12.5 Dynamic Modulus Testing Setup and Results.

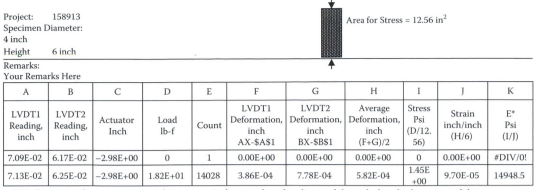

Project:　　158913
Specimen Diameter:
4 inch
Height　　　6 inch

Area for Stress = 12.56 in^2

Remarks:
Your Remarks Here

A	B	C	D	E	F	G	H	I	J	K
LVDT1 Reading, inch	LVDT2 Reading, inch	Actuator Inch	Load lb-f	Count	LVDT1 Deformation, inch AX-A1	LVDT2 Deformation, inch BX-B1	Average Deformation, inch (F+G)/2	Stress Psi (D/12.56)	Strain inch/inch (H/6)	E* Psi (I/J)
7.09E-02	6.17E-02	−2.98E+00	0	1	0.00E+00	0.00E+00	0.00E+00	0	0.00E+00	#DIV/0!
7.13E-02	6.25E-02	−2.98E+00	1.82E+01	14028	3.86E-04	7.78E-04	5.82E-04	1.45E+00	9.70E-05	14948.5

NOTE: Determine the average stress and average strain for a number of readings and then calculate the dynamic modulus

FIGURE 12.6 Example Calculation of Dynamic Modulus.

Generally a (log) T (in Rankine) versus log log viscosity plot is generated from asphalt test data obtained at a few temperatures, and then A (intercept) and VTS (slope) of the plot are determined to develop a regression equation relating viscosity to temperature. This equation is then used to predict the viscosity at temperatures spanning the relevant range. An example is shown in Table 12.1.

Note that it is important to consider both temperature and loading time because of the fact that asphalt mixes are sensitive to those factors, as well as the fact that the temperature does vary over the months of a year (with changing seasons) at any specific depth and, over the depths, at any specific time. The loading time (which can be represented as the "width" of the loading pulse, in a time scale) varies with the depth of the pavement for any selected design speed of the vehicle. An example of variation of temperature with depth and months (for a pavement location in the state of Maine) is shown in Table 12.2, and the corresponding dynamic modulus values are shown in Table 12.3 (for a frequency of 10 Hz). Note that for practical purposes, all of the surface-layer (and all of the base-layer) HMA could be considered as one single layer, and dynamic modulus values are determined at a range of temperature that spans the highest and lowest temperatures that can be expected in that layer.

TABLE 12.1
Change in Viscosity with a Change in Temperature

T_c	T_r	log log η	η (Poise)	η (10^{-6} Poise)
14.21	517.25	0.97687	30285832.77	30.28583277
15.65	519.85	0.96938	20852148.05	20.85214805
19.86	527.42	0.94776	7357533.81	7.35753381
24.93	536.56	0.92211	2281143.336	2.281143336
30.42	546.43	0.89486	707609.29	0.70760929
34.18	553.19	0.87649	334804.3767	0.334804377
36.45	557.28	0.8655	217146.7636	0.217146764
35.74	556.01	0.86889	247867.73	0.24786773
31.82	548.96	0.88797	532454.8361	0.532454836
27.08	540.42	0.9114	1427172.935	1.427172935
22.67	532.47	0.93353	3808745.994	3.808745994
17.18	522.59	0.9615	14179717.73	14.17971773
		A	VTS	
		10.312	−3.44	

TABLE 12.2
Temperature in Different Months (in °C)

Month	Air Temperature (°C)	Surface HMA Layer (h = 75 mm)			HMA Base Layer (h =125 mm)		
		Top	1/3h	2/3h	Top	1/3h	2/3h
January	−11.05	11.07	22.90	20.99	19.46	17.52	15.92
February	−9.11	13.01	24.73	22.74	21.15	19.11	17.44
March	−3.44	18.68	30.08	27.84	26.05	23.76	21.87
April	3.38	25.51	36.54	33.99	31.95	29.36	27.22
May	10.77	32.90	43.52	40.64	38.34	35.41	33.00
June	15.83	37.96	48.29	45.19	42.72	39.56	36.96
July	18.88	41.01	51.18	47.94	45.36	42.06	39.35
August	17.94	40.07	50.28	47.09	44.54	41.29	38.61
September	12.66	34.79	45.30	42.34	39.98	36.96	34.48
October	6.27	28.40	39.27	36.59	34.45	31.72	29.48
November	0.33	22.46	33.65	31.24	29.31	26.85	24.83
December	−7.05	15.07	26.67	24.59	22.92	20.80	19.05

Temperatures in pavements can be determined from available data, thermocouples, and/or equations, such as those shown in Table 12.4.

12.1.2 Models

Fatigue cracking models relating the number of load repetitions to failure have been developed by different agencies, as shown in Table 12.5. Note that the models use different parameters, and each one is applicable for the specific definition of failure, based on which model has been developed.

TABLE 12.3
Dynamic Modulus for Different Months (Dynamic Modulus in psi)

Month	Air Temperature (°C)	Surface HMA Layer (h = 75 mm)			HMA Base Layer (h = 125 mm)		
		Top	1/3h	2/3h	Top	1/3h	2/3h
January	−11.05	9,166,279	4,452,691	5,029,908	5,537,231	6,249,742	6,892,441
February	−9.11	8,200,535	3,958,450	4,501,614	4,982,925	5,664,571	6,284,646
March	−3.44	5,821,436	2,794,771	3,236,174	3,636,507	4,217,280	4,758,755
April	3.388	3,763,643	1,837,941	2,167,159	2,473,542	2,930,351	3,368,561
May	10.77	2,323,931	1,182,623	1,414,822	1,636,158	1,974,936	2,309,144
June	15.83	1,678,406	887,065	1,068,462	1,243,758	1,516,244	1,789,631
July	18.88	1,384,429	750,503	906,627.6	1,058,590	1,296,773	1,537,953
August	17.94	1,465,851	788,345	951,551.9	1,110,067	1,357,902	1,608,160
September	12.66	2,056,430	1,060,497	1,272,352	1,475,392	1,788,054	2,098,531
October	6.27	3,117,721	1,543,560	1,831,574	2,102,279	2,510,245	2,906,070
November	0.33	4,582,654	2,214,560	2,591,529	2,938,459	3,449,581	3,933,799
December	−7.05	7,253,065	3,486,796	3,992,318	4,444,117	5,089,618	5,682,087

TABLE 12.4
Temperature Equations for Asphalt Pavements

Temperature at the surface	$T_{surf} = T_{air} - 0.00618 \, lat^2 + 0.2289 \, lat + 24.4$ where T is expressed in °C and the latitude is in degrees.	Solaimanian and Kennedy (1993), Huber (1994), and Solaimanian and Bolzan (1993)
Temperature at different depths	$T(d) = T(surf) \, (1 - 0.063 \, d + 0.007 \, d^2 - 0.0004 \, d^3)$ where T(d) and T(surf) are in °F and the depth, d, is in inches.	

TABLE 12.5
Fatigue Cracking Models

The Asphalt Institute (AI; 1991) fatigue cracking model is as follows:

$N_f = 0.00432 \, C \, (\varepsilon_t)^{-3.291} \, (E)^{-0.854}$

$C = 10^M$

$$M = 4.84 \left(\frac{V_b}{V_a + V_b} - 0.69 \right)$$

where:

N_f = number of load repetitions to failure, which is defined as fatigue cracking over 20% of the entire pavement area

C = correction factor

ε_t = tensile strain at the bottom of the asphalt mix layer

V_a = air void (%) in the asphalt mix

V_b = asphalt content (%) by volume in the asphalt mix

E = dynamic modulus of the asphalt mix, psi

The shift factor, needed to transform the laboratory fatigue data to field data, for the AI equation is 18.4.

For a standard asphalt mix with 5% air voids and 11% asphalt (by volume), the AI equation reduces to the following:

$N_f = 0.0796 \, (\varepsilon_t)^{-3.291} \, (E)^{-0.854}$

The Shell equation (Shell, 1978) is as follows:

$N_f = 0.0685 \, (\varepsilon_t)^{-5.671} \, (E)^{-2.363}$

where:

N_f = number of load repetitions at constant strain to failure

ε_t = tensile strain at the bottom of the asphalt mix layer

E = dynamic modulus of the asphalt mix, psi

Shell equations for allowable tensile strains are as follows:

For constant stress:

$\varepsilon_t = (4.102 \, PI - 0.205 \, PI \, V_b + 1.094 V_b - 2.7807) * E_{mix}^{-0.36} * N^{-0.2}$

For constant strain:

$\varepsilon_t = (0.300 \, PI - 0.015 \, PI \, V_b + 0.080 \, V_b - 0.198) * E_{mix}^{-0.28} * N^{-0.2}$

where:

N = number of equivalent single-axle loads

ε_t = allowable permissible tensile strain, mm/mm

PI = Penetration Index

V_b = bitumen content, % by volume of the mix

E_{mix} = stiffness modulus of the asphalt mix, N/m²

(Continued)

TABLE 12.5 (CONTINUED)
Fatigue Cracking Models

Modified Shell equation:

$$N_f = A_f \ F'' K_{1\sigma} \left(\frac{1}{\varepsilon_t} \right)^5 E^{-1.4}$$

$$F'' = 1 + \frac{F}{1 + exp^{(1.354 hac - 5.408)}}$$

where:

N_f = number of load repetitions to cause fatigue cracks

ε_t = tensile strain at critical location

E = dynamic modulus of asphalt mix, psi

$K_{1\sigma}$ = laboratory calibration parameter

A_f = laboratory to field adjustment factor

h_{ac} = thickness of the HMA layer, inches

Modified Shell equation considering total strain:

$$Nji = F K_\alpha^5 \left(\frac{1}{\varepsilon_t} \right)^5 \left(\frac{1}{E} \right)^{1.4} Nj$$

where:

$$F = \frac{1 + (13908 E^{-0.4}) - 1}{1 + e^{1.354 hac - 5.408}}$$

$K_\alpha = [0.0252 \ PI + 0.00126 \ PI \ (V_{be}) - 0.0167]$

ε_t = tensile strain

E = dynamic modulus of asphalt mix, psi

h_{ac} = thickness of asphalt mix layer, inches

PI = Penetration Index of binder

V_{be} = effective binder content, % by volume

For the surface sublayer (top and bottom):

$\varepsilon_t = \varepsilon_{tl} + \varepsilon_{tt}$

where:

ε_t = total tensile strain

ε_{tl} = tensile strain due to the load

ε_{tt} = tensile strain due to temperature drop

The FHWA cost allocation (Rauhut et al., 1984) study model is as follows (note that this model uses the resilient modulus instead of the dynamic modulus):

$N_f = K_1 (\varepsilon_t)^{K2}$

where:

$$K_1 = K_{1R} \left[\frac{E_r}{E_{Rr}} \right]^{-4}$$

$K_2 = 1.75 - 0.252 \ [Log \ (K_1)]$

K_{1R} = coefficients determined from fatigue tests at a reference temperature of 70°F, 7.87 * 10^{-7}

E_r = resilient modulus (total) from indirect tensile test, at a specific test temperature, psi

E_{Rr} = reference resilient modulus (total) for a test temperature of 70°F, 500,000 psi

TABLE 12.5 (CONTINUED)
Fatigue Cracking Models

The Transportation and Road Research Laboratory (Powell et al., 1984) model:

$N_f = 1.66* 10^{-10} (\varepsilon_t)^{-4.32}$

Maupin and Freeman (1976):

$N_f = K_1 (\varepsilon_t)^{-n}$

where:

$n = 0.0374(\sigma_t) - 0.744$

$Log (K_1) = 7.92 - 1.122 (\sigma_f)$

σ_f = indirect tensile strength at 70°F, psi

Illinois procedure (Thompson, 1987):

$N_f = 5.0* 10^{-6} (\varepsilon_t)^{-3.0}$

where:

N_f = number of strain repetitions to failure (initiation of fatigue cracks) in terms of 18-kip ESALs

$N_{18} = 5.6* 10^{11} (\delta_{18})^{-4.6}$

N_{18} = number of 80 kN (18-kip) single-axle loads to fatigue failure

δ_{18} = surface deflection for a moving 80 kN (18-kip) single-axle load (mils)

Belgian Road Research Center (Verstraeten et al., 1977):

$N_f = 4.92* 10^{-14} (\varepsilon_t)^{-4.76}$

Probabilistic distress models for asphalt pavements (PDMAP; Monismith et al., 1972; Finn et al., 1973):

$$LogN_f (Laboratory) = 14.82 - 3.291Log\left(\frac{\varepsilon_t}{10^{-6}}\right) - 0.854Log\left(\frac{E}{10^3}\right)$$

$$LogN_f (fatigue\ cracks\ over\ 10\%\ of\ wp) = 15.947 - 3.291\ Log\left(\frac{\varepsilon_t}{10^{-6}}\right) - 0.854\ Log\left(\frac{E}{10^3}\right)$$

$$LogN_f (fatigue\ cracks\ over\ 45\%\ of\ wp) = 16.086 - 3.291Log\left(\frac{\varepsilon_t}{10^{-6}}\right) - 0.854Log\left(\frac{E}{10^3}\right)$$

where:

N_f = load applications of constant stress to cause fatigue failure

ε = initial horizontal tensile strain on the bottom of the HMA layer, in./in.

E = HMA modulus, psi

MICH-PAVE program (Baladi, 1989):

$$Log(ESAL) = -2.544 + 0.154h_{AC} + 0.0694h_{EQ} - 2.799Log\delta_0 - 0.261V_a + 0.917LogE_{base}$$

$$+ 0.0000269M_R - 1.0694Log\varepsilon_t + 1.173Log\varepsilon_v - 0.001\eta_K + 0.064ANG$$

$$h_{EQ} = h_{base} + h_{subbsase}\left(\frac{E_{subbase}}{E_{base}}\right)$$

where:

ESAL = number of 80 kN (18-kip) equivalent single-axle loads to failure

h_{AC} = thickness of the HMA layer, in.

h_{EQ} = equivalent thickness of base materials, in.

h_{base} = actual thickness of base materials, in.

$h_{subbase}$ = actual thickness of subbase materials, in.

(Continued)

TABLE 12.5 (CONTINUED)
Fatigue Cracking Models

E_{base} = resilient modulus of the base materials, psi

$E_{subbase}$ = resilient modulus of the subbase, psi

δ_0 = surface deflection, in.

V_a = air voids (%) in the mix

M_R = resilient modulus of subgrade, psi

ε_t = tensile strain at the bottom of the HMA layer, in./in.

ε_v = compressive strain at the bottom of the HMA layer, in./in.

η_K = kinematic viscosity at 135°C (275°F), Cst

ANG = aggregate angularity (4 for 100% crushed material, 2 for 100% rounded river-deposited material, and 3 for a 50% mix of crushed and rounded aggregate)

Dense graded (Monismith, 1967) base courses:

Controlled stress: $N_f = 2.0 * 10^{-7}(\varepsilon_t)^{-3.38}$

Controlled strain: $N_f = 7.5 * 10^{-6}(\varepsilon_t)^{-2.79}$

Asphalt-treated base (Kallas and Puzinauskas, 1972; test temp = 70°F):

$N_f = 2.520 * 10^{-9}(\varepsilon_t)^{-3.58}$

Relationships between log failure strain and log resilient modulus have been developed, and a longer fatigue life is indicated by a relatively larger tensile strain at failure for a specific modulus value.

The models shown in Table 12.5 are based on empirical test methods and parameters. The second approach for characterizing fatigue behavior of asphalt mix is with the use of the concept of dissipated energy under dynamic loading, using either a flexural center-point or a third-point fatigue test. The following equation can be used (Tangella et al., 1990):

$$N_f = \left(\frac{W}{A} \right)^{-z}$$

where:

W = total dissipated energy = $W_{total} = \Sigma_{i=1}^{n} W_i$

A, Z = mixture characteristic constants

Dissipated energy, ith load cycle:

$$w_i = \pi * \sigma_0 * \varepsilon_0 * \sin \phi_0$$

where:

σ_0 = amplitude of initial stress

ε_0 = amplitude of initial strain

ϕ_0 = phase angle, indicating lag between stress and strain in viscoelastic materials

Another method is the use of concepts of *fracture mechanics*, in which stress and fracture toughness are utilized. The steps are based on the original work by Paris, Gomez, and Anderson (1961), and are as follows.

1. Rate of growth of crack is proportional to stress intensity factor.

$$\frac{dc}{dN} = A(K)^n$$

where:

　　c = length of crack

　　N = number of repetitions of load

　A, n = constants related to fracture behavior of the material

　　K = stress intensity factor = $\sigma\sqrt{\pi a}$, where σ = stress and a = radius of the cylindrical specimen

2. Lytton et al.'s (1993) modified model for crack propagation (note that the crack initiation can be determined from laboratory beam fatigue tests, as described earlier).

$$N_{fp} = \left(\frac{1}{A}\right)\int_{c_0}^{h_{ac}} \frac{dc}{K^n}$$

where:

　N_{fp} = number of load applications to propagate a crack of length 8 mm (0.3 in.) to the surface

　h_{ac} = asphalt mix layer thickness

　c_o = initial crack length

　K = stress intensity factor

　A = asphalt mix properties

Majidzadeh et al.'s (1970) equation for determination of A:

$$A * 10^9 = 0.23213 + 2613(\sigma_t/E^*) - 3.2334\ (K_{IC})$$

where:

　σ_t = indirect tensile strength of the mix

　E^* = complex modulus of the mix

　K_{IC} = fracture toughness

Molenaar's (1983) equation for determination of A:

$$|LogA| = 0.977 + 1.628(n)$$

$$LogA = 4.389 - 2.52(LogE) * (\sigma_t)(n)$$

where:

　σ_t = indirect tensile strength of the asphalt mix at a given loading condition, MPa

　E = modulus of mix at a given loading condition, MPa

Schapery's (1986) equation for determination of A:

$$n = 2\left(\frac{1}{m} + 1\right)$$

$$n = \frac{2}{m}$$

$$n = 0.8\left(\frac{1}{m} + 1\right)$$

where:

　m = slope of the log creep compliance curve.

12.1.3 DEFINITION OF FAILURE

Note that the number of repetitions at which the pavement will fail in the field is significantly different from the number of repetitions that is indicated in the laboratory, because of discrepancies between field and laboratory loading conditions. These discrepancies include lateral wandering of the traffic in the field versus loading over the same spot in the laboratory and rest periods in the field being much longer than those in the laboratory. Also, while crack initiation and a resulting decrease in modulus, for example, are taken as failures in the laboratory, it is the visual manifestation of cracks that is considered as a sign of failure in the field. To be visible on the surface, the cracks after initiation at the bottom need to propagate to the surface, and the process takes time. A fundamental discrepancy could also be that while according to elastic theory the tensile stresses/strains are always maximum at the bottom of the asphalt mix layer, due to rapid cooling of the surface and formation of a rigid layer over a relatively soft layer, the cracks could initiate somewhere in the middle of the layer in the field.

Whatever the discrepancies may be, actually observed N_f could be 10 to 100 times that of laboratory-obtained N_f. Hence the laboratory-obtained N_f values are multiplied by an appropriate "shift factor" derived from field experience. One question at this point is as follows: to determine N_f in the field (which is required for predicting the shift factor for a specific location and class of asphalt mixes used by an agency), how would one determine whether a pavement has failed or not—in other words, what is the criterion for failure?

Fatigue failure has been defined earlier. Structurally, a pavement is said to have failed by fatigue when one or more of the following conditions are noted within the scope of the project. Note that the more conservative criteria are used for interstate and freeway pavements, whereas the less conservative ones are used for primary or secondary routes.

1. Over 45% of the wheelpath shows fatigue cracking and crack deterioration around the edges.
2. Over 20% of the wheelpath area has severely deteriorated cracks, with pieces moving under traffic.
3. 20% of the entire pavement area has fatigue cracks.
4. 37% of the wheelpath area has fatigue cracks.

12.1.4 USE OF MODELS

The next step is to utilize the fatigue equations in design processes. Note that with changing seasons, critical properties such as moduli of layers change, causing a change in the tensile stresses/strains in the asphalt mix layer, and hence causing a change in the way damage occurs. Furthermore, if the traffic is different in different seasons, then the relative proportion of damage throughout a year differs also. To consider all these factors, the concept of cumulative damage (Miner, 1945) is used. It is assumed that accumulation of damage causes ultimate "failure" of a pavement, damage by fatigue failure is cumulative, and a cumulative damage factor or index is defined such that when it reaches a value of 1, the pavement is assumed to "fail."

The damage factor for the ith loading is defined as the number of actual repetitions (n_i) of load divided by the "allowable" repetitions (N_i) that would cause failure. The Total Damage Index, or the Cumulative Damage Factor (CDF), for the parameter is given by summing the damage factors over all of the different loadings in the different seasons.

Mathematically, this can be expressed as follows:

$$\text{Total damage index (or DI)} = \sum_{t=1}^{m} \Delta DI_{ij}$$

$$\Delta DI_{i,j} = \left(\frac{n_i}{N_f} \right)_{i,j}$$

where:

n_i = number of actual load repetitions in a specific season (specific season being defined in terms of temperature and moisture conditions), i, for load of j, for a specific year

$N_{f(ij)}$ = allowable number of load repetitions for a specific season (i) and load (j) determined from fatigue equations

ΔDI_{ij} = Damage Index for a specific year

The concept of endurance limit has been proposed, which assumes that under a very small repetitive strain, an asphalt mix layer can sustain infinite or a very large number of loads (such as > 10 million). The concept of long-life or perpetual pavement is based on this theory. Suggested values of such low strains include 100 and 65 microstrains. (Currently, an NCHRP study is underway to evaluate this concept of endurance limit; see http://www.trb.org/TRBNet/ProjectDisplay.asp?ProjectID=972; Transportation Research Board, n.d.)

12.1.5 RELATIONSHIP BETWEEN MIX DESIGN AND FATIGUE PERFORMANCE

The fatigue characteristics of an asphalt mix are dictated by its materials—types and relative proportions, as well as mix volumetric properties—and are reflected in its mechanical properties. The different materials and properties are discussed below.

Aggregate: The effect of percentage of material passing the No. 200 sieve is significant—this material increases the fatigue resistance up to an optimum content and then reduces it beyond that content. The more angular the fine aggregate is, the more resistant it is to fatigue failure.

Asphalt: Both asphalt properties and amount (asphalt content in the mix) are significant. As discussed in Chapter 9 ("Asphalt and Emulsions"), the stiffness and the phase angle are controlled ($(G^*) * (Sin\delta)$) to maintain sufficient fatigue resistance, through limiting the stiffness (G^*) and the phase angle (δ). A higher asphalt content provides greater fatigue resistance.

Air voids/density: A high density (low air voids) is better for fatigue cracking resistance.

Note that some of the properties also affect the aging of the binder and hence affect its fatigue properties indirectly, since a higher rate of aging increases stiffness and reduces the resistance against fatigue failure. For example, a higher asphalt content and/or a higher density reduces the potential of aging of the binder and hence reduces the potential of deterioration of fatigue resistance.

12.1.6 RELATIONSHIP BETWEEN PAVEMENT STRUCTURE AND FATIGUE PERFORMANCE

The fatigue resistance of the asphalt mix layer is also affected by the structure of the asphalt pavement—the thickness of the asphalt mix and base course layer. Depending on the magnitude of the load and tire pressures, a very thin asphalt mix layer (for example, less than 2 in.) will not generate tensile stresses/strains and will not be susceptible to bottom-up fatigue failure; a moderately thin asphalt layer will be susceptible to fatigue failure; whereas a thick asphalt pavement (for example, > 8 in.) will definitely be very resistant to fatigue failure, because of the low tensile stress/strain. Both thickness and modulus of the base course are important, since they contribute to the overall stiffness of the pavement structure—a high modulus and a thicker layer improve fatigue resistance.

Finally, the wheel load is an important factor—a higher load naturally increases the potential of fatigue cracking.

12.1.6.1 Steps for Avoiding Premature Fatigue Cracking

To consider all of the important factors, the following steps can be taken, in general, to provide adequate resistance to fatigue cracking:

1. Adopt a standard maximum load.
2. Select appropriate materials for different layers so that they have desirable fatigue resistance properties.
3. Design the mixes (for the base as well as asphalt mix layers, for example) such that the volumetric properties are at the desirable level for adequate fatigue resistance.
4. Considering the mechanical properties of the designed mixes, design the pavement structure such that the overall stiffness of the structure is adequate.
5. Construct the pavement such that the controllable properties are brought up/down to desirable levels, as determined in Step 3.

12.2 THERMAL CRACKING

Thermal cracking occurs in two ways: (1) when the thermal stresses due to a drop in temperature exceed the fracture strength of the material (low temperature cracking), and (2) when, due to repeated thermal cycles, the strain in the asphalt layers causes thermal fatigue cracking. The potential of thermal cracking of a mix is evaluated by evaluating the mix stiffness and fracture strength characteristics with respect to temperature and time of loading.

The tensile strengths of asphalt mixes change with a change in temperature. As the temperature in the pavement drops (from the surface, which is in contact with the cold air, which causes the drop in temperature), the tensile strength keeps increasing, but beyond a point, due to micro-cracks at the aggregate-asphalt interface, it does not increase anymore and actually begins to decrease (Figure 12.7). On the other hand, as the temperature drops the asphalt mix layer tends to shrink, but is restrained by friction with the layers underneath. If the stress builds up relatively slowly, then it can be dissipated through stress relaxation in the viscoelastic asphalt mix at temperatures above 70°F. However, at lower temperatures, depending on the rate at which temperature drops and the properties of the asphalt mix components and the mix, the built-up stress exceeds the tensile strength (at that specific temperature), causing a crack on the surface.

Similar to load-related fatigue cracking, repetitive alternate cycles of high stress (due to a drop in temperature at night) and low stress (due to warmer temperatures during the day) can cause cracking (thermal fatigue cracking). The number of stress cycles required to cause failure depends on the magnitude of the repeated high stress and the thermal fatigue–related properties of the mix, which include the effects of viscoelastic properties as well as aging.

12.2.1 MATERIAL CHARACTERIZATION

For low-temperature cracking, the drop in temperature and the rate at which it drops are both important. The effect can be quantified with the creep modulus parameter—a higher value indicates a lower relaxation of stress, and hence a higher potential of buildup of thermal stress and hence cracking.

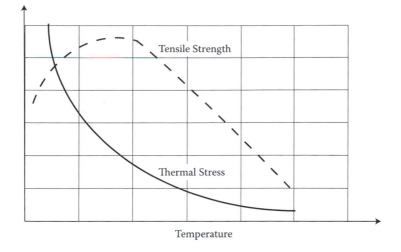

FIGURE 12.7 Concept of Change in Tensile Strength and Stress with Temperature.

The creep modulus is affected primarily by the low-temperature properties of the asphalt binder, and appropriate tests have been discussed in Chapter 9 ("Asphalt and Emulsions"). The tests include the bending beam rheometer test (to evaluate stiffness and rate of relaxation) and direct tension test (to evaluate strain at failure).

Asphalt mix properties: The low temperature indirect tensile creep test is used to determine the two properties that are used, the indirect tensile strength and the creep modulus. The test is run by applying a load at the rate of 1.25 mm per minute, at a temperature ranging from 0 to –20°C, for 100 to 3600 seconds.

Indirect tensile strength/creep compliance test: This test (AASHTO T-315, T-322) is conducted to determine the tensile strength (as discussed above) as well as the indirect tensile creep compliance, which is defined as the ratio of time-dependent strain to the applied stress. The sample setup is shown in Figure 12.8. In the test for creep compliance, a stress is applied quickly and then maintained constantly throughout a period of time. The applied

FIGURE 12.8 Measurement of Deformation with Extensometers on the Asphalt Mix Sample.

load and the resulting vertical and horizontal deformations are measured (as a function of time), and the creep compliance D(t) is calculated as follows:

$$D(t) = \frac{\Delta X_t * d * b}{P * GL} * C_{cmpl}$$

where:
  ΔX_t = mean horizontal deflection, mm
  d = diameter of the sample, mm
  b = thickness of the sample, mm
  P = creep load, kN
  GL = gage length over which deformation is measured, mm
  C_{cmpl} = compliance factor

$$C_{cmpl} = 0.6354 \left(\frac{X}{Y}\right)^{-1} - 0.332$$

where

$\frac{X}{Y}$ = ratio of horizontal to vertical deflection.

The limits of C_{cmpl} are as follows:

$$\left[0.704 - 0.213\left(\frac{b}{d}\right)\right] \leq C_{cmpl} \leq \left[1.566 - 0.195\left(\frac{b}{d}\right)\right]$$

$$0.20 \leq \frac{b}{d} \leq 0.65$$

The compliance, D(t), is used to calculate the relaxation modulus, E, which is used for predicting thermal stress in the asphalt mix. Paris' law is used then to determine the development of cracks.

The coefficient of *thermal contraction* of asphalt mixes can be estimated from the following equation (Jones et al., 1968):

$$B_{mix} = \frac{VMA * B_{AC} + V_{AGG} * B_{AGG}}{3 * V_{TOTAL}}$$

where:
  B_{mix} = linear coefficient of thermal contraction of the asphalt mix (1/°C)
  B_{AC} = volumetric coefficient of thermal contraction of the asphalt binder in the solid state (1/°C)
  B_{AGG} = volumetric coefficient of thermal contraction of the aggregate (1/°C)
  VMA = voids in mineral aggregate (% by volume)
  V_{AGG} = volume of aggregate in the mix (%)
  V_{TOTAL} = 100%

Typical values of coefficient of thermal contraction of asphalt mixes have been reported as 1.17 to 2.05 * 10^{-5}/°F (2.11 to 3.69 * 10^{-5}/°C).

12.2.2 MODELS

The thermal cracking prediction models are based on the approach of predicting a critical temperature below which an asphalt mix layer is expected to crack, and most methodologies combine the effect of low temperature and thermal fatigue cracking. The use of these methods requires the determination of environmental conditions (for the project site) and asphalt mix properties, some which are determined, while the others are commonly taken from available values in the literature.

12.2.2.1 Environmental Conditions

The pavement temperature, in response to air temperature, could be determined with the help of the enhanced integrated climatic model (EICM; developed by Larson and Dempsey, 1997), based on the integrated climatic model (ICM; developed by Lytton et al., 1990). The model is based on the balance of heat transfer between the pavement and the environment, considering the three different modes of heat transfer—conduction, convention, and radiation. The equation relating these factors is as follows:

$$T_{(i,t+\Delta t)} = T_{(i,t)} + \left(\frac{k\Delta t}{\gamma_d C \Delta z^2} \right) [T_{(i+1,t)} + T_{(i-1,t)} - 2T_{(i,t)}]$$

where:
  T = temperature, °F
  z = axis along the depth of the pavement
  K = thermal conductivity, BTU/(hr-ft^2-°F)
  C = mass specific heat in BTU/(hr-°F)
  γ_d = dry density

   Using the predicted conditions and the asphalt mix properties, either regression equations or fracture mechanics–based equations could be used to predict the extent of low-temperature-related cracking in an asphalt mix layer.

12.2.2.2 Regression Equation Approach (Hajek and Haas, 1972)

$$I = 30.3974 + (6.7966 - 0.8741h + 1.338a)\log(0.1S_{bit}) - 2.15165d - 1.2496m + 0.06026S_{bit} \log d$$

where:
  I = Cracking Index (≥ 0); number of full cracks plus one-half of the half-transverse cracks per 500-ft section of two-lane roads (does not consider cracks shorter than half-lane widths)
  S_{bit} = stiffness modulus, kg/cm^2, of the original asphalt; determined from Van der Poel (1954) nomograph with a loading time of 20,000 s and the winter design temperature. Penetration Index (PI) and Ring and Ball Softening Point determined from penetration at 25°C and kinematic viscosity at 275°F (135°C), according to McLeod's method (1970)
  h = total thickness of asphalt mix layer, in.
  a = age of pavement, years
  m = winter design temperature, without considering the "negative" sign, in °C
  d = code for subgrade type: sand, 5; loam, 3; and clay, 2

Haas et al.'s (1987) model (based on cracking in airfields):

$$TCRACK = 218 + 1.28 * ACTH\ 2.52\ MTEMP + 30 * PVN - 60 * COFX$$

where:

$\quad$ TCRACK = transverse crack average spacing, in meters

$\quad$ MTEMP = minimum temperature recorded on site, °C

$\quad$ PVN = dimensionless pen-vis number

$\quad$ COFX = coefficient of thermal contraction, mm/100 mm/°C

$\quad$ ACTH = thickness of asphalt mix layer, cm

12.2.2.3 Fracture Mechanics Approach: SHRP Thermal Cracking Model

There are three components of the thermal cracking distress model, which work on the basis of each other. The stress intensity model determines the stress at the tip of a local vertical crack, which is used to predict the depth of crack propagation, and based on that, the crack amount model predicts the number of thermal cracks per unit length of the pavement.

The time- and temperature-dependent relaxation modulus of asphalt mix, which reflects its viscoelastic property, dictates the thermal stress developed in an asphalt mix layer during temperature drop. Mathematically, this is expressed as follows:

$$\sigma(\xi) = \int_0^{\xi} E(\xi - \xi')\frac{d\varepsilon}{d\xi}d\xi'$$

where:

$\quad \sigma\,(\xi)$ = stress at reduced time ξ

$\quad E\,(\xi - \xi')$ = relaxation modulus at reduced time $(\xi - \xi')$

$\quad \varepsilon$ = strain at reduced time $(\xi - \xi') = \alpha(T(\xi') - T_0))$

$\quad \alpha$ = linear coefficient of thermal contraction

$\quad T(\xi')$ = pavement temperature at reduced time ξ'

$\quad T_0$ = pavement temperature when $\sigma = 0$

$\quad \xi'$ = variable of integration

The relaxation modulus of asphalt mix can be represented by the Maxwell model for viscoelastic materials:

$$E(\xi) = \sum_{i-1}^{N+1} E_i e^{\frac{-\xi}{\lambda_i}}$$

where:

$\quad \xi$ = reduced time = t/a

$\quad E(\xi)$ = relaxation modulus at reduced time

$\quad T$ = real time

$\quad a_T$ = temperature shift factor

$\quad E_i, \lambda_i$ = Prony series parameters

The creep test is performed at multiple temperatures to determine the creep compliance D(t), from which the relaxation modulus function is obtained.

The data are shifted from different temperatures to one continuous "master" creep compliance curve using a shift factor.

$$D(\xi) = D(0) + \sum_{i-1}^{n} D_i\left(1 - e^{\frac{-\xi}{\tau_i}}\right) + \frac{\xi}{\eta_v}$$

where:

$$D(\xi) = \text{creep compliance at reduced time } \xi$$
$$\xi = \text{reduced time} = \frac{t}{a_T}$$
$$a_T = \text{temperature shift factor}$$
$$D_{(0)}, D_i, \tau_i, \eta_v = \text{Prony series parameters}$$

A series of steps are needed to develop the master curve for creep compliance, using regression.

1. Creep tests are conducted for different times of loading and temperature, and the results are plotted as shown, with log D(t) and log(time), for the different temperatures (T).
2. A reference temperature is selected, and all of the curves at the other temperatures are shifted onto this curve to result in one smooth master curve. The determination of the shift factors (as a function of the temperature) as well as of Prony series parameters is done with regression, using a least-square criterion. Four Prony series coefficients are determined for data obtained at three temperatures. To complete the master curve, shift factors for temperatures at which tests were run are determined on the basis of the regression relationship between the shift factor and temperature.
3. An equation for the compliance curve is then developed to determine m, the slope of the linear portion of the log time versus log compliance curve:

$$D(\xi) = D_0 + jD_1\xi^m$$

where, D_0, D_1, and m are the coefficients of the equation, and m is the primary parameter that characterizes the low temperature cracking potential of an asphalt mix.

The next step is the determination of the relaxation modulus, which enables one to determine the stress for causing thermal cracks. While the relaxation modulus can be approximated as just the inverse of the creep compliance, the more precise relationship is the following:

$$sL[D(t)]*sL[E(t)] = 1$$

where:

$L[D(t)] = \text{Laplace transform of creep compliance, } D(t)$
$L[E(t)] = \text{Laplace transform of relaxation modulus, } D(t)$
$s = \text{Laplace parameter}$
$t = \text{time or reduced time, } \xi$

Using a computer program, the master relaxation modulus can be determined from the master creep compliance equation.

Boltzmann's superposition principle (1874) for linear viscoelastic materials is used to determine thermal stresses (as mentioned earlier):

$$\sigma(\xi) = \int_0^\xi E(\xi - \xi')\frac{de}{d\xi}\, d\xi'$$

where:

$$\sigma(\xi) = \text{stress at reduced time } \xi$$
$$E(\xi - \xi') = \text{relaxation modulus at reduced time, } \xi - \xi'$$
$$\varepsilon = \text{strain at reduced time } \xi\ (= \alpha(T(\xi') - T_0))$$
$$T(\xi') = \text{pavement temperature at reduced time } \xi'$$
$$T_0 = \text{pavement temperature when } \sigma = 0$$
$$\xi' = \text{variable of integration}$$

The assumptions are that the HMA material is a thermorheologically simple material, allowing the use of the time-temperature superposition principle, and that the layer behaves as a uniaxial rod.

The form of the equation with real time, t, is as follows:

$$\sigma(t) = \int_0^\xi E[(\xi(t) - \xi'(t)] \frac{de}{dt'} dt'$$

The use of Prony series representation of $E(\xi)$ results in the following finite difference solution:

$$\sigma(t) = \sum_{i=1}^{N+1} \sigma_i(t)$$

where:

$$\sigma_i(t) = e^{\frac{-\Delta}{\lambda_i}} \sigma_i(t - \Delta t) + \Delta E_i(\varepsilon) \frac{\lambda_i}{\Delta \xi} (1 - e^{\frac{-\Delta \xi}{\lambda_i}})$$

where:

$\Delta\varepsilon$ = change in strain over time $t - \Delta t$ to t

$\Delta\xi$ = change in reduced time $t - \Delta t$ to t

12.2.2.4 Models for Cracking

Stress intensity model 2 d FE model (Chang et al., 1976):

$$K = \sigma \left(0.45 + 1.99 C_0^{0.56} \right)$$

where:

K = stress intensity factor

σ = far-field stress from the pavement response model at depth of the crack tip

C_0 = current crack length

The change in depth of a local crack subjected to a cooling cycle is determined by the Paris law.

$$\Delta C = A(\Delta K)^n$$

where:

ΔC = change in crack depth due to a cooling cycle

ΔK = change in stress intensity factor due to a cooling cycle

A, n = fracture parameters for the specific asphalt mix

m = slope of the linear portion of the log compliance–log time curve is determined from the creep test

The stress intensity factor can be used, in Paris' equation, to determine the change in crack length due to a drop in temperature.

$$\Delta C = A(\Delta K)^n$$

A and n can be determined from fracture tests and the following equations, where m = slope of the linear portion of the log compliance–log time curve and is determined from the creep test.

$$n = 0.8\left[1 + \frac{1}{m}\right]$$

Then, using n and σ_m (which is determined from the indirect tensile strength test), A is determined using the following formula:

$$A = 10^{(\beta*(4.389-2.52*\log(E*\sigma_m*n)))}$$

where:
    E = modulus of the asphalt mix
    σ_m = tensile strength of the asphalt mix, psi
    β = parameter from calibration

The total crack depth is estimated by computing and summing up the change in crack depth (ΔC) with time. The HMA layer is subdivided into four layers. A relationship between the probability distribution of the log of crack depth to layer thickness ratio and the percentage of cracking is used to predict the degree of cracking.

$$C_f = \beta_1 * N\left(\frac{\log C/h_{ac}}{\sigma}\right)$$

where:
    C_f = observed amount of thermal cracking
    β_1 = coefficient of regression equation for field calibration
  $N(z)$ = standard normal distribution evaluated at (z)
    σ = standard deviation of the log of the depth of cracks in the pavement = 0.769
    C = crack depth
   h_{ac} = thickness of asphalt mix layer

12.2.3 CRACKING AND PROPERTIES OF ASPHALTS AND AGGREGATES

The resistance of an asphalt mix layer to thermal cracking can be controlled by a variety of parameters, which affect tensile strength and relaxation properties. Apart from the asphalt grade, which has the primary influence, parameters such as asphalt content, air voids, and aggregate type and gradation have significant influence.

The effect of asphalt type has been discussed in Chapter 9. A high asphalt content results in a low creep modulus and hence lower potential of thermal cracking. Note that the effective asphalt content can be lowered by an absorptive aggregate (if not considered properly during mix design) and hence increase the potential of cracking directly (or indirectly, by allowing the asphalt to age faster). The aging can also be accelerated by the presence of relatively high air voids, which can also decrease the tensile strength, and hence increase the potential of cracking. The increase in tensile strength can be achieved by using appropriate gradation as well as aggregate with crushed faces.

12.3   RUTTING OR PERMANENT DEFORMATION

Rutting is the result of repeated loading, which causes accumulation, and increase of permanent deformations. The one-dimensional densification-consolidation rutting, resulting from a decrease in air voids, occurs with volume change and is vertical deformation only (primary rutting), whereas the two-dimensional rutting is caused by shear failure and is accompanied by both vertical and lateral movement of the material (secondary and tertiary rutting).

The densification-consolidation rutting may be caused by the action of high stress load at high temperatures near the surface of the pavement, leading to a decrease in air voids of the asphalt mix, especially if they are too high compared to desirable air voids; and/or due to the densification of the underlying layers if they are at densities lower than the maximum dry density or desirable densities; or due to the consolidation of fine-grained soils in the subgrade, for example, with high levels of moisture. The one-dimensional rutting mechanism can be relatively easily simulated in the laboratory.

In the structural design of pavement, the rutting in a pavement is assumed to be caused by either or both of two causes—excessive strain in the subgrade or permanent deformation in any of the layers in the pavement.

12.3.1   MATERIAL CHARACTERIZATION

Any element in any layer in a pavement subjected to a moving load experiences a variety of stresses, which change as the load moves. Figure 12.9 shows the different stresses acting on an element—vertical, horizontal, and shear stress—as well as the phenomenon of stress reversal as the load changes position.

For example, when the load is "near" any point near the bottom of a stiff surface layer, the point experiences compressive stress in the horizontal direction—this stress changes to a tensile stress when the load is directly above it. Because of the variety of stresses and stress reversals, it is extremely difficult, if not impossible, to completely simulate such a stress state in the laboratory with any test procedure. As such, any one or more of different types of tests, varying in complexity and ability of simulation of in-place conditions, are used, depending on the level of sophistication required. Such tests may be "fundamental" or "simulative." Fundamental tests create in-place loading conditions with the help of complex loading equipment and allow the determination of fundamental parameters such as modulus, whereas simulative tests simulate in-place loading conditions, not necessarily with consideration of scale effects, but specifically to "compare" the performance between different asphalt mixes.

The relevant parameter that resists the assumed shear failure due to shear stress is the shear modulus:

$$G = \frac{\tau}{\gamma} = \frac{E}{2(1+\mu)}$$

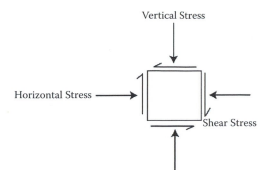

Different Stresses Acting on a Pavement Element Under Load

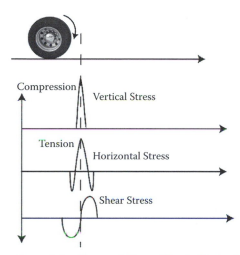

Stress Reversals in a Pavement Element Due to Moving Load

FIGURE 12.9 Different Stresses Due to a Moving Wheel Load.

where:
   G = shear modulus of an elastic material
   τ = shear stress
   γ = shear strain—angle, in radians, of deformation of the material due to the shear stress
   E = Young's modulus
   μ = Poisson's ratio

   Since G is affected by both time of loading (longer time of loading lower is G) and temperature (higher temperature lower is G), rutting is also affected by these parameters.

12.3.1.1 Creep Testing

The deformation of a pavement material in any of the layers begins with elastic deformation, then changes over to partly elastic and partly plastic (elasto-plastic) and then completely plastic deformation. In the case of asphalt mixes, which consist of asphalt, which is a viscoelastic material, the elastic and plastic deformations also consist of viscoelastic and viscoplastic deformations. The deformations resulting from the temperature and time of loading-dependent viscoelastic and visco-plastic behavior cause rutting, and can be simulated in the laboratory using either static or repeated

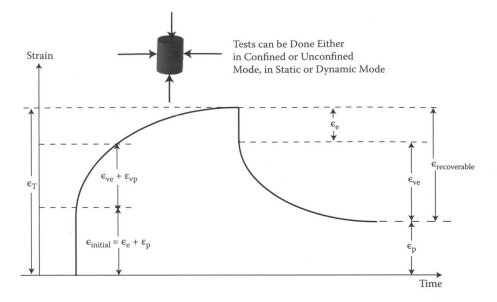

FIGURE 12.10 Schematic of Creep Testing.

load creep tests. In the creep test the vertical strain in an asphalt mix sample subjected to a constant load (or repeated load), at a specific temperature, is measured during and after the application of the load. The static creep tests can be done either in uniaxial mode or with triaxial mode, which applies confining stress.

The total strain, ε_T, consists of different components, as shown below (Figure 12.10):

$$\varepsilon_T = \varepsilon_e + \varepsilon_p + \varepsilon_{ve} + \varepsilon_{vp}$$

where:
  ε_T = total strain
  ε_e = elastic strain, recoverable and time independent
  ε_p = plastic strain, irrecoverable and time independent
  ε_{ve} = viscoelastic strain, recoverable and time dependent
  ε_{vp} = viscoplastic strain, irrecoverable and time dependent

Note that in creep tests, it is very important to have the loading faces of the sample smooth, parallel, and perpendicular to the loading head. The creep test can be run with one load-unload cycle or with incremental load-unload cycles, and the results provide the different deformations mentioned earlier. From the creep test, the creep modulus is calculated to determine the increase in elastic and plastic deformations at specific loading times. The creep test is used specifically for asphalt mixes (and not for other layer materials such as soils) because only the asphalt mixes generally have viscous components with time-dependent stiffness and deformation properties.

$$E_c(t) = \frac{\sigma_c}{\varepsilon_T(t)}$$

where:
  $E_c(t)$ = compressive creep modulus at time t
  σ_c = compressive stress
  $\varepsilon_T(t)$ = compressive strain at time t

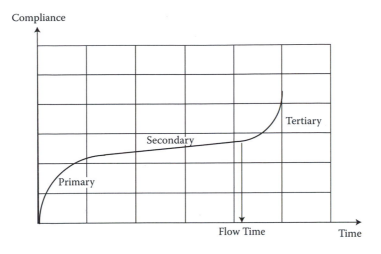

FIGURE 12.11 Different Parts of the Compliance Curve.

Also measured is the percentage of recovery of the deformation in the elastic-plastic region of the deformation, for which a higher number indicates lower permanent deformation or lower potential of rutting.

$$X_R = \frac{\varepsilon_t(t)}{\varepsilon_r}$$

where
  ε_r = recoverable strain.

For viscoelastic material, the term *compliance* D(t) is used rather than *modulus*, since D(t) allows the separation of the different deformation components. D(t) is approximately equal to the inverse of the modulus. Just like the deformation, the D(t) can be divided, with respect to time, into a primary, secondary, and tertiary component (Figure 12.11).

The secondary part of the curve can be modeled as follows, using a power regression curve:

$$D' = D(t) - D_0 = a(t)^m$$

where:
  D' = viscoelastic compliance
  $D(t)$ = total compliance at time t
  D_0 = instantaneous compliance
   t = loading time
  a, m = coefficients for specific asphalt mix

The coefficients a and m are used to predict the mix's rutting potential. A low a value is desirable since a large a value means a large D value and lower modulus value, and also, for a specific a, the higher the m value, the higher is the potential of permanent deformation.

12.3.1.2 Triaxial Test

From the triaxial repeated load tests, resilient modulus, E_R; total plastic strain, ε_p; plastic strain per load cycle, $\varepsilon_{p(N)}$; and strain ratio, $\varepsilon_p/\varepsilon_r$, can be obtained. The resilient strain, ε_r, is the recoverable strain during the rest period of the load cycle. The resilient shear modulus is the ratio of compressive stress to resilient axial strain. The strain ratio is the ratio of plastic strain to resilient strain.

In the triaxial shear strength test, a modification of AASHTO T-167 can be used with confining stress to develop a Mohr-Coulomb failure envelope, which is expressed as follows:

$$\tau = c + \sigma(\tan \phi)$$

where:
   τ = shear stress
   σ = normal stress
   c = cohesion
   ϕ = angle of internal friction, or the slope of the failure envelope

This simulation is based on the assumption that the shear strength of HMA develops from cohesion (c) from the asphalt binder and interlock of the aggregates matrix, which is characterized by the angle of internal friction (ϕ). The relative contributions of c and ϕ toward the shear strength of the mix are dependent on the time of loading, temperature, and volumetric properties of the mix.

The results of triaxial tests can also be used for characterizing the behavior of the asphalt mix according to the Drucker-Prager failure envelope. In this model, the combined effect of failure stresses is represented in terms of the first invariant stress tensor and the second invariant deviatoric stress tensor:

$$J_2 = \gamma^{\frac{1}{2}}I_1 + k$$

$$J_2^{\frac{1}{2}} = \left(\frac{1}{3}\right)(\sigma_1 - \sigma_3)$$

$$I_1 = \sigma_1 + 2\sigma_3$$

where:
   I_1 = first invariant stress tensor
   $J_2^{\frac{1}{2}}$ = second invariant deviatoric stress tensor
   k = intercept
   $\gamma^{1/2}$ = slope

12.3.2 MODELS

The use of the *subgrade strain model* ensures that there is sufficient cover over the subgrade to resist plastic deformation of the subgrade and, hence, resulting "ride-on" deformation in all of the layers above it. This is based on the assumption that the layers above the subgrade are designed and constructed properly, so that there is no rutting in any of those layers. The form of the model is as shown below:

$$N_f = b_1\beta(\varepsilon_v)^{-b_2}(M_R)^{b_3}$$

where:
   N_f = number of load repetitions causing failure that is causing deformation exceeding a specific surface deformation
   M_R = design resilient modulus of the subgrade soil (resilient modulus has been discussed in Chapter 7, "Soil")

TABLE 12.6
Examples of Rutting Models Based on Subgrade Strain

Asphalt Institute (1984):

N_f for 13 mm (0.5 inch rutting) = $1.365 * 10^{-9} (\varepsilon_v)^{-4.477}$

Shell (50% reliability; Shell, 1978):

N_f for 13 mm (0.5 inch rutting) = $6.15 * 10^{-7} (\varepsilon_v)^{-4.0}$

Shell (95% reliability):

N_f for 13 mm (0.5 inch rutting) = $1.05 * 10^{-7} (\varepsilon_v)^{-4.0}$

Transport Road Research Laboratory (TRRL, UK; 85% reliability; Powell, 1984):

N_f for 10 mm (0.4 inch rutting) = $6.18 * 10^{-8} (\varepsilon_v)^{-3.95}$

Belgian Road Research Center (Verstraeteu, 1977):

N_f for 10 mm (0.4 inch rutting) = $3.05 * 10^{-9} (\varepsilon_v)^{-4.35}$

U.S. Army Corps of Engineers (modified by Rauhut et al., 1984; Von Quintas et al., 1991):

N_f for 13 mm (0.5 inch rutting) = $1.259 * 10^{-11} (\varepsilon_v)^{-4.082} (M_R)^{0.955}$

ε_v = vertical compressive strain at the top of the subgrade

b_1, b_2, b_3 = coefficients derived from triaxial tests

β = field calibration factor

The predominant way of using this model is to limit the vertical subgrade strain to a specific value such that the N_f matches or exceeds the anticipated load (traffic) repetition. Different agencies have developed their own models based on this model—some of which do contain M_R, and the rest contain ε_v only. It is important to note that the models are developed for a specific range of site, materials, and environmental conditions, and that failure criteria (or, more specially, the specific depth of rutting which is called a rutting failure) are different for the different models, and one must use the same criteria for failure as those in the specific model that is being used. Examples of such models are shown in Table 12.6.

12.3.2.1 Consideration of Rutting in Asphalt Mix Only *environmental aspect*

Rutting in the asphalt mix layer is caused specifically by high temperature and standing traffic. At high temperatures, such as those above 60°C, the effect of repeated loading on the mix can be considered to be primarily dictated by the properties of the binder, such as viscosity. The influence of binder and the measures taken to select an appropriate binder with adequate resistance against rutting have been discussed.

The Penetration Index (PI) has been shown to relate (inversely) well with rutting. Rutting results from wheel-tracking tests have been correlated well with creep stiffness measured from a uniaxial unconfined creep test. The deformation is inversely proportional to the stiffness of the mix, which is proportional to the stiffness of the binder, which can be derived from the PI of the binder.

$$\text{Deformation} = k_1 \frac{1}{S_{mix}}$$

$$S_{mix} = k_2 (S_{asphaltbinder})^{k_3}$$

where k_3 has been suggested as 0.25. Increasing the PI results in a significant increase in viscosity and decrease in rutting potential.

For strains that cause rutting, the binder is assumed to remain in zero shear viscosity regime, with Newtonian fluid characteristics, in which stress and strain are linearly proportional to each other and the stiffness modulus is independent of the stress or strain level and the viscosity is independent of the shear rate. Such viscosity can be measured by a dynamic shear rheometer. A practical upper limit of zero shear viscosity has been proposed as 10^5 Pa s. The value of the zero shear viscosity or the shear complex modulus and elastic component of the binder, as represented by G*/sin delta, for example, can be used to evaluate the rutting potential of an asphalt binder.

The effect of asphalt mix properties can be modeled for rutting prediction by different ways, such as through the use of statistical (regression) equations, or through the use of material characterization parameters, as shown below.

12.3.2.2 Statistical Predictive Models on the Basis of Different Properties (Baladi, 1989)

$$Log\ RD = -1.6 + 0.067\ V_a - 1.4\ log\ h_{AC} + 0.07\ T_{avg} - 0.000434\ \eta_k + 0.15\ log\ ESAL - 0.4\ log\ M_R$$
$$- 0.50\ log\ E_{base} + 0.1\ log\ \delta_0 + 0.01log\ \varepsilon_v - 0.7\ log\ h + 0.09\ log\ (50 - h_{AC} - h_{EQ})$$

where:
RD = rut depth, in.
ESAL = number of 80 kN (18-kip) ESAL corresponding to the rut depth
T_{avg} = average annual temperature, °F
h_{AC} = equivalent thickness of base material, in.
E_{base} = effective resilient modulus of base materials, psi (*effective* means it is influenced by freezing index and seasonal variations)
δ_0 = surface deflection, in.
V_a = air voids (% in mix)
M_R = effective resilient modulus of subgrade, psi
ε_v = compressive strain at the bottom of the asphalt mix layer, in./in.
η_k = kinematic viscosity of the asphalt binder at 135°C (275°F), Cst

The Shell International Procedure (Shell, 1978) uses a compressive creep test:

$$\Delta h = C_m h_{AC}\left(\frac{\sigma_c}{E_{mix}}\right)$$

where:
Δh = reduction of thickness in asphalt mix layer, mm
C_m = correction factor for dynamic effect, depending on the type of mix
H_{AC} = thickness of asphalt mix layer, mm
σ_c = average vertical compressive stress in the asphalt mix layer, kPa
E_{mix} = modulus of the asphalt mix, kPa

12.3.2.3 The Layered Vertical Permanent Strain Approach

In the layered vertical permanent strain approach, the permanent strain in each layer is determined as a function of load repetitions, the deformation is calculated by multiplying the strain times the layer thickness, and then the deformation in different layers is added up to determine total rutting which is visible on the surface of the pavement. One such model recommended by Barenberg and Thompson (1992) is of the form:

$$Log(\varepsilon_p) = a + b[Log(N)]$$

where:

ε_p = permanent strain
a, b = empirical coefficients
N = number of load repetitions

As suggested by Huang (1993), the steps consist of (1) dividing the pavement structure into a number of layers, (2) estimating vertical and radial stresses at the mid-depth of each layer, (3) conducting repeated load tests in the lab using the estimated vertical stress and radial stress plus overburden stress as confining stress, (4) determining permanent strain, (5) computing vertical deformation by multiplying permanent strain by thickness, and (6) adding the permanent strains in all of the layers to obtain rutting at the surface.

Examples of such models are shown in Table 12.7.

12.3.2.4 Permanent Strain Rate Method

In this method, the strain per load application is expressed as a power model:

$$\frac{\partial \varepsilon_p}{\partial N} = \varepsilon_{pn} = \frac{\partial (aN^b)}{\partial N}$$

$$\varepsilon_{pn} = abN^{(b-1)}$$

$$\frac{\varepsilon_{pn}}{\varepsilon_r} = \left(\frac{ab}{\varepsilon_r}\right)N^{b-1}$$

$$\text{If } \mu = \frac{ab}{\varepsilon_r}$$

and

$$\alpha = 1 - b$$

then,

$$\frac{\varepsilon_{pn}}{\varepsilon_r} = \mu N^{-\alpha}$$

12.3.2.5 Plastic-Elastic Vertical Strain Ratio Method

In this method, the plastic strain is expressed as a function of the elastic or resilient strain.

$$\frac{\varepsilon_p}{\varepsilon_r} = \beta_r aN^b$$

ε_p = accumulated plastic strain after N repetitions of load
ε_r = resilient strain depending on mix properties
N = number of load applications
a, b = nonlinear regression coefficients
β_r = field adjustment factor

The different models are shown in Table 12.8.

TABLE 12.7

Examples of Layered Vertical Permanent Strain Approach Models

Asphalt Institute (May and Witczak, 1992), for pavements with > 3% air voids, deviator stress ≤ 90 psi; strain from other layers is negligible:

$$Log\varepsilon_p = -14.97 + 0.408Log(N) + 6.865Log(T) + 1.107Log(S_d) - 0.117Log(V)$$

$$+ 1.908LogP_{eff} + 0.971(V_v)$$

where:

ε_p = permanent axial strain

N = number of load repetitions to failure

T = temperature, °F

S_d = deviator stress, psi

V = viscosity at 21°C (70°F), Ps * 106

P_{eff} = % by volume of effective asphalt

V_v = air voids (%)

Allen and Deen (1986), for any layer:

$$Log\varepsilon_p = C_0 + C_1[Log(N)] + C_2[Log(N)]^2 + C_3[Log(N)]^3$$

where:

ε_p = permanent axial strain

N = number of load applications

C_0, C_1, C_2, C_3 = coefficients depending on the type of material

For asphalt mixes:

$C_0 = -0.000663T^2 + 0.1521T - 13.304 + (1.46 - 0.00572T) * log\Phi_1$

$C_1 = 0.63974, C_2 = -0.10392, C_3 = 0.00938$

For dense graded aggregate base:

$C_0 = -4.41 + (0.173 + 0.003w) * \Phi_1 - (0.00075 + 0.0029w) * \Phi_3$

$C_1 = 0.72, C_2 = -0.142 + 0.092 (log w), C_3 = 0.006 - 0.004 (log w)$

For subgrade:

$C_0 = -6.5 + 0.38w - 1.1 (log\Phi_3) + 1.86 (log \Phi_1)$

$C_1 = 10^{(-1.1+0.1w)}, C_2 = 0.018w, C_3 = 0.007 - 0.001w$

where:

T = temperature, °F

Φ_1 = deviator stress, psi

Φ_3 = confining stress, psi

w = moisture content (%)

FHWA VESYS model (Moavenzadeh et al., 1974; Brademeyer, 1988):

Axial repeated load creep test is conducted to determine the resilient strain, resilient modulus, and permanent deformation.

$$\varepsilon_p = IN^s$$

where:

I = linear intercept on the permanent strain axis

N = number of load applications

S = slope of the linear portion of the log log curve

The two derived parameters used to characterize the rutting potential of mixes are gnu and alpha:

$$Gnu \ (or \ \mu) = \frac{IS}{E_r}$$

alpha (or α) = I − S.

TABLE 12.8
Different Plastic-Elastic Vertical Strain Ratio Models

Leahy (1989):

$$\log\left(\frac{\varepsilon_p}{\varepsilon_r}\right) = -6.631 + 0.435 \log N + 2.767 \log T + 0.110 \log S + 0.118 \log \eta$$

$$+ 0.930 \log V_{beff} + 0.5011 \log V_a$$

where:

ε_p = accumulated permanent strain

ε_r = resilient strain

N = number of load repetitions

T = mix temperature (°F)

S = deviatoric stress (psi)

η = viscosity at 70°F (10^6 poise)

V_{beff} = effective asphalt content (% by volume)

V_a = air void content (%)

Ayress (1997):

$$\log\left(\frac{\varepsilon_p}{\varepsilon_r}\right) = -4.80661 + 2.58155 \log T + 0.429561 \log N$$

Kaloush and Witczak (2000):

$$\frac{\varepsilon_p}{\varepsilon_r} = 10^{-3.15552\beta_{r1}} \, T^{1.734*\beta_{r2}} N^{0.39937*\beta_{r3}}$$

where

$\beta_{r1}, \beta_{r2}, \beta_{r3}$ = factors obtained from calibration of rut model, and where the resilient strain can be determined as follows:

$$\varepsilon_{rz} = \frac{1}{|E^*|}(\sigma_z - \mu\sigma_x - \mu\sigma_y)$$

where

E* is the dynamic modulus, which is determined in the laboratory, as a function of time of loading and temperature using a master curve. E* can also be determined from a regression equation relating asphalt, aggregate and asphalt mix volumetric properties, and time of loading.

12.3.2.6 Rutting Rate Method (Majidzadeh, 1981)

$$\frac{\varepsilon_p}{N} = A(N)^m$$

where:

ε_p = permanent strain

N = allowable number of load applications

a, m = constants derived from experiments

12.3.2.7 Alternate Model Relating Tertiary Flow Characteristics to Mix Properties

Since many of the above equations use regression coefficients derived from the secondary rutting part of tests, the rutting resulting from tertiary rutting is ignored, resulting in a lower than actual rutting prediction. An alternative method is as follows.

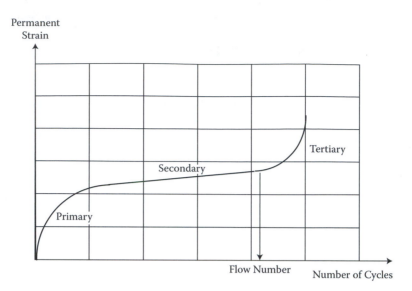

FIGURE 12.12 Concept of Flow Number.

In the tertiary zone, the material does not undergo any volume change, but there is a large increase in compliance, D(t). The starting point of tertiary deformation, known as *flow time* (F_T; see Figure 12.11; see also Witczak et al., 2000), has been shown to be a significant indicator of the mix's rutting potential. Flow time is defined as the time at which the shear deformation without any volume change starts.

Another parameter that has been used to relate to the mix's permanent deformation potential is the flow number, F_N (Figure 12.12), which is defined as the number of cycles at which tertiary permanent strain begins, when the permanent strain is plotted against the number of cycles. This type of plot is generated from data obtained in an unconfined or confined repeated load test, in which a haversine load is applied for, say, 3 hours of 10,000 cycles, with a loading time of 0.1 second and rest period of 0.9 seconds. The results of these tests are plotted on a log (N) versus log ε_p scale, and the linear portion is modeled as a power curve with intercepts as a and slope as b.

$$\varepsilon_p = aN^b$$

The number of cycles of loading at which tertiary flow would occur has been correlated to mix properties (Kaloush and Witczak, 2000):

$$F_N = (1.00788E5)\, T^{-1.6801}\, S^{-0.1502}\, \eta^{0.2179}\, V_{beff}^{-3.6444}\, V_a^{-0.9421}$$

where S = stress level.

12.3.2.8 Models for Unbound Materials

For unbound materials, the rutting models developed are shown in Table 12.9. The basic form is as follows (Tseng and Lytton, 1986):

$$\delta_a(N) = \left(\frac{\varepsilon_0}{\varepsilon_r}\right) e^{-\left(\frac{\rho}{N}\right)^{\beta}} \varepsilon_v h$$

TABLE 12.9
Rutting Models for Unbound Materials

Granular material:

$$\log\left(\frac{\varepsilon_0}{\varepsilon_r}\right) = 0.80978 - 0.06626W_c - 0.003077\sigma_\theta + 0.000003E_r$$

$$\log\beta = -0.9190 + 0.03105W_c + 0.001806\sigma_\theta - 0.0000015E_r$$

$$\log\rho = -1.78667 + 1.45062W_c + 0.0003784\sigma_\theta^2 - 0.002074W_c^2 - 0.0000105E_r$$

Subgrade material:

$$\log\left(\frac{\varepsilon_0}{\varepsilon_r}\right) = -1.69867 + 0.09121W_c - 0.11921\sigma_d + 0.91219\log E_r$$

$$\log\beta = -0.9730 - 0.0000278W_c^2\sigma_d + 0.017165\sigma_d - 0.0000338W_c^2\sigma_\theta$$

$$\log\rho = 11.009 + 0.000681W_c^2\sigma_d - 0.40260\sigma_d + 0.0000545W_c^2\sigma_\theta$$

where:
W_c = water content (%)
σ_d = deviator stress (psi)
σ_θ = bulk stress (psi)
E_r = resilient modulus of the layer/sublayer (psi)

where:

δ_a = permanent deformation for the layer/sublayer
N = number of traffic repetitions
ε_0, β, and ρ = material properties
ε_r = resilient strain imposed in laboratory tests to obtain material properties ε_0, β, and ρ
ε_v = average vertical resilient strain in the layer/sublayer as obtained from the primary response model
h = thickness of the layer/sublayer
β_1 = calibration factor for the unbound granular and subgrade materials

$\varepsilon_0/\varepsilon_r$ depend on the type of material, granular, or subgrade soil.

12.3.2.9 Ayres Combined Model for Subgrade and Granular Materials (NCHRP, 2004)

$$\log\left(\frac{\varepsilon_0}{\varepsilon_r}\right) = 0.74168 + 0.08109W_c - 0.000012157M_r$$

where:

$(\frac{\varepsilon_0}{\varepsilon_r})$ is multiplied by ADJ_Strain_Ratio
ADJ_Strain_Ratio = $1.2 - 1.39 * e^{-0.058*(\frac{M_r}{1000})}$
If $M_r/1000 < 2.6$, use 2.6
If ADJ_Strain_Ratio < 1e-7, use ADJ_Strain_Ratio = 1e-7
$\log\beta = -0.61119 - 0.017638W_c$
β is multiplied by 0.7 as a correction factor

$$\log \rho = 0.622685 + 0.541524 W_c$$
$$W_c = 51.712 * CBR^{-0.3586*GWT^{0.1192}}$$

$$CBR = \left(\frac{Mr}{2555}\right)^{(1/0.64)}$$

where:

W_c = water content (%)

CBR = CBR ratio of the unbound layer

GWT = groundwater table (ft)

M_r = resilient modulus of the layer/sublayer (psi)

Since the subgrade is often modeled as an infinite layer, it is not practical to divide it into sublayers, and it is better to adopt an alternative approach as suggested by Ayres:

$$\varepsilon_p(z) = (\varepsilon_{p,z=0})e^{-kz}$$

where:

$\varepsilon_p(Z)$ = plastic vertical strain at depth z, measured from the top of the subgrade, for an infinite subgrade

$(\varepsilon_{p\,Z=0})$ = plastic vertical strain at the top of the subgrade ($z = 0$)

k = regression constant

Total permanent deformation in the subgrade, $\delta = \int_0^{h_{bedrock}} \varepsilon_p(z)dz$.

Or:

$$\delta = \varepsilon_{p,z=0} * \int_0^{h_{bedrock}} e^{-kz} dz = \left(\frac{1 - e^{-kh_{bedrock}}}{k}\right) * \varepsilon_{p,z=0}$$

where:

δ = total plastic deformation of the subgrade

$h_{bedrock}$ = depth to bedrock

Cumulative damage concept—using subgrade vertical compressive strain, using Miner's law:

$$D_k = \sum_{j=1}^{k} \sum_{i=1}^{m} \frac{n_{ij}}{N_{fij}}$$

where:

D_k = subgrade distortion damage through season k; when it reaches 1, the pavement is assumed to have reached a state where it needs rehabilitation

n_{ij} = actual number of load repetitions for load class I (I = 1 ... m) during season j

N_{fij} = allowable number of load repetitions for load class I and season j to reach failure

The equivalent annual resilient modulus for unbound materials can be used to reduce the number of response computations.

12.3.2.9.1 For Aggregate Base and Subbase

$$M_{R(Aggregate)} = \frac{\Sigma[(M_{RA})_i * (UF)_i]}{[\Sigma UF_i]}$$

$$UF_i = 1.885 * 10^3 [M_{Ri}]^{-0.721}$$

where:

$M_{R(Aggregate)}$ = equivalent annual resilient modulus for unbound aggregate base and subbase materials, psi

UF_i = damage factor for unbound aggregate base and subbase materials in season i

M_{RAi} = resilient modulus measured in the laboratory for a moisture content in season i, psi

12.3.2.9.2   For Subgrade Soil

$$M_{R(Soil)} = \frac{\Sigma[(M_{RS})_i * (US)_i]}{[\Sigma US_i]}$$

$$US_i = 4.022*10^7[M_{RSi}]^{-1.962}$$

where:

$M_{R(Soil)}$ = equivalent annual or design resilient modulus of the subgrade soil, psi

M_{RSi} = resilient modulus of subgrade soil measured in the laboratory for a moisture content or physical condition within season i, psi

US_i = Damage factor for the subgrade soil in season i

The AASHTO 1993 method to determine a representative resilient modulus of soil, considering the damage due to different seasons, was presented in Chapter 11.

12.3.2.10   Equivalent Temperature Concept

Models are used to convert traffic (number of load applications) at a given temperature to an equivalent traffic at a reference temperature.

$$\sum TEF_i * ESAL_i = \text{equivalent } ESAL_i$$

where:

TEF_i = temperature equivalency factor for the ith temperature interval

$ESAL_i$ = design equivalent single-axle loads that accumulate during the temperature interval i

The critical location for rutting is selected close to the surface, such as 5 cm (2 in.) from the surface. The temperature gradient is given as follows:

$$G = \frac{(T_B - T_{2''})}{D}$$

where:

T_B = temperature at the bottom of the asphalt mix layer

$T_{2''}$ = temperature at a depth of 5 cm (2 in.)

D = the thickness of the asphalt mix layer less than 5 cm (2 in.)

A model relating number of load repetitions to a distortion failure to the temperature of interest is as follows:

$$\ln(N) = A_1 + A_2*T + A_3*G + A_4*T^2 + A_5*G^2 + A_6*T*G$$

where:
  N = number of load repetitions to failure
  T = temperature at critical pavement location
  G = temperature gradient
  A_x = coefficients form regression

Equivalency factors:

$$TEF_i = \frac{N_s}{N_i}$$

where:
  N_s = number of load repetitions to failure at a standard reference temperature
  N_i = number of load repetitions to failure at the ith temperature

The TEF_i is then used in the following equation mentioned earlier:

$$\sum TEF_i * ESAL_i = \text{equivalent } ESAL_i$$

12.3.2.11 El-Basyouny and Witczak Model (NCHRP, 2004)

Form:

$$\frac{\varepsilon_p(N)}{\varepsilon_r} = \left(\frac{\varepsilon_0}{\varepsilon_r}\right) e^{-\left(\frac{\rho}{N}\right)^\beta}$$

$$\text{For traffic level, } N = 1, \frac{\varepsilon_p(1)}{\varepsilon_r} = a_1 E^{b_1}$$

$$\text{For traffic level, } N = 2, \frac{\varepsilon_p(10^9)}{\varepsilon_r} = a_9 E^{b_9}$$

Assuming values for $\frac{\varepsilon_p}{\varepsilon_r}$ at different E to calculate the a and b coefficients,

$$\left(\frac{\varepsilon_0}{\varepsilon_r}\right) = \frac{(e^{(\rho)^\beta} * a_1 * E^{b_1}) + (e^{\left(\frac{\rho}{10^9}\right)^\beta} * a_9 * E^{b_9})}{2}$$

where:

$$\text{Log } \beta = -0.61119 - 0.017638 W_c$$

$$\rho = 10^9 \left(\frac{C_0}{[1-(10^9)^\beta]}\right)^{\frac{1}{\beta}}$$

$$C_0 = \left(\ln \frac{a_1 * E^{b_1}}{a_9 * E^{b_9}}\right)$$

The equilibrium moisture content of an unbound material can be predicted from the groundwater table depth and in-place modulus, using regression models which were developed on the basis of variables used with the enhanced integrated climatic model (EICM). The EICM uses the soil water characteristic curve (SWCC; variation of water storage capacity within the macro and micro pores of a soil, with respect to suction), generally plotted as variation of water content with soil suction. The SWCC prediction can be made by the use of equations proposed by Fredlund and Xing (1994). To avoid direct measurement of soil suction, the parameters of the Fredlund and Xing equation were fitted to soil index properties—percentage passing the No. 200 sieve (P_{200}), D_{60}, and Plasticity Index (PI). When the PI of a soil > 0, the SWCC parameters are correlated with P200 * PI; if the PI is zero, the parameters correlated with D_{60}. Other important properties are specific gravity of solids (G_s) and saturated hydraulic conductivity (k_{sat}), both of which can be estimated from P_{200} * PI and D_{60} values if no field or laboratory value is available.

The steps in the calculation of equilibrium moisture corresponding to different CBR and GWT depths content are shown in Table 12.10 (NCHRP, 2004).

TABLE 12.10
Steps in Determination of Equilibrium Moisture Content

1. $Mr = 2555(CBR)^{0.64}$

2. $P_{200}PI = \dfrac{\left(\frac{75}{CBR}\right)-1}{0.728}$

$D_{60} = \left(\dfrac{CBR}{28.091}\right)^{2.792516}$

3. Specific gravity, $G_s = 0.041(P_{200}PI)^{0.29} + 2.65$

4. Degree of saturation at optimum moisture content:

$S_{opt} = 6.752(P_{200}PI)^{0.147} + 78$

5. Gravimetric moisture content at optimum condition:

$W_{opt} = 1.3(P_{200}PI)^{0.73} + 11$

When PI = 0:

$w_{opt(T99)} = 8.6425(D_{60})^{-0.1038}$

For base courses:

$w_{opt} = w_{opt(T99)} - \Delta w_{opt}$

$\Delta w_{opt} = 0.0156[w_{opt(T99)}]^2 - 0.1465w_{opt(T99)} + 0.9$

For nonbase course layers:

$w_{opt} = w_{opt(T99)}$

6. Maximum compacted dry unit weight,

$\gamma_{d\,max\,comp} = \dfrac{G_s\gamma_{water}}{1+\frac{w_{opt}G_s}{S_{opt}}}$

7. Volumetric water content at optimum condition:

$\theta_{opt} = \dfrac{w_{opt}\gamma_{d\,max}}{\gamma_{water}}$

(Continued)

TABLE 12.10 (CONTINUED)
Steps in Determination of Equilibrium Moisture Content

8. Saturated volumetric water content:

$$\theta_{sat} = \frac{\theta_{opt}}{S_{opt}}$$

9. Volumetric water content is calculated from Fredlund and Xing (1994) equation:

$$\theta_w = C(h) * \left[1 - \frac{\ln\left(1 + \frac{h}{h_r}\right)}{\ln\left(1 + \frac{1.45*10^5}{h_r}\right)} \right]$$

10. The SWCC coefficients are calculated as follows:

$$a_f = \frac{0.00364(P_{200}PI)^{3.35} + 4(P_{200}PI) + 11}{6.895} psi$$

$$\frac{b_f}{c_f} = -2.313(P_{200}PI)^{0.14} + 5$$

$$c_f = 0.0514(P_{200}PI)^{0.465} + 0.5$$

$$\frac{h_r}{a_f} = 32.44e^{0.0186(P_{200}PI)}$$

11. The equilibrium gravimetric moisture content:

$$w_{equ} = \frac{\theta_w \gamma_{water}}{\gamma_{d\,max}}$$

Using the above calculations, the final regression equation was obtained as follows:

$$w_{equi}, in \% = 51.712 * CBR^{0.3586*GWT^{0.1192}}$$

12.3.3 Definition of Failure

With consideration of serviceability, an asphalt pavement is said to have failed by rutting when the rut depth at the surface exceeds 13 mm or 0.5 in.

The concept of incremental rutting in each layer and summing up rutting in all of the layers to get the total rutting has been used in several models. The steps include determination of the resilient strain, then the plastic strain, and then the rut depth at each depth, and finally summing up the ruts through the layer to get the total rutting in the layers.

1. $\varepsilon_{rz} = \frac{1}{|E^*|}(\sigma_z - \mu\sigma_x - \mu\sigma_y)$
2. $\varepsilon_p = f(\varepsilon_{rz})$
3. $\Delta R_{d_i} = \varepsilon_{pi}.\Delta h_i$
4. $R_d = \Sigma_{i=1}^n \Delta R_{d_i}$

12.4 SMOOTHNESS CONSIDERATION

Smoothness is a measure of ride quality as perceived by the user. Smoothness is affected significantly by rutting, rut depth variance, and fatigue cracking. The smoothness of the initial (newly constructed) pavement is an indicator of the future smoothness, since it indicates the overall quality of construction of the pavement as well as the rate of deterioration of smoothness over time. Different researchers have come up with different lists of factors affecting smoothness. They include initial smoothness, ESAL, age, base thickness, freezing index, initial International Roughness Index (IRI), type of subgrade, thickness of overlay, maximum and minimum temperatures, and annual number of wet days. Researchers have also listed the various distresses that affect smoothness. They include rut depth, potholes, depression and swells, transverse cracks, standard deviation or variance of rut depth, patching, and fatigue cracking.

The different models for smoothness are shown in Table 12.11.

TABLE 12.11
Different Models for Considering Smoothness

AASHO Road Test model:

$$PSR = 5.03 - 1.91 Log(1 + SV) - 0.01(C + P)^{0.5} - 1.38RD^2$$

where:

PSR = present serviceability rating (mean rating of the panel)

SV = slope variance

$$SV = \frac{\Sigma Y^2 - \left(\frac{1}{n}\right)(\Sigma Y^2)}{n - 1}$$

Y = difference between two elevations 9 in. apart

n = number of elevation readings

C = major cracking in ft per 1000 sq ft area

P = asphalt mix patching in sq ft per 1000 sq ft area

RD = average rut depth of both wheelpaths in in., measured at the center of a 4 ft span in the most deeply rutted part of the wheel path

Zero-Maintenance Pavements Study model (Darter and Barenberg, 1976)

$$PSR = 4.5 - 0.49RD - 1.16RDV^{0.5}(1 - 0.087RDV^{0.5}) - 0.13Log(1 + TC) - 0.0344(AC + P)^{0.5}$$

where:

RD = rut depth in both wheelpaths of the pavement, in.

RDV = rut depth variance, in.^{2} * 100

AC = Class 2 or Class 3 alligator or fatigue cracking, ft^2/1000 ft^2

TC = transverse and longitudinal cracking, ft^2/1000 ft^2

P = patching, ft^2/1000 ft^2

World Bank HDM III model (World Bank, 1995):

$$\Delta RI = 134e^{mt}MSNK^{-0.05}\Delta NE4 + 0.1114\Delta RDS + 0.0066\Delta CRX$$

$$+ 0.003h\Delta PAT + 0.16\Delta POT + mRI_t\Delta t$$

where:

ΔRI = increase in roughness over time period t, m/km

MSNK = factor related to pavement thickness, structural number, and cracking

$\Delta NE4$ = incremental number of equivalent standard-axle loads in period t

ΔRDS = increase in rut depth, mm

ΔCRX = increase in area of cracking (%)

ΔPAT = increase in surface patching (%)

ΔPOT = increase in total volume of potholes, m^3/lane km

M = environmental factor

RI_t = roughness at t years

(Continued)

TABLE 12.11 (CONTINUED)
Different Models for Considering Smoothness

Δt = incremental time period for analysis, years

t = average age of pavement or overlay, years

h = average deviation of patch from original pavement profile, mm

FHWA/Illinois Department of Transportation Study (Lee and Darter, 1995) model:

PSR = 4.95 − 0.685D − 0.334P − 0.051C − 0.211RD

where:

D = number of high-severity depressions (number per 50 m)

P = number of high-severity potholes (number per 50 m)

C = number of high-severity cracks (number per 50 m)

RD = average rut depth, mm

Models Suggested in the 2002 Design Guide:

The model suggested by work conducted under NCHRP 1-37A (Development of the 2002 Guide for Design of New and Rehabilitated Pavements) on the basis of LTPP data was used to modify existing models.

$IRI = IRI_0 + \Delta IRI_d + \Delta IRI_F + \Delta IRI_S$

where:

IRI_0 = initial IRI

ΔIRI_d = IRI due to distress

ΔIRI_F = IRI due to frost heave potential of the subgrade

ΔIRI_S = IRI due to swell potential of the subgrade

The models were then separated for asphalt pavements with different types of bases and subbases, as follows.

Unbound aggregate base and subbase:

$$IRI = IRI_0 + 0.0463\left[SF\left(e^{\frac{age}{20}} - 1\right)\right] + 0.00119(TC_L)_T + 0.183(COV_{RD})$$

$$+ 0.00384(FC)_T + 0.00736(BC)_T + 0.00155(LC_{SNWP})_{MH}$$

where:

IRI_0 = initial IRI

SF = site factor

$$SF = \left(\frac{(R_{SD})(P_{0.075} + 1)(PI)}{2*10^4}\right) + \left(\frac{\ln(FI+1)(P_{02}+1)[\ln(R_m+1)]}{10}\right)$$

R_{sd} = standard deviation of the monthly rainfall, mm

$P_{0.075}$ = % passing the 0.075 mm sieve

PI = % plasticity index of the soil

FI = average annual freezing index

P_{02} = % passing the 0.02 mm sieve

R_m = average annual rainfall, mm

$e^{\frac{age}{20}}$ = age term

$(FC)_T$ = fatigue cracking in wheelpath

COV_{RD} = coefficient of variation of rut depths = $\frac{SD_{RD}}{RD}$

SD_{RD} = standard deviation of rut depth along the pavement

RD = average rut depth

$(TC_L)_T$ = length of transverse cracks

$(BC)_T$ = area of block cracking as a % of the total lane area

$(LCS_{NWP})_{MH}$ = length of sealed longitudinal cracks outside wheelpath, ft/mile (moderate and high severity)

TABLE 12.11 (CONTINUED)
Different Models for Considering Smoothness

Asphalt-treated base:

$$IRI = IRI_0 + 0.0099947(Age) + 0.0005183(FI) + 0.00235(FC)_T + 18.36 \left[\frac{1}{(TC_S)_H} \right] + 0.9694(P)_H$$

where:

$(TC_S)_H$ = transverse crack spacing, m (high severity level)

$(P)_H$ = area of patching as a % of total area (high severity)

Cement or pozzolanic-treated bases:

$$IRI = IRI_0 + 0.00732 (FC)_T + 0.07647 (SD_{RD}) + 0.0001449 (TC_L)_T + 0.00842 (BC)_T + 0.0002115 (LC_{NWP})_{MH}$$

where:

$(LC_{NWP})_{MH}$ = length of longitudinal cracks outside wheelpath, m/km (moderate and high severity)

Model to predict SD_{RD}:

$$SD_{RD} = 0.665 + 0.2126 (RD)$$

where:

SD_{RD} = standard deviation of rut depths, mm

RD = mean rut depth, mm

HMA overlays placed on flexible pavements:

$$IRI = IRI_0 + 0.011505 (Age) + 0.0035986 (FC)_T + 3.4300573 \left(\frac{1}{(TC_s)_{MH}} \right)$$
$$+ 0.000723 (LC_s)_{MH} + 0.0112407 (P)_{MH} + 9.04244 (PH)_T$$

where:

$(TC_S)_H$ = average spacing of medium and high severity transverse cracks, m.

(LC_S) MH = medium and high severity sealed longitudinal cracks in the wheel path, m/km.

$(P)_{MH}$ = area of medium and high severity patches, percent of total lane area, %.

$(PH)_T$ = pot holes, percent of tatal lane are, %.

HMA overlays placed on rigid pavements:

$$IRI = IRI_0 + 0.0082627 (Age) + 0.0221832 (RD) + 1.33041 \left(\frac{1}{(TC_S)_{MH}} \right)$$

where:

RD = average rut depth, mm

12.5 TOP-DOWN CRACKING

Top-down cracking initiates at the surface, generally in the wheelpath. It can extend up to 1 inch or the depth of the top layer, and leads to ingress of water and raveling and formation of potholes.

The primary cause of top-down cracking is the failure of the surface mix as a result of excessive surface shear and tensile stresses from tires (generally truck tires). The excessive stress can be due to one or many factors, including anomalies in load distribution, axle configuration, axle loadings, tire types, and rigidity as well as tire condition and pressures. The failure could be accelerated because of poor mix qualities which could result from many factors, including segregation, inadequate density, and premature aging of asphalt binder.

The best approach to avoid such problems, where high tensile and shear stresses on the surface are expected, is to use HMA mixes with adequate resistance against such stresses—such as those with polymer-modified binders. Top-down cracks, once found, should be sealed as soon as possible, to avoid ingress of moisture and further rapid deterioration of the layer.

A fracture mechanics–based design method to consider top-down cracking is being developed at this time (NCHRP Project 01-42A, http://www.trb.org/trbnet/projectdisplay.asp?projectid=228; Transportation Research Board, n.d.).

QUESTIONS

1. Determine the indirect tensile strength of a sample from the following test data:

   Thickness of the sample: 2.9 in., breaking load = 4908 lb, diameter = 6 in.

2. Determine the resilient modulus of a sample from the following test data:

   Load = 250 lb, total deformation = $1.9567 * 10^{-4}$ in., thickness = 2 in.

3. Compute the dynamic modulus values at different frequencies and temperatures for the following mix:
   Asphalt test data:

T, C	η (10^{-6} Poise)
−28	237870351.6
−22	11729449.93
−16	783062.0737
−10	68238.2326
−4	7521.773921
2	1020.675635
8	166.5412517
14	32.01494597
20	7.122800176
26	1.805759608
32	0.514587666
38	0.162871888
44	0.056654775
50	0.021457668
56	0.008776083
62	0.003847795
64	0.002965245

Mix test data:

Air Void Content (%)	Effective Binder Content, % by Volume	Cumulative % Retained on the 3/4 In. Sieve	Cumulative % Retained on the 3/8 In. Sieve	Cumulative % Retained on the No. 4 Sieve	% Passing on the No. 200 Sieve
5.8	12	0	3	37	6

4. Using the Asphalt Institute (AI) and Shell equations, estimate the number of load repetitions to fatigue cracking failure for a HMA pavement with a tensile strain of 250 microstrain and a modulus of 3500 MPa. Air voids = 5%, asphalt content = 11% (by volume).

5. Using the Transportation Research Laboratory (TRL) and the U.S. Army Corps of Engineers models, determine the number of repetitions to subgrade rutting failure for a pavement with a vertical subgrade strain of 200 microstrain, with a resilient modulus of 7252 psi.

13 Distress Models and Material Characterization for Concrete Pavements

The major distresses of concrete pavement against which the pavement is designed consist of cracking, faulting, pumping, and punchouts. Degradation of smoothness and ride quality over time will severely affect the functionality of the pavement. These distresses are commonly caused by excessive stresses, strains, and deflections. The goal of the mechanistic-empirical (M-E) approach to pavement design is to reduce these stresses, strains, and deflections and maintain the pavement below a "critical" threshold to minimize deterioration and prevent failure.

To prevent cracking, the stresses are limited by a certain value. Stresses develop as a combination of factors such as expansive stresses due to temperature and moisture changes, or curling and warping, or such as fatigue stresses caused by repeated loading. To counter the cracking problem, some of the major test properties required are modulus of elasticity, the static and dynamic flexural strength, and coefficients of thermal expansion. The tests are discussed in the second part of this chapter.

Faulting is prevented by providing a stiff base with relatively low deformation under load. Minimizing curling and warping also helps in preventing faulting. Pumping is prevented by providing a free draining base, preventing moisture intrusion into the joints, and minimizing curling. Tests on fresh concrete are conducted for quality control and achieving consistency in the mix. The purposes are also to ensure good placement and finishing of the concrete.

Hence, the pavement design features that require the appropriate selection of slab thickness, base type and thickness, joint spacing, dowel design, and drainage, all contribute to reducing induced stresses, strains, and deflections and minimizing observed distress.

13.1 DISTRESSES AND MODELS

13.1.1 CRACKING

Cracking in concrete pavements can occur due to excessive static stresses or strains due to fatigue stress/strain failure. The high stresses that can cause cracking can be induced by a combination of factors, including restraint forces that are developed due to temperature and/or moisture volume changes that cause warping and curling and traffic loads. Figure 13.1 shows the deflected pavement shape at the onset of cracking with a combination of attributes that contribute to cracking.

13.1.1.1 Fatigue Cracking in JPCP

For JPCP, contraction joints are provided at close intervals to prevent the development of midslab cracks resulting from excessive stresses induced by loading, temperature, and moisture changes. The consequences of these stresses is an accumulation of damage in a critical portion of the slab. Enough accumulated stress will eventually initiate cracking, and further propagation of cracks will manifest at the bottom or surface of the PCC slab over time. Depending on the critical factors involved in crack propagation, fatigue cracking in JPCP can be divided into four major categories:

- Bottom-up transverse cracks
- Top-down transverse cracks

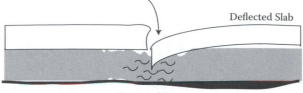

FIGURE 13.1 Deflections in Concrete Pavement.

- Longitudinal cracks
- Corner breaks

Bottom-up transverse cracks occur when a critical combination of loading and temperature curling creates summative stresses. For example, when truck axles are applying loads near the longitudinal edge of a slab and midway between the transverse joints, a critical tensile bending stress occurs at the extreme bottom fiber of the slab. This bending stress increases greatly when coupled with a positive thermal gradient that produces downward curling. The process initiates microcracks that eventually grow and propagate with repeated loading stresses that manifest into visible transverse cracks that affect the pavement performance. Bottom-up cracking models must therefore account for accumulation of fatigue damage caused by every truck axle load within a specified time increment, including appropriate thermal stresses and summed over the total design period.

Top-down transverse cracks also occur when a critical combination of loading and temperature curling creates summative stresses. In this case, when the truck front axle (steering axle) approaches a transverse joint within 10 to 20 ft, a large tensile stress is induced in the top of the slab between axles. This top tensile stress increases greatly when there is a negative thermal gradient through the slab, a built-in negative gradient from the slab construction, or significant drying shrinkage at the top of the slab. *Negative thermal gradient* is defined when the top of the slab is cooler than the bottom and upward curling occurs. Similarly to the case above, repeated loading stresses will eventually manifest into visible transverse cracks that affect the pavement performance. Models for top-down transverse cracks should account for similar stresses as presented for bottom-up transverse cracks.

Longitudinal cracks and corner breaks also occur due to combined stresses. The mechanism for this type of distress is similar to that of top-down transverse cracking except for the location of the critical stresses that develop. The critical induced stresses occur along both the longitudinal and transverse joints but near the corner.

Assessment of damage accumulation is essential for structural design to properly consider all the different load magnitudes that are applied to highway pavements. The fatigue damage concept as presented by Miner (1945) allows the designer to sum the fatigue damage from different loads of various magnitudes and to compute the resulting combined fatigue damage, as shown below.

$$\text{Fatigue Damage} = \sum_{ikj=1}^{p} \frac{n_{i,k,j}}{N_{i,k,j}}$$

where:

n_{ijk} = number of actual load applications under conditions represented by i, j, and k
N_{ijk} = number of allowable load applications under conditions represented by i, j, and k

The subscripts i, j, and k represent factors such as PCC strength and modulus, load, axle type, temperature gradient, location of applied load, and k, among others.

PCC pavements almost never fail due to a single application of load. Field and laboratory experience has shown that fatigue failure of PCC slabs and beams occurs due to repeated flexural stresses. Representative modeling of fatigue stresses should include more than loading, and this includes critical fatigue location, critical traffic load stresses, critical curling and warping stresses, seasonal variation in support, load transfer, warping and curling, concrete strength over time, and appropriate use of the concept of accumulated damage.

Based on numerous published data, it has been observed that the concrete will not reach failure by fatigue if the stress ratio (applied stress divided by maximum flexural stress) is less than 50% after approximately 20 million cycles or repetitions (Mindess, 1981, Neville, 1981).

Since concrete continues to gain strength with age, the flexural strength of normal concrete reaches only about 70% of its long-term potential strength at 28 days and approximately 90% at 90 days. Based on Wood (1991), a 20% increase in flexural strength above the 28-day strength can be used for analysis purposes.

Fatigue damage in PCC pavements can be determined using Miner's equation when the allowable number of load repetitions is known. Various models have been developed to determine the allowable number of load repetitions as a function of the stress ratio $(\frac{\sigma}{S_c})$. These models are presented next.

13.1.1.2 Zero-Maintenance Design Fatigue Model

The zero-maintenance design model presents the relationship between the number of stress applications to failure and the stress ratio in JPCP. The model was developed by Darter and Barenberg (1976, 1977), and is presented in the following equation.

$$\log N_f = 17.61 - 17.61 \left(\frac{\sigma}{S_c} \right)$$

where:

N = number of stress applications

$(\frac{\sigma}{S_c})$ = stress ratio

σ = applied cyclic stress

S_c = maximum static flexural strength

This relationship represents a 50% probability of failure, which is too high a risk for design purposes. Another equation was later developed that reduced the probability of failure to 24%. Developing an expression with a lower probability of failure in fatigue life was not possible due to the inherent variation in strength.

$$\log N_f = 16.61 - 17.61 \left(\frac{\sigma}{S_c} \right)$$

13.1.1.3 Calibrated Mechanistic Design Fatigue Model

This model was developed from data obtained from the U.S. Army Corps of Engineers' full-scale accelerated traffic tests and from the AASHTO Road Test. The model presents the relationship between the number of edge stress repetitions, the stress ratio, and the probability level (P) as presented by Salsilli et al. (1993).

$$\log N = \left[\frac{-\left(\frac{\sigma}{S_c} \right)^{-5.367} \log(1 - P)}{0.0032} \right]^{0.2276}$$

13.1.1.4   ERES-COE Fatigue Model

This model was developed by Darter (1988) using the Corps of Engineers' test results from full-scale field sections. This model was originally developed for airport pavements but was used for other applications later.

$$\log N = 2.13 \left(\frac{\sigma}{S_c} \right)^{-1.2}$$

13.1.1.5   PCA Fatigue Model

The PCA fatigue model is similar to the zero-maintenance model. However, the PCA model assumes that no fatigue failure will occur below a stress ratio of 45% even at an infinite number of cycles. The PCA equations as presented by Packard and Tayabji (1985) are as follows:

$$\text{For } \frac{\sigma}{S_c} \ge 0.55 : \; \log N_f = 11.737 - 12.077 \left(\frac{\sigma}{S_c} \right)$$

$$\text{For } 0.45 < \frac{\sigma}{S_c} \ge 0.55 : N_f = \left[\frac{4.2577}{\left(\frac{\sigma}{S_c} \right) - 0.4325} \right]^{3.268}$$

$$\text{For } \frac{\sigma}{S_c} \le 0.45 : \; N_f = \text{unlimited}$$

13.1.1.6   ARE Fatigue Model

This model was developed based on the AASHTO Road Test data from all sections that developed Class 3 and 4 cracking. The traffic loads were converted to 18-kip ESALs, and the maximum mid-slab stresses were computed using elastic layer theory.

$$N_f = 23,440 \left(\frac{\sigma}{S_c} \right)^{-3.21}$$

13.1.1.7   Vesic Distress Model

The Vesic and Saxena (1969) model was also based on the AASHO Road Test using a Serviceability Index of 2.5 as the failure criterion.

$$N_f = 225,000 \left(\frac{\sigma}{S_c} \right)^{-4}$$

13.1.1.8   RISC Distress Function

The RISC model is also dependent on the AASHTO Road Test, and failure was defined as the number of 18-kip ESALs required to reach a terminal Serviceability Index of 2.0.

$$N_f = 22,209 \left(\frac{\sigma}{S_c} \right)^{-4.29}$$

13.1.1.9 Transverse Cracking

Transverse cracks occur in both JPCP and JRCP. The cracking mechanism is, however, different for each type. Transverse cracks in JPCP are commonly caused by thermal curling or fatigue loading induced by traffic loads. However, transverse cracks in JRCP are commonly caused by thermal curling and shrinkage. As described in the previous section, the incremental damage induced by traffic and environmental conditions is converted to physical pavement distress such as transverse cracks, longitudinal cracks, and corner breaks using a calibrated damage-to-distress correlation model. Some of the models developed for predicting transverse cracking are presented next.

13.1.1.9.1 Transverse Cracking Prediction Models

13.1.1.9.1.1 RIPPER

The RIPPER model is based on a fatigue life depletion approach similar to the one used in the zero-maintenance design model. The maximum number of allowable repetitions "N" was computed using fatigue damage. The cumulative fatigue damage (n/N) is then computed and plotted against the corresponding percentage of slabs cracked. The model is as follows:

$$P = \frac{1}{0.01 + 0.03 \left[20^{-\log(^n/_N)} \right]}$$

where:
P = percentage of slabs cracked
n = actual number of 18-kip ESALs at slab edge
N = allowable 18-kip ESALs from the stress ratio from the ERES-COE fatigue model

Calibrated mechanistic design:

$$P = \frac{1}{0.01 + 0.0713[2.5949^{-\log(FD)}]}$$

where:
P = percentage of slabs cracked
FD = fatigue damage

13.1.1.9.1.2 RIPPER 2

This model is the follow-up to RIPPER and was developed by Yu et al. (1996) from the largest national-scale PCC pavement performance study. The approach was similar to that of RIPPER. In addition, the effect of wander was given more emphasis. A ratio termed the *pass-to-coverage (p/c) ratio* was defined, which gives the number of traffic passes needed to produce the same amount of fatigue damage at the critical location as one pass that causes the critical loading condition (for example, edge location). For example, if the (p/c) has a value of 100, this means that it takes 100 traffic passes to cause the same amount of damage as one load placed directly at the edge. The (p/c) is represented mathematically as follows:

$$^p/_c = \frac{FD_{Dii}}{\sum P(COV_{Dj})FD_{Dij}}$$

where:
FD_{Dii} = fatigue damage at a location D_i due to load at D_j
$P(COV_D)$ = probability that the load will pass through location D_j
FD_{Dij} = fatigue damage at a location D_i due to load at D_j

The results from RIPPER 2 show that fatigue cracking is sensitive to slab thickness, joint spacing, and shoulder type. A reduction in fatigue cracking in JPCP is achievable by increasing the slab thickness and/or reducing the joint spacing. A tied PCC shoulder and wider slabs can significantly reduce the amount of slab cracking. A stabilized bonded base can also reduce slab cracking. Hotter climates can cause more cracking because of the greater temperature gradients induced in the slab. The flexural strength or modulus of rupture (MOR) of the PCC will affect cracking. A higher MOR will result in less slab cracking. The modulus of subgrade reaction has a minor role, except for extreme conditions.

13.1.2 Transverse Joint Faulting in Jointed Plain Concrete Pavements

Transverse joint faulting is the difference in elevation between adjacent slabs across a joint. Excessive faulting reduces the performance of JPCP by increasing roughness and user discomfort and resulting in ultimate cracking and corner breaks. Faulting is caused by the erosion beneath the leave slab and the buildup on the approach slab of base fines by the action of pumping (Figure 13.2). For faulting to occur, the right combination of factors must exist, and they include heavy axle loading, poor load transfer across joints, base materials that are erodible, and free water in the base (water is the medium to transport fines under the slab). Faulting is commonly measured using a faultmeter (such as the Georgia Digital Faultmeter).

Although it is commonly agreed upon that pumping is the basic mechanism that leads to faulting, other factors can still influence faulting. Faulting increases in the cold months since joints open up due to PCC contraction and aggregate interlock is reduced. Joints with adequate dowel support increase joint load transfer and decrease faulting. However, if doweled joints are deficient or become so, faulting can result and has been observed in doweled joints. In addition, the base can usually have a higher moisture content that contributes to pumping. Conversely, in the warmer months the concrete expands and the joints are tighter, increasing aggregate interlock. Upward curling can contribute to slab corner deflection when subjected to heavy axle loads.

13.1.2.1 Models to Predict Faulting

M-E faulting models tend to incorporate the following parameters that contribute to faulting: slab corner deflections, base or subbase erodibility, and free water within the pavement structure. For example, a typical model is as follows:

$$\text{Fault} = C_1 DE^{\alpha} EROD^{\beta} WTR^{\gamma}$$

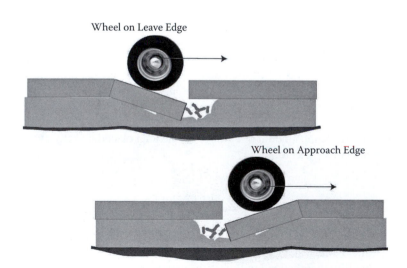

FIGURE 13.2 Faulting in Jointed Plain Concrete Pavements without Dowels.

where:
    Fault = average joint faulting, in. (or total joint faulting per mile)
    DE = a factor representing slab corner deflection
    EROD = a factor representing base/subbase erodibility
    WTR = a factor representing the amount and frequency of free water within the pavement system
    $C_1 \alpha \beta \gamma$ = regression constants

The models developed for faulting use mechanistic principles to predict slab corner deflections; however, base or subbase erodibility and moisture conditions are estimated using empirical methods.

13.1.2.2 Slab Corner Deflections

13.1.2.2.1   NAPCOM Approach

The Purdue method for analysis of rigid pavements (PMARP) uses energy methods to determine the volume of materials pumped from beneath a loaded slab given a deformation energy induced by wheel loads.

$$TE = \sum_{i-1}^{n} k_i A_i \Delta_i^2$$

where:
    TE = deformation energy imposed by an axle load, psi/in.
    K_i = modulus of subgrade reaction, psi/in.
    A_i = area associated with node i, in.^{2}
    Δ_1 = deflection at node i, in.

The deformed area under the loaded PCC slab with the potential for pumping is shown in Figure 13.3.

To model the energy that contributes to water movement under a slab, the differential deformation energy was defined as the difference in energy (pore water pressure) experienced at the interface of adjacent slabs beneath the joint.

$$DE = E_L - E_{UL}$$

where:
    DE = differential elastic deformation energy per unit area between the loaded and the unloaded slab, kN/mm
    E_L = elastic deformation energy per unit area imposed on the loaded slab, KN/mm
    E_{UL} = elastic deformation energy imposed on the unloaded slab, kN/mm

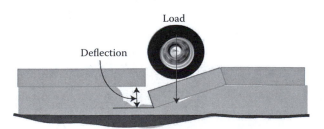

FIGURE 13.3 Deflection under Load.

The expression for DE can be rewritten as follows:

$$DE = \tfrac{1}{2}k(W_L + W_{UL})(W_L - W_{UL})$$

where the term $(W_L + W_{UL})$ is the differential deflection between the loaded and unloaded slabs. The higher the difference, the higher the probability of movement of material from beneath the leave slab to beneath the approach slab.

DE is used as the mechanistic parameter for computing the allowable number of axle load applications, N.

$$LogN_{ijk} = \alpha_1 - \alpha_2 Log(DE + 1)$$

where:

N_{ijk} = allowable axle load repetitions of axle type I and load j for foundation condition k
α_1, α_2 = regression constants
DE = differential slab deflections due to axle type I and load j for foundation condition k

The damage due to repeated differential slab deflections and pumping can be calculated using Miner's damage equation:

$$DAMAGE_{ijk} = \sum_{ijk} \frac{n_{ijk}}{N_{ijk}}$$

where:

$DAMAGE_{ijk}$ = damage for axle type i and load j for foundation condition k
n_{ijk} = expected number of axle load repetitions for each axle group i and load j for foundation condition k
N_{ijk} = allowable number of repetitions for each axle group i and load j for foundation condition k to a maximum faulting

13.1.2.2.2 PCA Model

In the PCA model, erosion damage is related to power. Power is defined as the rate of work each axle pass over the PCC slab joint exerts on the underlying pavement base/subbase materials. The PCA defines power exerted by each axle pass at the slab corner as the product of corner deflection (w) and pressure (p) at the slab base–subbase interface divided by the length of the deflection basin, which is a function of the radius of relative stiffness. The equation is as follows:

$$P = 268.7 \frac{p^2}{hk^{0.73}}$$

where:
P = power (rate of work)
p = pressure at slab-foundation interface, psi
h = slab thickness, in.
k = modulus of subgrade reaction, psi/in.

The allowable axle load repetitions for a given axle type and foundation condition can then be determined using the following:

$$N_{ijk} = 10^{(14.52 - 6.77[C_1 P_{ijk} - 9]^{0.103})}$$

where:

N_{ijk} = allowable axle load repetitions of axle type i and load j for foundation condition k

$C_1 = 1 - (k/2000 * 4/h)^2$ (approximately equal to 1 for granular materials and 0.9 for high-strength nonerodible materials)

P = power of axle type i and load j for foundation condition k.

Erosion damage can be calculated using Miner's equation as follows:

$$\text{EROSION}_{ijk} = \sum_{ijk} \frac{100 C_2 n_{ijk}}{N_{ijk}}$$

where:

EROSION_{ijk} = percentage of erosion damage for axle type i and load j for foundation condition k

C_2 = 0.06 for pavements without a shoulder and 0.94 for pavements with a tied concrete shoulder

n_{ijk} = expected number of axle load repetitions for each axle group i and load j for foundation condition k

N_{ijk} = allowable number of repetitions for each axle group i and load j for foundation condition k to a maximum faulting

13.1.2.2.3 Yao et al. (1990) Model

Prior to determining the number of repetitions to failure, the Foundation Damage Index is first determined and is expressed as follows:

$$DI = \frac{W^2}{hk^{1.27}}$$

where:

DI = Foundation Damage Index

w = corner deflection, in.

h = pavement thickness, in.

k = modulus of subgrade reaction, psi/in.

For DI > 1.862×10^{-4}:

$$N_{ijk} = \frac{0.5064}{DI_{ijk}^{0.312}}$$

where:

N_{ijk} = allowable number of repetitions for each axle group i and load j for foundation condition k

DI_{ijk} = deflection index for axle type i and load j for foundation condition k

For DI < 1.862×10^{-4}:

$$N_{ijk} = \frac{3.75 * 10^{-22}}{DI_{ijk}^{5.96}}$$

Using Miner's damage equation, the damage at a slab corner is as follows.

$$DAM_{ijk} = \sum_{ijk} \frac{n_{ijk}}{N_{ijk}}$$

where:

DAM_{ijk} = total damage for axle type i and load j for foundation condition k

n_{ijk} = expected number of axle load repetitions for each axle group i and load j for foundation condition k

N_{ijk} = allowable number of repetitions for each axle group i and load j for foundation condition k

The damage calculated in Yao et al.'s model may be used to assess the potential for erosion, pumping, and the loss of slab support due to void creation beneath the PCC slab. This damage index, however, was not correlated to field faulting data to predict transverse joint faulting. Yao et al.'s model shows that corner deflection could be used as a predictor for joint damage, erosion, and pumping that could ultimately induce transverse faulting.

13.1.2.2.4   NCHRP 1-26 (RIPPER Model)

This model was developed for transverse joint faulting in doweled jointed pavements. This was done since it is the belief that previously introduced joint-faulting models are not appropriate for doweled joints. Faulting in doweled joints is attributed to free vertical movement between the dowel and the contiguous or surrounding concrete. This can be attributed to excessive deformation of the concrete when its localized bearing strength is exceeded. The bearing stress between dowels and concrete has been introduced earlier and is presented here for completeness.

Dowel deflection is defined as follows:

$$D_o = \frac{P_t(2+\beta z)}{4\beta^3 E_d I_d}$$

where:

D_o = the deformation of the dowel at the face of the joint

P_t = shear force acting on the dowel (load on one dowel)

z = the joint width

E_d = modulus of elasticity of the dowel

$I_d = \frac{\pi d^4}{64}$ = the moment of inertia of the dowel

$$\beta = \sqrt[4]{\frac{Kd}{4E_d I_d}}$$

The bearing stress σ_b is proportional to the deformation:

$$\sigma_b = KD_o = \frac{KP_t(2+\beta z)}{4\beta^3 E_d I_d}$$

An estimation of the amount of load transferred by the dowel is commonly selected as 1,500,000 psi/in. (as adopted by NCHRP 1-26). For the NCHRP 1-26 study, the critical pavement response parameter which is the maximum bearing stress at the concrete dowel interface was directly correlated with measured transverse joint faulting.

13.1.2.2.5   Byrum et al. (1997) Model

Byrum et al. developed a mechanistic-based transverse faulting model for nondoweled jointed pavement. This model is based on the NAPCOM and PCA approaches and coupled with thermal gradients and shrinkage to accentuate the faulting due to upward slab curvature. Byrum proposed a curvature index to quantify the slab curvature due to thermal gradients and shrinkage. There was a good correlation between CI and transverse joint faulting. Hence, this approach may have a good potential for predicting joint pumping and faulting in the future. (Byrum, 1999):

$$CI = 0.5(LCI + RCI)$$

where:
    CI = slab curvature index
    LCI = left curvature index
    RCI = right curvature index

Curvature is defined as follows:

$$\text{Curvature} = \frac{\text{Slope}_{n+1} - \text{Slope}_n}{X_{n+1} - X_n}$$

where:
    Curvature = slab curvature for a given interval or segment
    X_n = horizontal distance along a given slab

Slope is defined as follows:

$$\text{Slope} = \frac{Y_{n+1} - Y_n}{X_{n+1} - X_n}$$

where:
    Slope = slope along the slab surface for a given interval or segment
    Y_n = vertical elevation along a given slab
    X_n = horizontal distance along a given slab

13.1.3 EROSION CHARACTERIZATION OF BASE/SUBBASE

The erodibility of a base is an important factor contributing to the potential pumping and faulting that may occur. Understanding, measuring, and predicting foundation material erodibility is, therefore, critical for developing accurate pumping and faulting models. Many factors may contribute to quantifying erodibility; they include soil compaction, shearing, friction, water pressure and mobility, and treatment for bound materials.

Characterization of material erodibility was attempted by many. PIARC (the World Road Association) developed specifications for characterizing the erosion potential of base and subbase material into five classes. (PIARC, 1987). This is presented in Table 13.1.

Other methods for developing more reliable predictors for erodibility include the *rotational shear device*, which uses water for erosion after applying a critical shear stress (this does not evaluate cohesive materials); the *jetting erosion device*, which is a modified rotational shear device that can evaluate cohesive soils; the *brush test*, which assesses the erosion due to a standard brush and load of a soil by measuring weight loss; and the *South African erosion test*, which uses a loaded wheel on a submerged confined soil specimen, and quantifies erodibility using the Erosion Index (L).

13.1.4 CHARACTERIZING FREE WATER WITHIN A PAVEMENT STRUCTURE

The enhanced integrated climatic model (EICM) is a one-dimensional couple heat and flow program that is used for analyzing pavement soil systems with climatic conditions. EICM can model numerous climate-related parameters such as rainfall, solar radiation, cloud cover, wind speed, and

54345

270   Pavement Engineering: Principles and Practice

TABLE 13.1
PIARC Recommendations for Erosion Potential of Base/Subbase Material

Class	Description
A: extremely erosion resistant	Class A materials are extremely resistant to erosion. The typical material of this category is lean concrete with at least 7 to 8% cement (or 6% with special addition of fines) or bituminous concrete with a bitumen content of at least 6%.
B: erosion resistant	Class B materials are five times more erodible (on average) than class A materials, but they still offer good guarantees of erosion resistance because they are far from the threshold at which erodibility increases exponentially. The typical material of this category is a granular material that is cement treated in the plant and contains 5% cement.
C: erosion resistant	Class C materials are five times more erodible (on average) than those in class B, and they are close to the threshold under which erodibility increases very rapidly in inverse proportion to the amount of binder. The typical material of this category is a granular material that is cement treated in the plant and contains 3.5% cement, or a bitumen-treated granular material with 3% bitumen.
D: fairly erodible	Class D materials are five times more erodible (on average) than those of Class C. Their low binder content makes their erosion-resistance properties highly dependent on construction conditions and the homogeneity of the material. The typical material of this category is a granular material treated in place with 2.5% cement. Also falling within this category are fine soils treated in place, such as cement-treated silt-lime and cement-treated sand. By extension, clean, well-graded, good-quality granular materials would also fall in this category.
E: very erodible	Class E materials are over five times more erodible (on average) than those of class D. Class E materials are untreated or very poorly treated mixes. The typical material of this category is an unprocessed treated material rich in fine elements, and especially untreated silt.

Source: PIARC (1987).

air temperature, but the prediction of moisture at the interface between the PCC slab and base is essential to modeling pumping and faulting.

13.1.5 PRS M-E Transverse Joint-Faulting Prediction Model

The PRS M-E model estimates JPCP transverse joint faulting in two ways: estimating faulting without considering the amount of consolidation around dowels, and estimating faulting while considering the amount of consolidation around dowels.

The model is presented as follows:

$$FAULT = DAMAGE^{0.275} * [0.1741 - 0.0009911 * DAYS90 + 0.001082 * PRECIP]$$

where:

FAULT = average transverse joint faulting per joint, in.
DAMAGE = n/N
    n = actual number of applied cumulative 18-kip ESALs
    N = allowable number of applied cumulative 18-kip ESALs
DAYS90 = number of days per year with the maximum temperature greater than 32°C (90°F)
PRECIP = average annual precipitation, in.

Statistics:

N = 511 (JPCP sections from LTPP and FHWA RIPPER databases).
R^2 = 56%
SEE = 0.029 in. (0.74 mm)

The following equation is used to compute allowable ESALs (N):

$$Log(N) = 0.785983 - Log(EROD) - 0.92991 * (1 + 0.40 * PERM * (1 - DOWEL)) *$$
$$Log(DE * (1 - 1.432 * DOWELDIA + 0.513 * DOWELDIA^2))$$

where:

N = allowable number of applied cumulative million 18-kip ESALs
$EROD$ = base erodibility factor for the base (value between 1 and 5)
$PERM$ = base permeability (0 = not permeable, 1 = permeable)
$DOWEL$ = presence of dowels (1 if dowels are present, 0 if dowels are not present)
$DOWELDIA$ = dowel diameter, in. (maximum allowed is 1.50 in.)
DE = differential energy density at a corner

DE at a corner is defined as the energy difference in the elastic base/subgrade deformation under the loaded leave slab and the unloaded approach slab. Another parameter used in the calculation of DE is that for the nondimensional aggregate interlock stiffness (AGG*) factor. When the percentage of consolidation around dowels is not considered, AGG* is computed using the following equation:

$$AGG* = (AGG/kL) = 2.3 * Exp(-1.987 * JTSPACE/L + DOWELDIA^{2.2})$$

where:

$AGG*$ = nondimensional aggregate interlock stiffness
AGG = aggregate load transfer stiffness, psi
k = dynamic modulus of subgrade reaction (dynamic k value, approximately twice the static value), psi/in.
L = slab's radius of relative stiffness, in.
$\quad = \left[E_{PCC} * h_{PCC}^3 \right]/(12 * (1 - v^2) * k)]^{0.25}$
E_{PCC} = PCC modulus of elasticity, psi
h_{PCC}^3 = slab thickness, in.
v = PCC Poisson's ratio (assumed to be equal to 0.15)
$JTSPACE$ = slab length (joint spacing), ft
$DOWELDIA$ = dowel diameter, in. (maximum allowed is 1.50 in.)

The computation of DE involves completing a multistep process in which maximum corner deflections are computed for loaded and unloaded conditions.

13.1.6 PUNCHOUTS IN CONTINUOUSLY REINFORCED CONCRETE PAVEMENTS

Punchouts are the main structural distress of continuously reinforced pavements (CRCP). Punchouts initiate as transverse cracks. Continued traffic loading on the pavement edge, coupled with loss of support for the slab, will eventually lead to a loss of load transfer at the transverse cracks. This scenario results in a section of the PCC slab between the transverse cracks to behave like a cantilever beam when subjected to wheel loads at the slab edge. When excessive stresses build up, micro-cracking develops, and eventually top-down cracking leads to punchouts (Figure 13.4).

Load transfer efficiency (LTE) at the transverse crack is critical to potential tensile stresses that develop at the cantilever joint due to wheel edge loading. If adequate aggregate interlock exists at the transverse crack, the progression of punchouts could be retarded. Adequate load transfer is provided by aggregate at the crack interface, sufficient steel reinforcement in the CRCP, and PCC slab thickness that may increase the surface area of aggregate interlock faces.

Loss of foundation support will increase the shear at the crack faces, reduce aggregate interlock, and increase the tensile stresses at the cantilever section, resulting in increased potential for

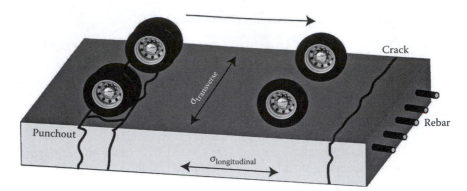

FIGURE 13.4 Punchout Resulting from Cantilever Effect at Slab Edge.

punchouts. Therefore, it is critical to maintain a nonerodible base. This can be accomplished by using cement or asphalt-treated bases.

13.1.6.1 Development of CRCP Punchout Models

Punchouts are commonly formed between transverse cracks spaced less than 2 ft apart. This, coupled with tensile stress development, LTE, and edge loading, could be used as a framework for developing CRCP punchout models. The following process is recommended:

- Assemble relevant pavement properties.
- Determine transverse crack spacing.
- Estimate number of crack spacing less than 2 ft.
- Estimate transverse crack width and mean LTE over time.
- Compute tensile stress at top of slab.
- Compute tensile stress-to-strength ratio and allowable number of load repetitions.
- Use Miner's damage model to compute fatigue damage for 2 ft crack spacing.
- Calibrate cumulative fatigue damage with field punchout distress data to predict punchouts.

Procedure for determining transverse crack development (after Abou-Ayyash, 1974):

- Determine tensile strength of CRCP slab at time t_n.
- Compute drying shrinkage Z_n, change in temperature ΔT_1 at time t_1.
- Calculate maximum tensile stress generated in CRCP.
- Compare computed CRCP stress to concrete tensile strength; cracking will occur if actual concrete slab stress exceeds concrete tensile strength.
- Increase time to t_2, and repeat the process. If failure occurs between time n and (n + 1), then use iterative tools to determine the exact time of cracking between intervals.
- Repeat the process at a later time and search for additional cracks, modifying the slab configuration to account for initial cracks.

Crack spacing in CRCP is affected by numerous factors such as concrete properties, age, environmental conditions, reinforcement content, placement and steel properties, slab geometry and boundary conditions, and construction factors, among others.

The bending tensile stress at the cantilever joint in the transverse direction is used to compute the critical pavement response parameter stress ratio (SR), and is calculated as follows:

$$SR = \frac{\sigma}{M_R}$$

George Green Library - Issue Receipt

Customer name: Singh Thakur, Kulbir

Title: Pavement engineering : principles and practice / Rajib B. Mallick, Tahar El-Korchi.
ID: 1005549916
Due: 04 Dec 2015 23:59

Total items: 1
09/10/2015 14:53

All items must be returned before the due date and time.
The Loan period may be shortened if the item is requested.

WWW.nottingham.ac.uk/library

where:

σ = top tensile stress in the transverse direction along the longitudinal crack for CRCP
M_R = PCC flexural strength

The number of allowable load repetitions is computed using the stress ratio by the following:

$$LogN_{ij} = \forall(SR)^{\Omega}$$

where:

SR = stress ratio
$\Omega\forall$ = regression constants

Fatigue damage is then determined using Miner's damage accumulation equation and can account for incremental damage due to traffic, concrete properties, age, foundation, and seasonal and climate variables. The cumulative fatigue damage due to all transverse crack spacings can be accumulated over the entire crack-spacing distribution using Miner's approach.

$$FD = \sum_{i=1}^{n} \frac{n_i}{N_i}$$

where:

FD = accumulated fatigue damage for ith crack spacing
n_i = number of applied ESALs
N_i = number of allowable ESALs
i = counter for crack spacing

13.1.6.2 Punchout Distress Model

An example of an M-E-based distress model that predicts punchouts as a function of accumulated fatigue damage due to bending in the transverse direction is given by the following:

$$PO = 100e^{-\left(\frac{FD}{\alpha}\right)^{\beta}}$$

where:

PO = total predicted number of punchouts per mile
FD = accumulated fatigue damage (due to slab bending in the transverse direction) assuming ith crack spacing
α β = calibration constants for punchout prediction

It can be seen that punchouts occur due to the deterioration of transverse cracks that develop in CRCP slabs. The transverse cracks are induced by climate-related tensile stresses and built-in curling and warping stresses, and deteriorate due to traffic loading, loss of base support, and loss of load transfer through the transverse crack. Fatigue cantilever stresses due to repeated wheel edge loads will eventually fracture the slab and cause it to settle into a moderate to severe punchout. Punchouts are influenced to a certain degree by the following parameters: PCC slab thickness, PCC stiffness or elastic modulus, base/subbase support, amount and depth of reinforcing steel, coefficient of thermal expansion for steel, shrinkage of PCC, PCC creep characteristics, slab-base friction, age, temperature, and humidity during construction.

13.1.7 SMOOTHNESS CONSIDERATIONS

Smoothness or roughness is a rider-perceived judgment of pavement ride quality. A pavement's functionality is based more on ride quality and user comfort than on structural performance.

Road roughness does not have one standard definition. Road roughness can be described as the lack of smoothness. The American Society of Testing and Materials (ASTM) defines roughness as follows: "The deviations of a pavement surface from a true planar surface with characteristic dimensions that affect vehicle dynamics, ride quality, dynamic loads, and drainage, for example, longitudinal profile, transverse profile, and cross slope." Or it has been described as the "irregularities in the pavement surface that affect the smoothness of the ride or induce vibration in traveling vehicles."

Road roughness is caused by either longitudinal or transverse distortions in the roadway, where longitudinal is in the direction of travel and transverse is perpendicular to the direction of travel. Longitudinal distortions can occur in either long wavelengths or short wavelengths. Long wavelengths are caused by consolidation of the pavement's foundation material and usually occur with high amplitudes but low frequency. Transverse distortions are usually the result of rutting or settlement within the road. Roadways with various frequencies and amplitudes of distortions cause oscillations in vehicles traveling over the pavement, and these oscillations are referred to as *roughness*.

To determine roughness or smoothness, a pavement profile must be obtained. A profile can be defined as the elevation profile representation of a two-dimensional slice of pavement surface. To obtain a profile, one can use a rod and level, a Dipstick™ (walking elevation measuring device), or an inertial profiler.

An inertial profilometer is a vehicle that is equipped with an accelerometer mounted to an axle and provides inertial reference by measuring the vertical acceleration of the vehicle. A sensor mounted to the vehicle's body measures the distance from a set point on the vehicle to the pavement. This is the vehicle's relative elevation displacement. Different types of distance-measuring sensors are commonly used such as laser, optical, and ultrasound. Since the sensors are digital instruments, they only measure on a set interval. This interval, expressed in terms of the longitudinal vehicle motion between readings, is known as *sample spacing*.

To determine the profile, the accelerations are first integrated twice to determine the vertical displacement of the axle. The change between the height and relative displacement is the pavement profile. The longitudinal distance is measured using a distance-measuring device (DMI).

There are two different types of inertial profilers—high speed and lightweight. High-speed devices are those which are usually van mounted, and operate at highway speeds. Lightweight inertial profilers are mounted on a smaller vehicle, similar to a golf cart. The high-speed profilers are typically used for network condition assessment, while the lightweights are used for pavement construction quality acceptance.

The International Roughness Index (IRI) is an index computed from the cumulative elevation changes over a distance, as determined from a longitudinal road. This is correlated to a standard quarter-car (one wheel's suspension) model, and measures the cumulative output of the suspension travel. It was developed for the World Bank in 1982 to establish a relevant, transportable, and time-stable measure of road roughness.

The IRI is an important ride quality statistic used for M-E smoothness modeling. The M-E design procedures include performance prediction models to predict the progression of certain key pavement distresses over the design life of a pavement.

Pavement smoothness or ride quality has been used since the AASHO Road Test. The present serviceability rating (PSR) concept was a subjective assessment of ride quality, while the PSI was developed as an objective quantitative assessment. The longitudinal portion of the PSI contributed greatly to the PSI value. And, therefore, smoothness has become used as the parameter that most closely relates to serviceability.

The first model to relate serviceability and distress was developed for the AASHO Road Test (as developed by Carey and Irick, 1960). The PSR was defined as follows:

$$PSR = C + A_1 R_1 + B_1 D_1 + B_2 D_2$$

where:

$\quad$ C, A_1,B_1,B_2 = regression coefficients

$\qquad\quad$ R_1 = function of longitudinal profile

$\qquad\quad$ D_1 = function of transverse profile

$\qquad\quad$ D_2 = function of surface distress

The rigid pavement model used for the AASHO Road Test was given as follows:

$$\text{PSR} = 5.41 - 1.78\log(1+\text{SV}) - 0.09(\text{C}+\text{P})^{0.5}$$

where:

$\quad$ PSR = present serviceability rating or panel mean rating

$\quad$ SV = slope variance

$\qquad$ C = major cracking in ft per 1000 sq ft area

$\qquad$ P = bituminous patching in sq ft per 1000 sq ft area

Slope variance is defined as follows:

$$\text{SV} = \frac{\sum Y^2 - \frac{1}{n}(\sum Y)^2}{n-1}$$

where:

$\quad$ Y = difference between two elevations 9 in. apart

$\quad$ n = number of elevation readings

Correlation statistics show that PSR is highly correlated with smoothness and distress for rigid pavements. It was also observed that distress greatly affects smoothness or profile measurements. Other researchers have studied and developed various models to correlate smoothness to pavement distress features. Unfortunately, the models vary widely based on the type and extent of distress. Increasing quantities and severities of distresses such as cracking, faulting, joint spalling, crack and joint deterioration, and punchouts will greatly influence smoothness degradation in pavements. Hence, the development of M-E pavement distress models will be an initial step to the development of smoothness models. Therefore, input parameters for distress models, such as traffic, construction, and environment over time, will also be applicable for smoothness predictor models.

Initial smoothness was found to be highly correlated to future smoothness and degradation of smoothness. Pavements that are built with a smoother profile tend to last longer and will remain smoother longer. This probably relates to the quality of the construction, where smoother pavements are a result of good construction practices. In addition, smoother pavements will be subject to less truck-dynamic loads than a rougher pavement surface (NCHRP 1-31). This can be seen in the incorporation of smoothness specifications that are part of many state highway agencies' new pavement construction projects. Smoothness of new pavements has become so important to the owners and riding public that many new construction specifications have included a bonus/penalty clause. If the contractor exceeds expectations, he or she is paid a bonus. If the contractor has not met smoothness specifications, he or she must pay a penalty or reconstruct the pavement if it is below a certain threshold.

Any factors that will contribute to changing the pavement profile will contribute to smoothness loss. This includes foundation movement (consolidation and swelling of soils, and frost-susceptible soils), maintenance activities (such as patching and crack filling), and initial smoothness. Therefore these conditions must be addressed when developing predictor models for pavement smoothness.

A model to predict smoothness could be structured as follows:

$$S(t) = S_0 + a_i D(t)_i + b_i M_j + c_i SF$$

where:

$S(t)$ = pavement smoothness over time (IRI, m/km)

B_0 = initial smoothness over time (IRI, m/km)

a_i b_j c_i = regression constants

$D(t)_i$ = ith distress at a given time

M_j = maintenance activities that significantly influence smoothness (e.g., patching and crack filling)

SF = site factors, which are defined as follows:

$$SF = AGE(SC_1 * SC_2 * SC_3)$$

A model example using the proposed structure above was developed under a FHWA study (Hoerner, 1998). Input parameters include initial IRI, cracking, spalling, and faulting. A site factor (SITE) was added to account for the effects of age, Freezing Index, and percentage of subgrade passing the 0.075 mm (No. 200) sieve. The model that predicts smoothness in IRI is given as follows:

$$IRI = IRI_0 + 0.013 * \%CRACKED + 0.007 * \%SPALL + 0.001 * TFAULT + 0.03 * SITE$$

where:

IRI_i = initial smoothness measured in m/km

%CRACKED = percentage of slabs with transverse cracking and corner cracks (all severities) expressed as a number between 0 and 100

%SPALL = percentage of joints with spalling (medium and high severities) expressed as a number between 0 and 100

TFAULT = total joint faulting cumulated per km, mm

SITE = site factor = $AGE (1 + FI)^{1.5} (1 + P_{0.075}) * 10^{-6}$

AGE = pavement age since construction

FI = Freezing Index, °C days

$P_{0.075}$ = percentage of subgrade material passing the 0.075 (No. 200) sieve, expressed as a number between 0 and 100

This model can predict future smoothness in IRI as a function of initial IRI, transverse slab cracking, transverse joint spalling, transverse joint faulting, and pavement site condition.

13.2 TESTS FOR CONCRETE

13.2.1 FLEXURAL STRENGTH TESTS

The flexural strength (also called the *modulus of rupture*, or MOR) is a specified design parameter for rigid pavements. This is due to the fact that pavements are stressed in bending during traffic loading and under curling stresses. The flexural strength can be determined using two different tests: a center-point load (Figure 13.5) and a third-point load. The beam size is typically $6 \times 6 \times 20$ in. for an aggregate maximum size less than 2 in. The difference between the two test methods is that the third-point method provides pure moment in the middle third of the span with no shear. In the center-point load, the point of fracture is unknown and there is both moment and shear contributing to the failure. The standard flexural strength test is given by AASHTO T-97 and ASTM C-78, (Flexural Strength of Concrete [Using Simple Beam with Third-Point Loading]), and AASHTO T-177 and ASTM C-293, (Flexural Strength of Concrete [Using Simple Beam with Center-Point Loading]).

The MOR of rupture is calculated using the following equation:

$$MOR = \frac{Mc}{I}$$

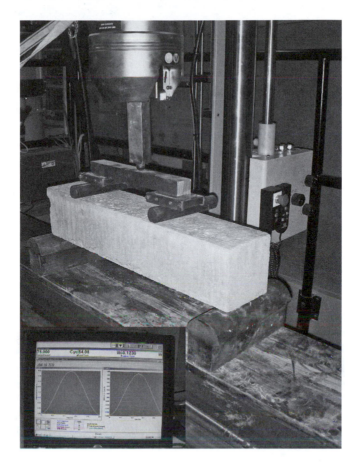

FIGURE 13.5 Center-Point Flexural Strength Test.

where:
  M = moment
  c = distance from neutral axis to the extreme fiber
  I = moment of inertia

Usually, mix designs are typically tested for both flexural and compressive strength; they must meet a minimum flexural strength, which is then correlated to measured compressive strengths so that compressive strength (an easier test) can be used in field acceptance tests.

13.2.2 COMPRESSIVE STRENGTH

The compressive strength test is performed using a 6×12 in. cylinder after 28 days of curing. The test is shown in Figure 13.6 and is conducted in accordance with AASHTO T-22 and ASTM C-39 (Compressive Strength of Cylindrical Concrete Specimens). Although the MOR is the specified strength for structural design in rigid pavements, the compressive strength is a much easier test to conduct and can be correlated reliably to the MOR. Hence, compressive strength is used by certain agencies for acceptance criteria. The compressive strength test is also less prone to testing variation than the flexural strength tests.

13.2.3 TENSILE STRENGTH

PCC tensile strength is important in pavement applications. Tensile strength may be used as a PCC performance measure for pavements because it best simulates tensile stresses at the bottom of the

FIGURE 13.6 Compressive Strength Test.

PCC surface course during traffic loading. Because of secondary stresses generated during gripping and pulling of a sample, a true direct tensile strength is difficult to conduct. Therefore, tensile stresses are typically measured indirectly using a splitting tension test.

A splitting tension test (AASHTO T-198 and ASTM C-496, Splitting Tensile Strength of Cylindrical Concrete Specimens) uses a standard 6 × 12 in. cylinder laid on its side. A diametral compressive load is then applied along the length of the cylinder until it fails (see Figure 13.7). Because PCC is much weaker in tension than compression, the cylinder will typically fail due to horizontal tension and not due to a vertical compression.

The modulus of elasticity for normal weight concrete (Ec), ranges between 2,000,000 and 6,000,000 psi. It can be determined by calculating the stress-to-strain ratio in the elastic range in accordance with ASTM C-469. Alternatively, it can be approximated using the ACI empirical equations by Ec = 57,000$\sqrt{\text{f'c}}$ (f'c in psi) or Ec = 5,000$\sqrt{\text{f'c}}$ (f'c in SI units).

The value of Poisson's ratio for concrete will range between 0.15 to 0.25, but is commonly selected as 0.20 to 0.21.

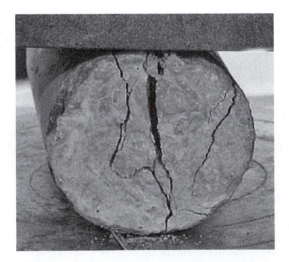

FIGURE 13.7 Tensile Strength Test.

13.2.4 Coefficient of Thermal Expansion Test

In this test (AASHTO TP-60), a concrete sample is placed in a temperature-controlled chamber, and its expansion is monitored at different temperatures. The values range from 6 to $13 * 10^{-6} / °C$.

13.2.5 Fatigue Testing for PCC

In a fatigue test for PCC, a repeated flexural loading is applied on a prismatic beam specimen measuring $3 \times 3 \times 15$ in. Loading is usually applied at the third points with a rate of 1 to 2 repetitions per second (2 Hz) for a duration of 0.1 s.

The stress (σ) in the extreme fiber of the beam is calculated using the following equation:

$$\sigma = \frac{3aP}{bh^2}$$

where:
  a = the distance from the edge support to the inner support
  P = the total applied load
  b = base of the beam
  h = height of the beam

Note: when the third points are equidistant and a = L/3, then the stress equation reduces to PL/bh^2.

The results are plotted on a semilog plot where the number of cycles to failure are plotted on the log scale—(log N_f) is plotted versus the flexural stress ratio (the quotient of the applied cyclic stress, σ, and the modulus of rupture, S_c, as determined by ASTM C78-84).

Based on numerous published data, it has been observed that the concrete will not reach failure by fatigue if the stress ratio is less than 50% after approximately 20 million cycles or repetitions (Mindess, 1981; Neville, 1981). Based on data published by others, Huang (1993) found that the relationship representing a 50% probability of failure is expressed as follows:

$$\log N_f = 17.61 - 17.61\left(\frac{\sigma}{S_c}\right)$$

Similarly, the Portland Cement Association recommends the use of the following fatigue equations:

$$\text{For } \frac{\sigma}{S_c} \geq 0.55 : \log N_f = 11.737 - 12.077\left(\frac{\sigma}{S_c}\right)$$

$$\text{For } 0.45 < \frac{\sigma}{S_c} \geq 0.55 : N_f = \left(\frac{4.2577}{\sigma/S_c - 0.4325}\right)^{3.268}$$

$$\text{For } \frac{\sigma}{S_c} \leq 0.45 : N_f = \text{unlimited}$$

QUESTIONS

1. A 12-in. long concrete cylinder with a 6-in. diameter failed at a load of 122,000 lb. Determine the compressive strength and approximate modulus of elasticity.
2. For a concrete slab of dimension 40 ft (length) by 12 ft (width) by 6 in. (thickness), determine the stresses developed due to a temperature drop of 20°C. Use concrete property values from question 1.

14 Mix and Structural Design of Asphalt Mix Layers

Structural and mix design of asphalt pavements are interlinked. The mix design step develops the most optimum combination of materials to produce a mix, and the structural design step utilizes the properties of the designed mix to determine the most optimum thickness and number of layers. Even though it seems that the structural design step should come after mix design, in many cases preliminary structural design is done prior to mix design. This is because, for most agencies, a database of information on relevant properties of typical mixes and materials exists.

Mix design is conducted mostly on the basis of consideration of volumetric properties of asphalt mixes. The materials, aggregates and asphalts, are selected on the basis of empirical and/or mechanistic test properties. These properties are considered to ensure that the mix will be resistant against the major distresses—thermal cracking, fatigue cracking, and rutting. Empirical and, in some cases, calibrated mechanistic models are available which relate volumetric properties to levels/amounts of distress. These models are considered to determine appropriate optimum levels of the volumetric and other properties during mix design.

Structural design of asphalt pavements can be done primarily in two ways—either through the use of charts and nomographs that have been developed on the basis of experiments and/or experience (empirical method), or part through the use of principles of mechanics and part through results obtained from experiments and/or experience (mechanistic-empirical).

Note that while the tendency for structural design has been to move from purely empirical to mechanistic-empirical methods (to reduce uncertainties regarding extrapolation of data), mix design has more or less remained primarily dependent on empirical relationships between volumetric properties and distress levels in asphalt mixes (although more sophisticated models relating volumetric properties to mechanistic properties have been developed). The two major advancements made in mix design are the development and adoption of better mix compaction methods and mix evaluation methods in the laboratory. For structural design, the major advancements have been in terms of better testing methods, sophisticated data analysis techniques, and field calibration of models through instrumentation and accelerated loading and testing.

14.1 PHYSICAL AND VOLUMETRIC PROPERTIES OF ASPHALT MIX

Physical and volumetric properties are used extensively in mix design, as they help us to convert weight to volume and vice versa, and as they have been shown to have significant correlations with the performance of asphalt mixes. Correlations between volumetric properties and structural properties are also used in structural design, when testing is not possible or actual test data are not available.

Figure 14.1 shows a simplified schematic of a compacted asphalt mix (that is, what we find in pavements). It consists of three primary components—the aggregate, the asphalt binder, and some air voids. Before the asphalt is added to the aggregates, there were only aggregates and some voids within the aggregate structure. Part of the total voids is filled with asphalt binder, and the rest remains as air voids. Of the asphalt binder, part of it is absorbed by the aggregate, and part remains as effective asphalt binder on the aggregate surface. The relative proportions of the volume of the air voids with respect to the total volume and with respect to the volume of the total voids, and the relative proportion of the volume of the effective asphalt binder with respect to the volumes of the total voids, are important volumetric parameters—those that have been shown to have significant effects on the performance of the asphalt mix. In addition, the proportion of mass of the effective asphalt with respect to the total mass is also important, since it tells us how much of the asphalt binder that we add to the aggregates is actually being effective in providing all the good properties that the asphalt binder is suppose to provide.

The different parameters discussed above can be separated into the following (as explained in Figure 14.1).

 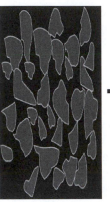 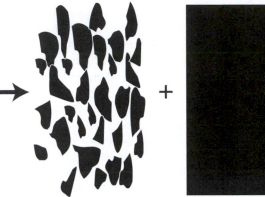

Actual   Conceptual   Aggregate + Air + Asphalt

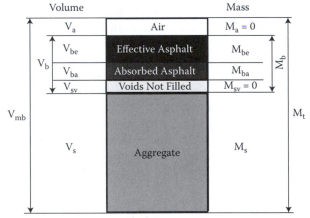

Block Diagram

Definitions

$$VTM,\% = \frac{V_a}{V_{mb}}*100$$

$$P_{ba},\% = \frac{M_{ba}}{M_s}*100$$

$$VMA,\% = \frac{V_a+V_{be}}{V_{mb}}*100$$

$$P_{be},\% = \frac{M_{be}}{M_t}*100$$

$$VFA,\% = \frac{V_{be}}{V_a+V_{be}}*100$$

Calculation of Volumetric Parameters Requires the Determination of Density or Specific Gravity Values of the Different Components

FIGURE 14.1 Conceptual Block Diagram.

Voids in total mix (VTM): Total volume of pockets of air in between the *asphalt-coated aggregates* in a compacted asphalt mix, expressed as a percentage of the total volume of the mix. Neither a high nor a very low VTM is desirable in a compacted asphalt mix—a VTM of 3–5 is the optimum range for adequate stability and durability.

Absorbed asphalt (P_{ba}): The mass of asphalt binder that is absorbed by the aggregates, expressed as a percentage of the mass of the aggregate. There is no specific limit on absorption except the fact that highly absorptive aggregates (> 2% absorption) may not be used for economical reasons (would need a relatively high amount of asphalt binder). However, if aggregates with relatively high absorption are used, the "aging" process during mix design in the laboratory needs to be modified to accurately simulate the expected high absorption during production in the plant.

Effective asphalt (P_{be}): The mass of the asphalt binder that is not absorbed by the aggregate (total—absorbed), expressed as a percentage of the mass of the total mass of the mix.

Voids in mineral aggregate (VMA): Total volume of the void spaces in between the *aggregates* in a compacted asphalt mix, part of which is filled with effective asphalt binder, and part with air, expressed as a percentage of the total volume of the mix. Generally, a minimum VMA is desirable to provide space for adequate asphalt binder for durability, and air voids for stability, although a very high VMA would lead to unstable mixes as well. Generally, for surface and base course paving mixes, the specified minimum VMA values range from 12 to 16, depending on the nominal maximum aggregate size (NMAS) of the mix—the higher the NMAS, the lower is the minimum VMA. This is based on the principle that a gradation with a higher NMAS has a relatively lower surface area (of aggregates) and hence would require a relatively lower amount of asphalt to provide an optimum film thickness of asphalt binder.

Voids filled with asphalt (VFA): The part of the total void spaces in between the aggregates in a compacted mix that is filled with asphalt binder, expressed as a percentage. A low VFA indicates an inadequate amount of asphalt binder or too high VTM, which are undesirable for durability, whereas a high value indicates a relatively low VTM, which can cause stability problems, Hence a range, generally 65–80, is specified on the basis of traffic levels—a higher volume of traffic would need a mix with lower asphalt content for stability, and hence a lower VFA. Note that the volumetric properties need the determination of density or specific gravity.

Next consider Figure 14.2, in addition to Figure 14.1. They show definitions of the different parameters discussed above, in equations, and point out the properties that we need from tests to determine these volumetric properties. These are listed below.

14.1.1 Bulk-Specific Gravity of Compacted Asphalt Mix (G_{MB})

This test determines the ratio of weight in air of a compacted asphalt mix specimen at a certain temperature to the weight of an equal volume of water at the same temperature. For dense graded asphalt paving mixes, the test can be run according to AASHTO T-166 (Bulk Specific Gravity of Compacted Hot Mix Asphalt Using Saturated Surface Dry Specimens). The test consists of measuring the weight of the sample in air, in water (after immersing for 3–5 minutes), and then measuring the saturated surface dry (SSD) weight (after patting the surface dry). The bulk-specific gravity is determined from the following formula.

$$G_{mb} = \frac{A}{B - C}$$

Working Formula

$$VTM,\% = \left(1 - \frac{G_{mb}}{G_{mm}}\right)*100$$

$$VMA,\% = \left(1 - \frac{G_{mb}(1 - P_b)}{G_{sb}}\right)*100$$

$$VFA,\% = \left(\frac{VMA - VTM}{VMA}\right)*100$$

$$P_{ba} = 100*\left(\frac{G_{se} - G_{sb}}{G_{sb}G_{se}}\right)*G_b$$

$$P_{ba},\% = P_b - \frac{P_{ba}}{100}\,P_s$$

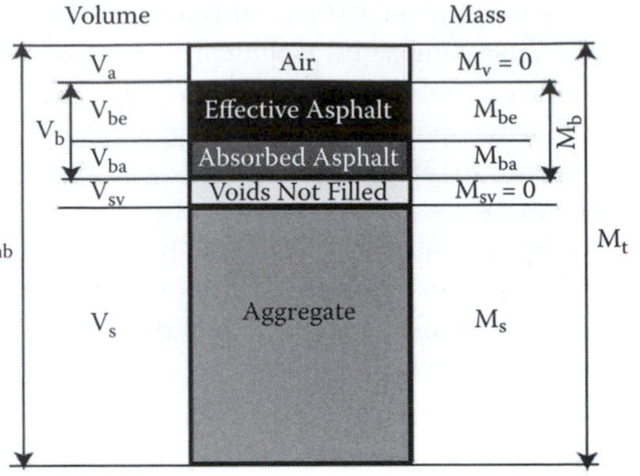

G_{mb} = Bulk Specific Gravity of Compacted Mix

$$G_{mb} = \frac{Mass\ of\ Aggregate\ and\ Asphalt}{Volume\ of\ Aggregate,\ Asphalt\ and\ Air} = \frac{M_t}{V_{mb}}$$

G_{mm} = Maximum (Mix) Specific Gravity

$$G_{mm} = \frac{Mass\ of\ Aggregate\ and\ Asphalt}{Volume\ of\ Aggregate\ and\ Asphalt} = \frac{M_s + M_b}{V_{mb} + V_a}$$

P_b = Percentage of Asphalt Binder by Weight of Mix, or Asphalt Content, %

$$P_b = \left(\frac{M_b}{M_t}\right)*100$$

$$G_{se} = \frac{Mass\ of\ Aggregate}{Volume\ of\ Aggregate,\ Im\,permeable\ and\ permeable\ voids\ in\ the\ agregate-volume\ of\ absorbed\ asphalt}$$

Calculated as Follows:

$$G_{se} = \frac{100 - P_b}{\dfrac{100}{G_{mm}} - \dfrac{P_b}{G_b}}$$

FIGURE 14.2 Different Parameters Required for Calculation of Volumetric Properties.

where:
  A = weight in grams of the specimen in air
  B = weight in grams of the saturated surface dry specimen in air
  C = weight in grams of the specimen in water

The bulk-specific gravity is multiplied by the density of water to obtain bulk density.

Specimens that contain open or interconnected voids or absorb more than 2% water (or both) can be tested by using either ASTM D-1188 (Bulk Specific Gravity and Density of Compacted Bituminous Mixtures Using Coated Samples) or AASHTO TP-69 (Bulk Specific Gravity and Density of Compacted Asphalt Mixtures Using Automatic Vacuum Sealing Method). Note that for testing of field cores which have been obtained by sawing or coring and which contain a significant amount of water, the procedure is the same, except that the dry mass (mass in grams of the specimen in air) is obtained last after drying the specimen at 110°C to constant mass.

14.1.2 THEORETICAL MAXIMUM DENSITY (TMD) OR MAXIMUM SPECIFIC GRAVITY OF THE MIX (G_{MM})/RICE SPECIFIC GRAVITY (AFTER JAMES RICE)

This test determines the ratio of weight of a unit volume of an uncompacted asphalt mix sample to the weight of an equal volume of water at the same temperature. This test can be run according to ASTM D-2041 (Theoretical Maximum Specific Gravity and Density of Bituminous Paving Mixtures) or AASHTO T-209 (Theoretical Maximum Specific Gravity and Density of Hot-Mix Asphalt Paving Mixtures). A specified amount of mix (ranging in weight from 1.1 to 5.5 lb for a range of maximum size of aggregate of #4 to 1 in.) is warmed in an oven and taken out, and the particles are separated from each other. The mix is weighed in air, and then placed in a pycnometer with sufficient water to cover it. A vacuum pump is then connected to the pycnometer and turned on to remove entrapped air from the mix. The mix is agitated during this procedure to help remove the air. At the end of this step, the pycnometer is submerged in a water bath, and its weight (along with the mix inside it) is noted. The following expression is used for the calculation of G_{mm}.

$$G_{mm} = \frac{A}{A - B}$$

where:
  A = weight in grams of sample in air
  B = weight in grams of sample in water

Note that there is a check for making sure that porous aggregates without an adequate amount of asphalt binder cover are not getting saturated with water during the vacuum procedure.

The different specific gravities of aggregates and asphalt binder have been discussed in Chapters 8 and 9, respectively. The importance of the three specific gravities of aggregate is discussed in the following paragraphs.

The need for the determination of the different specific gravities of aggregate arises from the fact that all aggregates absorb water (and hence asphalt as well) to some extent. Determination of the different specific gravities allows the estimation of such absorbed water, and then the estimation of absorbed asphalt. Whatever the specific gravity might be, the testing procedure always has the basic fundamental objective of determination of weight (or, more precisely, mass) divided by volume. The weight is either dry (for G_{sb}) or saturated surface dry (for $G_{sb\text{-}SSD}$), and the volume is (volume of aggregate + surface voids) for G_{sb}, and volume of aggregate only for apparent specific gravity (G_{sa}; Figure 14.3). Using G_{sb} and G_{sa}, as shown in Figure 14.2, one can determine the absorption.

Note that when the specific gravities of different size fractions are used (for a blend of aggregates), the average specific gravity value of the blend needs to be calculated and used in the determination of the volumetric properties of the mix. The procedure is explained in Figure 14.3.

The concept of effective specific gravity arises in an asphalt-coated aggregate. In this case, for the mass divided by volume parameter, the mass is still the dry aggregate, but the volume consists of the volume of the aggregate plus the volume of the surface voids that are not filled with asphalt. This definition is important since the "effective" aggregate volume is used to calculate the absorbed asphalt and hence the effective asphalt binder content. Effective specific gravity is a calculated parameter, from G_{mm}, as shown in Figure 14.2.

A worked-out example of the use of the different parameters is shown in Figure 14.4.

Bulk Specific Gravity

$$G_{sb}, Dry = \frac{Mass\ of\ Dry\ Aggregate}{Volume\ of\ Aggregate\ and\ Surface\ Voids}$$

$$G_{sb}, SSD = \frac{Mass\ of\ Dry\ Aggregate\ and\ Water}{Volume\ of\ Aggregate\ and\ Surface\ Voids}$$

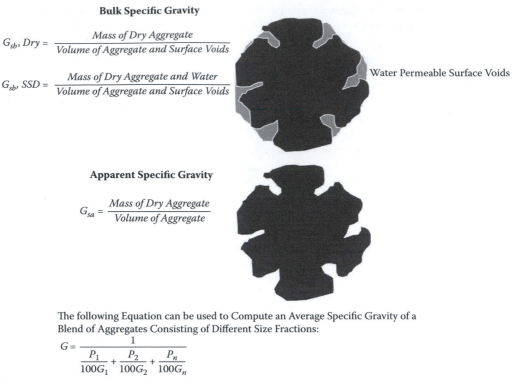

Water Permeable Surface Voids

Apparent Specific Gravity

$$G_{sa} = \frac{Mass\ of\ Dry\ Aggregate}{Volume\ of\ Aggregate}$$

The following Equation can be used to Compute an Average Specific Gravity of a Blend of Aggregates Consisting of Different Size Fractions:

$$G = \frac{1}{\dfrac{P_1}{100G_1} + \dfrac{P_2}{100G_2} + \dfrac{P_n}{100G_n}}$$

Where
G = Average specific gravity
G_1, G_2, ...G_n = Appropriate specific gravity of each size fraction
P_1, P_2, ...P_n = Percentage (by mass) of each fraction in the total sample

FIGURE 14.3 Bulk and Apparent Specific Gravity of Aggregates.

14.2 MIX DESIGN METHODS

In the following paragraphs, three different mix design systems are presented. Note that all of the mix design systems have the same objectives of selecting the most desirable aggregates and asphalts and combining them in the most appropriate ratio to obtain the optimum volumetric properties.

14.2.1 Hveem Method (ASTM D-1560, D-1561)

The Hveem method was developed by Francis Hveem of the California Division of Highways in the 1920s. It is intended for mixes with a maximum aggregate size of 1 in. and has been used primarily for the design of dense graded mixes. Currently, it is used mainly in some of the Western states in the United States.

The primary steps in this procedure are the following:

1. Select suitable aggregates, gradation, and asphalt binder for the project.
2. Determine the approximate asphalt (binder) content as a percentage of dry aggregate using the centrifuge kerosene equivalent (CKE) test procedure and aggregate gradation.
3. Prepare specimens, 2.5 in. high and 4 in. in diameter, for a range of asphalt contents at, below, and above the estimated asphalt content from Step 1, using a kneading compactor, at recommended temperatures (for specific asphalt grades).

Example: Determine the different volumetric properties of the mix, for which the following information is provided: $G_{mb} = 2.335$, $G_{mm} = 2.530$, $G_b = 1.02$, $P_b = 5.5\%$ (by mass of total mix), G_{sb} (blend of aggregate) = 2.650

Steps

1. Determine Density of the Mix: Density = $G_{mb} * Y_w = 2.335*1g/cm^3 = 2.335\ g/cm^3$;

2. Determine Masses: Assume 1 cm^3 of Compacted HMA; Total Mass = 2.335 g;
   $M_a = 0$; $M_b = (5.5/100)*2.335 = 0.128$ g; $M_s = (95/100)*2.335 = 2.218$ g;

3. Determine Total Volume of Asphalt Binder: $V_b = M_b/(G_b * Y_w) = 0.128/(1.02*1.00) = 0.125\ cm^3$

4. Determine Volume of Bulk Aggregate: $V_{sb} = M_s/(G_{sb} * Y_w) = 2.218/(2.650*1.00) = 0.837\ cm^3$

5. Determine Volume Associated with G_{mm}: $V_{mm} = M_t/(G_{mm}*Y_w) = 2.335/(2.530*1.00) = 0.923\ cm^3$

6. Determine Volume of Air: $V_a = V_{mb} - V_{mm} = 1.00 - 0.923 = 0.077\ cm^3$

7. Determine Volume of Effective Asphalt Binder: $V_{be} = V_{mm} - V_{sb} = 0.923 - 0.837 = 0.086\ cm^3$

8. Determine Volume of Absorbed Asphalt Binder: $V_{ba} = V_b - V_{be} = 0.125 - 0.086 = 0.039\ cm^3$

9. Determine Volume of Effective Aggregate: $V_{se} = V_{sb} - V_{ba} = 0.837 - 0.039 = 0.798\ cm^3$

10. Determine Mass of Effective Asphalt Binder: $M_{be} = V_{be} * G_b * Y_w = 0.086*1.02*1.00 = 0.088$ g

11. Determine Mass of Absorbed Asphalt Binder: $M_{ba} = V_{ba} * G_b * Y_w = 0.039*1.02*1.00 = 0.039$ g

12. Determine Volumetric Properties: VTM, % = $100\% * V_a/V_{mb} = 100\%*0.077/1.00 = 7.7\ \%$;

    VMA, % = $100\%*(V_a + V_{be})/V_{mb} = 100\%*(0.077+0.086)/1.00 = 16.3\%$;

    VFA, % = $100\%*V_{be} (V_{be} + V_a) = 100\% * 0.086/(0.086+0.077) = 52.76\%$;

    Effective Specific Gravity of Aggregate, $G_{se} = M_s/(V_{se} * Y_w) = 2.218/(0.798*1.00) = 2.779$;

    Percent Absorbed Asphalt Binder: P_{ba}, % = $100\%*M_{ba}/M_s = 100\%*0.039/2.218 = 1.76\%$

    Percent Effective Asphalt Binder: P_{be}, % = $100\%*M_{be}/M_t = 100\%*0.088/2.335 = 3.77\%$

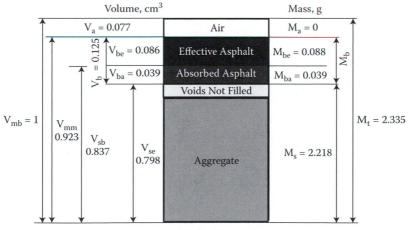

FIGURE 14.4 Worked-Out Example on Calculation of Different Volumetric Properties of a Compacted Hot Mix Asphalt (HMA).

4. Test using a stabilometer and a swell test apparatus.
5. Compute stabilometer value, swell, density, and air voids (using maximum specific gravity).
6. Determine the optimum asphalt content by selecting the maximum asphalt content that meets the minimum air voids and stabilometer value criteria. The other criteria that must be met are those for maximum swell and observed flushing of asphalt.

The key steps are illustrated in Figure 14.5, with the relevant charts in Figure 14.6. A kneading compactor is shown in Figure 14.7. The criteria are shown in Table 14.1.

14.2.2 MARSHALL METHOD

The Marshall method was developed by Bruce Marshall of the Mississippi Highway Department in the 1930s, and was refined by the U.S. Army Corps of Engineers in the 1940s and 1950s. Although intended for use of mixes with a maximum aggregate size of 1 in., the procedure has since been modified to accommodate large-size aggregates (up to 1 1/2 in.). This method continues to be used widely in airport pavement design jobs, and has been used for designing asphalt emulsion mixes, as well as for compaction and testing of quality control samples during construction.

Following are the main steps in this method.

1. Select aggregates using available test methods, and gradation, considering VMA; select appropriate viscosity-grade asphalt, conduct viscosity tests at different temperatures, and determine mixing and compaction temperature from temperature versus viscosity data.
2. Prepare specimens using a Marshall compactor (handheld or mechanical) with an appropriate number of blows per side (considering the traffic level of the project) at the mixing and compaction temperature determined in Step 1.
3. Determine bulk-specific gravity.
4. Condition the specimens, and run a Marshall stability and flow test.
5. Compute VTM, VMA, and VFA, using maximum specific gravity of the mix and specific gravity of aggregates, and stability and flow from test data obtained in Steps 3 and 4.
6. Determine the optimum asphalt content using criteria from the appropriate method, on the basis of VTM, VMA, VFA, stability, and flow values, for the expected traffic level in the project.

The method is explained in Figure 14.8, with the relevant charts and criteria in Figure 14.9. The fabrication of a Marshall specimen consists of placing a specified amount of mix in a mold assembly (with a protection paper disk at the bottom and top) and compacting each side of the mix inside the mold using a drop weight hammer. Generally 1,200 grams of HMA material are used to fabricate a Marshall specimen, which is 4 in. in diameter and 2 1/2 in. in height, or 509 to 522 cc in volume. The Marshall mold assembly consists of the collar (2 3/4 in. high) on the mold (3 7/26 in. high) which sits on a base (4 in. in diameter). The compaction hammer has a 10 lb sliding weight with a free fall height of 18 in., which could be raised automatically by a mechanism or manually (Figure 14.10). The mold assembly sits on a compaction pedestal which consists of an 8 in. by 8 in. by 18 in. wooden post capped with a 12 in. by 12 in. by 1 in. steel plate. The post is secured to a concrete floor by brackets, and the cap should be level. The mold assembly and compaction hammer face are heated to 300°F before compaction. The mix is transferred carefully inside the mold, with the collar on top of it, and spaded vigorously around the perimeter and at the center. After a paper disk is put on top of the smoothed surface of the mix in the mold, the mold assembly is transferred to the pedestal, and compaction starts when the temperature of the mix reaches a temperature at which the asphalt binder has a specified viscosity, for example, 280 ± 30 Cs. Each side of the specimen is compacted by the specified number of blows, the paper disks are peeled off the surface, and the specimen is left to cool down to a temperature that can be handled easily, after which the specimen is extruded out of the mold with an extruder.

1. Compute Equivalent Surface Area of the Aggregate Sample for the Selected Aggregate Blend

Sieve Size (inch)	Percent Passing (Column B)	Surface Area Factor, ft²/lb (Column C)	Surface Area, ft²/lb Column B*Column C
Maximum Size	100	2	
No. 4		2	
No. 8		4	
No. 16		8	
No. 30		14	
No. 50		30	
No. 100		60	
No. 200		160	
		Total Surface Area	Sum of Above Cells

2. Determine Centrigauge Kerosene Equivalent (CKE) of Fine Aggregates from the CKE Test and Percent Oil Retained for Coarse Aggregates from Surface Capacity Test

$$CKE_{corrected} = CKE^* \frac{SG}{2.65}$$

$$CKE = \frac{W_w - W_D}{W_D} *100$$

SG = Specific Gravity of Fine Aggregates

W_w = Weight of Aggregate + Absorbed Kerosene

W_D = Weight of Aggregate

$$Percent\ Oil\ Retained_{corrected} = Percent\ Oil\ Retained^* \frac{SG}{2.65}$$

$$Percent\ Oil\ Retained = \frac{W_w - W_D}{W_D} *100$$

SG = Specific Gravity of Coarse Aggregates

W_w = Weight of Aggregate + Absorbed Oil

W_D = Weight of Aggregate

3. Determine K_f from Chart 1, and K_c from Chart 2

4. Determine the Correction to K_f Using Chart 3

5. Calculate K_m as: $K_m = K_f$ + Correction (Correction is Negative if $K_c - K_f$) is Negative, Otherwise Positive

6. Determine the Oil Ratio for a Cutback Asphalt from Chart 4

7. Determine Oil Ratio for the Paving Grade Asphalt Binder from Chart 5 and Compute Asphalt Content as follows:

$$Asphalt\ Content = 100^* \frac{Oil\ Ratio}{100 + Oil\ Ratio}$$

8. Mix Asphalt and Aggregates Using Mixing Temperature as Follows:

Grade	Temperature Range, °F
AC 2.5, AR-1000, or 200–300 Pen	225–250
AC-5, AR 2000, or 120–150 Pen	250–275
AC-10, AR 4000, or 85–100 Pen	275–300
AC-20, AR 8000, or 60–70 Pen	300–325
AC-40, AR 16000, or 40–50 Pen	300–325

Prepare 4 specimens - One at Asphalt Content Determined in Step 7, 2 at 0.5 % increments Above, and One at 0.5 % Increment Below - for Stabilometer Test; 2 at Asphalt Content Determined in Step 7 for Swell Test

9. After Curing for 15 Hours at 140F, Reheat the Mix to 230F, and then Compact Using a Kneading Compactor

10. Conduct Stabilometer Test; Determine Stabilometer Value (S) as folows:

$$S = \frac{22.2}{\frac{P_h D_2}{P_v - P_h} + 0.222}$$

D_2 = Displacement on Specimen

P_v = Vertical Pressure (Typically 400 psi)

P_h = Horizontal Pressure, from Stabilometer Pressure Gage Reading Obtained at the Instant P_v is 400 psi

Correct the Stabilometer Value for Specimen Height Using Chart 6

11. Conduct Swell Test and Determine Swell as Change in Height of The Specimen After 24 Hours of Conditioning

12. Determine Air Voids in % in the Specimen Mixes

13. Determine the Optimum Asphalt Content using Chart 7

FIGURE 14.5 Key Steps in the Hveem Mix Design Method.

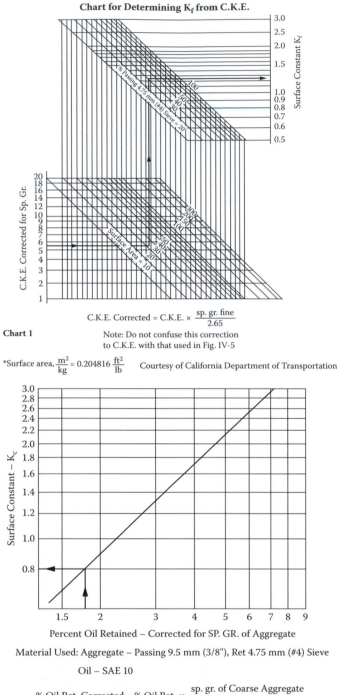

Chart for Determining K_f from C.K.E.

$$\text{C.K.E. Corrected} = \text{C.K.E.} \times \frac{\text{sp. gr. fine}}{2.65}$$

Chart 1

Note: Do not confuse this correction to C.K.E. with that used in Fig. IV-5

*Surface area, $\frac{m^2}{kg} = 0.204816 \frac{ft^2}{lb}$ Courtesy of California Department of Transportation

Percent Oil Retained – Corrected for SP. GR. of Aggregate

Material Used: Aggregate – Passing 9.5 mm (3/8"), Ret 4.75 mm (#4) Sieve

Oil – SAE 10

$$\% \text{ Oil Ret. Corrected} = \% \text{ Oil Ret.} \times \frac{\text{sp. gr. of Coarse Aggregate}}{2.65}$$

Courtesy of California Department of Transportation.

Chart 2

FIGURE 14.6 Charts for the Hveem Mix Design Method.
Source: Reprinted from Hot Mix Asphalt Materials, Mixture Design, and Construction by permission of National Asphalt Pavement Association Research and Education Foundation.

Chart for Combining K_f and K_c to Determine K_m

If ($K_c - K_f$) is neg., corr. is neg.
If ($K_c - K_f$) is pos., corr. is pos.
$K_m - K_f$ + corr. to K_f

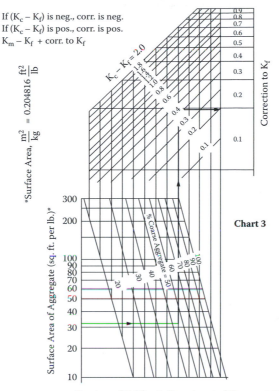

Chart 3

Courtesy of California Department of Transportation.

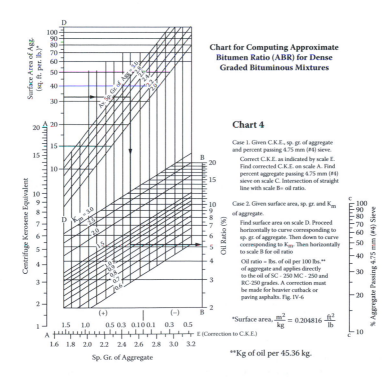

Chart for Computing Approximate Bitumen Ratio (ABR) for Dense Graded Bituminous Mixtures

Chart 4

Case 1. Given C.K.E., sp. gr. of aggregate and percent passing 4.75 mm (#4) sieve.

Correct C.K.E. as indicated by scale E. Find corrected C.K.E. on scale A. Find percent aggregate passing 4.75 mm (#4) sieve on scale C. Intersection of straight line with scale B= oil ratio.

Case 2. Given surface area, sp. gr. and K_m of aggregate.

Find surface area on scale D. Proceed horizontally to curve corresponding to sp. gr. of aggregate. Then down to curve corresponding to K_m. Then horizontally to scale B for oil ratio

Oil ratio = lbs. of oil per 100 lbs.** of aggregate and applies directly to the oil of SC - 250 MC - 250 and RC-250 grades. A correction must be made for heavier cutback or paving asphalts. Fig. IV-6

*Surface area, $\dfrac{m^2}{kg} = 0.204816 \dfrac{ft^2}{lb}$

**Kg of oil per 45.36 kg.

FIGURE 14.6 (*Continued*).

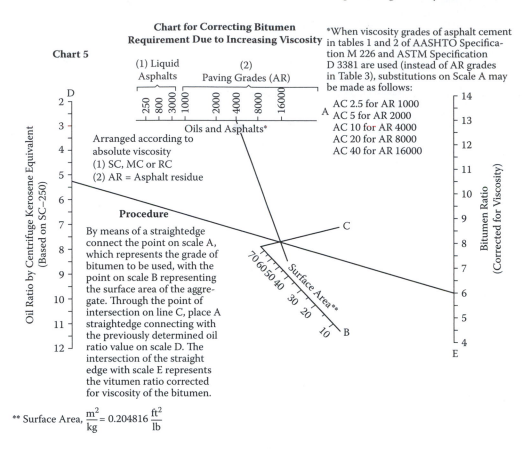

FIGURE 14.6 (*Continued*).

The procedure for checking the stability and flow starts with conditioning the specimen in a 140°F water bath for 30–40 minutes. Testing is conducted using a stability and flow tester (Figure 14.11) at a temperature of 70–100°F. The peak load at which the sample breaks and the flow are recorded. Stability correlation ratios are provided in AASHTO T-245 to adjust the values of the stability if a specimen with height different from 2 1/2 in. is used.

14.2.3 SUPERPAVE METHOD

The Superpave (*Su*perior *Per*forming *Pave*ment) method was developed as a result of the Strategic Highway Research Program (SHRP) research effort in the United States during the late 1980s and early 1990s. The system includes the newly developed performance-graded (PG) asphalt binder classification system, aggregate properties identified as "consensus" properties, new mix compaction, and mix analysis procedures. The key steps in the mix design method are the following:

1. Select asphalt binder and aggregates on the basis of environmental conditions and traffic.
2. Select aggregate gradation and estimate preliminary asphalt binder contents by considering volumetric properties—VTM, VMA, and VFA—and by considering traffic levels
3. Compact specimens at a range of asphalt contents (enveloping the estimate content from Step 3) using a Superpave gyratory compactor, with the number of gyrations commensurate with the traffic level expected.
4. Determine bulk density, and compute VTM, VMA, and VFA (using the theoretical maximum density of mixes and bulk-specific gravities of the aggregates and asphalt).

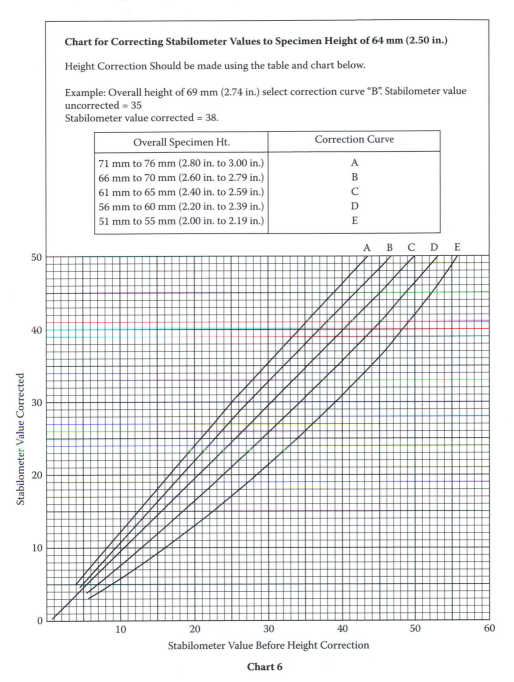

Chart for Correcting Stabilometer Values to Specimen Height of 64 mm (2.50 in.)

Height Correction Should be made using the table and chart below.

Example: Overall height of 69 mm (2.74 in.) select correction curve "B". Stabilometer value uncorrected = 35
Stabilometer value corrected = 38.

Overall Specimen Ht.	Correction Curve
71 mm to 76 mm (2.80 in. to 3.00 in.)	A
66 mm to 70 mm (2.60 in. to 2.79 in.)	B
61 mm to 65 mm (2.40 in. to 2.59 in.)	C
56 mm to 60 mm (2.20 in. to 2.39 in.)	D
51 mm to 55 mm (2.00 in. to 2.19 in.)	E

Chart 6

FIGURE 14.6 (*Continued*).

5. Determine the optimum asphalt content on the basis of optimum VTM and allowable VMA and VFA.
6. Prepare specimens at construction voids with optimum asphalt content.
7. Subject one set of moisture-conditioned and another set of dry specimens to indirect tensile strength tests.
8. Compute the tensile strength ratio, and ensure adequate resistance against moisture susceptibility.

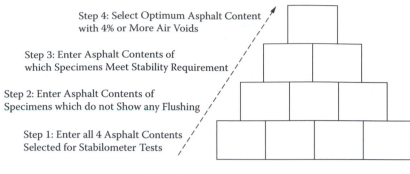

Step 4: Select Optimum Asphalt Content
with 4% or More Air Voids

Step 3: Enter Asphalt Contents of
which Specimens Meet Stability Requirement

Step 2: Enter Asphalt Contents of
Specimens which do not Show any Flushing

Step 1: Enter all 4 Asphalt Contents
Selected for Stabilometer Tests

Chart 7

FIGURE 14.6 (*Continued*).

The method is illustrated in Figure 14.12, with the relevant plots and criteria in Figures 14.13 and 14.14, respectively.

AASHTO 312 specifies the method for using a Superpave gyratory compactor for compaction of 150-mm diameter HMA specimens for subsequent determination of bulk-specific gravity. The equipment consists of a loading system within a frame, which applies a specified load for compaction of the mix in a mold, which sits between the loading ram and a rotating base. The loading ram applies a stress of 600 kPa, and the base rotates with a tilt of 1.25° (Figure 14.15), for a specified number of gyrations (depending on the level of compaction, which is dependent on the design traffic as well as high temperature), at 30 gyrations per minute. The final height is approximately 115 mm, and during compaction the height is continuously measured to 0.1 mm. Generally mix in the amount of approximately 5,000 gram is obtained following the AASHTO-specified quartering

Compacting Ram
with Triangular Face

Rotating Sample
Holder Base

FIGURE 14.7 Kneading Compactor.

TABLE 14.1
Requirements of Hveem Mix Design Mixes

Traffic Category	Heavy	Medium	Light
Stabilometer value	≥ 37	≥ 35	≥ 30
Swell	< 0.030 in.		

1. Conduct Viscosity Tests at Different Temperatures and Plot Viscosity versus Temperature; Select Mixing Temperature Corresponding to 170 ± 20 Centistokes Select Compaction Temperature Corresponding to 280 ± 30 Centistokes

2. Mix 18 Specimens - Six at Each of Three Asphalt Contents, Optimum (from Guidelines), 0.5 % Above Optimum, 0.5 % Below Optimum

3. Compact Specimens Using a Marshall Hammer, with Numbers of Blows As Shown in the Table in Figure 14.9, After Curing Mix, If Required

4. Determine Bulk Specific Gravity and Compute VTM, VMA and VFA

5. Test Specimens for Stability and Flow Using the Marshall Stability-Flow Equipment

6. Plot Asphalt content Versus Unit Weight, VMA, Stability, VFA, Flow and VTM, as Shown in Figure 14.9

7. Select the Optimum Asphalt Content on the Basis of Criteria for Specific Method Used; National Asphalt Pavement Association (NAPA) and Asphalt Institute (AI) Guidelines are Provided Below

NAPA Procedure

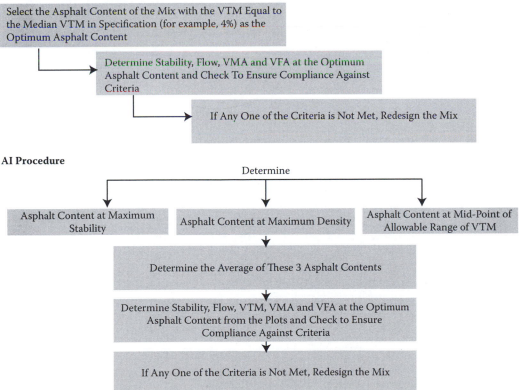

AI Procedure

Note: Change Aggregate Gradation and Redesign the Mix if VMA Criterion is Not Met

FIGURE 14.8 Key Steps in the Marshall Mix Design Method.

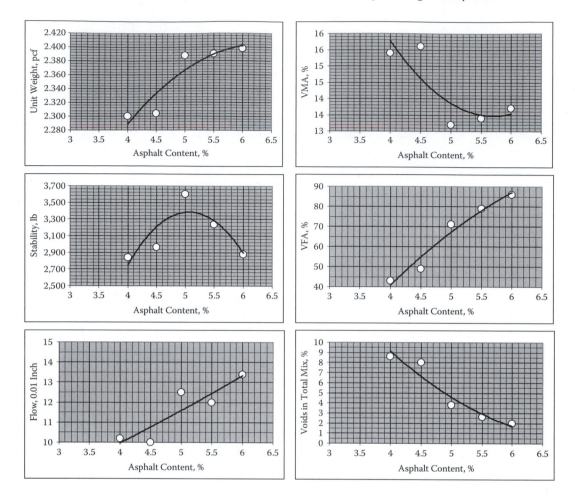

Traffic	Heavy	Medium	Light
Compaction, Number of Blows on Each Side	75	50	35
Stability, lb	Minimum: 1,800	Minimum: 1,200	Minimum: 750
Allowable Range			
Flow, 0.01 inch	8-14	8-16	8-18
Voids in Total Mix, %	3-5	3-5	3-5

Nominal Maximum Aggregate Size, inch	Minimum VMA, %		
	Voids in Total Mix, 3.0%	Voids in Total Mix, 4.0%	Voids in Total Mix, 5.0%
2.5	9.0	10.0	11.0
2.0	9.5	10.5	11.5
1.5	10.0	11.0	12.0
1.0	11.0	12.0	13.0
3/4	12.0	13.0	14.0
1/2	13.0	14.0	15.0
3/8	14.0	15.0	16.0
No. 4	16.0	17.0	18.0
No. 8	19.0	20.0	21.0
No. 16	21.5	22.5	23.5

FIGURE 14.9 Plots Required in the Marshall Mix Design Method.

Hammers

FIGURE 14.10 Marshall Hammer.

procedure from fresh samples, and is used for compaction. The mix should be compacted at a temperature at which the asphalt binder in the mix possesses a viscosity of 0.28 + −0.03 Pa s. The mix is transferred to the preheated (to compaction temperature) steel mold (150 mm high with an inside diameter of 150 mm) with a base plate. Protective paper disks must be placed at the bottom of the mold before placing the mix, and at the top of the mix before placing it inside the gyratory compactor. The number of gyrations is then set, and the compaction started. The number of compactions could be set at N_{max} or N_{design}, depending on the agency-specific requirements. The number of gyrations versus specimen height data is generally stored to a disk via the equipment computer and/or printed out at the end of the compaction process. The specimen is then extruded out of the mold, and the paper discs are removed as quickly as possible. The specimens are then left to cool down for subsequent tests.

The height data at the final compaction are used to estimate the volume of the sample, and hence the bulk-specific gravity, which is then corrected on the basis of the actual bulk-specific gravity, which is determined by testing of the specimen. The correction factor is then used to determine the corrected bulk-specific gravities at all gyrations. The corrected bulk-specific gravity can be used along with the theoretical maximum density (TMD) or maximum Rice gravity (G_{mm}) to calculate

Asphalt Mix
Sample

FIGURE 14.11 Marshall Stability Tester.

1. Select Asphalt Binder on the Basis of Climate and Traffic (See Chapter 9) from Performance Graded Classifications According to AASHTO MP1

2. Select Aggregates on the Basis of Consensus and Source Property Requirements as Shown in Figure 14.14.

3. Compact mixes with Three Trial Gradations Using an Estimated Asphalt Content, Using the Superpave Gyratory Compactor, with Number of Gyrations Recommended for Specific Traffic Levels and Average Design High Temperature. Use Mixing and Compaction temperature from Temperature versus Viscosity Plot Obtained by Testing Viscosities at 135 and 165° C. Select Mixing temperature Corresponding to Viscosity Range of 0.15 and 0.19 Pa-s and Compaction Temperature Corresponding to Viscosity Range of 0.25 and 0.31 Pa-s. Use 2 or 4 Hour Aging at 135°C Prior to Compaction by Spreading in a Pan at 21 - 50 Kg per Square Meter.

4. Select Gradation that Gives Mix with the Best Overall Match with Specified VTM, VMA and VFA

5. Mix and Compact 8 Specimens with the Selected gradation - 2 Each at 4 Asphalt Contents Enveloping the Initial Asphalt Content

6. Determine Bulk Specific Gravity and Compute VTM, VMA and VFA

7. Select the Asphalt Content That Gives Mixes with 4% VTM as the Optimum Asphalt Content. Check Whether or Not the Mix at 4% VTM Meets Other Criteria, as Shown in Figure 14.14.

8. Prepare Six Specimens at Optimum Asphalt Content at Construction (7%) VTM. Subject Three to Moisture Condtioning. Determine Tensile Strengths of three Conditioned and Three Unconditioned Specimens. Determine Tensile Strength Ratio (TSR), as Shown in in Figure 14.14. Check if TSR is Equal to or Greater than Minimum Specified Value. If not, Consider Adding Antistripping Agent, such as Lime to the Mix.

FIGURE 14.12 Key Steps in the Superpave Mix Design Method.

the air voids. Figure 14.13 shows a complete example of all calculations. The densities at $N_{initial}$, N_{design}, and N_{max} are of primary importance, and the determined values are checked against specified values for compliance with mix design. The data can also be used for calculation and checking of voids in mineral aggregates (VMA) and voids filled with asphalt (VFA).

A test for resistance against moisture damage (AASHTO T-283) is then conducted on samples compacted with the optimum asphalt content. In this test, the resistance to moisture damage is evaluated by measuring the tensile strength of a dry set of HMA samples and another set of wet/conditioned samples, and comparing the two. The tensile strength ratio (strength of conditioned specimens divided by the strength of the unconditioned/dry specimens) is checked against a specified minimum limit (such as 80%). Note that the conditioning phase is supposed to simulate the destructive action of moisture/freezing/thawing, through which cohesive bonds inside the specimen are destroyed, and the tensile strength is lowered. This conditioning can be simply soaking and drying, or it can be freezing or thawing, depending on the environmental conditions that are expected in the location of the pavement being constructed. Specimens used in this test are compacted after conditioning the mix for 16 hours in an oven at 60°C. The mix is compacted at the specified compaction temperature such that the resulting specimens have 7 ± 1% air voids. These specimens need to be stored for 72–96 hours at room temperature before the next step.

Based on the air voids (determined from bulk-specific gravity and maximum specific gravity, which are measured from an additional sample), the specimens are divided into two groups (each

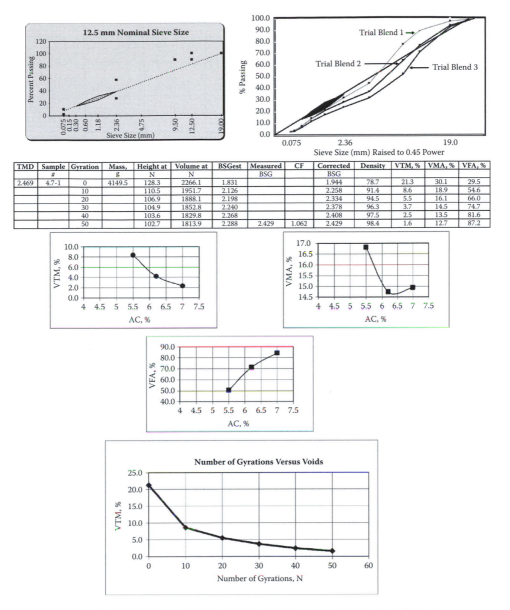

TMD	Sample	Gyration	Mass,	Height at	Volume at	BSGest	Measured	CF	Corrected	Density	VTM, %	VMA, %	VFA, %
	#		g	N	N		BSG		BSG				
2.469	4.7-1	0	4149.5	128.3	2266.1	1.831			1.944	78.7	21.3	30.1	29.5
		10		110.5	1951.7	2.126			2.258	91.4	8.6	18.9	54.6
		20		106.9	1888.1	2.198			2.334	94.5	5.5	16.1	66.0
		30		104.9	1852.8	2.240			2.378	96.3	3.7	14.5	74.7
		40		103.6	1829.8	2.268			2.408	97.5	2.5	13.5	81.6
		50		102.7	1813.9	2.288	2.429	1.062	2.429	98.4	1.6	12.7	87.2

FIGURE 14.13 Example Calculations and Plots Used in the Superpave Mix Design Method.

group with three samples) such that the average air voids of the two groups are approximately the same. Thickness and diameter measurements are made. The dry subset samples are placed inside a water bath maintained at 25°C after placing them in leak-proof plastic bags. After 2 hours, the samples are removed from the bags and tested for their tensile strength. Load is applied at a rate of 50 mm per minute. The peak load is recorded, and is used along with the diameter and the thickness of the sample to calculate the tensile strength. The wet subset samples are vacuum saturated to at least 55–80% of saturation (which should be checked by measuring the SSD weight), after which they are covered with plastic wrap and placed in a plastic bag with 10 ml of water inside the bag. The bag is then placed inside a freezer at –18°C for 16 hours, after which the specimen is placed in

Traffic Level	Coarse Aggregate Angularity		Fine Aggregate Angularity	
	<100 mm	> 100 mm	<100 mm	> 100 mm
<0.3	75/—	50/—	40	40
0.3 to < 3.0	85/80	60/—	45	40
3.0 to < 30.0	95/90	80/75	45	40
>30.0	100/100	100/100	45	45

Traffic Level	Sand Equivalent, %	Flat and Elongated, %
<0.3	40	—
0.3 to < 3.0	45	10
3.0 to < 10.0	45	10
10 to < 30.0	45	10
>30.0	50	10

$$0.6 \leq \frac{\%\,Weight\,of\,Material\,Passing\,the\,0.075\,mm\,Sieve}{\%\,Weight\,of\,Effective\,Asphalt\,Binder} \leq 1.2$$

Traffic Level, ESALs, Million	Compaction, Number of Gyrations			
	$N_{initial}$		N_{design}	$N_{maximum}$
	Gyrations	Density, % G_{mm}		
<0.3	6	<91.5	50	75
0.3 – <0.3	7	<90.5	75	115
3.0 – <3.0	8	<89.0	100	160
>30.0	9	<89.0	125	250

Slow/Standing Traffic: Increase N_{design} by 1 Level

Nominal Maximum Aggregate Size, mm	Minimum VMA, %
9.5	15
12.5	14
19	13
35	12
37.5	11

Traffic (ESAL, Million)	Range of VFA, %
<0.3	70–80
1 – 3	65–78
>3.0	65–75

$$Tensile\,Strength\,Ratio = \frac{Average\,Tensile\,Strength\,of\,Conditioned\,Specimens}{Average\,Tensile\,Strength\,of\,Unconditioned\,Specimens}$$

FIGURE 14.14 Requirements in the Superpave Mix Design Method.

Controls
with Disk Drive

Compactor
Printer

Mix Put in the Mold

Filled Mold with Paper Disc on Top and Bottom

Mold Placed in Compactor

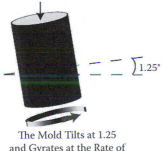

Compacted Sample

1.25°

The Mold Tilts at 1.25
and Gyrates at the Rate of
30 Revolutions Per Minute

Schematic of Mold Gyration

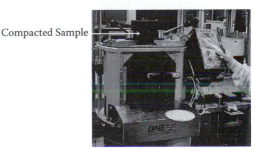

Compacted Sample at The End of
Gyrations

FIGURE 14.15 Compaction Using the Superpave Gyratory Compactor.

a 60°C water bath for 24 hours. Then the specimen is conditioned at 25°C water, and then tested for tensile strength.

After breaking, the aggregates inside the sample could be examined to determine the degree of stripping or loss of asphalt, if any. Some agencies (such as the Georgia DOT) specify a method of evaluating the degree of stripping by this method. Note that one freezing and thawing cycle is described above—agencies could also require multiple freeze-thaw tests, depending on the severity of the environment in which the pavement is expected to serve. Figure 14.16 shows a typical agency job mix formula, resulting from a mix design. In more and more cases, designed mixes are being "proof" tested for such properties as resistance against rutting and/or moisture effects, or compared amongst each other, with the help of simulative tests. The Asphalt Pavement Analyzer (APA), shown in Figure 14.17, is such a test procedure, in which gyratory-compacted samples can be subjected to a pressure of 100 psi and a load of 100 lbs in a temperature-controlled chamber, with or without moisture. The steel wheels apply the load through pressurized rubber hoses, and the

DEPARTMENT OF Maine TRANSPORTATION	Ref. No. **3566**	**HMA DESIGN** SUBMITTAL DATA	**100 GYRATIONS**

Date Submitted: **5/6/2003**　　Submitter:　　　　　　　　　Item No.: **404.40**

Mix Grading: **HMA MIX-12.5MM(SMA)**　　RAP, %:　　Plant:

PLANT DATA	Make	Size	Type	Location
		3 TON	BATCH	PORTLAND

BITUMEN DATA	Grade	Refiner	Supplier	Temp. Range (Aim ± 20°F [10°C])	
				Plant Aim, °F [°C]	Street Aim, °F [°C]
	PG70-28	BITUMAR CA	BOUCHARD	310 [155]	290 [143]

SUPERPAVE DESIGN DATA

ESAL's:　**3 to <30**　　Gsb:　**2.65**　　% Binder: **6.2**

Nominal Size: **12.5mm**　　Gmm: **2.440**　　Fines to Eff. Binder Ratio: **1.2**

Gmb, weight, g

AGGREGATE DATA

Size	Type	Original Source/Owner	Original Source/Location
12.5	QUARRY		WELLS
9.5	QUARRY		WELLS
DRY STONE SCREENINGS	QUARRY		WELLS

Stockpile Gradation (Percentages Passing Sieve Sizes)

% Used	1½ in. 37.5 mm	1 in. 25.0 mm	¾ in. 19.0 mm	½ in. 12.5 mm	⅜ in. 9.5 mm	No. 4 4.75 mm	No. 8 2.36 mm	No. 16 1.18 mm	No. 30 0.600 mm	No. 50 0.300 mm	No. 100 0.150 mm	No. 200 0.075 mm
65			100	94	37	4	3	2	2	2	1	1
15				100	97	29	7	4	3	3	2	1.5
12.6					100	99	78	54	39	26	17	10.5
7.4										100	99	94

	37.5 mm	25.0 mm	19.0 mm	12.5 mm	9.5 mm	4.75 mm	2.36 mm	1.18 mm	0.600 mm	0.300 mm	0.150 mm	0.075 mm
Aim			100	96	59	27	20	16	14	12	11	9
Lower			100	89	52	20	16	12	11	10	9	7
Upper			100	100	66	34	24	20	17	14	13	11
Spec.			100	90 - 100	26 - 78	20 - 28	16 - 24					8 - 10
Restricted Zone												

Status: **PENDING**

JMN: **PII-PD3-100-12**

COMMENTS:

MAINE DOT USE ONLY

JOB MIX SPECIFICATIONS

	Pb, %	%Passing No. 200 0.075mm Sieve
Aim	6.2	
Lower	5.8	
Upper	6.6	

DOT Consensus Qualities

Fractured, 1 Face (ASTM D 5821), %	99
Fractured, 2 Face (ASTM D 5821), %	99
F.A. Angularity (T 304), %	45
Flat/Elongated (ASTM D 4791), %	2
Micro-Deval (TP58 99), %	6
Sand Equivalent (T 176)	66
Washington Degradation (MeDOT)	
Combined Agg, Gsb (PP 28)	

Authorized by:　　　　　　　　　　　Date:

Paper Copy: Lab File　　*Electronic: Area Supervisor; Resident; Contractor*

FIGURE 14.16 Results of a Mix Design in a Standard Format (job mix formula).
Courtesy: Rick Bradbury, Maine Department of Transportation.

deformation in the HMA samples after a specific number of "passes" of the wheels is used as the indicator of strength against rutting.

14.3　STRUCTURAL DESIGN

The basic methods of consideration of the major distresses have been discussed in Chapter 12. Recall that some of the models are based on statistical regression with empirical data, whereas

Wheels Running on Hose During Testing

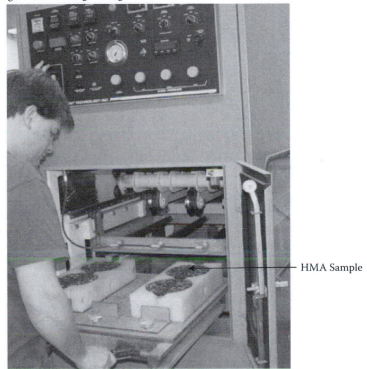

Rutted Samples after Testing

FIGURE 14.17 The Asphalt Pavement Analyzer (APA).

others are based on mechanics. In an empirical structural design method, only empirical models are used, whereas in a mechanistic-empirical method, both empirical as well as mechanistic expressions are utilized. The following paragraphs present examples of both types of design method.

14.3.1 EMPIRICAL METHODS

The empirical methods include mostly those based on the soil classification system and subgrade type (U.S. Bureau of Public Roads) and those based on road tests conducted by AASHO. In both cases the design methods are based on experience—either from in-service pavements or from controlled experiments. The advantages of such methods are that sufficient data are available and design procedures are relatively simple and less time consuming. The primary disadvantage is that the design cannot accommodate traffic, environmental conditions, and materials for which experience and/or results of experiments do not exist.

14.3.1.1 California Bearing Ratio (CBR) Method

This method is based on charts relating the required thickness of pavement over the subgrade to the strength of the subgrade—the strength being represented in terms of a parameter called the *California Bearing Ratio* (CBR; ASTM D-1883-99, AASHTO T-193). The CBR concept and test method were originally developed by the California Division of Highways in the 1920s, later modified by the U.S. Army Corps of Engineers, and finally adopted by the American Society of Testing and Materials (ASTM) and the American Association of Highway and Transportation Officials (AASHTO).

The CBR test is a penetration test that uses a standard piston (3 sq in.) which penetrates the soil at a standard rate of 0.05 in. per minute. A unit load is recorded at several penetration depths, typically 0.1 and 0.2 in. The CBR value is computed by dividing the recorded unit load by a standard load that is required for penetration for a high-quality crushed stone material. The CBR test is conducted on a soaked sample of soil—the soaking in water (for 96 hours) is conducted to simulate the worst (saturated) condition under which the pavement would perform.

Figure 14.18 shows the original curves for CBR versus thickness for two different traffic levels, and an example to illustrate its use.

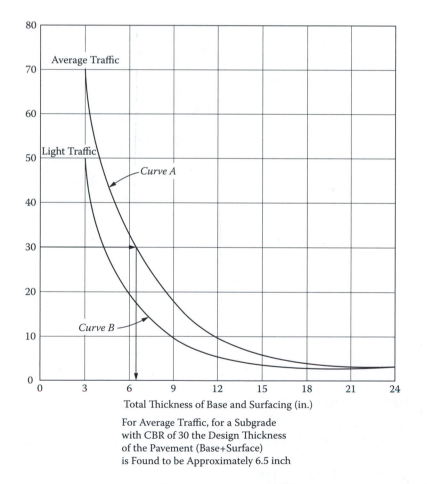

For Average Traffic, for a Subgrade
with CBR of 30 the Design Thickness
of the Pavement (Base+Surface)
is Found to be Approximately 6.5 inch

FIGURE 14.18 Example of the Use of a California Bearing Ratio (CBR) Curve for Pavement Design. *Source:* From ASCE, 1950, and Monismith and Brown, 1999, reprinted with kind permission from the Association of Asphalt Paving Technologists.

14.3.1.2 American Association of State Highway and Transportation Officials (AASHTO) Method

The AASHTO design procedure introduced a number of important design concepts. The design procedure came out of statistical analysis of data from an extensive road test carried out between 1955 and 1960 near Ottawa, Illinois. Traffic, consisting of trucks with different loads and axles, were run to failure over different pavement sections, consisting of different materials and thickness. Two important concepts—serviceability and reliability—were utilized in the development of the design process. This design procedure has been subsequently revised to add more mechanistic components over the years. The latest version is the AASHTO *Guide for Design of Pavement Structures*, published in 1993.

The procedure started with the AASHO Road Test. The concept of Present Serviceability Index (PSI) was developed, and an equation relating traffic to thickness was developed. This equation was mapped into a nomograph, which is often used in this method. The tests were carried out in Ottawa, Illinois, between the years of 1958 and 1961. In total there were 6 test tracks, with both HMA and reinforced concrete sections. For the test location, the mean temperatures in January and July were 27°F and 76°F, respectively. The annual average rainfall and frost depth were 34 and 28 in., respectively. The subgrade soil consisted of A6 (silty clay), with 82% of the material passing the No. 200 sieve, and an average CBR of 2.9 and optimum moisture content of 13%. The subbase and base courses consisted of a sand/gravel mixture (CBR = 28–51) and crushed limestone (average = 108), respectively. The HMA was made up of 85–100 penetration grade asphalt.

There were multiple HMA sections with subbase courses varying in thickness from 0 to 16 in., base courses 0–9 in., and HMA 1 to 6 in. The majority of the HMA sections failed, with most failing due to the effect of spring thaw, which was found to decrease with the use of a thicker base and subbase.

The complete equation relating traffic and thickness is as follows.

$$\log W_{18} = Z_R \times S_0 + 9.36 \log(SN+1) - 0.20 + \frac{\log\left(\frac{\Delta PSI}{4.2-1.5}\right)}{0.40 + \frac{1094}{(SN+1)^{5.19}}} + 2.32 \log(M_r) - 8.07$$

The use of this equation requires the selection of reliability levels in terms of Z_R and S_0, initial and terminal PSI or change in PSI, and determination of resilient modulus, M_r. The following paragraphs explain the various parameters in the equation.

14.3.1.2.1 log W18

Base 10 logarithm of the predicted number of ESALs over the lifetime of the pavement. The logarithm is taken based on the original empirical equation from the AASHO Road Test.

14.3.1.2.2 SN

Structural number. An abstract number expressing the structural strength of a pavement required for given combinations of soil support (M_r), total traffic (ESALs), and allowable change in serviceability over the pavement life (ΔPSI).

The structural number is used to determine layer depths by using a parameter called *layer coefficient*, such that

$$SN = a_1 D_1 + a_2 D_2 m_2 + a_3 D_3 m_3 + \cdots$$

where:
a_i = layer structural coefficient
D_i = layer depth (in.)
m_i = layer drainage coefficient

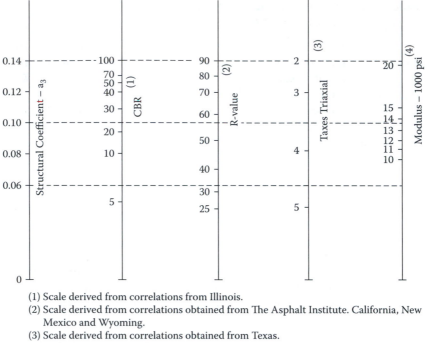

(1) Scale derived from correlations from Illinois.
(2) Scale derived from correlations obtained from The Asphalt Institute. California, New Mexico and Wyoming.
(3) Scale derived from correlations obtained from Texas.
(4) Scale derived on NCHRP project (3).

FIGURE 14.19 Variation in the Granular Subbase Layer Coefficient.
Source: From AASHTO Guide for Design of Pavement Structures © 1993, by the American Association of State Highway and Transportation Officials, Washington, D.C. Used by permission.

Typical values of layer coefficients are as follows. HMA (surface layer): 0.44; base course (crushed stone): 0.14; stabilized base course material: 0.30–0.40; subbase (crushed stone): 0.11. Values of drainage coefficients can range from 0.4 (slow-draining, saturated layers) to 1.4 (fast-draining layers that do not get saturated). The term can be neglected by using m = 1. An example chart relating the various stiffness/strength measuring for subbase materials is shown in Figure 14.19.

The concept of reliability is based on the assumption that the distribution of variables such as stress, resulting from uncontrollable factors such as loading and the environment, and strength/stiffness of materials/layers, resulting from controllable factors such as variations in construction quality and materials, can be assumed to be of the normal distribution type. *Reliability* refers to the probability that the predicted design life will exceed the required design, or, more specifically, the strength of a material will exceed the stress on the material. Two factors are considered for reliability:

Z_R = standard normal deviate
S_0 = combined standard error of the traffic prediction and performance prediction

Table 14.2 shows the different values of Z_R for the different levels of reliability, which could be selected on the basis of guidelines that have been provided in the AASHTO Design Guide.

14.3.1.2.3 S_0

The typical values for a flexible pavement are 0.40 to 0.50. S_0 cannot be calculated from actual traffic or construction numbers, so it is almost always assumed to be 0.50.

TABLE 14.2
Z_R Values for Different Reliabilities

Reliability	Z_R
99.99	−3.750
99.9	−3.090
99	−2.327
95	−1.645
90	−1.282
80	−0.841
75	−0.674
70	−0.524
50	0

14.3.1.2.4 Solution of the AASHTO Equation

The equation is solved using an iterative process. The equation is often solved assuming ESAL values. The process consists of the following steps.

1. Determine the total ESALs.
2. Determine M_r, or, using the following formula, convert existing CBR values to M_r.
   AASHTO method: M_r, psi = 1500 * (CBR)
   NCHRP 1-37 method: M_r, psi = 2555(CBR)$^{0.64}$
3. Select reliability (selecting a reliability means selecting a specific value of Z_R).
4. Select p_o, p_t.
5. Decide on a basic structure, and select the M_r of the materials of the layers.
6. Solve the equation for two layers (combine lower two layers and upper two layers, for example); solve each layer using the M_r of the layer directly underneath.
7. Determine total SN: SN required from lower two layers (combined) and from upper two layers.
8. Use guidelines and experience to select thickness of the different layers to satisfy the SN requirement.

14.3.1.2.5 Correlation with Layer Coefficient

The AASHTO standard guide for design of pavement provides relationships between the resilient modulus and layer coefficients, whereas the relationship between the CBR and resilient modulus is provided by the UK Transportation Research Laboratories (TRL). The relationships are as follows:

$$M_r = 2,555 \ (CBR)^{0.64}$$

where M_r, resilient modulus, is in psi.
   And:

$$M_r = 30,000 * \frac{a_i}{0.14}$$

where M_r is in psi and a_i is the AASHTO layer coefficient.
   The nomograph used in the AASHTO design method is shown in Figure 14.20.

Nomograph Solves:

$$\log_{10} W_{18} = Z_R {}^* S_o + 9.36 {}^* \log_{10}(SN+1) - 0.20 + \dfrac{\log_{10}\left[\dfrac{\Delta PSI}{4.2 - 1.5}\right]}{0.40 + \dfrac{1094}{(SN+1)^{5.19}}} + 2.32 {}^* \log_{10} M_R - 8.07$$

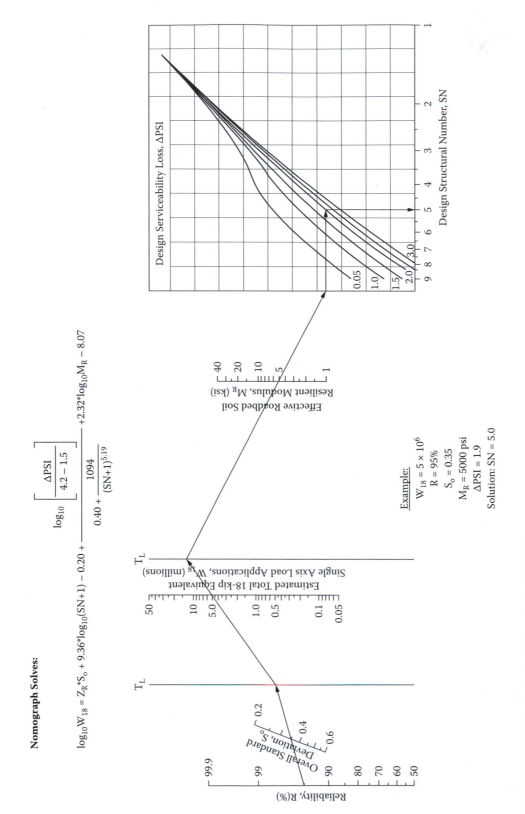

Example:

$W_{18} = 5 \times 10^6$

R = 95%

$S_o = 0.35$

M_R = 5000 psi

ΔPSI = 1.9

Solution: SN = 5.0

FIGURE 14.20 Nomograph for Solving the AASHTO Equation.

Source: From AASHTO Guide for Design of Pavement Structures © 1993, by the American Association of State Highway and Transportation Officials, Washington, D.C. Used by permission.

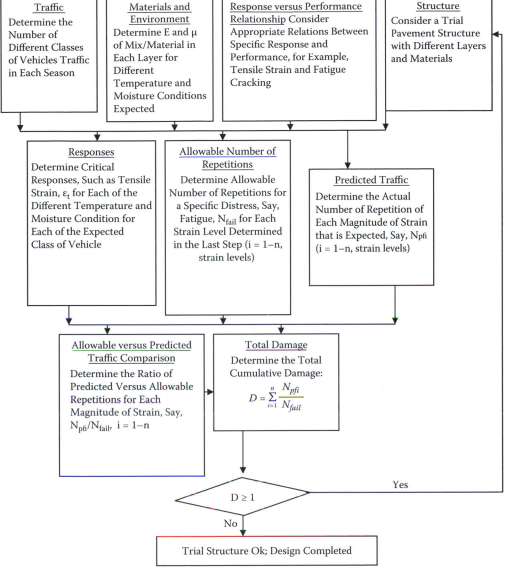

FIGURE 14.21 Flowchart in a Typical Mechanistic-Empirical (M-E) Design.

Example

Determine the thicknesses for two layers in an asphalt pavement, where a total structural number of 4.5 is required. Layer 1 is a HMA layer with a layer coefficient of 0.44, and layer 2 is a granular base course with a layer coefficient of 0.13. Consider the drainage coefficient or the base as 0.9, and the minimum thickness of the HMA layer as 2 in.
Consider the minimum thickness of the HMA first, say, 2 in.

$$SN = a_1 D_1 + a_2 D_2 m_2$$

where:

  a_i = layer structural coefficient
  D_i = layer depth (in.)
  m_i = layer drainage coefficient

$$SN_{required} = 4.5 = (0.44)(2) + (0.13)(D_2)(0.95)$$

where D_2 = thickness of the base = 29.3 in., approximately 29.5 in.

If a 4–in. HMA layer is used instead, the required base thickness is approximately 22 in.

14.3.2 Mechanistic-Empirical (ME) Methods

A typical mechanistic-empirical design procedure basically consists of two parts—a mechanistic part or an analytical part which helps in the determination of pavement response under load, and an empirical part that relates the response to pavement performance or (more specifically) to distress, such as rutting. The complete flow chart for a mechanistic-empirical design is shown in Figure 14.21.

A typical schematic of a layered pavement structure with material properties relevant for ME design is shown in Figure 14.22. The layers are shown with typical values for a specific location. The layer thicknesses as well as the material properties are then changed within reasonable ranges. The effects of these changes on a critical response, the tensile strain at the bottom of the HMA surface layer, are also shown in Figure 14.22.

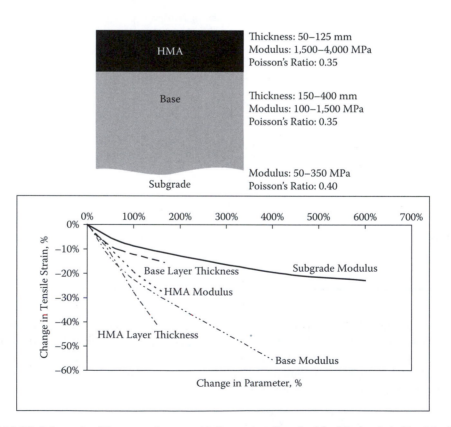

FIGURE 14.22 Schematic of Pavement Layers with Parameters Required for Mechanistic-Empirical Design and Example of Sensitivity of Response to Change in Parameters.

Once the critical responses are determined, they are used in *relationships* between such responses and critical performance conditions or distresses, to predict the number of load repetitions that the pavement can sustain before reaching the threshold/critical values of the distress conditions. Alternatively, if the number of load repetitions is given for a specific design period, the cumulative "damage" to the pavement for the given number of load repetitions is determined. This cumulative "damage" is then checked against the appropriate criteria, and if found unsuitable, the layer thicknesses and/or material properties are changed, and the process is repeated until an acceptable "damage" is obtained—and this represents the end of the "structural design" step.

In the design methods, the damage from each group of traffic (for each season) is considered, and the damage factor for the ith loading is defined as the number of repetitions (N_{actual}) of a given response parameter divided by the "allowable" repetitions ($N_{allowable}$) of the response parameter that would cause failure. The Cumulative Damage Factor (CDF) is obtained by summing the damage factors over all the loadings in the traffic spectrum. If the CDF reaches 1.0, then the pavement is presumed to have reached the end of its design life. Hence the objective of the design process is to change the different factors until the computed CDF is found to be close to but less than 1.0.

A simple application of the two steps is illustrated in the following example.

Example

Check the design of a pavement with the following information.

TRAFFIC

1 million repetitions of a standard 18-kip dual-tire axle.

LOAD

Considering one side of the axle only, there are two tires, each supporting a load of 4500 lb, with a tire pressure of 100 psi, 13 in. apart, center to center.

ASSUMED PAVEMENT STRUCTURE

The pavement is made up of four layers: HMA surface and binder (3 in.), stabilized base (6 in.), crushed gravel subbase (10 in.), and subgrade.

PAVEMENT MATERIALS
Properties

$$E_{HMA} = 3,500 \text{ MPa} = 507 \text{ ksi}; \ \upsilon_{HMA} = 0.35$$

$$E_{stabilized\ base} = 1,500 \text{ MPa} = 217 \text{ ksi}; \ \upsilon_{stabilized\ base} = 0.35$$

$$E_{subbase} = 1,50 \text{ MPa} = 21.7 \text{ ksi}; \ \upsilon_{subbase} = 0.35$$

$$E_{subgrade} = 50 \text{ MPa} = 7.2 \text{ ksi}; \ \upsilon_{subbase} = 0.4$$

Performance Models

$$N_{HMA} = \left[\frac{0.005889}{tensilestrain} \right]^5$$

$$N_{subgrade} = \left[\frac{0.003900}{verticalstrain} \right]^{7.10}$$

RESULTS

Maximum horizontal tensile strain at the bottom of the HMA = 75 με.

$$N_{HMAallowable} = \left[\frac{0.005889}{0.000075}\right]^5 = 2,984 * 10^6$$

$$CDF_{HMA} = \sum_{i=1}^{N} \frac{N_{actual}}{N_{allowable}} = \frac{10^6}{2984 * 10^6} = 0.00033$$

Maximum vertical compressive stress on top of the subgrade = 390 με.

$$N_{subgrade} = \left[\frac{0.003900}{0.000390}\right]^{7.10} = 12.59 * 10^6$$

$$CDF_{subgrade} = \sum_{i=1}^{N} \frac{N_{actual}}{N_{allowable}} = \frac{10^6}{12.59 * 10^6} = 0.0079$$

Since the CDFs are both less than 1, the pavement is sufficiently strong to withstand the expected design traffic. In fact, since the CDFs are very low compared to the limiting value of 1, there is scope for redesigning the pavement for the design traffic.

Note that in this case fatigue cracking and rutting, and relatively simple models relating tensile strains and compressive strains to them (respectively), have been used. A design could involve more sophisticated models and/or other distresses such as thermal cracking, as discussed in Chapter 12.

14.3.2.1 Example of Structural Design Procedure Using Mechanistic Principles

The Asphalt Institute procedure (Asphalt Institute, 1991) contains multilayer elastic theory; is built on the basis of data from the AASHTO Road Test, the Western Association of State Highway Organizations (WASHO) Road Test, and other different state and local road tests; and considers rutting and fatigue cracking as the distresses. For rutting the vertical compressive strain (ε_c) at the top of the subgrade is considered, whereas for fatigue cracking the horizontal tensile strain (ε_t) at the bottom of the lowest asphalt mix layer is considered. Rutting in the asphalt mix layer only is considered to be due to consolidation due to improper compaction of the layer during construction. Design charts are the basis of this design procedure. The design charts were generated with the use of the computer program DAMA, originally developed in 1983 (Asphalt Institute, 1983), which determined the largest of the two strains, and recommended the thickness of the layer to meet the higher (or more critical) of the two strains, on the basis of inputs which include resilient modulus of the subgrade, selection of type of surface and base layers, traffic information, and mean annual air temperature.

The procedure is illustrated with an example in Figure 14.23. In the AI method, a minimum thickness of asphalt mix surface layers is provided for different types of bases—emulsion treated and untreated. Charts are provided for different mean annual air temperatures—45°F, 60°F, and 75°F. Limits of CBR, Atterberg Limits, and percentage passing the No. 200 sieve for subbase/base are also provided.

14.3.2.1.1 Staged Construction

When adequate funds are not available for constructing a pavement to meet the demands of a full design period, the pavement can be constructed in stages. The initial thickness can accommodate

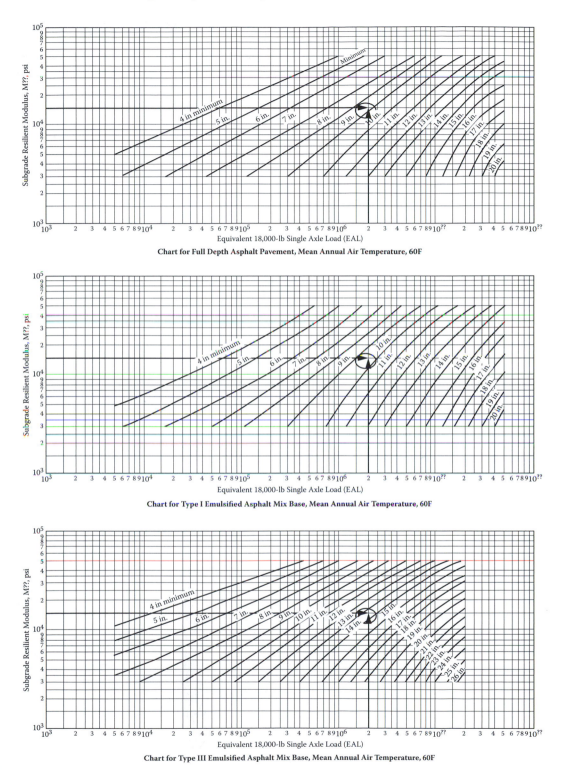

FIGURE 14.23 Example of Use of the Asphalt Institute (AI) Method.
Source: Reprinted, with permission, from Thickness Design: Asphalt Pavements for Highways and Streets Manual Series No. 1 (MS-1) Ninth Edition 1999, Copyright the Asphalt Institute, Inc.

a certain number of years of traffic, and the second construction adds the thickness required for sustaining the remaining traffic for the full design period.

The principle of design in staged construction is that the second construction will take place (say, as an overlay) before the first pavement has reached the end of its life. Based on the concept of cumulative damage (discussed earlier), it is often assumed that the second overlay will be applied when the first pavement has at least 40% of its life left (that is, damage is a maximum of 60%).

If:

n_1 = actual accumulated ESAL for the first stage

N_1 = allowable number of ESAL for the stage 1 pavement thickness

D_1 = percentage of life of the pavement expended right before the second stage

$\frac{n_1}{N_1} = D_1 = 0.6$

$N_1 = \frac{n_1}{0.6} = 1.67 n_1$

Determine h_1 on the basis of N_1.

If:

n_2 = actual accumulated ESAL for the second stage

N_2 = allowable number of ESAL for stage 2 pavement thickness

$\frac{n_2}{N_2} = (1 - D_1) = 0.4$

$N_2 = \frac{n_2}{0.4} = 2.5 n_2$

Determine h_2 on the basis of N_2, then thickness to be added in stage 2 = $(h_2 - h_1)$.

14.3.2.2 NCHRP 1-37A Mechanistic-Empirical Design Guide

The NCHRP 1-37A Mechanistic Empirical Design Method (commonly referred to as the *2002 Design Guide*) was developed in NCHRP Project 1-37A (Development of the 2002 Guide for the Design of New and Rehabilitated Pavement Structures). The *Guide* presents a very comprehensive method, with considerations of appropriate sophisticated models, and uses state-of-the-art testing procedures. The software and detailed instructions of the design guide can be obtained from www.trb.org/mepdg/ (NCHRP, n.d.).

The *Guide* suggests three different levels of design: Level 1, which requires very basic information and the use of correlations; Level 2, which requires some test data; and finally Level 3, for which most of the data needed to be input are obtained from testing. The basic premises of this *Guide* are the same as the mechanistic principles discussed in Chapter 12. Note that the models have been calibrated against national data (in the United States and Canada)—it is extremely important that those models be calibrated against locally available data before the *Guide* is adopted by any specific state or local transportation department. In the meantime, the *Guide*/software serves as a very useful tool for checking the sensitivity of the performance/distress of the pavements to the different control variables. The most powerful feature of the guide is the ability to cycle through variations in material properties as a function of changes in seasonal conditions, and compute the incremental damage due to cumulative traffic loading through the different seasons for long design periods. This concept is illustrated with an example output in Figure 14.24. Note that it shows three relevant factors—cumulative traffic, changes in modulus of asphalt mix, and total rutting during the entire 20-year design period. Such plots can also be generated for other types of distresses.

The different parameters that need to be input are shown in Table 14.3, and the steps in running the software (for a Level-1 design) for the guide are presented thereafter. A brief description of the different steps is provided in the following paragraphs.

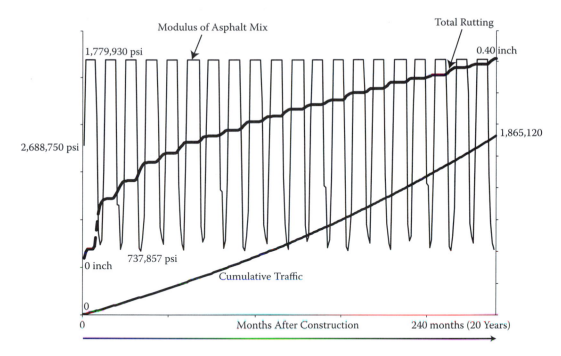

FIGURE 14.24 Example of Information Obtained from M-E Design.

14.3.2.2.1 Overview of the Software

The NCHRP 1-37A Mechanistic Empirical Pavement Design (commonly referred to as the Design Guide 2002, or DG2002) software is a Windows-based application for the simulation of pavement structures. It can be used to simulate pavement structures under different traffic and climatic conditions. The software has the ability to predict the amount of damage that a structure will display at the end of its simulated design life. In addition to using laboratory test results, predicting properties for pavement mixes can be done by using correlations. Values (either default or entered by the user) that represent structural, climatic, or traffic properties are used in calculating distresses. The following example demonstrates how to provide the different inputs for a design project. The project is based in Guilford, Maine, and is designed for a 20-year life.

14.3.2.2.2 Creating a New Project

To create a new project, first open the Design Guide 2002 software. Click on **File,** then **New**; a window, as shown in Figure 14.25, will pop up. Here you can specify the name of the project, the folder to create it in, and the measurement system (currently the software can only handle U.S. Customary). After this is complete, clicking **OK** will bring you into the main software console (shown in Figure 14.26). The red boxes next to each section indicate that information needs to be entered; yellow boxes indicate that there are default values for that section which need to be either accepted or changed.

14.3.2.2.3 Specifying Project Information

On the top left corner of the console, double click on **General Information** (see Figure 14.27). This will bring up a window where you can specify the Design Life, Base/Subgrade Construction Month, Pavement Construction Month, and Traffic Open Month (see Figure 14.28). Then choose the Type of Pavement.

TABLE 14.3
Inputs from Test Results for NCHRP 1-37A Design Guide

| | Inputs from Tests | | |
| | Levels | | |
Material	1	2	3
Soil	Resilient Modulus from AASHTO T-307; Plasticity Index, AASHTO T-90; Gradation, AASHTO T-27; Specific Gravity, Maximum Dry Unit Weight, Optimum Moisture Content, AASHTO T-99, T-180; Saturated Hydraulic Conductivity, AASHTO T-215	Resilient Modulus from Correlation with CBR, AASHTO T-193; Layer Coefficient, AASHTO 1993; PI/Gradation, AASHTO T-27/T-90, DCP, ASTM D-6951	Resilient Modulus from CBR from Soil Classification, AASHTO M-145/ASTM D-2487 (default values available)
Asphalt Binder	Performance Grading Tests, AASHTO T-315; or Viscosity, AASHTO T-201; Penetration, AASHTO T-49; Specific Gravity, ASTM D-70; Softening Point, ASTM D-36	Performance Grading Tests, AASHTO T315; or Viscosity, AASHTO T201; Penetration, AASHTO T49; Specific Gravity, ASTM D70; Softening Point, ASTM D36	A and VTS
Aggregates	Coefficient of Thermal Contraction, ASTM D-4535	Gradation: retained on ¾ in., 3/8 in., and No. 4 sieve, and passing No. 200 sieve, AASHTO T-30	Gradation: retained on ¾ in., 3/8 in., and No. 4 sieve, and passing No. 200 sieve, AASHTO T-30
Asphalt Mix	Dynamic Modulus, AASHTO T-P62; Volume of Air Voids, AASHTO T-269; Unit Weight, AASHTO T-166; Indirect Tension Test at 3 Temperatures, AASHTO T-322; Voids in Mineral Aggregates, AASHTO R-35	Volume of Air Voids, AASHTO T-269; Unit Weight, AASHTO T-166; Indirect Tension Test at 1 Temperature, AASHTO T-322; Voids in Mineral Aggregates, AASHTO R-35	Volume of Air Voids, AASHTO T-269; Unit Weight, AASHTO T-166; Voids in Mineral Aggregates, AASHTO R-35
Existing Pavement for Rehabilitation Concrete	Unbound layers: resilient modulus from falling weight deflectometer (FWD) tests, rutting from trench	Unbound layers: resilient modulus from correlations with other tests, rutting from user	Unbound layers: resilient modulus from soil classification, rutting from user
	Stabilized base layers: maximum modulus, existing modulus from FWD	Stabilized base layers: maximum modulus and % fatigue cracking	Stabilized base layers: typical maximum modulus and pavement condition
	Fractured concrete: modulus from FWD		Fractured concrete: crack spacing
	Asphalt mix layer: existing modulus from FWD, volumetric, and binder properties	% alligator cracking, volumetric, and binder properties	Pavement rating, estimates of mix properties

In the main console, double click on **Site/Project Information**. In the window that pops up, enter the Location, and the Project ID shown in Figure 14.29 (for the date and traffic direction, defaults are used here and can be changed if needed). Click **OK**.

Double click on Analysis Parameters in the main console. These are the values that will determine at the end of the design life whether that aspect of the pavement structure has failed; these are shown in Figure 14.30. For example, the limit for AC Thermal Fracture is 1000 feet per mile. In the

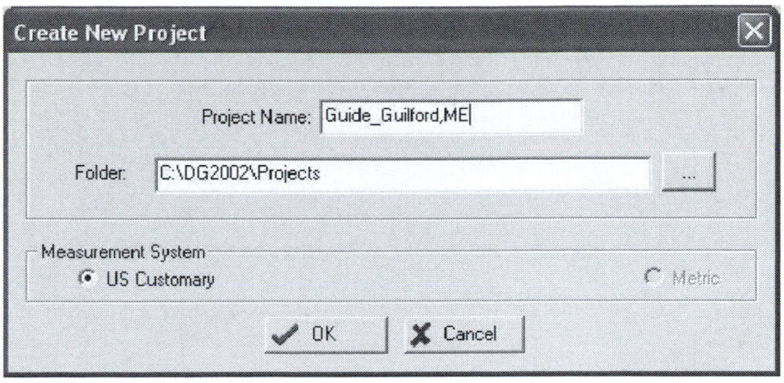

FIGURE 14.25 Creating a New Project.

final summary, if the simulation shows that the structure has thermal cracking in excess of 1000 feet per mile, then it has failed. Figure 14.30 shows the default values; accept them by clicking **OK**.

14.3.2.2.4 Specifying Traffic Inputs

On the main console under **Inputs**, double click on **Traffic**. Figure 14.31 shows the traffic window where the specific inputs are shown for Guilford; enter these values as shown. After these are finished, click on the **Edit** button next to Traffic Volume Adjustment. This will pull up a window like the one in Figure 14.32. Going through each tab:

- *Monthly adjustment*: these are default values.
- *Vehicle class distribution*: enter 100 for **Class 9**, and set all others to 0.
- *Hourly distribution*: these are default values.
- *Traffic growth factors*: the compound growth should be set to 4%. All other parameters have been set already.

Press **OK**.

Next, click on the **Edit** button next to Axle Load Distribution Factor. The window that pops up (see Figure 14.33) will contain default values; press **OK**.

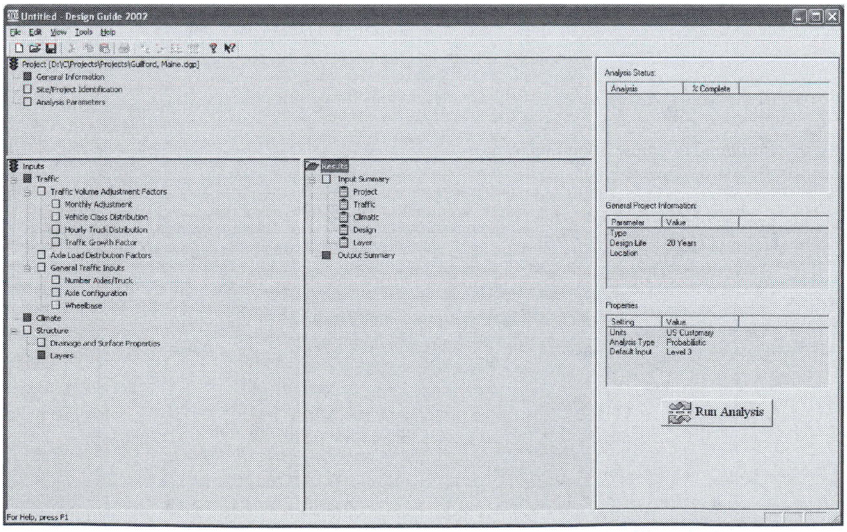

FIGURE 14.26 Main Software Console.

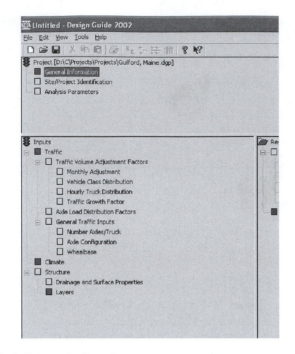

FIGURE 14.27 Upper Left Corner of a Console.

General Traffic Inputs is the last parameter set that needs to be specified; click **Edit**. All the parameters in this window (see Figure 14.34) under each tab are default values; click on each tab until all are green, and then press **OK**. Now the original Traffic window is displayed; click **OK** to return to the main console.

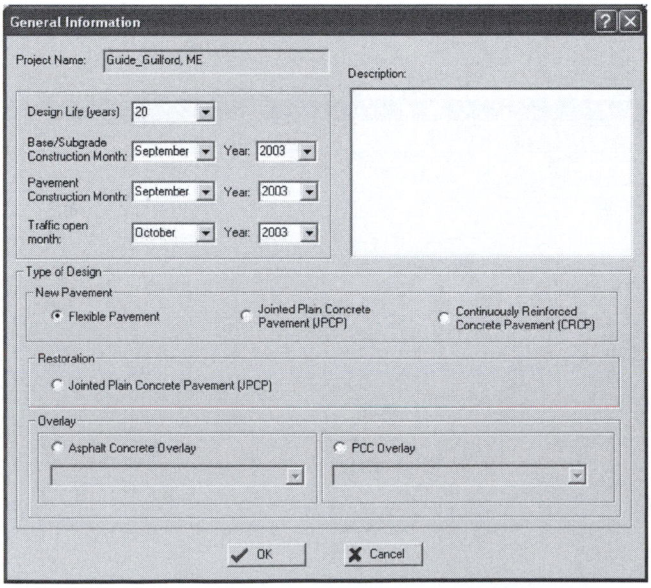

FIGURE 14.28 General Information.

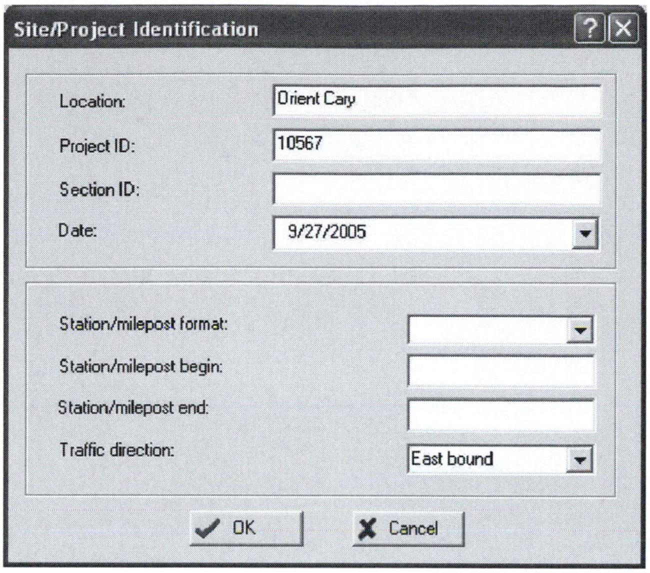

FIGURE 14.29 Site/Project Identification.

14.3.2.2.5 Specifying Climate

In the main console, double click on **Climate**. A window will pop up where you can designate climate and water table information (see Figure 14.35). For the **Depth of Water Table**, enter 10. Next, click on **Generate**. Figure 14.36 shows the window that pops up; click on **Interpolate Climatic**

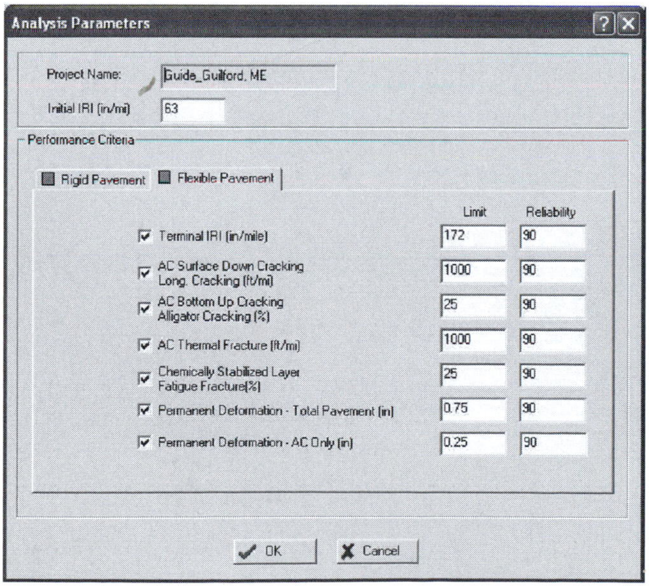

FIGURE 14.30 Analysis Parameters.

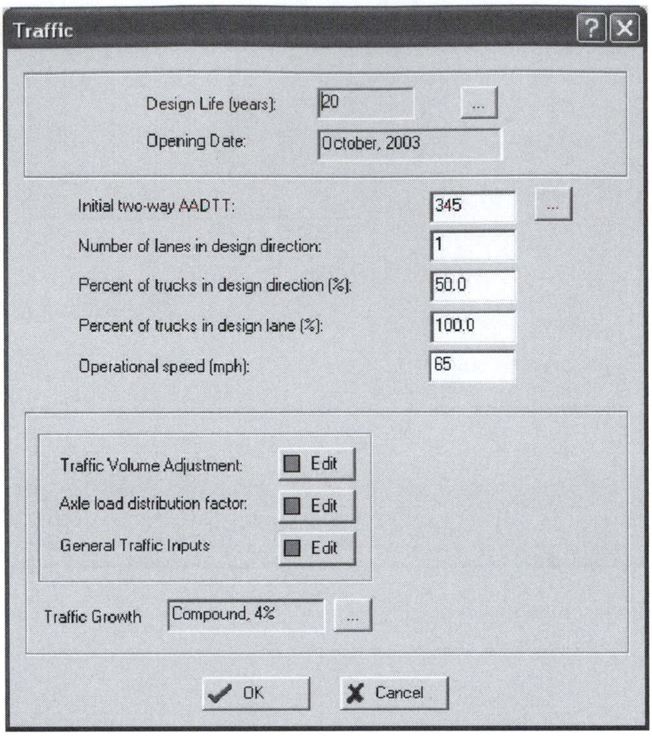

FIGURE 14.31 Traffic Parameters.

Data for Given Station. Enter the latitude, longitude, and elevation values for Guilford that are shown in Figure 14.36. Select the first three options (Millinocket, Greenville, and Bangor) to interpolate weather data for Guilford. Click **Generate**. Save the generated weather file; this will make it available for use in future projects.

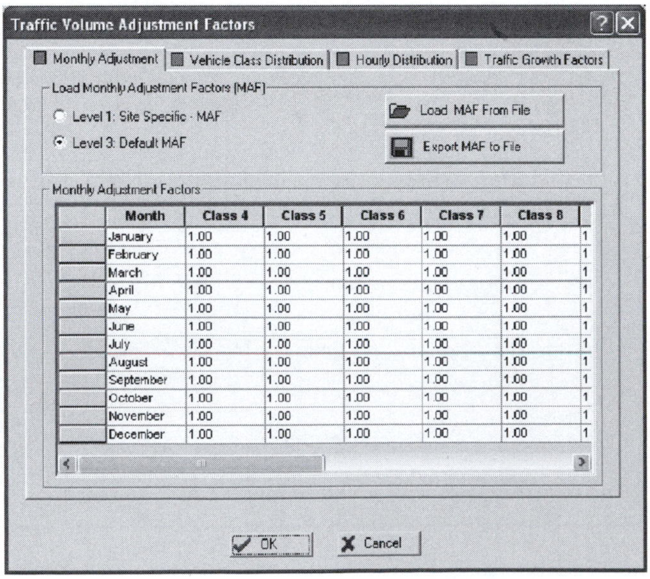

FIGURE 14.32 Traffic Volume Adjustment Factors.

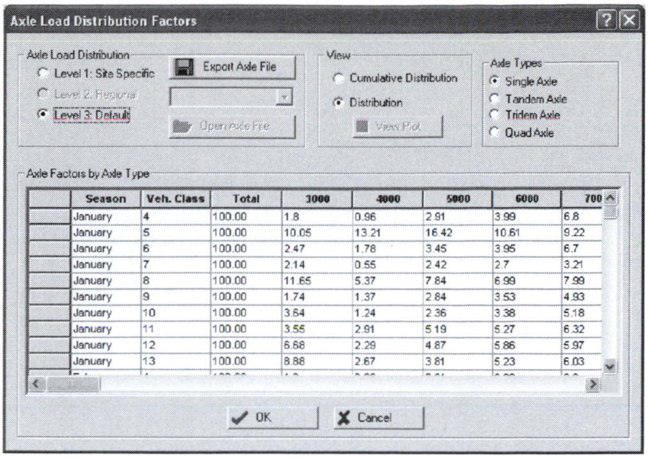

FIGURE 14.33 Axle Load Distribution Factors.

14.3.2.2.6 Defining Pavement Structure

14.3.2.2.6.1 Hot Mix Asphalt Layer

On the main console, double click on **Structure**. Figure 14.37 shows the window that pops up. Highlight the Asphalt Concrete layer, and click on **Edit**. Here you can enter and select the properties of the HMA layer.

- Under the Asphalt Mix tab, enter the gradations shown in Figure 14.38. The layer thickness should be 3 in., and be sure that Level 3 is selected.

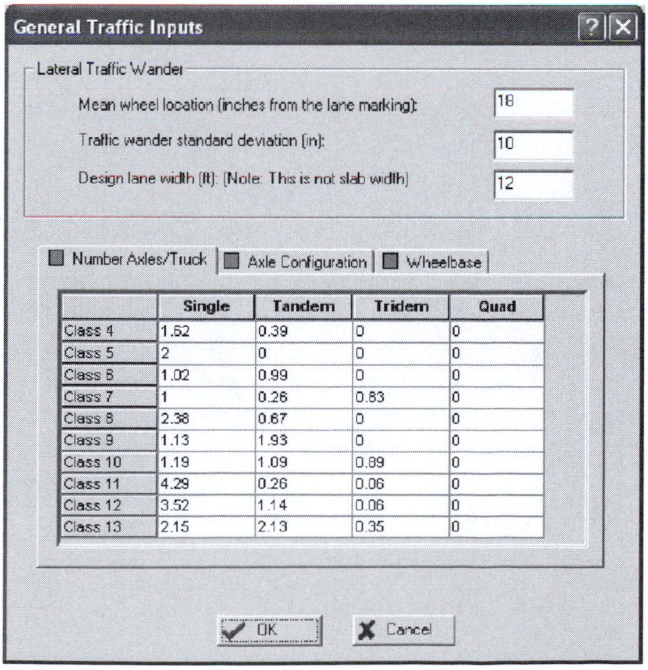

FIGURE 14.34 General Traffic Inputs.

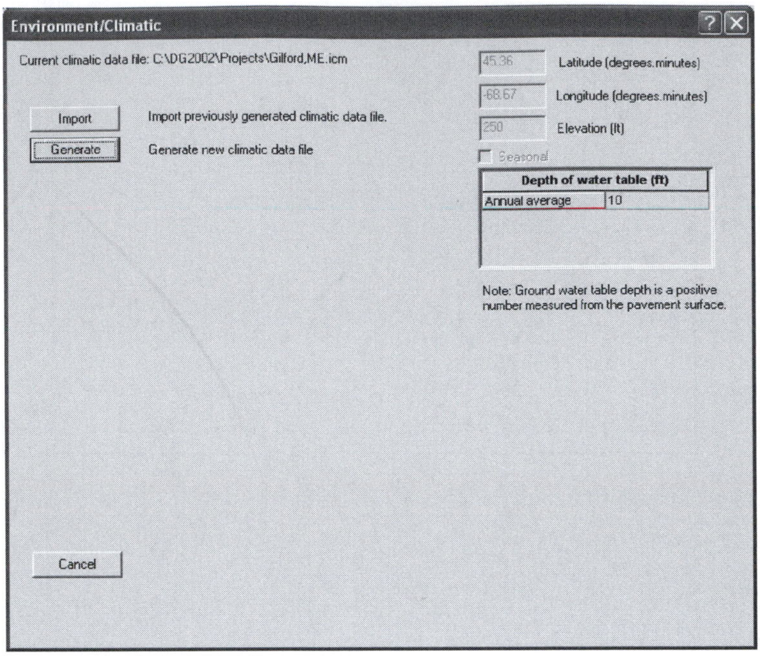

FIGURE 14.35 Environment/Climatic Window.

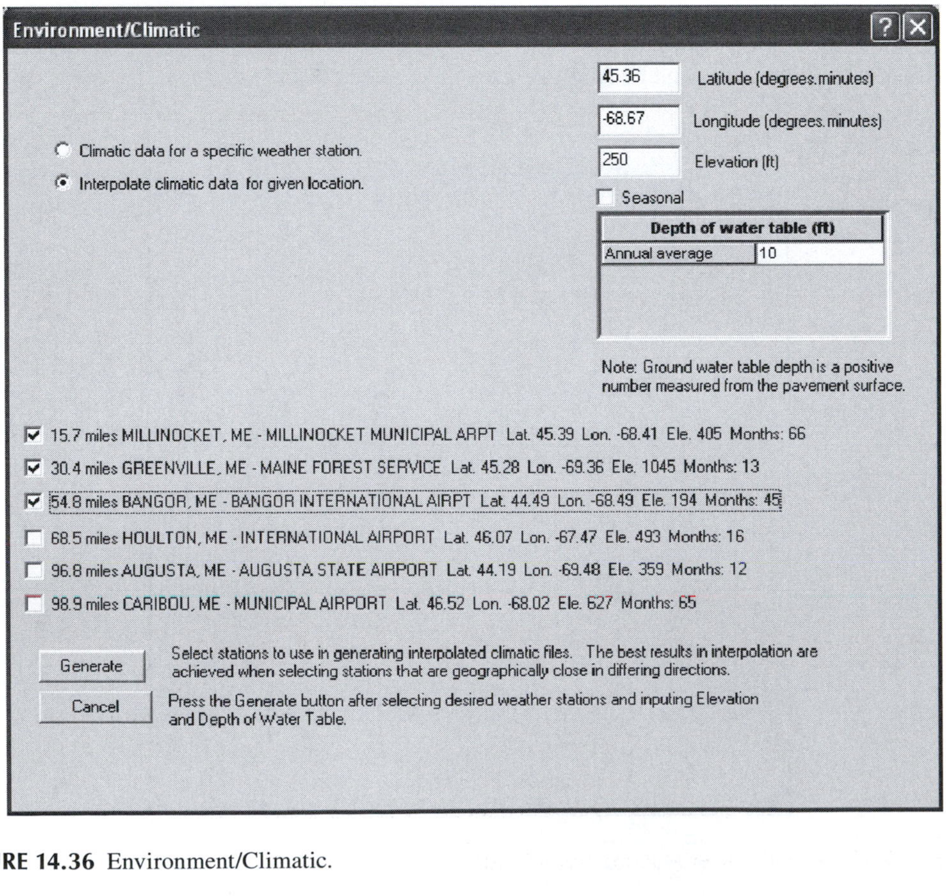

FIGURE 14.36 Environment/Climatic.

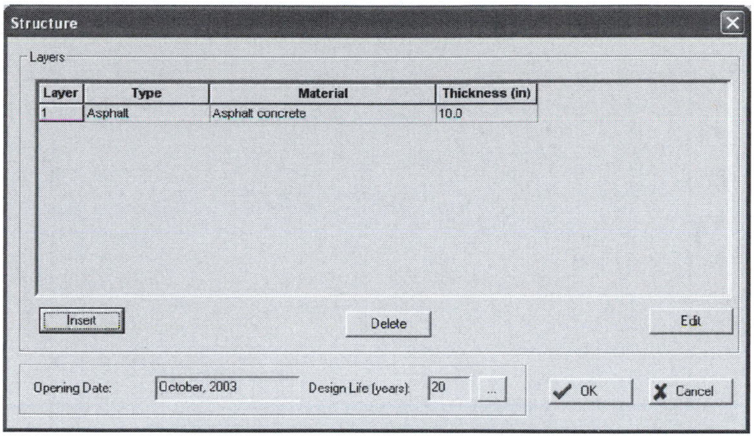

FIGURE 14.37 Structure.

- Under the Asphalt Binder tab, select Superpave binder grading. Choose 64-28 for the binder (high and low temperature, respectively; shown in Figure 14.39).
- Under the Asphalt General tab, change the values for Effective Binder Content and Air Voids, as shown in Figure 14.40.

Click **OK** to return to the Structure window.

14.3.2.2.6.2 Granular Base Layers

Highlight the HMA Layer that you just created, and click **Insert**. Refer to Figure 14.41 for the information that needs to be entered. Click **OK**.

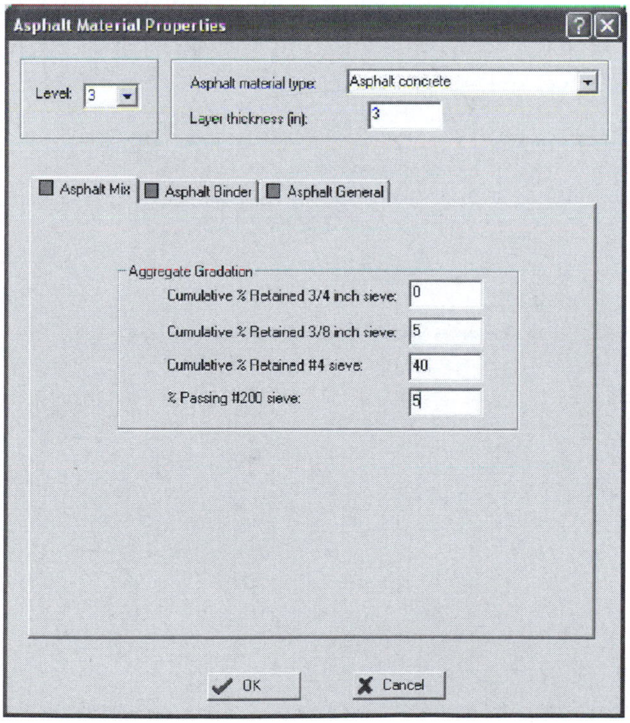

FIGURE 14.38 Asphalt Material Properties—Gradation.

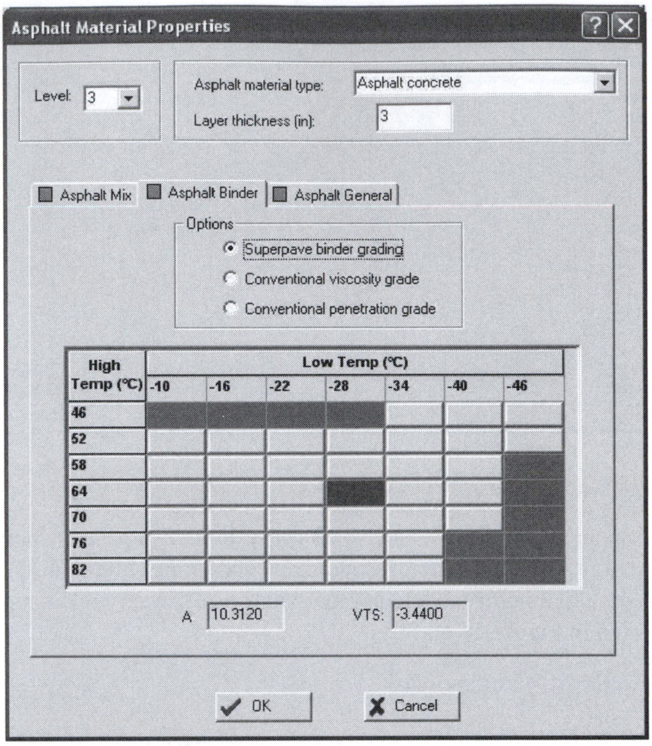

FIGURE 14.39 Asphalt Material Properties—Binder.

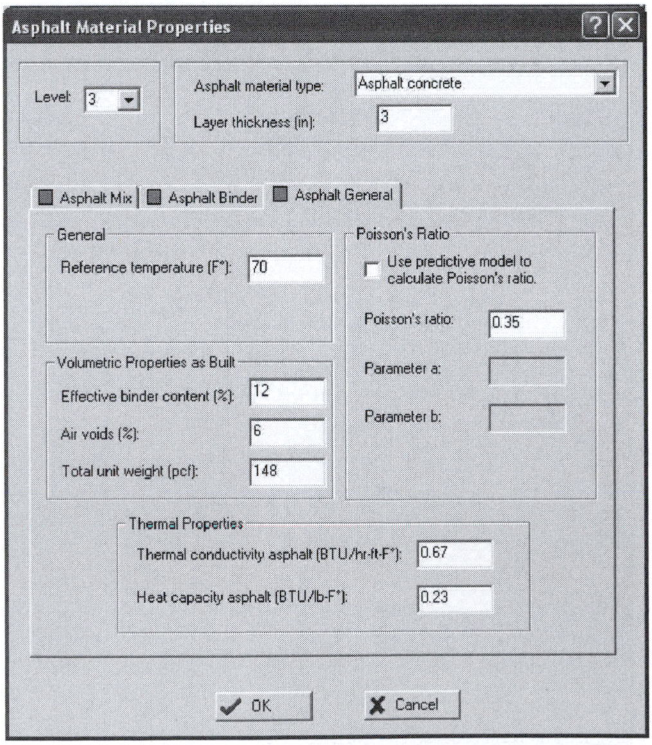

FIGURE 14.40 Asphalt Material Properties—Volumetric Properties.

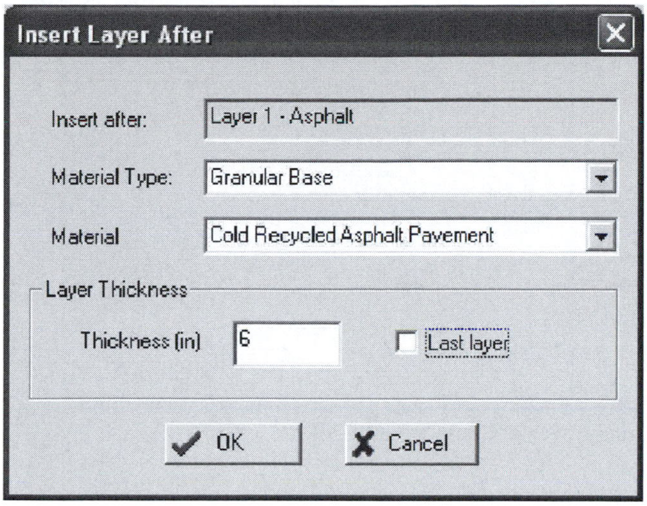

FIGURE 14.41 Inserting a New Layer.

Highlight the Granular Base that you just created, and click **Edit**. Be sure that the Modulus (psi) button is selected, and enter 51,428 as shown in Figure 14.42. Note: the software is able to calculate modulus values based on other material properties if you are using a level 1 or 2 design.

Now, repeat the process for a second granular base layer beneath the one you just created. Follow Figures 14.43 and 14.44 for relevant inputs.

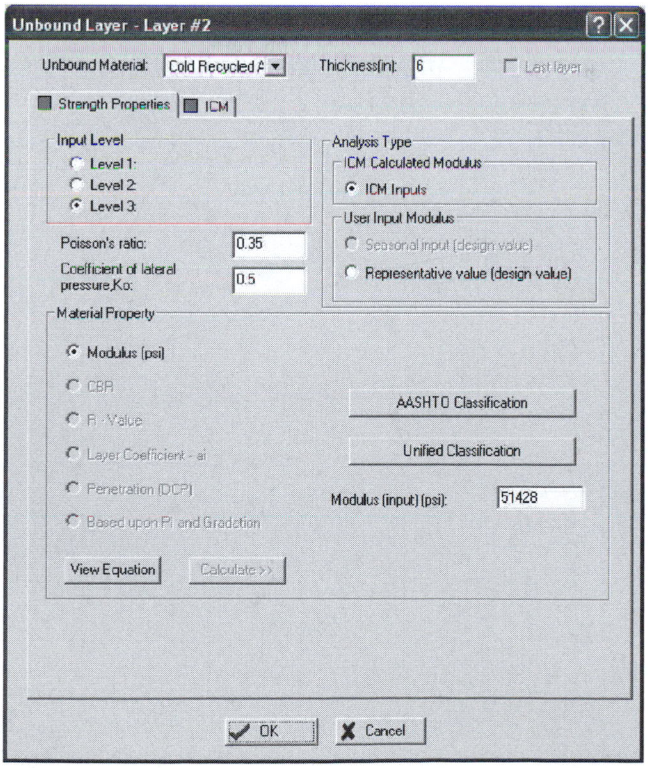

FIGURE 14.42 Granular Layer Strength Properties.

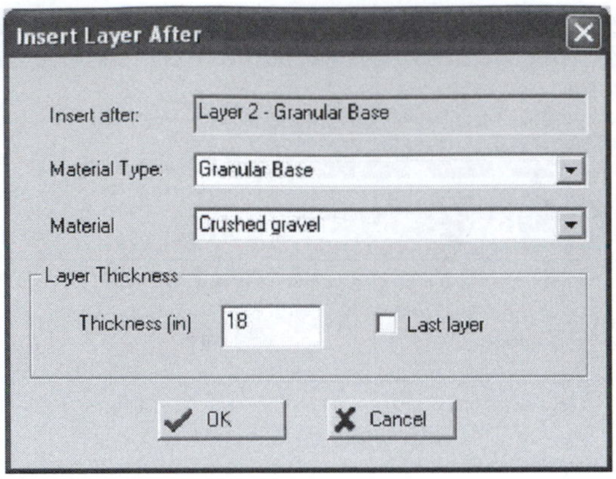

FIGURE 14.43 Inserting a Second Granular Layer.

14.3.2.2.6.3   Subgrade

To complete the structure, highlight the second granular layer and click **Insert**. Refer to Figures 14.45 and 14.46 for subgrade inputs. Once finished, return to the main console.

> *Thermal cracking*: on the main console, double click on **Thermal Cracking**. Figure 14.47 shows the window that pops up; the values are calculated by the software and are default. Click **OK**.
>
> *Distress potential*: on the main console, double click on **Distress Potential**. Figure 14.48 shows the window that appears. Click **OK** since these are default values.

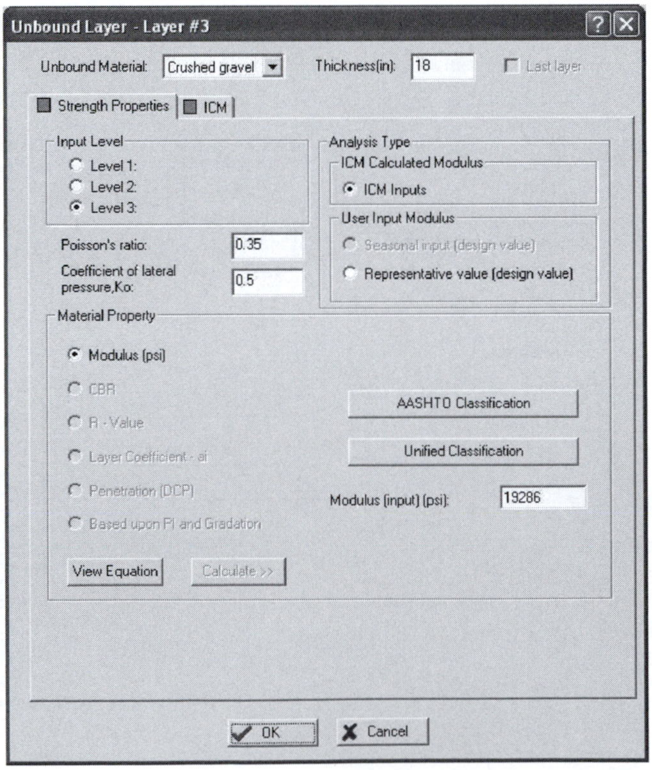

FIGURE 14.44 Second Granular Layer Strength Properties.

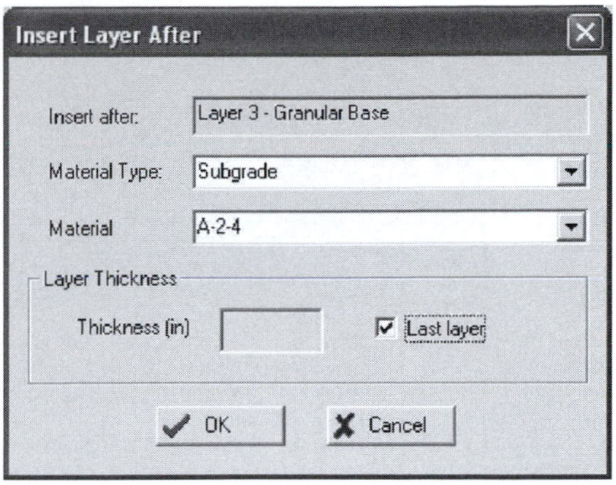

FIGURE 14.45 Inserting a Subgrade Layer.

14.3.2.2.7 Running the Analysis

Before running the analysis, there are two simple steps that need to be taken care of:

- Check all values that you entered. This ensures that there will be no surprises or inaccuracies on your part when the simulation is finished.
- Save your work before running the simulation. If anything happens to the computer or the software, you will not have to reenter all of your work.

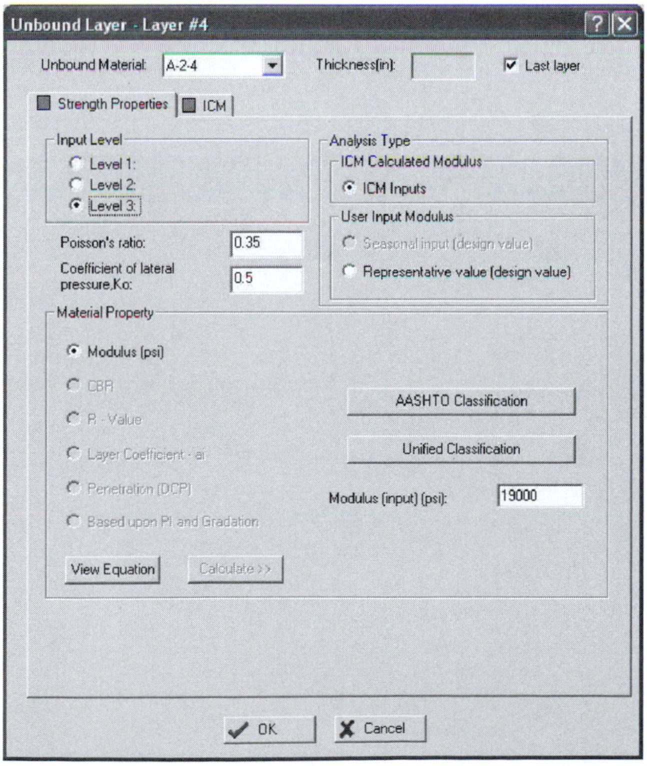

FIGURE 14.46 Subgrade Layer Strength Properties.

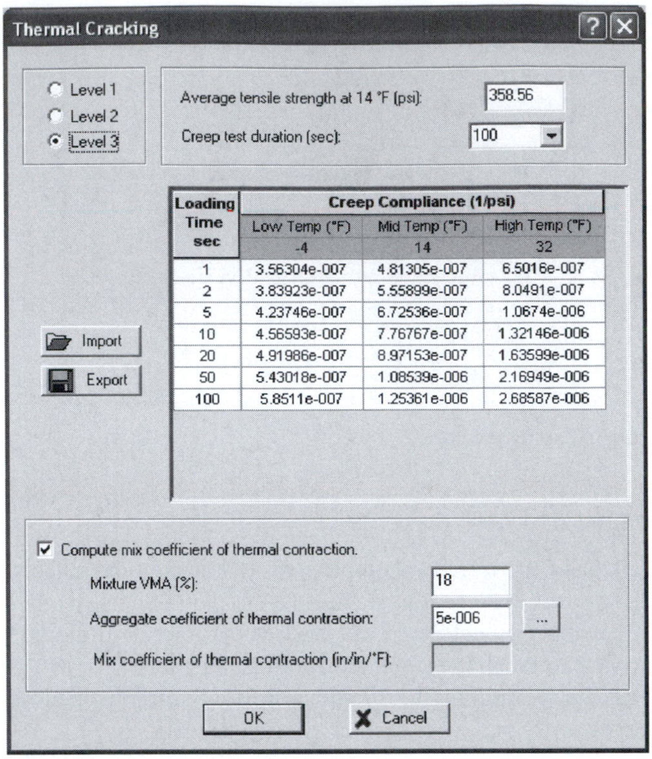

FIGURE 14.47 Thermal Cracking Values.

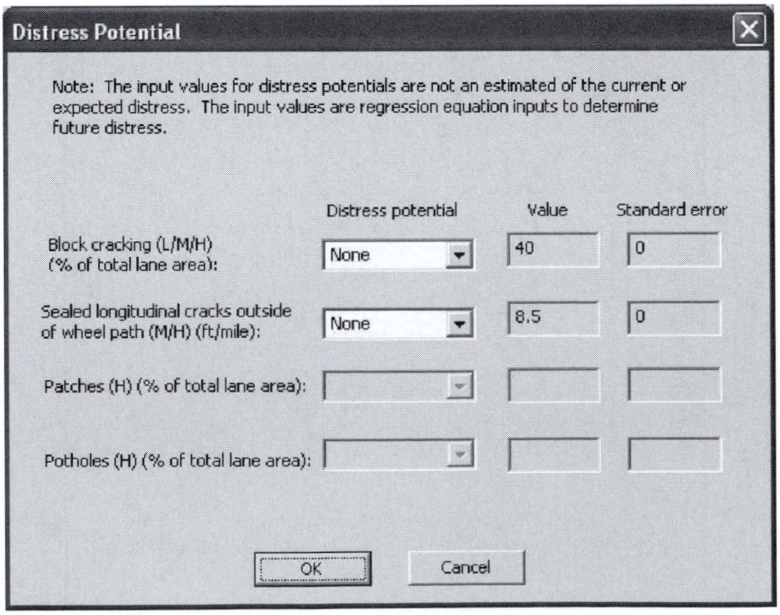

FIGURE 14.48 Distress Potential Window.

Once you have verified your inputs and saved the program, run the simulation by clicking **Run Analysis**. Depending on the speed of your computer and the design life that you have entered, the software can take anywhere from 30 to 90 minutes or more (based on runs on a laptop computer, Dell Inspiron 8100, with Microsoft Windows XP 2002 Professional, 1200 MHz Intel Pentium III, 256 MB of RAM, and 20 GB hard drive).

14.3.2.2.8 Practical Issues

While using the DG2002, some glitches and problems could be encountered, since the software is still relatively new and is most likely prone to compatibility problems. The user is strongly advised to participate in the Highway Community Exchange Online Forum (Highway Community Exchange, n.d.) to discuss any problems.

QUESTIONS

1. Determine the VTM, VMA, and VFA of a compacted HMA sample with the following test data: bulk-specific gravity = 2.351, theoretical maximum density = 2.460, and aggregate bulk-specific gravity = 2.565, asphalt content = 5.6%, and specific gravity of asphalt binder = 1.03.

2. Determine the optimum asphalt content for a mix, on the basis of VTM, from the following data:

AC (%)	VTM
4.5	6.7
5	5.2
5.5	3.4
6	2.4

3. Using the AASHTO procedure, determine the thickness required for a base and a surface course, over a subgrade. The structural number (SN) required for the pavement is 4.5. The available materials are a granular base with a layer coefficient of 0.13 and an HMA surface with a layer coefficient of 0.40. The drainage coefficient (m_2) of the base course can be considered to be 0.9.

4. Design a four-lane pavement for a 15-year design period, with the following traffic, layers, and material properties, using the mechanistic-empirical design method. Use rutting and fatigue failure criteria. Calculate the tensile strain at the bottom of the HMA surface and vertical strain on top of the subgrade layer using any layered analysis program. Use Asphalt Institute models for rutting and fatigue failure.

Layer	Modulus (MPa)	Poisson's Ratio
HMA surface	3500	0.35
Recycled base course	1500	0.35
Gravel subbase	300	0.4
Subgrade	40	0.4

Note: Traffic: initial two-way annual average daily truck traffic (AADTT) = 345. % of trucks in design direction: 50; % of trucks in design lane: 95%; traffic growth rate: 4%.

Assume 80 kN tandem axles for all trucks, and tire pressure of 690 kPa.

15 Mix Design and Structural Design for Concrete Pavements

15.1 MIX DESIGN

The absolute volume method detailed in ACI 211 (American Concrete Institute, ACI Committee 211, 1991) is commonly used to proportion the local concrete materials to achieve the desired fresh and ultimate hardened concrete properties. This includes selecting the water-cement ratio, coarse and fine aggregate, cement and pozzolans, water, and admixture requirements. The effects of all of these volumetric properties are interrelated. Their selection depends on many factors, including the compressive strength, which is the simplest and easiest to measure.

15.1.1 CONCRETE STRENGTH

According to ACI 318 (American Concrete Institute, ACI Committee 318, 2002), a concrete strength test is defined by the average compressive strength of two concrete cylinders. The running average of three consecutive tests must be equal to or greater than the specified compressive strength (f_c') at 28 days. ACI 318 also requires the f_c' to be at least 2500 psi (17.5 MPa), and no individual test (average of two cylinders) can be more than 500 psi (3.5 MPa) below the specified strength (ACI 318). To allow for variability in materials, batching, mixing, placing, finishing, curing, and testing the concrete, the ACI recommends a higher target strength that would assure the designer to achieve the minimum specified strength, which is called f_{cr}' (see Table 15.1, Table 15.2, and Table 15.3).

15.1.2 WATER-TO-CEMENTITIOUS MATERIALS RATIO

The water-to-cementitious materials ratio (or water-cementitious ratio; W/CM), is the mass of water divided by the mass of all cements, blended cements, and pozzolanic materials such as fly ash, slag, silica fume, and natural pozzolans. In most literature, *water-cement ratio* (W/C) is used synonymously with W/CM.

Assuming that a concrete is made with clean sound aggregates, and that the cement hydration has progressed normally, the strength gain is inversely proportional to the water-cementitious ratio by mass. The paste strength is proportional to the solids volume or the cement density per unit volume (Abrams, 1918) and is shown schematically in Figure 15.1. It should be mentioned, however, that the cement hydration process is more complex in its microstructural development of calcium-silicate-hydrates and pore structure. But even at a given water-cementitious ratio, differences in strength may result based on the influence of aggregate gradation, shape, particle size, surface texture, strength, and stiffness, or factors associated with cement materials (e.g., different sources, chemical composition, and physical attributes), the amount of entrained air, and the effects of admixtures and curing. A typical relationship between W/C, compressive strength, and air-entrainment is shown in Figure 15.2.

15.1.3 SELECTION OF THE WATER-TO-CEMENTITIOUS MATERIALS RATIO

The selection of the W/CM is influenced by strength requirements and exposure conditions. The W/CM will govern the permeability or watertightness that is necessary for preventing aggressive

TABLE 15.1

Modification Factor for Standard Deviation When Less Than 30 Tests Are Available

Number of Tests*	Modification Factor for Standard Deviation**
Less than 15	Use Table 15.2/Table 15.3
15	1.16
20	1.08
25	1.03
30 or more	1.00

* Interpolate for intermediate numbers of tests.

** Modified standard deviation to be used to determine required average strength, f'_{cr}.

Source: Reprinted from Design and Control of Concrete Mixtures, Engineering Bulletin 001, 14th edition, with kind permission from The Portland Cement Association.

salts and chemicals from entering the paste pore system and causing deleterious effects in the concrete or reinforcing steel. For different exposure conditions, ACI recommendations for a maximum W/CM and minimum design compressive strengths are provided in Table 15.4 and Table 15.5. For pavement design, flexural strength is recognized as the more appropriate strength characteristic, and is specified for pavement design. However, the ACI uses only compressive strength in its specifications because of the lower variability associated with compressive strength testing.

When durability does not govern, the W/CM should be selected based on compressive strength. In this case adequate knowledge about strength and mix proportions should be obtained through trial mixes or field data. For example, Figure 15.2 and Table 15.6 and Table 15.7 show the relationship between the W/CM and compressive strength for a given concrete mix and curing conditions. It should be noted that the ACI-recommended values given in subsequent tables and figures should be a starting point from which to initiate the mix design process. Variations in mix ingredients may require a few iterations to optimize the desired properties.

TABLE 15.2

(Metric) Required Average Compressive Strength When Data Are Not Available to Establish Standard Deviation

Specified Compressive Strength, f'_c, MPa	Required Average Compressive Strength, f'_{cr}, MPa
Less than 21	$f'_c + 7.0$
21 to 35	$f'_c + 8.5$
Over 35	$1.10\,f'_c + 5.0$

Source: Reprinted from Design and Control of Concrete Mixtures, Engineering Bulletin 001, 14th edition, with kind permission from The Portland Cement Association.

TABLE 15.3
(Inch-Pound) Required Average Compressive Strength When Data Are Not Available to Establish Standard Deviation

Specified Compressive strength, f'_{cr}, psi	Compressive Strength, f'_{cr}, psi
Less than 3000	$f'_c + 1000$
3000 to 5000	$f'_c + 1200$
Over 5000	$1.10\,f'_c + 700$

Source: Reprinted from Design and Control of Concrete Mixtures, Engineering Bulletin 001, 14th edition, with kind permission from The Portland Cement Association.

TABLE 15.4
Maximum Water-Cement Ratios and Minimum Design Strengths for Various Exposure Conditions

Exposure Condition	Maximum Water-to-Cementitious Materials Ratio by Mass for Concrete	Minimum Design Compressive Strength, f'_c, MPa (psi)
Concrete protected from exposure to freezing and thawing, application of deicing chemicals, or aggressive substances	Select water-to-cementitious materials ratio on the basis of strength, workability, and finishing needs	Select strength based on structural requirements
Concrete intended to have low permeability when exposed to water	0.50	28 (4000)
Concrete exposed to freezing and thawing in a moist condition, or deicers	0.45	31 (4500)
For corrosion protection for reinforced concrete exposed to chlorides from deicing salts, saltwater, brackish water, seawater, or spray from these sources	0.40	35 (5000)

Source: Reprinted from Design and Control of Concrete Mixtures, Engineering Bulletin 001, 14th edition, with kind permission from The Portland Cement Association.

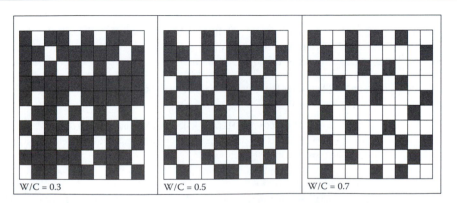

FIGURE 15.1 Schematic Showing the Effect of the Water-to–Cementitious Materials Ratio (W/CM) on Solids Density per Unit Volume. (Dark = Solid; White = Voids.)

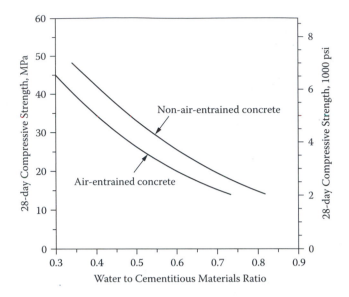

FIGURE 15.2 Typical Relationship between Compressive Strength and W/CM for Concrete Using 19 mm to 25 mm Nominal Maximum Aggregate Size Coarse Aggregates.
Source: Reprinted from Design and Control of Concrete Mixtures, Engineering Bulletin 001, 14th edition, with kind permission from The Portland Cement Association.

TABLE 15.5
Requirements for Concrete Exposed to Sulfates in Soil of Water

Sulfate Exposure	Water-Soluble Sulfate (SO₄) in Soil, % by Mass*	Sulfate (SO₄) in Water, ppm	Cement Type**	Maximum Water-to–Cementitious Materials Ratio, by Mass	Minimum Design Compressive Strength, f'_c, MPa (psi)
Negligible	Less than 0.10	Less than 150	No special type required	—	—
Moderate†	0.10 to 0.20	150 to 1500	II, MS, IP(MS), IS(MS), P(MS), I(PM)(MS), I(SM)(MS)	0.50	28 (4000)
Severe	0.20 to 2.00	1500 to 10,000	V, HS	0.45	31 (4500)
Very severe	Over 2.00	Over 10,000	V, HS	0.40	35 (5000)

* Tested in accordance with the Method for Determining the Quantity of Soluble Sulfate in Solid (Soil and Rock) and Water Samples (Bureau of Reclamation, 1977).

** Cement Types II and V are in ASTM C-150 (AASHTO M-85), types MS and HS in ASTM C-1157, and the remaining types are in ASTM C-595 (AASHTO M-240). Pozzolans or slags that have been determined by test or service record to improve sulfate resistance may also be used.

† Seawater.

Source: Reprinted from Design and Control of Concrete Mixtures, Engineering Bulletin 001, 14th edition, with kind permission from The Portland Cement Association.

15.1.4 AGGREGATES

Aggregates have an important role in proportioning concrete. Aggregates have a great effect on the workability of fresh concrete. The aggregate particle size and gradation, shape, and surface texture will influence the amount of concrete that is produced with a given amount of paste (cement plus water). The selection of the maximum size aggregate is governed by the thickness of the slab and by the closeness of the reinforcing steel. The maximum size aggregates should not be obstructed and should flow easily during placement and consolidation.

Guidelines provided by ACI state that the maximum size coarse aggregate should not exceed one-fifth of the narrowest dimension between the sides of forms, or three-fourths of the clear space between individual reinforcing bars or the clear spacing between reinforcing bars and form walls. For unreinforced slabs on grade, the maximum aggregate size should not exceed one-third of the slab thickness. For pavements, typically 1–1.5 in. maximum size aggregate is used.

The most desirable gradation for fine aggregate will depend on the properties of the concrete the designer is trying to seek and will depend on coarse aggregate volume and maximum size, and amount of paste. Figure 15.3 and Table 15.8 provide the recommended bulk volume fraction of coarse aggregate per unit volume of concrete based on the maximum nominal size of the coarse aggregate and the fineness modulus of the sand or fine aggregate. PCC for pavements may have a stiffer consistency and therefore coarse aggregate volumes may be increased by 10%.

15.1.5 AIR CONTENT IN CONCRETE

The purposeful entrainment of air in concrete provides tremendous protection against freezing and thawing action and against deicing salts. Hydraulic pressure is generated when water in the paste pore structure freezes and pushes against the unfrozen water. The tiny entrained air bubbles act as relief valves for this developed hydraulic pressure. Figure 15.4 shows the recommended relationship between the target percentage air content, the maximum nominal aggregate size, and the severity of exposure.

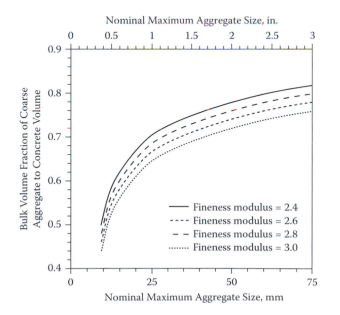

FIGURE 15.3 Bulk Volume of Coarse Aggregate per Unit Volume of Concrete.
Note: Bulk volumes are based on aggregates in a dry-rodded condition.
Source: Reprinted from Design and Control of Concrete Mixtures, Engineering Bulletin 001, 14th edition, with kind permission from The Portland Cement Association.

TABLE 15.6
(Metric) Relationships between the Water-to-Cementitious Materials Ratio and the Compressive Strength of Concrete

Compressive Strength at 28 Days, MPa	Water-to-Cementitious Materials Ratio by Mass	
	Non-Air-Entrained Concrete	Air-Entrained Concrete
45	0.38	0.30
40	0.42	0.34
35	0.47	0.39
30	0.54	0.45
25	0.61	0.52
20	0.69	0.60
15	0.79	0.70

Note: Strength is based on cylinders moist-cured 28 days in accordance with ASTM C-31 (AASHTO T-23). Relationship assumes a nominal maximum size aggregate (NMSA) of about 19 to 25 mm.

Source: Reprinted from Design and Control of Concrete Mixtures, Engineering Bulletin 001, 14th edition, with kind permission from The Portland Cement Association.

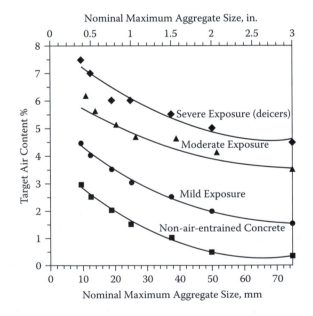

FIGURE 15.4 Target Total Air Content Requirements for Concrete Using Different Sizes of Aggregate. *Note:* The air content in job specifications should be specified to be delivered within −1 to +2 percentage points of the target value for moderate and severe exposure. *Source:* Reprinted from Design and Control of Concrete Mixtures, Engineering Bulletin 001, 14th edition, with kind permission from The Portland Cement Association.

TABLE 15.7
(Inch-Pound Units) Relationships between Water-to-Cementitious
Material Ratio and Compressive Strength of Concrete

Compressive Strength at 28 Days, psi	Water-to-Cementitious Materials Ratio by Mass	
	Non-Air-Entrained Concrete	Air-Entrained Concrete
7000	0.33	—
6000	0.41	0.32
5000	0.48	0.40
4000	0.57	0.48
3000	0.68	0.59
2000	0.82	0.74

Note: Strength is based on cylinders moist-cured 28 days in accordance with ASTM C-31 (AASHTO T-23). Relationship assumes a nominal maximum size aggregate (NMSA) of about ¾ in. to 1 in.

Source: Reprinted from Design and Control of Concrete Mixtures, Engineering Bulletin 001, 14th edition, with kind permission from The Portland Cement Association.

15.1.6 SLUMP

Fresh concrete must have the appropriate workability, consistency, and plasticity suitable for construction conditions. Workability is a measure of the ease of placement, consolidation, and finishing of the concrete. Consistency is the ability of freshly mixed concrete to flow, and plasticity assesses the concrete's ease of molding. If the concrete is too dry and crumbly, or too wet and soupy, then it lacks plasticity. The slump test is a measure of consistency and indicates when the characteristics of the fresh mix have been changed or altered. However, the slump is indicative of workability when assessing similar mixtures. Different slumps are needed for different construction projects. Table 15.9 provides recommendations for slump for different types of construction. For pavements, state DOTs may use different slump specifications for fixed-form and slip-form construction.

15.1.7 WATER CONTENT

The water demands for a concrete mix depend on many variables, including mix proportions, air content, aggregate gradation, angularity and texture, and climate conditions. Table 15.10 and Table 15.11 and Figure 15.5 present the relationship between mix-water, different slumps, and nominal maximum size aggregates (NMSAs). The recommendations given in Table 15.10 and Table 15.11 and Figure 15.5 are for angular aggregates, and the water estimates may be reduced by 20 to 45 lb for subangular to rounded aggregates, respectively.

15.1.8 CEMENTING MATERIALS CONTENT AND TYPE

The cementing materials content is usually determined based on the specified water-to-cementitious materials ratio. However, usually a minimum amount of cement is also specified to ensure satisfactory durability, finishability, and wear resistance of slabs even though strength needs may be satisfied at lower cement contents, as shown in Table 15.12. However, economic mixes are obtained with the least amount of cement. To accomplish this, the concrete should be as stiff as possible, should use the largest practical maximum coarse aggregate, and should use the optimum ratio of coarse to fine aggregate.

TABLE 15.8
Bulk Volume of Coarse Aggregate per Unit Volume of Concrete

Nominal Maximum Size of Aggregate, mm (in.)	Bulk Volume of Dry-Rodded Coarse Aggregate per Unit Volume of Concrete for Different Fineness Moduli of Fine Aggregate*			
	2.40	**2.60**	**2.80**	**3.00**
9.5 (3/8)	0.50	0.48	0.46	0.44
12.5 (1/2)	0.59	0.57	0.55	0.53
19 (3/4)	0.66	0.64	0.62	0.60
25 (1)	0.71	0.69	0.67	0.65
37.5 (1 1/2)	0.75	0.73	0.71	0.69
50 (2)	0.78	0.76	0.74	0.72
75 (3)	0.82	0.80	0.78	0.76
150 (6)	0.87	0.85	0.83	0.81

* Bulk volumes are based on aggregates in a dry-rodded condition as described in ASTM C-29 (AASHTO T-19).

Source: Reprinted from Design and Control of Concrete Mixtures, Engineering Bulletin 001, 14th edition, with kind permission from The Portland Cement Association.

For durability, the type of cement used is also important. When the concrete will be exposed to significant amounts of sulfates, cements with low C_3A contents should be used, as presented in Table 15.5. When the potential for corrosion of reinforcing bars exists due to a favorable chloride condition, then blended cements and a low W/CM should be considered. ACI 357R should be consulted for designing for corrosion protection. Table 15.13 shows limits on the amounts of pozzolans and mineral admixtures to be used when designing concrete exposed to deicing chemicals. It should be noted that these ACI recommendations should be used as a starting point for trial mixes. However, local experience should be consulted whenever it exists.

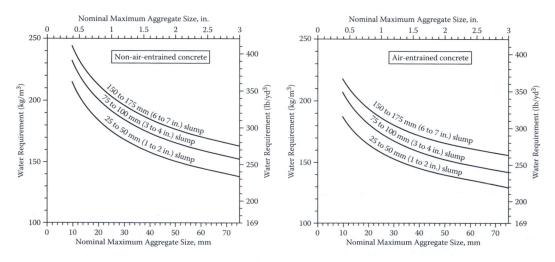

FIGURE 15.5 Approximate Water Requirement for Various Slumps and Crushed Aggregate Sizes for (left) Non-Air-Entrained Concrete and (right) Air-Entrained Concrete.
Source: Reprinted from Design and Control of Concrete Mixtures, Engineering Bulletin 001, 14th edition, with kind permission from The Portland Cement Association.

TABLE 15.9
Recommended Slumps for Various Types of Construction

Concrete Construction	Slump, mm (in.)	
	Maximum*	Minimum
Reinforced Foundation		
Walls and footings	75 (3)	25 (1)
Plain footings, caissons, and substructure walls	75 (3)	25 (1)
Beams and reinforced walls	100 (4)	25 (1)
Building columns	100 (4)	25 (1)
Pavements and slabs	75 (3)	25 (1)
Mass concrete	75 (3)	25 (1)

*May be increased 25 mm (1 in.) for consolidation by hand methods, such as rodding and spading.

Note: Plasticizers can safely provide higher slumps.

Source: Reprinted from Design and Control of Concrete Mixtures, Engineering Bulletin 001, 14th edition, with kind permission from The Portland Cement Association.

15.1.9 ADMIXTURES

Admixtures are used in concrete to enhance the desirable characteristics in the fresh and hardened concrete. This includes reducing the W/CM, entraining air, retarding or accelerating the set time, increasing the slump without increasing the W/CM, inhibiting corrosion, and self-leveling. Some admixtures are calcium chloride based and may affect steel corrosion. ACI 318 places limits on the maximum chloride-ion content for corrosion protection, as presented in Table 15.14.

The design of a concrete mix using the ACI method is explained with the following example.

15.1.10 EXAMPLE OF MIX DESIGN

Concrete is specified for a 10-in. pavement slab that will be exposed to severe freeze-thaw conditions and a moist environment. The required 28-day compressive strength is 3500 psi.

Cement: Type II, Sp. Gr. 3.15.
Coarse aggregate: well-graded gravel containing some crushed particles. The maximum nominal size is 3/4 in., with a specific gravity of 2.65 and an oven dry-rodded unit weight of 100 pcf; the moisture content for the trial batch is 3%; and the absorption is 0.7%.
Fine aggregate: a natural sand with a specific gravity of 2.60; moisture content for the trial batch is 1.2%, the absorption is 2.8%, and the fineness modulus is 2.50.

The required W/CM is determined by strength (or structural design) and durability requirements. Our strength requires a 3500 psi concrete. Since no statistical data exist, Table 15.3 is used to establish a target strength, $f'_{cr} = f'_c + 1200$ psi. Our new target compressive strength is (3500 + 1200 = 4700 psi). For an environment with moist freezing and thawing, the maximum W/CM is 0.45 (Table 15.4). From Figure 15.2, the recommended W/CM = 0.42 for a compressive strength of 4700 psi. Or it could be interpolated from Table 15.7. Table 15.4 suggests a W/CM of 0.45 to protect against moisture and freeze-thaw conditions. However, since the lower W/CM governs, the W/CM = 0.42 is based on strength.

Select slump when it is not specified according to the desired application from Table 15.9. For a pavement slab, choose a slump of 3 in. Note that you can add 1 in. to the maximum values when placing by hand or rodding. For slip forming construction, the slump could be as low as zero.

TABLE 15.10

(Metric) Approximate Mixing Water and Target Air Content Requirements for Different Slumps and Nominal Maximum Sizes of Aggregate

Slump, mm	Water, Kilograms per Cubic Meter of Concrete, for Indicated Sizes of Aggregate*							
	9.5 mm	12.5 mm	19 mm	25 mm	37.5 mm	50 mm**	75 mm**	150 mm**
Non-Air-Entrained Concrete								
25 to 50	207	199	190	179	166	154	130	113
75 to 100	228	216	205	193	181	169	145	124
150 to 175	243	228	216	202	190	178	160	—
Approximate amount of entrapped air in non-air-entrained concrete (%)	3	2.5	2	1.5	1	0.5	0.3	0.2
Air-Entrained Concrete								
25 to 50	181	175	168	160	150	142	122	107
75 to 100	202	193	184	175	165	157	133	119
150 to 175	216	205	197	184	174	166	154	—
Recommended Average Total Air Content (%) for Level of Exposure[†]								
Mild exposure	4.5	4.0	3.5	3.0	2.5	2.0	1.5	1.0
Moderate exposure	6.0	5.5	5.0	4.5	4.5	4.0	3.5	3.0
Severe exposure	7.5	7.0	6.0	6.0	5.5	5.0	4.5	4.0

* These quantities of mixing water are for use in computing cementitious material contents for trial batches. They are maximums for reasonably well-shaped angular coarse aggregates graded within the limits of accepted specifications.

** The slump values for concrete containing aggregates larger than 37.5 mm are based on slump tests made after the removal of particles larger than 37.5 mm by wet screening.

† The air content in job specifications should be specified to be delivered within −1 to +2 percentage points of the table target value for moderate and severe exposures.

Source: Reprinted from Design and Control of Concrete Mixtures, Engineering Bulletin 001, 14th edition, with kind permission from The Portland Cement Association.

Specified maximum coarse aggregate size: 3/4 in. This could be dictated by availability and the ACI, which states that maximum coarse aggregate size should be less than one-third of the slab thickness or not more than three-fourths of the distance between reinforcements, and not more than one-fifth of form dimensions.

The water content requirement depends upon the coarse aggregate maximum size, particle shape and grading of aggregates, and amount of air entrainment. From Table 15.11 for a 3-in. slump, a 3/4-in. maximum coarse aggregate, and an air-entrained concrete, we need 305 lb of water per cu yd. Since our aggregate is a gravel with some crushed particles, the water content may be reduced by 35 lb. Therefore, (305 − 35 = 270).

Air content: for a severe freeze-thaw exposure, Table 15.11 recommends a target air content of 6%. Since it is difficult to achieve a target value for air, a range of +2% and −1% is acceptable, hence a range of 5–8%. Use 7% air.

Cement content: calculate based on the maximum W/CM.

Cement content = (Water Requirement) / (W/CM) = (270) / (0.42) = 643 lb/cu yd

Coarse aggregate content: estimate the quantity of 3/4-in. maximum size coarse aggregate with a sand fineness modulus of 2.5.

TABLE 15.11

(Inch-Pound Units) Approximate Mixing Water and Target Air Content Requirements for Different Slumps and Nominal Maximum Sizes of Aggregate

Slump (in.)	Water, Pounds per Cubic Yard of Concrete, for Indicated Sizes of Aggregate*							
	³/₈ In.	½ In.	¾ In.	1 In.	1¼ In.	2 In.**	3 In.**	6 In.**
Non-Air-Entrained Concrete								
1 to 2	350	335	315	300	275	260	220	190
3 to 4	385	365	340	325	300	285	245	210
6 to 7	410	385	360	340	315	300	270	—
Approximate amount of entrapped air in non-air-entrained concrete (%)	3	2.5	2	1.5	1	0.5	0.3	0.2
Air-Entrained Concrete								
1 to 2	305	295	280	270	250	240	205	180
3 to 4	340	325	305	295	275	265	225	200
6 to 7	365	345	325	310	290	280	260	—
Recommended Average Total Air Content (%) for Level of Exposure†								
Mild exposure	4.5	4.0	3.5	3.0	2.5	2.0	1.5	1.0
Moderate exposure	6.0	5.5	5.0	4.5	4.5	3.5	3.5	3.0
Severe exposure	7.5	7.0	6.0	6.0	5.5	5.0	4.5	4.0

* These quantities of mixing water are for use in computing cement factors for trial batches. They are maximums for reasonably well-shaped angular coarse aggregates graded within the limits of accepted specifications.

** The slump values for concrete containing aggregates larger than 1½ in. are based on slump tests made after the removal of particles larger than 1½ in. by wet screening.

† The air content in job specifications should be specified to be delivered within −1 to +2 percentage points of the table target value for moderate and severe exposures.

Source: Reprinted from Design and Control of Concrete Mixtures, Engineering Bulletin 001, 14th edition, with kind permission from The Portland Cement Association.

TABLE 15.12
Minimum Requirements of Cementing Materials for Concrete Used in Flatwork

Nominal Maximum Size of Aggregate, mm (in.)	Cementing Materials, kg/m³ (lb/yd³)*
37.5 (1½)	280 (470)
25 (1)	310 (520)
19 (¾)	320 (540)
12.5 (½)	350 (590)
9.5 (3/8)	360 (610)

* Cementing materials quantities may need to be greater for severe exposure. For example, for deicer exposures, concrete should contain at least 335 kg/m³ (564 lb/yd³) of cementing materials.

Source: Reprinted from Design and Control of Concrete Mixtures, Engineering Bulletin 001, 14th edition, with kind permission from The Portland Cement Association.

TABLE 15.13
Cementitious Materials Requirements for Concrete Exposed to Deicing Chemicals

Cementitious Materials*	Maximum % of Total Cementitious Materials by Mass**
Fly ash and natural pozzolans	25
Slag	50
Silica fume	10
Total of fly ash, slag, silica fume, and natural pozzolans	50†
Total of natural pozzolans and silica fume	35†

* Includes portion of supplementary cementing materials in blended cements.

** Total cementitious materials include the summation of Portland cements, blended cements, fly ash, slag, silica fume, and other pozzolans.

† Silica fume should not constitute more than 10% of total cementitious materials, and fly ash or other pozzolans shall not constitute more than 25% of cementitious materials.

Source: Reprinted from Design and Control of Concrete Mixtures, Engineering Bulletin 001, 14th edition, with kind permission from The Portland Cement Association.

From Table 15.8 and Figure 15.3, an amount of 0.65 cu yd may be used. Since our coarse aggregate weighs 100 pcf, the required dry weight per cu yd is as follows:

$$100 \text{ pcf} \times 27 \text{ cu ft/cu yd} \times 0.65 = 1755 \text{ lb (dry)}$$

Admixture content: for a 7% air content, the air-entraining admixture manufacturer recommends a dosage of 0.9 fl oz per 100 lb of cement.

$$0.9 \times 643/100 = 5.8 \text{ fl oz per cu yd}$$

Fine aggregate content: at this point, all ingredients of the mix have been estimated except for the fine aggregate. We shall use the method of absolute volume to compute the fine aggregate.

TABLE 15.14
Maximum Chloride-Ion Content for Corrosion Protection

Type of Member	Maximum Water-Soluble Chloride Ion (Cl^-) in Concrete, % by Mass of Cement*
Prestressed concrete	0.06
Reinforced concrete exposed to chloride in service	0.15
Reinforced concrete that will be dry or protected from moisture in service	1.00
Other reinforced concrete construction	0.30

*ASTM C-1218.

Source: Reprinted from Design and Control of Concrete Mixtures, Engineering Bulletin 001, 14th edition, with kind permission from The Portland Cement Association.

Volume = mass/(Sp. Gr. × density of water)
Water volume = (270)/(1 × 62.4) = 4.33 cu ft per cu yd
Cement volume = (643)/(3.15 × 62.4) = 3.27 cu ft per cu yd
CA volume = (1755)/(2.65 × 62.4) = 10.61 cu ft per cu yd
Air volume = 0.07 × 27 = 1.89 cu ft
Total volume of known ingredients = 20.1 cu ft

(Note that the admixture volume is insignificant to add to the water volume.)

Solid volume of "fine aggregate" required is the difference between the unit volume and total volume of known ingredients.

$$27 - 20.1 = 6.9 \text{ cu ft}$$

Weight of dry "fine aggregate"

$$6.9 \times 2.60 \times 62.4 = 1119 \text{ lb}$$

So far, the mixture proportions are as follows (per 1 cu yd):

Water	270 lb
Cement	643 lb
Coarse aggregate (dry)	1755 lb
Fine aggregate (dry)	1119 lb
Air entraining admixture	5.8 fl oz
Total weight	3787 lb

Slump	3 in. (± 3/4 in. for trial batch)
Air content	7% (± 0.5% for trial batch)
Estimated density (using SSD aggregates)	270 + 643 + (1755 × 1.007) + (1119 × 1.012)/27
	= 3812 / 27 =141 lb per cu ft.

Moisture corrections: since the mix water is based on oven-dried aggregate, the batch weights must be corrected for absorbed and free moisture in aggregates. For the aggregates with the indicated moisture content (MC), the wet batch weights are as follows:

Coarse aggregate (MC = 3%): 1755 × 1.03 = 1808 lb/cu yd
Fine aggregate (MC = 1.2%) 1119 × 1.012 = 1132 lb/cu yd

Note: absorbed water should be excluded from mix water, and surface moisture should be included in mix water.

Coarse aggregate absorption = 0.7%
Fine aggregate absorption = 2.8%
Coarse aggregate contribution = 3% – 0.7% = +2.3 moisture (subtract from mix water)
Fine aggregate contribution = 1.2% – 2.8% = –1.6 % moisture (add to mix water)

Corrected mix water:

$$270 - 1755 \times (0.023) + 1119 \times (0.016) = 247 \text{ lb/cu yd}$$

The revised batch weights for 1 cu yd are as follows:

Water to be added to mix	247 lb
Cement	643 lb
Coarse aggregate (wet with MC = 3%)	1808 lb
Fine aggregate (wet with MC = 1.2%)	1132 lb
Total	3830

Note: Air-entraining admixture = 5.8 fl oz.

Trial batch: at this point, trial batches are used to check if the target parameters such as slump, air, and workability are met. In addition, concrete cylinders for strength testing at 28 days are made at this point. For a laboratory batch, mixing a smaller volume is adequate. For this example, a 2 cu ft or 2/27 cu yd batch will be made.

Water to be added to mix	$247 \times (2/27) = 18.30$ lb
Cement	$643 \times (2/27) = 47.63$ lb
Coarse aggregate (wet with MC = 3%)	$1808 \times (2/27) = 133.93$ lb
Fine aggregate (wet with MC = 1.2%)	$1132 \times (2/27) = 83.85$ lb
Total	283.6

Air-entraining admixture: $5.8 \times (2/27) = 0.43$ fl oz (Note: usually admixtures are batched in milliliters). Therefore, 5.8×29.573 ml/oz = 171.52 ml $\times (2/27) = 12.71$ ml.

After mixing the trial batch, the measured slump was 4 in., the air content was 8%, and the fresh concrete density was 140.3 pcf. During mixing, not all of the mix water was used. The net water used was $18.30 - 0.5 = 17.8$ lb. Therefore, the updated batch weights are as follows:

Water to be added to mix	17.8 lb
Cement	47.63 lb
Coarse aggregate (wet with MC = 3%)	133.93 lb
Fine aggregate (wet with MC = 1.2%)	83.85 lb
Total	283.21 lb

The yield for the trial batch is as follows:

$$283.21 / 140.3 = 2.018 \text{ cu ft}$$

The mixing water content is determined from the added water plus the free water on the aggregates, and is determined as follows:

Water added	17.8 lb
Free water on coarse aggregate: $133.93/1.03 \times (3\% - 0.7\%)/100$	2.998
Free water on fine aggregate: $83.85/1.012 \times (1.2\% - 2.8\%)/100$	−1.325
Total	19.47 lb

(The negative sign indicates that the fine aggregate is drier than the absorption moisture content and therefore additional water should be added to the mix to compensate for it.)

The mix water required for a cubic yard of the same slump concrete as the trial batch is as follows:

$$19.47 \times 27/2.018 = 260.5 \text{ lb}$$

Batch adjustments: the measured slump of the trial mix was 4 in. This is higher than the target slump of 3 in. and is unacceptable. The measured air content was also too high (more than target of $7\% \pm 0.5\%$), and the yield was higher too. Therefore, we shall adjust the yield, air entrainment, and water content to obtain a 3 in. slump.

When adjusting the mix design from the trial batch to the mix proportions for the large batch, use the following rule of thumb: increase the water by 5 lb for each 1% air reduction to maintain a similar slump, and reduce the water by 10 lb for each reduction in slump by 1 in. Therefore, the adjusted mix water for the reduced slump and air is as follows:

$$(5 \times 1) - (10 \times 1) + 247 = 242 \text{ lb per cu yd}$$

Now that the water is reduced, the cement content should decrease for a constant W/CM:

$$\text{cement} = 242/0.42 = 576 \text{ lb}$$

Since the workability is adequate, the coarse aggregate content will not be readjusted.

The new adjusted batch weights are as follows:

Water to be added to mix	$242/(1 \times 62.4) = 3.87$ cu ft
Cement	$643/(3.15 \times 62.4) = 3.27$ cu ft
Coarse aggregate (wet with MC = 3%)	$1755/(2.65 \times 62.4) = 10.61$ cu ft
Air content	$7\% \times 27 = 1.89$ cu ft
Total	19.61 cu ft
Fine aggregate volume	$27 - 19.61 = 7.39$ cu ft

Fine aggregate (dry weight basis) $= 7.39 \times 2.60 \times 62.4 = 1199$ lb

Air-entrainment admixture: use 0.8 fluid oz per 100 lb cement to obtain a 7% air content.

$$\text{AEA} = 0.8 \times 643/100 = 5.14 \text{ fl oz}$$

Adjusted batch weights per cubic yard of concrete are as follows:

Water	242 lb.	240 lb.
Cement	643 lb.	643 lb.
Coarse aggregate (wet with MC = 3%)	1755 lb. (dry)	1767 lb. (SSD)
Fine aggregate (wet with MC = 1.2%)	1199 lb. (dry)	1233 lb. (SSD)
Total	5606 lb.	5652 lb.

15.2 STRUCTURAL DESIGN

15.2.1 AASHTO METHOD (AASHTO, 1993)

The rigid pavement design guide was developed similarly to the flexible pavement design guide based on the AASHO Road Test's results and knowledge gained.

346

Pavement Engineering: Principles and Practice

The basic equations developed for rigid pavements are similar to and of the same form as the flexible pavements but with different regression equation constants. Over the years, other modifications were made to the original equation to take into consideration factors such as different subgrades and conditions other than those that existed during the AASHO Road Test. In 1972 the AASHTO adopted a modification developed using the Spangler (1942) equation to consider corner loading and to extend the original AASHO Road Test equation to other conditions. Other modifications include the drainage coefficient C_d, $Z = E_c/k$, and the reliability term $Z_R S_o$, replacing the term $(4.5-p_t)$ by ΔPSI.

When the AASHO equations were developed in the 1960s and the use of computers was not yet common, nomographs were developed to aid in solving the complex equations. Figure 15.6 shows the nomograph for solving the following AASHO equation:

$$\log W_{18} = Z_R \times S_0 + 7.35 \log(D+1) - 0.06 + \frac{\log\left[\frac{(\Delta PSI)}{4.5-1.5}\right]}{1 + \frac{1.624 \times 10^7}{(D+1)^{8.46}}} + (4.22 - 0.32 p_t)$$

$$\log\left\{\left\{\left[\frac{S_c C_d (D^{0.75} - 1.132)}{215.63 J\left[D^{0.75} - \frac{18.42}{\left(\frac{E_c}{k}\right)^{0.25}}\right]}\right]\right\}\right\}$$

where:
  S_c = flexural strength of concrete
  D = slab thickness
  E_c = modulus of elasticity of concrete
  k = modulus of subgrade reaction
  J = joint transfer coefficient: 3.2 for plain jointed and jointed reinforced pavements, 2.9–3.2 for CRCP for doweled but not tied pavements, and 3.6–4.2 for doweled and tied pavements

The rest of the parameters are the same as those used in the asphalt pavement thickness design.
  The following example will be used to demonstrate the use of the nomograph.

Example

Determine the thickness D given the input parameters.
  Given:

$K = 72$ pci (19.5 MN/m³)
$E_c = 5 \times 10^6$ psi (34.5 GPa)
$S_c = 650$ psi (4.5 MPa)
$J = 3.2$
$C_d = 1.0$
$\Delta PSI = 4.2 - 2.5 = 1.7$
$R = 95\%$
$S_o = 0.29$
$W_t = 5.1 \times 10^6$

Solution

1. Starting with $k = 72$ pci (19.5 MN/m³) in Figure 15.6a, draw a vertical line until it connects with $E_c = 5 \times 10^6$ psi (34.5 GPa). Draw a horizontal line until the end of the k versus E_c figure is reached. Draw a line from this end point that goes through Sc = 650 psi (4.5 MPa) and that ends at the turning line (T_L). Similar lines are drawn through J = 3.2 and $C_d = 1.0$ until the end line hits the value of 74 on the Match Line.

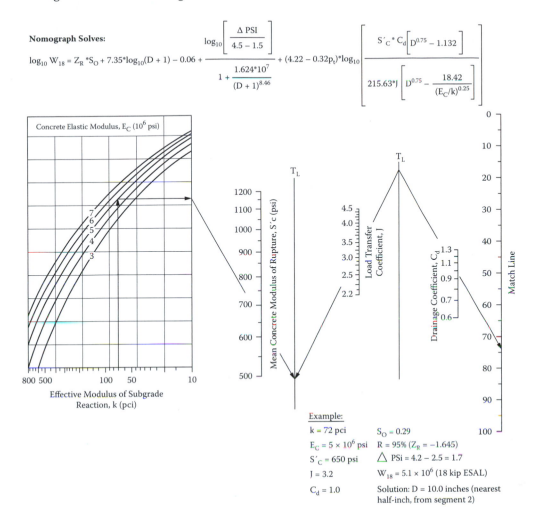

Nomograph Solves:

$$\log_{10} W_{18} = Z_R {}^*S_O + 7.35{}^*\log_{10}(D + 1) - 0.06 + \dfrac{\log_{10}\left[\dfrac{\Delta PSI}{4.5 - 1.5}\right]}{1 + \dfrac{1.624{}^*10^7}{(D + 1)^{8.46}}} + (4.22 - 0.32p_t){}^*\log_{10}\dfrac{S'_C {}^* C_d\left[D^{0.75} - 1.132\right]}{215.63{}^*J\left[D^{0.75} - \dfrac{18.42}{(E_C/k)^{0.25}}\right]}$$

Example:

k = 72 pci	$S_O = 0.29$
$E_C = 5 \times 10^6$ psi	R = 95% ($Z_R = -1.645$)
$S'_C = 650$ psi	$\triangle$ PSi = 4.2 − 2.5 = 1.7
J = 3.2	$W_{18} = 5.1 \times 10^6$ (18 kip ESAL)
$C_d = 1.0$	Solution: D = 10.0 inches (nearest half-inch, from segment 2)

FIGURE 15.6A Design Chart for Rigid Pavements. (Segment 2 in Figure Refers to Figure 15.6B.) *Source:* From AASHTO Guide for Design of Pavement Structures © 1993, by the American Association of State Highway and Transportation Officials, Washington, D.C. Used by permission.

2. Starting at 74 on the Match Line in Figure 15.6b, a line is drawn through $\Delta PSI = 1.7$ until it reaches the vertical axis.
3. Using the reliability axis, start with R = 95% and draw a line through $S_o = 0.29$, and then draw a line through $W_t = 5.1 \times 10^6$ until it intersects the horizontal axis.
4. A horizontal line is draw from the last step in step 2. A vertical line is drawn from the last step in step 3. The intersection of these two lines gives D = 9.75 in. (246 mm).

This nomograph can also be used to determine the amount of traffic that will produce a certain degradation in the present Serviceability Index.

The property of the subgrade used for rigid pavement design is the modulus of subgrade reaction k instead of the resilient modulus M_r. Therefore, it is necessary to convert M_r to k. Values of k are also season dependent and will vary. Therefore, the relative damage caused by the change in k needs to be evaluated. If the PCC slab is placed directly on the subgrade without the use of

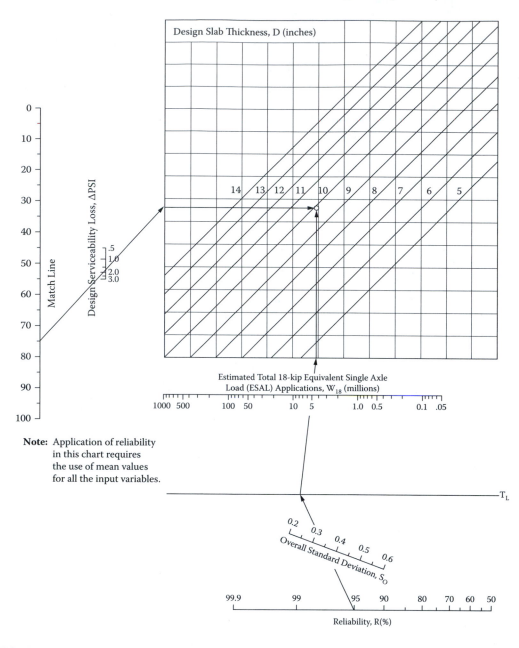

FIGURE 15.6B Design Chart for Rigid Pavements.
Source: From AASHTO Guide for Design of Pavement Structures © 1993, by the American Association of State Highway and Transportation Officials, Washington, D.C. Used by permission.

a subbase, the AASHTO recommends the use of the following theoretical relationship between k-values from a plate-bearing test and M_r.

$$k = \frac{M_r}{19.4}$$

where k = pci, and M_r is in psi.

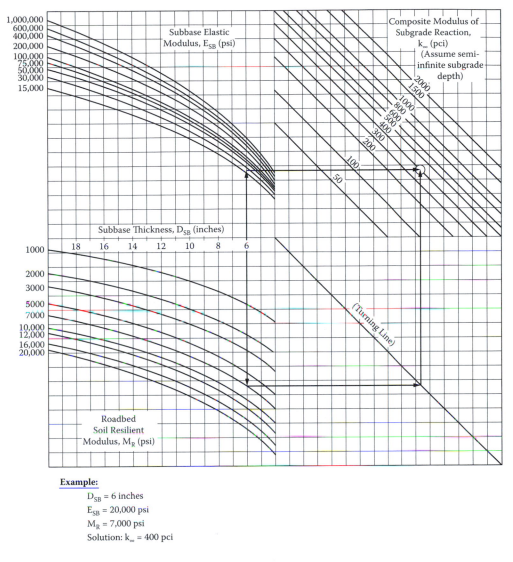

FIGURE 15.7 Estimating the Composite Modulus of Subgrade Reaction.
Source: From AASHTO Guide for Design of Pavement Structures © 1993, by the American Association of State Highway and Transportation Officials, Washington, D.C. Used by permission.

If a subbase exists between the slab and the subgrade, the composite modulus of subgrade reaction can be determined from Figure 15.7 (AASHTO, 1993). The modulus is based on a subgrade of infinite depth, k_∞. The chart was developed using the same method as for a homogeneous half-space except that the 30-in. plate is applied on a two-layer system. Hence the k values obtained from the chart are too large and are not representative of values obtained in the field.

Example

Given a subbase thickness D_{SB} of 6 in. (152 mm), a subbase resilient modulus of E_{SB} of 20,000 psi (138 psi), and a roadbed soil resilient modulus M_r of 7000 psi (48 MPa), determine the composite modulus of subgrade reaction k_∞.

Solution: the composite of the subgrade reaction can be determined as follows.

Using Figure 15.7, draw a vertical line from $D_{SB} = 6$ in. to the curve of $E_{SB} = 20,000$ psi. The same line is drawn downward until it intersects with $M_r = 7000$ psi, and then the line is turned horizontally until it intersects with the turning line.

A horizontal line is drawn from the point in step 1, and a vertical line from the point on the turning line in step 2. The intersection of these two lines gives a k_∞ of 400 pci.

15.2.1.1 Rigid Foundation at Shallow Depth

If the rigid foundation lies below the subgrade and the subgrade depth to rigid foundation D_{SB} is smaller than 10 ft (3m), then the modulus of subgrade reaction must be modified using the chart in Figure 15.8 (AASHTO). This chart can be used to modify PCC slabs with or without a subbase.

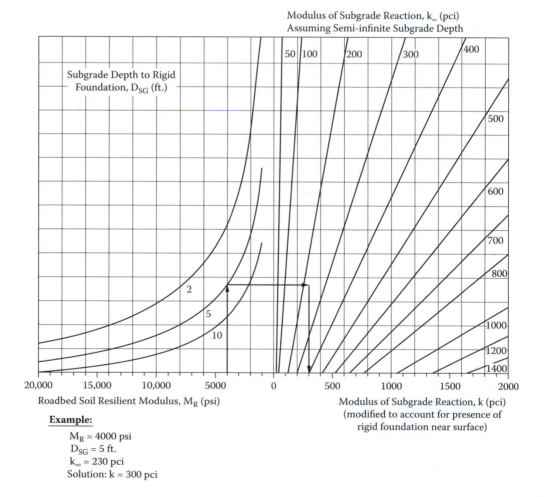

FIGURE 15.8 Modifying the Modulus of Subgrade Reaction Due to a Rigid Foundation near the Surface. *Source:* From AASHTO Guide for Design of Pavement Structures © 1993, by the American Association of State Highway and Transportation Officials, Washington, D.C. Used by permission.

Example: Determine k

Given:

$M_r = 4000$ psi
$D_{SG} = 5$ ft
$k_\infty = 230$ pci

Solution: using Figure 15.8, a vertical line is drawn from the horizontal scale with a $M_r = 4000$ psi until it intersects the curve with a $D_{SG} = 5$ ft. The line is then drawn horizontally until it reaches a point with $k_\infty = 230$ pci, and then vertically until a k of 300 pci is obtained.

15.2.1.2 Effective Modulus of Subgrade Reaction

The effective modulus of subgrade reaction is an equivalent modulus that would result in the same damage if seasonal values were used throughout the year. The relative damage to rigid pavements u_r is given by the following equation and Figure 15.9.

$$u_r = (D^{0.75} - 0.3k^{0.725})^{3.42}$$

Example

Given $D = 9$ in. and $k = 540$ pci, determine the u_r using the equation and Figure 15.8.
Solution: From the equation $u_r = [9^{0.75} - 0.3(540)^{0.725}]^{3.42} = 60.3$ which is similar to the value obtained using Figure 15.9.

Table 15.15 shows the steps in computing the effective modulus of subgrade reaction. To account for a potential loss of support by foundation erosion or differential vertical soil movement, the effective modulus of subgrade reaction must be reduced by the loss of subgrade support (LS) factor. The LS factor is shown in Figure 15.10. Table 15.16 provides recommendations for LS values for different types of subbases and subgrades.

Table 15.15 shows an example for determining the effective modulus of a subgrade reaction for a slab thickness of 9 in. The slab is to be placed directly onto the subgrade with the monthly resilient moduli shown in Table 15.15. Note that the year is divided into 12 months, each with different moduli. The normal summer modulus is 7000 psi, and the maximum of modulus of 20,000 psi occurs in the winter months, when the subgrade is frozen (December–February). The k-values are obtained from the equation $K = M_r/19.4$.

The relative damage can be obtained from Figure 15.9 or the following equation:

$$u_r = (D^{0.75} - 0.3k^{0.725})^{3.42}$$

The sum of the relative damage = 7.25, and the average over the 12 months is 0.6, which is equivalent to an effective modulus = 540 pci.

15.2.2 Dowel Bars

Dowel bars are used as load transfer devices between adjoining slabs to reduce stresses and deflections of the loaded slabs. The design of the dowel bars (i.e., size and spacing) is dependent on the resulting bearing stresses between the concrete and dowel. Joint functionality and dowel bar efficiency can greatly affect joint faulting and pumping.

The design of dowels and joints is based on both experience and analytical methods. The size of the dowels selected depends on the thickness of the slab. The Portland Cement Association (PCA; 1991) recommended the use of 1.25-in. diameter dowels for highway pavements less than 10 in.

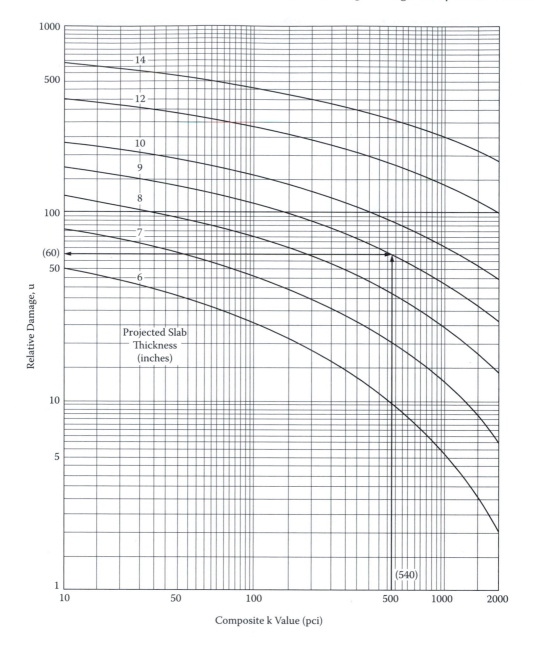

FIGURE 15.9 Estimating Relative Damage to Rigid Pavements.
Source: From AASHTO Guide for Design of Pavement Structures © 1993, by the American Association of State Highway and Transportation Officials, Washington, D.C. Used by permission.

thick and 1.5-in. diameter dowels for pavements 10 in. or thicker. A minimum dowel diameter of 1.25 in. to 1.5 in. is needed to control faulting by reducing the bearing stress in concrete.

15.2.2.1 Allowable Bearing Stress

Because concrete is much weaker than steel, the size and spacing of dowels required are governed by the bearing stress between the dowel and concrete. The allowable bearing stress can be determined

TABLE 15.15
Computation of Effective Modulus of Subgrade Reaction

	Trial subbase: type Granular Thickness (in.) 6 Loss of support (LS) 1.0		Depth to rigid foundation (ft) 5 Projected slab thickness (in.) 9		
(1)	(2)	(3)	(4)	(5)	(6)
Month	Roadbed Modulus, Mg (psi)	Subbase Modulus, E_{SB} (psi)	Composite k-Value (pci) (Figure 15.7)	k-Value (pci) on Rigid Foundation (Figure 15.8)	Relative Damage, u^r (Figure 15.9)
January	20,000	50,000	1100	1350	0.35
February	20,000	50,000	1100	1350	0.35
March	2500	15,000	160	230	0.86
April	4000	15,000	230	300	0.78
May	4000	15,000	230	300	0.78
June	7000	20,000	410	540	0.60
July	7000	20,000	410	540	0.60
August	7000	20,000	410	540	0.60
September	7000	20,000	410	540	0.60
October	7000	20,000	410	540	0.60
November	4000	15,000	230	300	0.78
December	20,000	50,000	1100	1350	0.35
				Summation: $\Sigma u_r =$	7.25

Average: $\bar{u}_r = \dfrac{\Sigma u_r}{n} = \dfrac{7.25}{12} = 0.60$

Effective modulus of subgrade reaction, k (pci) = 540

Corrected for loss of support: k (pci) = 170

Note: Reprinted from Design and Control of Concrete Mixtures, Engineering Bulletin 001, 14th edition, with kind permission from The Portland Cement Association.

by the following:

$$f_b = \frac{(4-d)}{3} f'_c$$

in which f_b is the allowable bearing stress in psi, d is the dowel diameter in in., and f'_c is the ultimate compressive strength of the concrete.

15.2.2.2 Bearing Stress on One Dowel

If the load applied to one dowel is known, the maximum bearing stress can be determined theoretically by assuming the dowel to be a beam and the concrete to be a Winkler foundation (Huang, 1993). Based on the original solution by Timoshenko (Timoshenko and Goodier, 1934/1951), Friberg (1938) developed a relationship for determining the maximum bearing stress. First, the deformation of concrete under the dowel can be expressed by the following equation:

$$y_0 = \frac{P_t(2+\beta z)}{4\beta^3 E_d I_d}$$

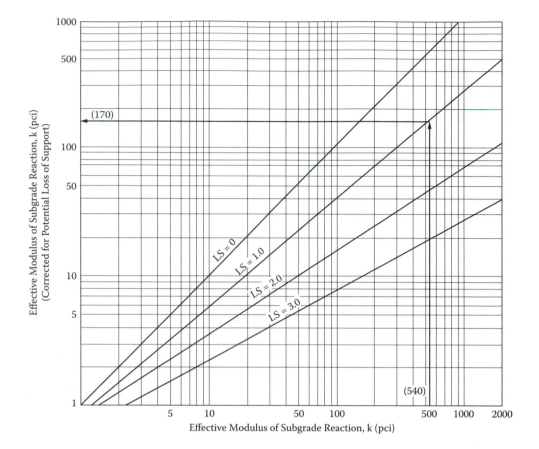

FIGURE 15.10 Correction of the Effective Modulus of Subgrade Reaction Due to a Loss of Foundation Contact. *Source:* From AASHTO Guide for Design of Pavement Structures © 1993, by the American Association of State Highway and Transportation Officials, Washington, D.C. Used by permission.

TABLE 15.16
Recommended Loss of Subgrade Support (LS) Factors

Type of Material	Loss of Support (LS)
Cement-treated granular base (E = 1,000,000 to 2,000,000 psi)	0.0 to 1.0
Cement aggregate mixtures (E = 500,000 to 1,000,000 psi)	0.0 to 1.0
Asphalt-treated base (E = 350,000 to 1,000,000 psi)	0.0 to 1.0
Bituminous stabilized mixtures (E = 40,000 to 300,000 psi)	0.0 to 1.0
Lime stabilized (E = 20,000 to 70,000 psi)	1.0 to 3.0
Unbound granular materials (E = 15,000 to 45,000 psi)	1.0 to 3.0
Fine-grained or natural subgrade materials (E = 3000 to 40,000 psi)	2.0 to 3.0

Note: E in this table refers to the general symbol for the elastic or resilient modulus of the material.

Source: Reprinted from Design and Control of Concrete Mixtures, Engineering Bulletin 001, 14th edition, with kind permission from The Portland Cement Association.

where:

y_0 = the deformation of the dowel at the face of the joint
P_t = the load on one dowel
z = the joint width
E_d = Young's modulus of the dowel
I_d = the moment of inertia of the dowel

$$I_d = \frac{1}{64}\pi d^4$$

β = the relative stiffness of a dowel embedded in concrete

$$\beta = \sqrt[4]{\frac{Kd}{4E_dI_d}}$$

K = the modulus of the dowel support, which ranges from 300,000 to 1,500,000 pci
d = the diameter of the dowel

The bearing stress σ_b is proportional to the deformation:

$$\sigma_b = Ky_0 = \frac{KP_t(2+\beta z)}{4\beta^3 E_d I_d}$$

The bearing stress obtained from this equation should compare with the allowable bearing stress in the equation given for f_b. If the actual bearing stress is greater than allowable, then dowel bars with a bigger area or a smaller dowel spacing should be used. Recent studies have shown that the bearing stress is related to the faulting of slabs. By limiting the bearing stress, the amount of faulting can be reduced to an acceptable level.

When a load is applied at the joint, the dowel bar immediately below the load carries a major portion of the load, while adjacent dowels will assume proportionally lesser amounts. Based on Westergaard's solutions, Friberg (1938) determined that the maximum negative moment for both interior and edge loadings occurs at a distance of $1.8l$ from the load, where l is defined earlier as the radius of relative stiffness. More recent research has shown the effective length to be at $1.0l$ (Heinrichs et al., 1989).

Example

A concrete pavement slab 8 in. thick has a joint width of 0.2 in., a modulus of subgrade reaction of 100 pci, and a modulus of dowel support of 1.5×10^6 pci. A load of 9000 lb is applied over the outermost dowel, which is located 6 in. from the edge. The dowels are 1 in. in diameter and spaced 12 in. apart center to center. Determine the maximum bearing stress between dowel and concrete.

Solution

If the dowel directly under the load is subjected to a shear force P_t, then the proportional forces on the dowels within a distance of $1.8l$, or 66 in., can be determined by assuming a linear relationship, as shown in Figure 15.11. The sum of all forces on the dowels is $3.27 P_t$, and is resisting half of the applied wheel load, assuming 100% joint efficiency or 50% of the wheel load is carried by each slab. $P_t = (9000/2)/3.27 = 1376$. $I_d = 0.049$, $\beta = 1.176$, and stress = 3615 psi. The allowable bearing stress for a 3500 psi concrete is 3500 psi. The actual bearing stress is approximately 3% higher than the allowable. At this stage, the designer should

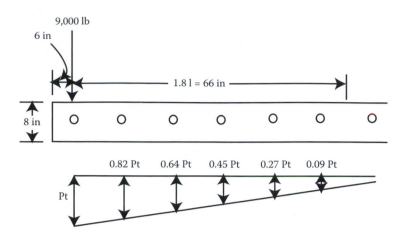

FIGURE 15.11 Shear Force Distribution in Pavement Dowels.

decide if a 3% difference is acceptable and within the variation of the materials and design parameters.

QUESTIONS

1. Determine the corrected batch weights for a 14 cu ft mix volume and for the following conditions.
   W/CM = 0.55.
   Air = 7%.
   Bulk volume of coarse aggregate = 59%.
   Coarse aggregate dry-rodded unit weight = 95 pcf.
   Estimated water = 324 lb per cu yd.
   Cement type = 1.
   Sand (moisture content = 5%; absorption = 3%) G_s = 2.60.
   Coarse aggregate (moisture content = 0%; absorption = 0.9%) G_s = 2.70.
   After mixing, the unit weight of the fresh concrete is 142 pcf.
   Calculate the yield, the gravimetric percentage of air content, the relative yield, and the cement factor (bags/cu yd).
2. Determine the thickness for a rigid pavement given the following:
   K = 100 pci
   E_c = 4 × 10⁶psi
   S_c = 650 psi
   J = 3.2
   C_d = 1.0
   ΔPSI = 4.2 − 2.5 = 1.7
   R = 95%
   S_o = 0.29
   W_t = 7 × 10⁶

16 Construction of Asphalt Pavements

16.1 OVERVIEW

The essential steps in the construction of asphalt pavements consist of selecting proper materials and conducting a proper mix design, ensuring drying and mixing of aggregates from properly maintained stockpiles, mixing aggregates with asphalt binder, delivering the mix in insulated trucks in a smooth way so as to match with paver speed, laying down with a paver, and using a properly selected combination of rollers of desirable weight and type, with correct vibration frequency and/ or tire pressure, in a correct rolling pattern. The rollers should be close to the paver and operating with desirable speed and manner of reversing and turning. All of the equipment should be in good operating condition. And, very importantly, good quality control must be ensured at every step in the plant and during laydown and rolling operations. There must be clear communication of any problems between the plant and the site, and if required, adjustments should be made in mix design, plant production, transportation, and laydown and rolling operations. Provisions should be made to use additives in the materials, such as natural rounded sand to reduce the harshness of a mix, and/or employ special equipment, such as a material transfer vehicle to reduce temperature-related segregation, if required.

16.1.1 PRODUCTION

The production of loose HMA starts in the HMA plant—which can be of either batch (Figure 16.1) or drum type. In both plants the steps consist of heating and drying aggregates and mixing them with heated liquid asphalt binder. In a batch plant the production happens in "batches"—one batch of aggregates of different sizes is dried in a drum, moved up in a hot elevator, and mixed with asphalt in a pugmill. In the case of a drum plant, the production is a continuous process—aggregates of different sizes are dried in the front part of the dryer drum and then mixed with the liquid asphalt binder in the back part of the drum. In a batch plant, aggregates are collected on cold feed conveyors, heated in a dryer drum, moved up a hot elevator, rescreened over a screen deck and separated into hot bins, and then mixed with asphalt in the pugmill. In the drum plant, the tower, consisting of the hot elevator, hot bins, and pugmill, is eliminated, and mixing is done in the drum. Hybrid plants, using some features of both batch and drum plants, are being used also.

Drum plants are continuous as opposed to batch plants where every batch of aggregate should be weighed and mixed individually. However, batch plants are also more efficient in cases of short production runs or delays in hauling, or when multiple types of mixes are required (such as paving for highways, parking lots, and driveways) in urban areas.

The production of HMA is a complex process and requires preparation and continuous supervision by experienced and trained people. Once the HMA is produced, it is either stored in a silo or put in trucks for transportation to the job site. The silo must be properly insulated to prevent oxidation and heat loss from the HMA. The trucks also need to be insulated (sometimes heated) to minimize heat loss from the HMA during transportation.

FIGURE 16.1 Hot Mix Asphalt Plant.
Courtesy: Matthew Teto, All States Asphalt.

16.1.2 TRANSPORTATION AND LAYDOWN

Once the trucks arrive at the job site, the HMA is transferred either directly to the paver or to a paver through a material transfer device (MTD). The role of the MTD is to store HMA such that it can be used when the paver hopper becomes empty and before more trucks arrive with new HMA. The MTD also helps keep the HMA heated and mixes it to maintain its uniformity and hence reduce the potential of segregation.

The paver lays down the HMA at a specific thickness with the help of the screed. It also provides some compaction to the mix. The remaining compaction is provided by the roller(s) following the paver. Generally rollers with these distinct features are used. A steel drum roller with one vibratory roller is used for breakdown rolling, a static steel drum or pneumatic roller is used for intermediate rolling, and usually a static steel drum roller is used for finishing.

Generally a mix will get compacted by 20–25%. Generally used thicker lifts are 2–4 in., and thinner lifts are 1–2 in. Thicker lifts are easier to compact since they retain heat longer and also help avoid the bridging tendency of thinner lifts. Thinner lifts, though, provide good results through better smoothness and help avoid shoving, which could happen in thicker lifts.

Tack coats are used to provide a bond between the old surface and the new asphalt mix layer. This is applied on the existing surface before the application of the HMA. The right kind and amount should be used, and adequate time should be given for curing before the HMA is laid down. The typical application rate is 0.05 gallons per square yard. Typically, it is applied in double lap with a distributor truck.

16.2 DESCRIPTION AND REQUIREMENTS OF COMPONENTS IN HOT MIX ASPHALT-PRODUCING PLANTS

16.2.1 AGGREGATE STOCKPILES

Aggregate stockpiles (Figure 16.2) should have uniform gradation of aggregates in all parts, and must be free from foreign materials and excessive amounts of moisture. The moisture content should be fairly constant throughout the depth of the stockpile.

FIGURE 16.2 Aggregate Stockpiles.
Courtesy: Matthew Teto, All States Asphalt.

Stockpiles should be constructed in the proper sequence using power equipment. Proper stockpiling should include remixing of materials, layering of remixed materials, and dispersion of newly delivered materials into existing piles. Front-end loaders are generally used. If trucks are used, then the pile should be built in layers, with each pile being one truck load of material. Manipulation by dozers should be limited. The stockpiles should be separated from each other and built on a surface with good drainage, and all parts of the stockpiles should be accessible and reachable from all sides.

16.2.2 COLD FEED BINS

The purpose of the cold feed bin system, which consists of multiple bins, is to provide a controlled distribution of aggregate in the conveyor system. The cold feed system (Figure 16.3) is controlled by manipulating the relative amounts of the different constituents as well as by the rate at which the material is fed into the drum. Cold bins contain the aggregates and direct them toward the cold feed through their sloping sides. The cold feed can be of the belt, vibratory, or apron type. The rate of flow from a bin into the cold feed is controlled by the adjustable gate at the bottom of the bins as well as by the rpm of the belt feeders or the frequency/amplitude of the vibratory feeders. The

FIGURE 16.3 Cold Feed Bins and Aggregate Feeder System.
Courtesy: Matthew Teto, All States Asphalt.

actual flow rate has to be calibrated for different aggregates, with difference in gradation, shape, and moisture content.

From the cold feeder the aggregate drops into a moving belt, which collects aggregates from multiple bins, which deliver the materials directly or through a scalping screen to the drum, as well as to sampling chutes. The final drop into the drum happens through an inclined belt, with scales for measuring the amount of aggregate, which dictates the amount of asphalt that should be added to the aggregates in the drum, if it is a drum plant.

The aggregate feeders need to be calibrated to determine the speed required to feed aggregates at a desired rate during production. The following steps are conducted to calibrate the feeders.

1. Close all feeders except the one being calibrated.
2. Run materials into the belt.
3. Determine the amount of material in a given length of the belt.
4. Adjust the feeder, and repeat steps 2 and 3.
5. Calibrate the next feeder using steps 1–4.
6. Determine the speed of the belt.

The relationship between the different parameters can be expressed as follows (NAPA, 1996).

$$R = \frac{1.8ws}{B}$$

where:
  R = rate of production in tons/hour
  w = weight of sample on B length (ft) of belt in lb
  s = speed of belt in ft/s

16.2.3 DRYER DRUM

In a batch plant the drum acts as a dryer. The drum is tilted, and the aggregates enter the upper end of the drum and flow toward the lower end, where the burner provides the heat for drying the aggregates (Figure 16.4). The burner works on oil and/or gas with a regulated flow of air from a

FIGURE 16.4 Dryer Drum with Burner and Inside of Drum.
Courtesy: Matthew Teto, All States Asphalt.

blower. The system for supplying fuel and air needs to be checked to ensure complete combustion of the fuel. The flow of hot air is further increased by an exhaust fan. The long thin flame of the burner as well as the hot gasses leaving the drum near the entry point of the aggregates dry off their moisture and heat them to a high temperature. The aggregates tumble through the flame through a set of flights inside the drum as it turns during the drying process. The time for which the aggregate remains in the dryer drum is known as the *dwell time*. It is important to use a proper dwell time, taking into consideration the amount of moisture in the aggregates, and the temperature to which it should be heated to mix with the asphalt binder. This temperature is equal to the temperature of the asphalt binder at which it has the optimum viscosity for mixing. For the proper dwell time, the moisture content of aggregates leaving the drum must not exceed 0.5%. Dwell time can be increased by slowing the drum rotation, lowering the tilt angle of the drum, rearranging the flights inside the drum, or increasing the total length of the dryer drum. If required, the amount of aggregate that is fed into the dryer, per unit time, should be decreased. Note that aggregates with higher moisture content will require more heat and hence more fuel to drive the moisture, and it is always advisable to store aggregates in enclosed storage areas.

16.2.4 Hot Elevator and Bins

The hot aggregates leave the dryer drum in a chute and are picked up by buckets in the hot elevator (Figure 16.5), which delivers it to the screen deck in the tower. The screen deck has a number of vibrating screen decks, arranged in a way such that the screen with the largest opening is at the top and the smallest opening is at the bottom. The screen sizes may range from 1½ in. to 5/32 in. The screened aggregates land in hot bins, which are placed beneath the screens. The screens help in separating the composite aggregate sample batch into different portions, depending on their size. Note that it is the final screening in this step that determines the blend of the aggregate in the mix, and hence the screens must operate efficiently without getting blinded due to processes such as high rates of loading or loading of improperly sized aggregates, or due to blinding of the screens themselves due to poor maintenance. The malfunctioning of screens causes the dumping of aggregates into inappropriate bins ("carryover"), and this affects the quality of the asphalt mix.

The function of the hot bins is to provide adequate temporary storage for the separate hot aggregates before mixing with the asphalt binder in the weigh hopper below them. Narrow hot bin gates open into the weigh hopper. Quality control samples should be taken at this point from hot bins to ensure proper functioning. To ensure a smooth flow of materials from the screen decks to the bins

FIGURE 16.5 Hot Elevator.
Courtesy: Matthew Teto, All States Asphalt.

FIGURE 16.6 Pugmill, View from the Bottom with Discharge Door Removed.
Courtesy: Matthew Teto, All States Asphalt.

and into the hopper and avoid segregation, it is important that the aggregates are supplied through the cold feed at a rate that is consistent with the demand for production. The plant operator could check on the relative amounts of materials in the different bins and, if needed, adjust the cold feed. A predetermined mass of aggregate from each hot bin (based on the mix design of the asphalt mix being produced) is dropped from the hot bins, one by one, to a hopper on a scale (weigh hopper). The batch is complete when all of the constituent aggregates have been dropped on the weigh hopper, and the cumulative mass is printed on a batch ticket, which is used as supporting documentation for verification of compliance with mix design.

16.2.5 PUGMILL

In the next step the weigh hopper empties the aggregates into the pugmill (Figure 16.6), where they are mixed with each other and with the asphalt binder with two mixing shafts with paddles, which rotate in opposite directions. It is important to make sure that the paddles are in good condition (and not worn out) and that the pugmill is filled to its optimum height (bottom to top of the paddles—the "live zone") to ensure good mixing. There are two stages of mixing in the pugmill: dry, when the aggregates are mixed with each other, typically 2 seconds or less; and wet, when the asphalt binder is introduced either by gravity or through a pressurized spray, typically 25–35 seconds. The total mixing time, which includes dry mixing time, wet mixing time, opening time of the pugmill gates below, discharge time of the mix, and closing time of the pugmill gates, affects the quality of the mix as well as the rate of production.

16.2.6 HAULING AND STORAGE

The mix from the pugmill is either carried away by hauling trucks to a job site or taken to a storage silo through a chute in a drag slat conveyor. The truck must be positioned underneath the pugmill to avoid spillage and one-sided drops, and ensure proper discharge, which is generally made by loading the front end of the truck first, then the back end, and finally the middle portion, to avoid segregation.

Drag slat conveyors (Figure 16.7) move the asphalt mix from the pugmill to either surge silos (Figure 16.7) for temporary storage (a few hours) or storage silos for long-term storage (overnight to several days). Drag slat conveyors have steel plates set at a 90° angle to the drag chain, and are typically enclosed to prevent any loss of heat from the mix.

Whenever there is movement of HMA mix, there is a tendency of coarser aggregate particles to roll down and get separated from the finer aggregates. This not only separates the aggregates of different sizes, but also creates an uneven distribution of asphalt binder in the mix, since it is the

FIGURE 16.7 Slat Conveyor and Silos.
Courtesy: Matthew Teto, All States Asphalt.

finer particle/matrix that contains more asphalt than the coarse aggregates. To minimize movement, sometimes a batcher is used for loading the silos from the slat conveyor. The batcher has a small hopper, which gets filled through a chute at the center, with a batch of mix. This batch of mix is then released to fall into the silo through gates at the bottom of the batcher. This way the silo is filled up in batches, and not in a continuous stream of HMA (thus limiting movement and hence segregation).

To minimize segregation from loading out from the silo, the trucks must be positioned properly beneath it, and loaded front first, then back, and finally the middle. Mix from the silo should not be used unless there is adequate mix inside the silo above the bottom cone portion of the silo. Also, if required, heated oil–carrying pipes throughout the shell of the silo could keep the mix warm, and inert gases could be used to reduce the oxidation of the mix inside.

16.2.7   DRUM PLANT

In a drum plant the aggregates are dried, heated, and mixed with the asphalt binder in the same drum, which is usually longer than the dryer drum used in a batch plant, and which is generally of the parallel flow type, in which the aggregates enter the drum in the same side as the burner flame. The drum has three distinct zones—the drying, heating, and mixing zones. The flame in the drum is shorter and wider, and the aggregates are kept in flights along the perimeter of the drum in the first zone, to prevent them from coming in direct contact with the flame. In the second zone, the flights cascade the aggregates through the center of the drum and through the hot burner gases, and as a result the aggregates get heated up to a high temperature. The configuration of the flights as well as the geometry of the drum could be varied in this zone to achieve this objective. Finally, in the mixing zone, flights are used to coat the aggregates with asphalt binder which is injected into the drum.

Although a typical drum is as it is describe above, numerous variations, as a result of continuous development processes, have resulted over the years, specifically to improve the efficiency of the heating process, and often to accommodate the use of recycled asphalt pavement (RAP) materials. Two such developments are the concepts of the double-barrel drum and the triple-barrel drum, which have outer "shells" that improve the heating process of aggregates, and at the same time cut down the potential of excessive oxidation of the asphalt binder–RAP materials. Another important development, in the context of paving jobs in remote areas, is the availability of portable drum plants, which can be transported from one site to another and readied for operation in a relatively short period of time. Note that counterflow drums are indeed used in many of these newly developed drum mix plants.

16.2.8 Dust Collection from HMA Plants

Dust is created during the drying, heating, and mixing of aggregates and asphalt. The dust primarily results from the existing fine portions of the aggregate; is created by abrasion during mixing, blown out by gases, or created by the air used for the burner; and comes from steam from the evaporation of water from the aggregates. The dust is actually captured through a collecting mechanism, and prevented from polluting the environment. Air quality codes from regulatory agencies direct the plants to test for particulate emissions in the exhaust gases from the plant to ensure the efficiency of the dust-collecting system.

The dust-collecting system consists of a primary and secondary mechanism. In the primary system, the relatively bigger size particles are separated from the exhaust gases. This is accomplished by either expanding the gases, reducing their velocity, and making the heavier particles drop out in the "knockout box" (expansion chamber process) or by speeding up the gases by directing them in a spiral flow through a chamber and separating the bigger particles by centrifugal force (cyclone method). The particles collected in the knockout box or the cyclone chamber are directed to a mineral filler silo, from where they could be reused for producing HMA.

In the secondary system of dust collection, either a baghouse (Figure 16.8) or a wet scrubber method is employed. In the baghouse method, bags stacked in vertical rows capture the particles as exhaust gases flow through them. The bags need to be cleaned periodically (generally by flexing, shaking, and blowing air pulses), and the recovered particles could be reused.

16.2.9 Asphalt Storage Tanks

Asphalt tanks (Figure 16.9) maintain asphalts in a plant at a required temperature, such that they can be supplied steadily to the pugmill or the drum at the desirable viscosity. Coiled pipes carrying hot oils around the periphery of the tanks maintain the high temperature. Delivery pipes are insulated to prevent loss of heat. It is important to note the grade and level of the binder before pumping new asphalt from a tanker to a storage tank.

FIGURE 16.8 Baghouse.
Courtesy: Matthew Teto, All States Asphalt.

FIGURE 16.9 Asphalt Storage Tanks.
Courtesy: Matthew Teto, All States Asphalt.

16.3 EQUIPMENT USED FOR TRANSPORTATION, LAYDOWN, AND COMPACTION

16.3.1 TRUCKS

For long-distance hauling, trucks should be insulated. A canvas should be used to hold heat, keep light rain showers off the HMA, and keep away dust. Truck beds should be clean, and trucks should be loaded properly to avoid segregation. Typically, trucks should be loaded in the front first, then in the back, and finally in the middle. Sometimes a material transfer vehicle is used to receive the mix from the truck, store it, remix it with augers, and then transfer it to the paver. This helps in maintaining the uniformity of the mix and also in maintaining constant paving operations. Note that prior to the placement of an HMA layer, a tack coat of emulsion (at the rate of 0.01 to 0.2 gallons per square yard, depending on the type of surface) is generally applied on the existing surface to help bond it with the new layer (Figure 6.10). The surface must be cleaned prior to the application of the tack coat. Figure 16.11 shows a view of an HMA paving site with the different equipment.

16.3.2 PAVERS

The HMA is placed and compacted to a certain degree with the paver (Figure 16.12). The paver consists of a hopper to receive the material from the truck or a material transfer vehicle, conveyors to send the material at the back, augers to distribute the mix transversely across the width of the pavement, and a screed (often heated and capable of vibrating), which ensures a specific depth of the material and preliminary compaction. Grade control is usually done with a laser.

16.3.3 ROLLERS

Rollers are used for breakdown and intermediate rolling, compaction, and finish rolling to remove roller marks. Generally a vibratory or static steel-wheel roller, with the first drum as the driver, is used for breakdown. The contact pressure that ensures compaction can be altered by changing the ballast load or dynamic force (in the case of a vibratory roller) or the tire pressure (for a rubber-tired roller). Rollers (Figure 16.13) can be rubber tired (these rollers are usually 15–20 tons in weight, and their tire pressure is generally around 90 psi, with 4500 lb per tire), a vibratory steel-wheel roller (usually 10 tons in weight, with high-frequency and low-amplitude vibrations), and a static

FIGURE 16.10 Application of Tack Coat.
Courtesy: Carolina Carbo, Road Recycling Council.

FIGURE 16.11 Different Equipment Used in Hot Mix Asphalt Paving.
Courtesy: Carolina Carbo, Road Recycling Council.

FIGURE 16.12 Paver.
Courtesy: Ed Kearney.

steel-wheel roller (10–12 tons). Recent developments in rollers include sensors for continuous monitoring of compaction of the mat.

16.4 IMPORTANT FACTORS

Several factors influence the asphalt pavement construction process, and hence their proper consideration is absolutely necessary to ensure good construction. The factors, with discussion, are listed below.

1. Meet specifications for aggregates and binders.
2. Control quality with properly trained and equipped personnel.
3. Maintain the flow of materials to the paver at a speed so as to limit the time the paver is kept waiting.

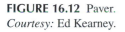

FIGURE 16.13 Steel-Wheel and Rubber-Tired Roller.
Courtesy: Ed Kearney and Mike Marshall, Wirtgen, GmbH.

4. Ensure the smooth transfer of materials from the truck to the paver, while the paver is moving.
5. All the mechanisms, such as the conveyor, hopper gate, and auger, in the paver should be working properly so as to maintain a good and consistent amount of HMA at the correct temperature in front of the screed.
6. Ensure paving with a minimum number of transverse joints.
7. The reference point for screed operation should not be readjusted too much—avoid constant/frequent readjustments of the paver's automatic controls.
8. Use proper guidelines to select the correct rolling techniques—the types and number of rollers, number of passes, and rolling pattern.
9. Make sure that the equipment, such as the paver and rollers, is in good operating condition.
10. During laydown or rolling, if a problem is identified as a material- or mix design–related problem, then it must be reported to the laboratory/plant as quickly as possible to ensure readjustments/corrective actions.

The different factors, along with their associated problems, are listed in Table 16.1.

TABLE 16.1
Steps, Potential Problems, and Solutions

Construction Step	Potential Problems/Suggested Solutions
Materials selection	Bleeding: check type of filler, grade of asphalt binder
	Blisters: ensure steel slag, if used, is cured; avoid aggregates with inorganic nitrogen or phosphorus, and quicklime, roots, or trapped volatiles from cutback asphalts
	Brown streaks: avoid highly absorptive aggregates; check quality of asphalt
	Checking under roller, tender mix: check quality of asphalt and its temperature susceptibility; check if source of asphalt has remained the same
	Tearing of mat: check maximum aggregate size
Mix design	Bleeding, poor-quality longitudinal joints: check amount of filler, VMA and asphalt content, voids, and laboratory compaction procedures
	Tender mix and checking under roller: check filler and asphalt content, asphalt binder temperature-viscosity characteristics, and residual moisture
	Segregation: check gradation
	Checking under roller, or tearing of mat: check asphalt content, VMA, asphalt content variation during production, and filler-metering system
	Transverse cracks: check asphalt content, filler content and asphalt binder temperature-viscosity characteristics, and residual moisture
	Poor-quality longitudinal joint: check filler and asphalt content, asphalt binder temperature-viscosity characteristics, residual moisture, and VMA
	Lean, brown, dull appearance of surface: check filler content
	Shoving of mat ahead of roller, or roller marks: check amounts of smooth, rounded aggregates, mix design temperature, asphalt content, and VMA
	Poor compaction: check for asphalt content and VMA
Storage	Bleeding: check stockpiling and drying techniques
	Segregation: check stockpiling
	Blisters: check drying techniques
	Shoving of mat ahead of roller, or lean, brown, dull appearance of surface: check for moisture in stockpiles

(Continued)

TABLE 16.1 (CONTINUED)
Steps, Potential Problems, and Solutions

Construction Step	Potential Problems/Suggested Solutions
Plant production	Bleeding, or lean, brown, dull appearance of surface: check fuel/ combustion/burner/nozzles
	Segregation: check storage bins
	Checking under roller: check mix temperature
	Poor-quality longitudinal/transverse joint: check filler-metering system and mix temperature
	Shoving of mat ahead of roller; lean, brown, dull appearance of surface; tearing of mat; or uneven thickness and quality: check mix temperature, asphalt cement type, and content; check filler-metering system, drying of aggregates, dwell time in dryer drum, and for proper quality control
	Poor compaction, or wavy surface: check mix temperature; asphalt viscosity-temp characteristics; asphalt properties, if overheated; filler-metering system; and for proper quality control at the plant
Tack/prime coating	Bleeding, transverse cracks, or shoving of mat ahead of roller: check type and application rates, and time available for curing of emulsion and/or penetration of prime coat into granular layers
Lift thickness	Checking under roller, shoving of mat ahead of roller, tearing of mat, or poor compaction: check proper lift thickness and number of lifts
Existing surface and environmental conditions	Bleeding: check for moisture in underlying layers; check if porous layer with water
	Blisters: check drainage
	Transverse cracks, checking under roller, shoving of mat ahead of roller, tearing of mat, or uneven thickness and quality: check for grading, drainage, density and weak areas, air temperature, and dust on existing surface
	Depression, or shoving of mat ahead of roller: check base preparation—density, drainage, and weak areas
	Poor-quality longitudinal joint: check air temperature
	Poor compaction: check preparation, ruts (if any), crack filling (if any), drainage and density, and existing air temperature
Transportation	Bleeding: check release agents on truck beds
	Poor-quality longitudinal/transverse joint; lean, brown, dull appearance of surface; tearing of mat; uneven thickness and quality; or poor compaction: check insulation in trucks
	Wavy surface: check for brakes in trucks
Transfer	Segregation, or poor-quality longitudinal/transverse joint: check for temperature variation across mat— consider using materials transfer vehicle
Paving	Bleeding, or segregation: check paver speed, and different parts such as auger and screed and mix in hopper
	Checking under roller, depression, or poor-quality longitudinal joint: check screed
	Poor-quality longitudinal/transverse joint: check screed, hopper and feed, and sensor, and check raking, bumping, and luting
	Tearing of mat: check paver speed and screed, feeder gate, kicker screws, and end plates
	Uneven thickness and quality, or wavy surface: check hopper gates, conveyor and spreading screw, sensor and auger, and screed pull point
	Poor compaction, or wavy surface: check for proper type of screed and hydraulics of screed operation
Rolling	Bleeding: check rolling pattern and number of passes
	Tender mix: change type of roller
	Checking under roller: check rolling pattern, speed, and number of rollers
	Depression, shoving of mat ahead of roller, roller marks, tearing of mat: check speed, reverse, turns, rolling pattern, parking of roller, tire pressure, roller weight, time of starting rolling, amplitude and frequency of vibration for vibratory roller, and pattern of rolling in superelevation
	Poor-quality longitudinal/transverse joint: check number and speed of rollers, and rolling direction
	Poor compaction: check for number, type and speed of rollers, frequency of vibrations, and pressure in tires of rubber-tired roller
Compacted surface	Brown streak: check for spilled gas or oil

Source: NAPA (2003).

16.5 SPECIFICATIONS

Specifications include description of construction materials, methods, and/or end product. There are primarily three types of specifications: performance-based or end product–based method specification and proprietary specification.

There are advantages and disadvantages of different kinds of specifications. For example, a performance- or end result–based specification could be shorter, but relatively difficult to prepare, and while it is easy to write a method specification, the results may not correlate well with performance, and while writing a proprietary specification is relatively easy, it does not allow the use of equal or better materials, and does not encourage innovations and competition.

An example of method-based specification, for compaction of an HMA pavement, is as follows:

> Use two passes with a vibratory roller weighing at least 10 tons. Vibrations per minute must exceed 1000, and the roller shall not travel faster than 4 miles per hour.

For the same purpose, a performance-based specification could be written as follows: compact the HMA to at least 98% of laboratory density. Most specifications combine aspects of both method- and performance-based specification.

Contracts for construction of asphalt pavements can be made through the low-bid system, best-bid system, or negotiated contract. In the low-bid system, generally the lowest bidder is selected. Although this system provides a fair and rational method of decision, it could lead to cutting corners. In the best-bid system, more emphasis is placed on the performance and experience of the contractor. In a negotiated contract, the owner negotiates the work with a specific contractor.

In the construction process, standards and specifications from ASTM and AASHTO are also followed. Most agencies would prepare a project specification from a list of guideline specifications from ASTM or AASHTO. Specifications and standards are particularly important with respect to sampling and testing for quality control and quality assurance.

A good specification must include a description of the following.

1. Lot size is the amount of materials that needs to be evaluated for acceptance or rejection. For example, 1 day of production of HMA.
2. Test properties indicate the properties that need to be tested. For example, density, aggregate gradation, asphalt content, and smoothness for an asphalt pavement.
3. Number of tests for each lot.
4. Point of sampling for tests. For example, aggregates from stockpiles and belt.
5. Method of sampling. For example, random sampling or representative sampling.
6. Number of tests to be reported. For example, an average of five tests should be reported.
7. Test method to be used for each property. For example, the density of the HMA layer should be measured with a nuclear density gage test.
8. The target value for each test property. For example, the target density is 95% of the laboratory density.
9. Tolerance of test results around the target value.
10. Actions to be taken if specification requirements are not met

A good specification should be free of vague and confusing words (such as "aggregate should be clean or to the satisfaction of the engineer"). Some other examples of words that should be avoided are as follows:

1. "…promptly and uniformly compacted…"—should state time of compaction, or that before the temperature drops to a certain value compaction should be completed.
2. "…heavy equipment or rollers…"—should state minimum weight of the roller.

An example of a specification for density of the HMA, including all of the above features, can be written as follows.

Example

The lot size of the construction for density evaluation is 1 day of production. The property to be tested is density. The lot will be divided into four equal sublots, and one random sample should be taken from each sublot. The in-place density should be measured. A 4-in. diameter core will be taken at each location. Use one test for each sample. Use ASTM D-2000 for determination of density. The average of the four core samples shall be 98–100% of laboratory density. When the density is below 98%, the contractor will be paid at the percentage shown in Table 16.2.

16.5.1 VARIABILITY OF MATERIALS

Test results are always variable because of errors of sampling and testing as well as material variability. Material variability will always be there, but the errors due to sampling and testing can be minimized by following specified methods of sampling, and using competent and qualified testing personnel, and appropriate testing equipment.

The variability of a number of test properties, such as density, can be approximated by a normal distribution. Using the different parameters, as listed below, allows the determination of variability of materials, and making decisions regarding acceptance or rejection in quality control and assurance.

The relevant parameters are as follows:

$$\text{Variance:} \quad s^2 = \frac{\Sigma_{i=1}^{n}(x_i - \overline{x})^2}{n-1}$$

$$\text{Standard deviation:} \quad s = \sqrt{\frac{\Sigma_{i=1}^{n}(x_i - \overline{x})^2}{n-1}}$$

$$\text{Coefficient of variation:} \quad s = \frac{s}{\overline{x}} * 100\%$$

$$\text{Standard error of mean:} \quad \overline{s} = \frac{s}{\sqrt{n}}$$

where:
  n = number of samples averaged to give each test result
  $\overline{x}$ = average
  x_i = individual test results

TABLE 16.2
Density Pay Factors

% of Density	% of Pay
Above 100	100
97.9–100	99
96.0–97.8	97
Below 96.0	Choice of paid 60%, or remove and replace

TABLE 16.3
Examples of Typical Standard Deviations

Property	Standard Deviation	Coefficient of Variation (%)
Subbase density	3.5	3.7
Asphalt content	0.2	0.03
Density of asphalt mix	1.02	1.0
Base course density	2.5	2.5

Typical standard deviation values have been determined as shown in Table 16.3.

In many specifications, the evaluation of a job for acceptance or rejection is based on the *percent within limits* (PWL) of results for each lot of the pavement. PWL (or *percent conforming*) is defined as the percentage of the lot falling above the lower specification limit (LSL), beneath the upper specification limit (USL), or between the USL and LSL. Although tests can be applied to verify it, in general, the population of most test results is assumed to be normally distributed, and the use of this procedure ensures the consideration of both average and variability of the test results for evaluation of the "quality" of the product. The PWL concept is based on the use of the area under a standard normal distribution.

A Z value, where $Z = \frac{y-\mu}{\sigma}$, can be used to determine the percentage of population that is within or outside a certain limit, using Table 16.4. Note that Z is the statistic that is to be used with the table, Y is the point within which the area is determined from the table, and μ and σ are the mean and standard deviation of the population, respectively. For example, consider the following test results for an in-place density measurement (as a percentage of theoretical maximum density).

Number of test results = 10; mean value = 94.0, standard deviation: 1.15. What percentage of the test results are expected to be below 92%?

$$Z = \frac{92 - 94}{1.15} = -1.74$$

From Table 16.4, (1 − 0.5 − 0.4591) * 100 = 4.1% of the test results are expected to be below 92%.

Now, in the case of the PWL procedure, the Quality Index value, Q, is used in instead of Z, and the sample mean, $\bar{x}$, and sample standard deviation, s, are used instead of the population mean and standard deviations. Also, instead of the "Y" value used for the calculation of Z, the USL and LSL values are used, as follows:

$$Q_L = \frac{\bar{x} - LSL}{s}$$

$$Q_U = \frac{USL - \bar{x}}{s}$$

where:

Q_L = Quality Index for lower specification limit, used when there is a one-sided lower specification limit

Q_U = Quality Index for upper specification limit, used when there is a one-sided upper specification limit

LSL = lower specification limit

USL = upper specification limit

$\bar{x}$ = sample mean

S = sample standard deviation

TABLE 16.4

Areas under the Standard Normal Distribution

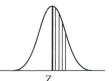 or

Z	0.00	0.01	0.02	0.03	0.04	0.05	0.06	0.07	0.08	0.09
0.0	0.0000	0.0040	0.0080	0.0120	0.0160	0.0199	0.0239	0.0279	0.0319	0.0359
0.1	0.0398	0.0438	0.0478	0.0517	0.0557	0.0596	0.0636	0.0675	0.0714	0.0753
0.2	0.0793	0.0832	0.0871	0.0910	0.0948	0.0987	0.1026	0.1064	0.1103	0.1141
0.3	0.1179	0.1217	0.1255	0.1293	0.1331	0.1368	0.1406	0.1443	0.1480	0.1517
0.4	0.1554	0.1591	0.1628	0.1664	0.1700	0.1736	0.1772	0.1808	0.1844	0.1879
0.5	0.1915	0.1950	0.1985	0.2019	0.2054	0.2088	0.2123	0.2157	0.2190	0.2224
0.6	0.2257	0.2291	0.2324	0.2357	0.2389	0.2422	0.2454	0.2486	0.2517	0.2549
0.7	0.2580	0.2611	0.2642	0.2673	0.2704	0.2734	0.2764	0.2794	0.2823	0.2852
0.8	0.2881	0.2910	0.2939	0.2967	0.2995	0.3023	0.3051	0.3078	0.3106	0.3183
0.9	0.3159	0.3186	0.3212	0.3238	0.3264	0.3289	0.3315	0.3340	0.3365	0.3389
1.0	0.3413	0.3438	0.3461	0.3485	0.3508	0.3531	0.3554	0.3577	0.3599	0.3621
1.1	0.3643	0.3665	0.3686	0.3708	0.3729	0.3749	0.3770	0.3790	0.3810	0.3830
1.2	0.3849	0.3869	0.3888	0.3907	0.3925	0.3944	0.3962	0.3980	0.3997	0.4015
1.3	0.4032	0.4049	0.4066	0.4082	0.4099	0.4115	0.4131	0.4147	0.4162	0.4177
1.4	0.4192	0.4207	0.4222	0.4236	0.4251	0.4265	0.4279	0.4292	0.4306	0.4319
1.5	0.4332	0.4345	0.4357	0.4370	0.4382	0.4394	0.4406	0.4418	0.4429	0.4441
1.6	0.4452	0.4463	0.4474	0.4484	0.4495	0.4505	0.4515	0.4525	0.4535	0.4545
1.7	0.4554	0.4564	0.4573	0.4582	0.4591	0.4599	0.4608	0.4616	0.4625	0.4633
1.8	0.4641	0.4649	0.4656	0.4664	0.4671	0.4678	0.4686	0.4693	0.4699	0.4706
1.9	0.4713	0.4719	0.4726	0.4732	0.4738	0.4744	0.4750	0.4756	0.4761	0.4767
2.0	0.4772	0.4778	0.4783	0.4788	0.4793	0.4798	0.4803	0.4808	0.4812	0.4817
2.1	0.4821	0.4826	0.4830	0.4834	0.4838	0.4842	0.4846	0.4850	0.4854	0.4857
2.2	0.4861	0.4864	0.4868	0.4871	0.4875	0.4878	0.4881	0.4884	0.4887	0.4890
2.3	0.4893	0.4896	0.4898	0.4901	0.4904	0.4906	0.4909	0.4911	0.4913	0.4916
2.4	0.4918	0.4920	0.4922	0.4925	0.4927	0.4929	0.4931	0.4932	0.4934	0.4936
2.5	0.4938	0.4940	0.4941	0.4943	0.4945	0.4946	0.4948	0.4949	0.4951	0.4952
2.6	0.4953	0.4955	0.4956	0.4957	0.4959	0.4960	0.4961	0.4962	0.4963	0.4964
2.7	0.4965	0.4966	0.4967	0.4968	0.4969	0.4970	0.4971	0.4972	0.4973	0.4974
2.8	0.4974	0.4975	0.4976	0.4977	0.4977	0.4978	0.4979	0.4979	0.4980	0.4981
2.9	0.4981	0.4982	0.4982	0.4983	0.4984	0.4984	0.4985	0.4985	0.4986	0.4986
3.0	0.4987	0.4987	0.4987	0.4988	0.4988	0.4989	0.4989	0.4989	0.4990	0.4980
3.1	0.4990	0.4991	0.4991	0.4991	0.4992	0.4992	0.4992	0.4992	0.4993	0.4993
3.2	0.4993	0.4993	0.4994	0.4994	0.4994	0.4994	0.4994	0.4995	0.4995	0.4995
3.3	0.4995	0.4995	0.4995	0.4996	0.4996	0.4996	0.4996	0.4996	0.4996	0.4997
3.4	0.4997	0.4997	0.4997	0.4997	0.4997	0.4997	0.4997	0.4997	0.4997	0.4998

Source: From Burati et al. (2003).

The use of the Quality Index is illustrated through the following example. Consider the in-place density data as follows:

n = number of samples = 10

$\bar{x}$ = sample mean = 97.5% of density of samples compacted by the Marshall procedure in the laboratory

S = sample standard deviation = 1.25

The specification states that in order to get 100% paid, 90% of the test results should be above 96.3% of density of the samples compacted by the Marshall procedure in the laboratory. That is, a 90% PWL is required for 100% pay. In this case, can the contractor expect 100% pay?

$$Q = \frac{97.5 - 96.3}{1.25} = 0.96$$

Referring to the Quality Index Table (Table 16.5), for n = 10, a Q value of 0.96 indicates a PWL of 83%, which is less than 90%. Hence, the contractor cannot expect to get paid 100% of the pay for the lot being tested.

Note that when two-sided specifications are used, the PWL is calculated as follows:

$$PWL_T = PWL_U + PWL_L - 100$$

where:

PWL_U = percentage below the upper specification limit (based on Q_U)

PWL_L = percentage above the upper specification limit (based on Q_L)

PWL_T = percentage within the upper and lower specification limits

Generally, pay adjustments, on the basis of pay factors, are made on the basis of the PWL of a lot. It is desirable that the pay factors are tied to the quality and resulting performance of the pavement on a rational basis. Generally, payments are reduced for substandard work and increased (above 100%, as a bonus) for superior-quality work. Both stepped and continuous pay factors are used, as shown in an example in Table 16.6.

Continuous payment factors:

$$PF = 55 + 0.5 * PWL$$

where:

PF = payment factor as a percentage of contract price

PWL = estimated percentage within limits

16.5.2 Use of Quality Control Charts

During construction, it is important to summarize and analyze the test results. A quality control chart is a plot of test results and numbers. The control chart can be used to summarize the data, identify trends, show specification limits and percentages within limits, and evaluate changes in any process.

Different data sets can be shown in quality control charts:

1. *Plot of individual values*: the advantage is that it shows all the data, and the disadvantage is that it is difficult to identify trends.
2. *Plot of running average*: it shows trends but does not show actual data.
3. *Plot of running variability*: it shows the variability of the mix.

Typical test properties that are plotted include the density and water content for the subgrade; the density, water content, and gradation for the subbase and base; and the gradation, asphalt content, in-place density, laboratory density, and voids for the HMA.

An example of a quality control chart for individual data values is shown in Figure 16.14.

TABLE 16.5

Example of Quality Index Value Table for Estimating PWL

PWL	$n = 3$	$n = 4$	$n = 5$	$n = 6$	$n = 7$	$n = 8$	$n = 9$	$n = 10$ to 11
100	1.16	1.50	1.79	2.03	2.23	2.39	2.53	2.65
99	–	1.47	1.67	1.80	1.89	1.95	2.00	2.04
98	1.15	1.44	1.60	1.70	1.76	1.81	1.84	1.86
97	–	1.41	1.54	1.62	1.67	1.70	1.72	1.74
96	1.14	1.38	1.49	1.55	1.59	1.61	1.63	1.65
95	–	1.35	1.44	1.49	1.52	1.54	1.55	1.56
94	1.13	1.32	1.39	1.43	1.46	1.47	1.48	1.49
93	–	1.29	1.35	1.38	1.40	1.41	1.42	1.43
92	1.12	1.26	1.31	1.33	1.35	1.36	1.36	1.37
91	1.11	1.23	1.27	1.29	1.30	1.30	1.31	1.31
90	1.10	1.20	1.23	1.24	1.25	1.25	1.26	1.26
89	1.09	1.17	1.19	1.20	1.20	1.21	1.21	1.21
88	1.07	1.14	1.15	1.16	1.16	1.16	1.16	1.17
87	1.06	1.11	1.12	1.12	1.12	1.12	1.12	1.12
86	1.04	1.08	1.08	1.08	1.08	1.08	1.08	1.08
85	1.03	1.05	1.05	1.04	1.04	1.04	1.04	1.04
84	1.01	1.02	1.01	1.01	1.00	1.00	1.00	1.00
83	1.00	0.99	0.98	0.97	0.97	0.96	0.96	0.96
82	0.97	0.96	0.95	0.94	0.93	0.93	0.93	0.92
81	0.96	0.93	0.91	0.90	0.90	0.89	0.89	0.89
80	0.93	0.90	0.88	0.87	0.86	0.86	0.86	0.85
79	0.91	0.87	0.85	0.84	0.83	0.82	0.82	0.82
78	0.89	0.84	0.82	0.80	0.80	0.79	0.79	0.79
77	0.87	0.81	0.78	0.77	0.76	0.76	0.76	0.75
76	0.84	0.78	0.75	0.74	0.73	0.73	0.72	0.72
75	0.82	0.75	0.72	0.71	0.70	0.70	0.69	0.69
74	0.79	0.72	0.69	0.68	0.67	0.66	0.66	0.66
73	0.76	0.69	0.66	0.65	0.64	0.63	0.63	0.63
72	0.74	0.66	0.63	0.62	0.61	0.60	0.60	0.60
71	0.71	0.63	0.60	0.59	0.58	0.57	0.57	0.57
70	0.68	0.60	0.57	0.56	0.55	0.55	0.54	0.54
69	0.65	0.57	0.54	0.53	0.52	0.52	0.51	0.51
68	0.62	0.54	0.51	0.50	0.49	0.49	0.48	0.48
67	0.59	0.51	0.47	0.47	0.46	0.46	0.46	0.45
66	0.56	0.48	0.45	0.44	0.44	0.43	0.43	0.43
65	0.52	0.45	0.43	0.41	0.41	0.40	0.40	0.40
64	0.49	0.42	0.40	0.39	0.38	0.38	0.37	0.37
63	0.46	0.39	0.37	0.36	0.35	0.35	0.35	0.34
62	0.43	0.36	0.34	0.33	0.32	0.32	0.32	0.32
61	0.39	0.33	0.31	0.30	0.30	0.29	0.29	0.29
60	0.36	0.30	0.28	0.27	0.27	0.27	0.26	0.26
59	0.32	0.27	0.25	0.25	0.24	0.24	0.24	0.24
58	0.29	0.24	0.23	0.22	0.21	0.21	0.21	0.21
57	0.25	0.21	0.20	0.19	0.19	0.19	0.18	0.18
56	0.22	0.18	0.17	0.16	0.16	0.16	0.16	0.16
55	0.18	0.15	0.14	0.14	0.13	0.13	0.13	0.13
54	0.14	0.12	0.11	0.11	0.11	0.11	0.10	0.10
53	0.11	0.09	0.08	0.08	0.08	0.08	0.08	0.08
52	0.07	0.06	0.06	0.05	0.05	0.05	0.05	0.05
51	0.04	0.03	0.03	0.03	0.03	0.03	0.03	0.03
50	0.00	0.00	0.00	0.00	0.00	0.00	0.00	0.00

Source: From Burati et al. (2003).

TABLE 16.6
Example of Pay Factors

Estimated PWL	Payment Factor (%)
95.0–100.0	102
85.0–94.9	100
50.0–84.9	90
0.0–49.9	70

Source: From Burati et al. (2003).

16.6 PREPARATION OF SUBGRADE, AND CONSTRUCTION OF BASE AND SUBBASE LAYERS

The subgrade can consist of relatively high-quality materials such as sand and gravel or poor-quality materials such as silt and clay. The pavement structure is designed to protect the subgrade from excessive deformation. However, the subgrade may need improvement before pavement construction begins. The improvement is done with compaction and/or stabilization.

Compaction can be measured by the nuclear density test, using a sand cone or a water balloon, or by testing samples recovered with cylinders. Density of the subgrade is specified as a percentage of Proctor or modified Proctor density or as a percentage of density of a control strip. The laboratory test data on optimum moisture content and maximum dry density are used.

Low density in subgrade can be due to the following factors.

1. Inadequate weight of rollers
2. Unsatisfactory roller patterns
3. Improper moisture content
4. Incorrect laboratory density due to wrong compactive effort and/or use of nonrepresentative samples
5. Poor testing, because of uncalibrated equipment or untrained personnel

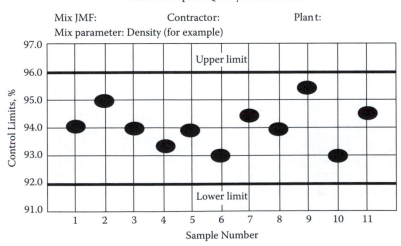

FIGURE 16.14 Example of a Control Chart.

The following tests could be run for characterizing the subgrade for performance.

1. California Bearing Ratio (CBR)
2. Plate bearing test
3. Unconfined compression test
4. Resilient modulus

Subbase and base materials generally consist of higher quality materials, such as clean, well-graded aggregates with fractured faces. Mixing is generally done in a portable pugmill, with close control on water content. Subbase and base materials are placed using a grader, spreader, or paver. Compaction of the subbase and base can be accomplished using a static and/or vibratory steel-wheel roller as well as rubber-tired rollers. Potential problems with the base and subbase include low density, segregation, grade control, and gradation of aggregates.

16.7 QUALITY CONTROL (QC) AND QUALITY ASSURANCE (QA)

An important part of the production process is the testing of plant-produced material as part of the quality control (QC) methods. QC sampling and testing are done on a regular basis to continuously evaluate the produced material, by the material producer. The results of the QC process can be fed back to the plant to correct or modify any setting, as required. Quality assurance (QA) tests are generally conducted by the owner or a consultant agency (or sometimes even by the producer) at longer time intervals to make sure that the produced material meets the specification.

The important tests and methods are presented below.

1. *Standard Test Method for Random Sampling (ASTM 3665)*: This standard outlines the process which can be used to eliminate bias during sampling. Sampling can be done either by random procedure or stratified random procedure. In the latter process, the lot is divided into sublots and then the random method is used to pick samples from each sublot. Tables for picking random numbers, by selecting any two numbers and rows and columns, are provided in ASTM 3665. Sampling should be done from different points, for example, aggregates from a belt or truck or from in-place paving material.
2. *Checking temperature of asphalt, HMA, and air*: HMA and asphalt temperatures are measured using dial- or digital-type thermometers (two thermometers) made with metal stems (which are inserted into the material) and armored glass. AASHTO T-245 specifies the requirements of the thermometers, which must be calibrated before use. Infrared thermometers, when calibrated against conventional thermometers, could be used. HMA temperatures are generally taken for each sample as well as at the start of the day, by pushing the stem to its full depth into the material, after waiting for 3 minutes or when there is no appreciable temperature difference between two readings at 1-minute intervals. The same type of thermometer could be used to check the temperature of the asphalt binder from sampling valves from the asphalt tank. For measuring air temperature, the thermometer should be kept out of direct sunlight.
3. *Preparing Marshall hammer compacted specimens*: AASHTO T-245 specifies the procedure for compacting specimens with a Marshall hammer and testing the specimens for stability and flow, after running bulk-specific gravity (AASHTO T-166). The bulk-specific gravity could be used along with the theoretical maximum density to determine the percentage of density and air voids.
4. *Preparing a specimen with a Superpave gyratory compactor*: AASHTO 312 specifies the method for using a Superpave gyratory compactor for the compaction of 150-mm diameter HMA specimens for subsequent determination of bulk-specific gravity.

5. *Determination of bulk-specific gravity of a compacted HMA specimen*: AASHTO T-166 specifies the procedure for the determination of bulk-specific gravity of a laboratory-compacted HMA specimen and in-place cores from pavement. Note that cores should be dried at 110°C to constant mass before running this test.

6. *Maximum specific gravity of HMA*: AASHTO T-209 specifies the method for determination of maximum specific gravity of HMA specimens, by determining the ratio of the weight of a void less the mass of HMA and the mass of an equal volume of water.

7. *Determination of thickness of HMA specimen*: ASTM D-3549 specifies the method of determination of thickness of a compacted HMA specimen. The devices that can be used are metal tape, a rule, a measurement jig, and a set of calipers. Required accuracy for measurements is specified in the standard. Measurements are made between the horizontal planes of the upper and lower surfaces of the specimens or between lines of demarcation (as in a thickness of a layer inside a full-depth core). The measurements are taken at four different points around the periphery, and the average of the four numbers is reported as the thickness of the specimen.

8. *Determination of asphalt content of HMA by ignition method (AASHTO T-308)*: In this test, an HMA sample (generally ranging in weight from 1.2 kg to 4 kg, for a range of nominal maximum aggregate size [NMAS] of 4.75 mm to 37.5 mm) is weighed, it is placed inside an oven maintained at 650°C, the asphalt binder is burnt off, the remaining aggregates are weighed, and the asphalt content is determined on the basis of the two weights. Aggregate calibration information, which accounts for loss of aggregates by burning, if any, can be input in the ignition oven computer. The ignition oven reports the weight of the sample to the nearest 0.1 g, taking into consideration the temperature correction, and prints out the asphalt content (using an onboard computer program) from the pre- and postignition weights, taking the aggregate calibration factor into consideration. Important items are provision of methods for reducing furnace emissions, availability of a self-locking oven door (which will not open until the end of the test), an audible alarm and lights for indication of completion of the test, as well as safety equipment such as a face shield and thermal gloves for the operator. The asphalt-free aggregates can be used for running a sieve analysis and checking gradation, although care must be taken to consider the breakdown of some aggregates and the creation of some fines (if any) during the ignition test of some types of aggregates.

9. *Determination of asphalt content by the nuclear method (AASHTO T-287)*: In this test the amount of asphalt is indicated by the amount of hydrogen in a mix as measured with a neutron source in the nuclear gage. The gage needs to be calibrated with both aggregate and mix samples before using. A pan filled with consolidated HMA is placed inside the nuclear gage, and the asphalt content is read off the equipment.

10. *Quantitative extraction of asphalt from HMA (AASHTO T-164)*: In this test a solvent is used to dissolve the asphalt binder out from an HMA sample, and the aggregates are recovered. The weight of the mix sample before and after the washing out of the asphalt is used to calculate the asphalt content, and the recovered aggregates can be used for sieve analysis. A special bowl is used to soak a specified amount of mix (ranging in weight from 0.5 kg to 4 kg, for an NMAS of 4.75 mm to 37.5 mm) in a solvent. A filter ring is placed on the base of the bowl, and the lid is clamped down. A centrifuge is used to spin the sample in the bowl at a speed of 3600 rpm (slowly increasing the speed to the maximum). The extraction fluid is run off the bowl, and the process is continued with fresh charges of solvent until all of the asphalt has been washed away. At the end of this step, the fines adhering to the filter paper can be scraped out on the remaining aggregate sample. The filter paper and the sample are then dried at 110°C to constant mass.

11. *Resistance of a compacted HMA specimen to moisture-induced damage (AASHTO T283)* as discussed in Chapter 14.

16.8 CONSTRUCTION OF LONGITUDINAL JOINTS

A longitudinal construction joint occurs when a lane of hot mix asphalt (HMA) is constructed adjacent to previously placed HMA. Longitudinal joints are inevitable in both highway and airfield pavements, unless paving is done in the echelon formation (which is generally not the case). Damaged longitudinal joints are of very serious concern in airfield pavements. Loose materials from such areas can cause foreign object damage (FOD) to aircrafts, leading to loss of life and equipment. Potential sharp edges along open longitudinal joints can also endanger aircraft. In addition, such joints can lead to ingress of moisture and undesirable materials and lead to premature failures in the subsurface and ultimately the entire pavement, leading to a cycle of costly and time-consuming repairs.

For the above reasons, engineers, consultants, and contractors have continuously tried to develop methods for constructing better performing longitudinal joints in pavements. Such methods include overlapping and luting operations; different types of rolling patterns for compacting the joint; and the use of special joint construction techniques and equipment, such as cutting wheels, restrained edge devices, notched wedge joints, and joint heaters. At the same time, many agencies have started using specifications that are written specifically for the construction of better joints, such as density requirements at joints.

There are several steps in the construction of a good-quality joint. These include paving the first lane in a uniform, unwavering line; compacting the unsupported edge of the first lane (cold lane) properly; controlling the height of the uncompacted HMA in the hot lane; and proper overlapping during the paving operation. Raking or luting at the longitudinal joint can be eliminated if the minimal overlapping is done. An excessive overlap will require the removal of extra material from the cold lane onto the hot lane; otherwise, the aggregate in the mix remaining on the compacted lane will get crushed, resulting in raveling. When that happens, the excessive overlapped material on the cold lane may be "bumped" with a lute onto the hot mat *just* across the joint. The bump should lie just above the natural slope or the wedge at the edge of the cold lane. Sometimes, there is a tendency to broadcast the raked material onto the HMA in the hot lane. This not only is undesirable for obtaining a good longitudinal joint but also affects the surface texture of the mat adversely.

Obtaining adequate compaction at the joint is the final key in obtaining a durable longitudinal joint. Joints with high densities generally show better performance than those with relatively low densities. By paying attention to construction details, it is possible to obtain a joint density within 1.5% of the mainline density.

Using proper specifications is another key requirement for obtaining good-quality joints. Joint density is best measured by obtaining a 6-in. (150-mm) diameter core centered on top of the visible line between the two lanes. It is recommended to make the compaction level at both the joint and mat based on theoretical maximum density (TMD) rather than the bulk-specific gravity of a daily compacted Marshall specimen, which is more variable.

Section 401-5.2 of Engineering Brief EB59A of the FAA, dated May 12, 2006, provides a good example of a specification of constructing longitudinal joints. The acceptance criteria are as follows.

Evaluation for acceptance of each lot of in-place pavement for joint density and mat density shall be based on PWL. The contractor shall target production quality to achieve 90 PWL or higher.

The percentage of material within specification limits (PWL) shall be determined in accordance with procedures specified in Section 110 of the General Provisions. The acceptance limits shall be as follows:

Mat density, % of TMD	92.8 minimum
Joint density, % of TMD	90.5 minimum

Notched Wedge Joint

Application of Rubberizd Asphalt Tack Coat at Joint

Joint Cutting

Infrared Heating of Joints

FIGURE 16.15 Different Longitudinal Joint Construction Techniques.
Courtesy: Prithvi S. Kandhal, the US Air Force, and Tom Allen (Ray- Tech Infrared Corp.).

16.8.1 Techniques of Constructing Good Longitudinal Joints

It is preferable to produce hot longitudinal joints by operating two or more pavers in echelon. If echelon paving is not possible, then any of the following good techniques can be used (Figure 16.15).

16.8.1.1 Combination of Notched Wedge Joint, Rubberized Asphalt Tack Coat, and Minimum Joint Density Requirements

Construct a notched wedge longitudinal joint. The unconfined edge of the first paved lane has a vertical notch at the edge generally ranging from ½ in. (13 mm) to 3/4 in. (19 mm) in height depending upon the nominal maximum aggregate size of the HMA mixture. Generally, a vertical notch of about ½ in. (13 mm) height is considered adequate for most surface course mixtures. It is recommended to end the taper with a minimal height such as 3/8 in. (9.5 mm) to avoid dragging of the material. Usually a loaded wheel, which is attached to the paver, is used to compact the taper. Typically, the roller weighs 100 to 200 lbs (45 to 91 kg) and is approximately 14 in. (356 mm) wide by 12 in. (305 mm) in diameter. There is no need to compact the taper with a conventional steel- or pneumatic-tired roller because it will simply destroy the vertical notch. The overlap layer of the adjacent paving lane is required to be placed and compacted within 24 hours unless delayed by inclement weather. The vertical notch and taper are tack-coated with rubberized asphalt binder prior to placing the overlap wedge, as described later. The notched wedge joint can be formed by using a homemade sloping steel plate or commercial devices attached to the inside corner of the paver screed extension. The notched wedge joint does not always work well for thinner HMA lifts. Ideally, it gives the best results with a minimum lift thickness of 1½ to 2 in. (37 to 51 mm). On the other hand, excessively thick lifts produce a long taper, which may not be desirable. In those cases, the length of the taper is generally restricted to 12 in. (305 mm). The top-course taper shall overlap and slope in the opposite direction of the lower-course taper.

After the first lane (cold lane) is paved with a notched wedge and compacted, a rubberized asphalt tack coat is applied on the face of the unconfined edge of the cold lane. The thickness of the tack coat is about 1/8 in. (3 mm) on the slope of the HMA edge. The rubberized asphalt tack coat need not be applied on the entire taper. It is considered adequate to apply it on the vertical notch and the top 3–4-in. (76–102-mm) wide band of the taper. Application excesses should not exceed more than ½ in. (13 mm) at the top of the joint. The sealant should preferably be applied within 4 hours of the time that the adjacent HMA lane is placed. The heat from the HMA in the adjacent lane and the roller pressure causes the sealant to adhere strongly along the joint face, resulting in a strong bond between the two lanes and providing a built-in sealer at the joint. After the rubberized asphalt tack coat is applied, the adjacent lane (hot lane) is placed. The height of the uncompacted HMA should be about 1 ¼ in. (32 mm) for each 1 in. (25 mm) of the compacted lift thickness in the cold lane. The end gate of the paver should extend over the top surface of the previously placed HMA by a distance of approximately 1 to 1 ½ in. (25 to 38 mm). The most efficient joint compaction method is to roll the longitudinal joint from the hot side, overlapping the cold lane by approximately 6 in. (150 mm). The steel-wheel roller can be operated in vibratory or static mode, preferably the vibratory mode to obtain better compaction.

For the following procedures, the best practices for paving and compacting the first lane, paving the second lane, and overlapping, raking and luting, and compacting the longitudinal joint, as well as minimum joint density and mat density requirements, are similar to those mentioned above.

16.8.1.2 Rubberized Asphalt Tack Coat and Minimum Joint Density Requirements

This practice is similar to that in Item 1 (above), except that no notched wedge joint is used. The first lane (cold lane) is paved as usual with the normal, unconfined edge slope. A rubberized asphalt tack coat is applied on the entire face of the unconfined edge of the cold lane using the procedure described in Item 1.

16.8.1.3 Notched Wedge Joint and Minimum Joint Density Requirements

This practice is similar to that in Item 1 (above) except that a conventional tack coat material (which is used on the main line) is applied to the entire face of the notched wedge joint in lieu of rubberized asphalt material.

16.8.1.4 Cutting Wheel and Minimum Joint Density Requirements

The cutting wheel technique involves cutting 1.5–2.0 in. (38–51 mm) of the unconfined, low-density edge of the first paved lane after compaction, while the mix is plastic. The cutting wheel is usually 10 in. (254 mm) in diameter, with the cutting angle about 10° from the vertical toward the mat to be cut and about 45° on the open side to push the trimmings away. The cutting wheel can be mounted on an intermediate roller or a motor grader. The HMA trimmings can be collected and recycled. A reasonably vertical face at the edge is obtained by this process, which is then tack-coated before placing the abutting HMA. It is important to restrict the overlap to about ½ in. (13 mm) while placing the adjacent lane. It is important to remove all low-density material at the edge of the first paved lane. Some contractors remove as much as a 3-in. (75-mm) strip to meet and exceed joint density requirements. It is very important to have a skilled cutting wheel operator, who must cut straight without wavering, and a skilled paver operator, who must closely match the cut line with minimal overlap.

16.8.1.5 Infrared Joint Heating and Minimum Joint Density Requirements

The objective of the infrared joint heating system is to obtain a hot joint during conventional longitudinal joint construction. The infrared joint heating system consists of two or three pre-heaters and one paver-mounted heater. The pre-heaters, which are connected in series, are towed with a small tractor over the joint approximately 100 ft (30 m) ahead of the paver. Propane cylinders are used to feed the infrared heaters. Both the pre-heaters and paver-mounted heaters are placed about 2–3 in. (51–76 mm) above the pavement surface and straddle the joint. The infrared heaters usually target a final surface temperature of 340°F (171°C) behind the paver-mounted heater. The target surface temperature can be achieved by changing the number of pre-heaters, the distance between the pre-heaters and paver, and the height of the heaters above the pavement surface. All these variables need to be adjusted, taking into account prevailing ambient conditions such as air temperature and wind velocity. The pre-heaters run continuously as long as the towing vehicle is operated at a speed high enough to prevent overheating the HMA at the joint. If the towing vehicle slows down too much, the pre-heater is designed to shut down automatically. The paver-mounted heater has also been designed to shut down once the HMA at the joint exceeds a specified temperature.

QUESTIONS

1. What are the different steps in the production of hot mix asphalt?
2. A contractor notices tender mix/shoving of the mix during compaction. What could be the probable reasons?
3. Ten tests were conducted to determine the density of a sample section of a new pavement. The densities were reported as a percentage of theoretical maximum density. Using a standard z distribution table, estimate the percentage of results that could be expected to fall below 92%. If the specification states that the PWL must be 90% to get full payment, and the lower limit for density is 91.0%, can the contractor expect to get paid 100%?

Test No.	Density (% of Theoretical Maximum)
1	95.0
2	92.1
3	91.2
4	93.3
5	92.5
6	95.0
7	95.5
8	92.2
9	92.6
10	93.2

4. Review a specification that is commonly used by your local or state department of transportation for pavement construction. Can you suggest some improvements?
5. What are some of the better techniques that could be utilized for constructing good-quality longitudinal joints?

17 Construction of Concrete Pavements

Sequentially, the construction process of concrete pavements consists of the following:

1. Concrete production
2. Preparation of subgrade and base
3. Presetting reinforcements such as dowel bars and continuous reinforcement
4. PCC slab construction—slipforming
5. Finishing
6. Jointing
7. Curing

17.1 CONCRETE PRODUCTION

All Portland cement concrete used for pavement applications is ready-mix concrete. ASTM C-94—AASHTO M-157 provides the standard specifications for the manufacture and delivery of freshly mixed concrete. Usually customers have a few options regarding the manner in which they can order the concrete. A performance-based option is where the customer specifies the strength, slump, and air, and the producer selects the mix proportions needed to achieve the desired performance. A prescribed option is where the customer specifies the mix proportions, whereas a combination option allows the producer to select the mix proportions with the minimum amount of cement and strength specified.

Batching is the process of combining the mix ingredients to produce the desired concrete. Batching is done by either volume or mass. Batching by mass, as specified in ASTM C-94—AASHTO M-157, is desirable and yields more accurate measurements since fine sands usually change apparent volume with a change in moisture content. Volumetric batching (ASTM C-685—AASHTO M-241) is used primarily for concrete mixed in continuous mixers. Mixing is the process of mixing all the ingredients thoroughly without overloading the equipment above its capacity. When a concrete is adequately mixed, it should have a uniform appearance, with all ingredients uniformly distributed. And when samples are taken from different portions of a batch, they should have the same slump, density, air content, and coarse aggregate content. Concrete can be mixed at the jobsite using a stationary mixer. For ready-mixed concrete, concrete is proportioned and mixed away from the construction site but delivered in a fresh and unhardened state. Ready-mixed concrete can be produced in a central mix plant, where it is completely mixed in a mixer and delivered in a concrete truck mixer (Figure 17.1).

17.2 PREPARATION OF SUBGRADE AND BASE

Prior to concrete placement and the construction of the PCC slab on grade, preparation of the base and subgrade is essential. This includes compacting, trimming, and wetting the subgrade. It is assumed that the subgrade has adequate bearing capacity, is suitable for structural loads, and is readied for construction. The subgrade should be moistened in dry conditions so that the dry subgrade soil does not draw too much moisture from the concrete. For base placing, the appropriate materials and compaction are important. For bound base, the material could be placed with a paver and compacted with rollers. Figure 17.2 shows different stages of subgrade and base preparation.

FIGURE 17.1 Concrete Plant with Mixer Truck.
Courtesy: Kim Franklin, Northeast Cement Shippers Association.

FIGURE 17.2 Preparations of Subgrade and Base.
Courtesy: Wouter Gulden, ACPA-SE Chapter.

17.3   PRESETTING REINFORCEMENTS SUCH AS DOWEL BARS AND CONTINUOUS REINFORCEMENT

The use of dowel bars for proper load transfer at PCC pavement joints is highly dependent on proper placement. Dowels can be placed either before PCC placement by using dowel baskets, or after PCC placement by using an automatic dowel bar inserter. Incorrect placement of dowels resulting in skewed alignment, shallow positioning, or excessively corroded dowels can cause failure and lead to faulting and cracking at the joints. The placement of dowel bars can be evaluated with nondestructive testing (magnetic induction tomography equipment, discussed in Chapter 20). To maintain long-term functionality and durability for PCC pavements, dowel bars must be kept in good functioning condition, including protection from corrosion. This can be accomplished by sealing joints and minimizing the amount of water that can seep through. However, this is impossible to achieve since water will seep through the concrete over time and, combined with deicing salts, may corrode unprotected steel bars. Typically dowel bars are protected from corrosion by the application of epoxy coating or stainless steel cladding. Additionally, dowels should be lightly coated with a lubricant such as grease or oil to prevent bonding with the PCC. Note that only half of a dowel must be greased.

Dowel baskets are used to hold dowel bars at the appropriate location and elevation before concrete placement. The baskets are simple truss structures fabricated from thick gauge wire and are left to remain as part of the PCC pavement structure. The FHWA recommends that a minimum of eight stakes be used to secure the baskets, and that the steel stakes must have a minimum diameter of 0.3 inches (8 mm) embedded at least 4 inches (100 mm) in stabilized bases, 6 inches (150 mm) in treated permeable bases, and 10 inches (250 mm) for untreated bases or subgrade. Figure 17.3 shows dowel bars in PCC pavements.

FIGURE 17.3 Dowels in PCC Pavements.
Courtesy: Wouter Gulden, ACPA-SE Chapter.

FIGURE 17.4 Slipform Paving with Tie Bar Inserter.
Courtesy: Wouter Gulden, ACPA-SE Chapter.

Dowel bar inserters are automated attachments to slipform pavers that allow the paver to insert transverse joint dowel bars as part of the paving process. Dowel bar insertion usually occurs after the vibrator but before the tamper bar. Dowel bars are placed on the fresh PCC surface, then pushed down to the correct elevation by forked rods. The rods are usually vibrated while they insert the dowel bar to facilitate insertion and consolidation around the dowels.

Tie bars are placed along the longitudinal joint to tie the two slabs together. Tie bars are not considered as load transfer devices and should not be designed as such. Typically, tie bars are about 0.5 inches (12.5 mm) in diameter and between 24 and 40 inches (0.6 and 1.0 m) long. Tie bars are typically inserted after concrete placement either by hand or using a tie bar inserter attachment in the slipforming process. When one lane at a time is paved, tie bars are inserted at midslab depth and bent back until the adjacent lane is ready to be paved. When using slipform pavers, tie bars are inserted through slab edges that will become future longitudinal joints. If two lanes are being paved in one paver operation, then the tie bars are inserted or pushed into a midslab area that will be cut as a future longitudinal joint similar to dowel bar construction. And similar to dowel bars, tie bars should be protected from corrosion. Figure 17.4 shows slipform paving with a tie bar inserter.

17.3.1 Reinforcing Steel (CRCP)

Proper placement of reinforcing steel is critical to the performance of CRCP. Failures in CRCP and inadequate performance are commonly associated with reinforcing steel issues and inadequacies. These include insufficient reinforcement bar lapping, inadequate consolidation of the fresh concrete around the steel, improper position of the steel in the slab, and extreme hot weather during construction. Reinforcing steel for CRCP can be placed by using the manual method or the mechanical method.

The manual method involves hand-placing the reinforcing steel before the fresh concrete is placed and finished. The rebar is placed in the proper location using supporting plastic or metal anchors called *chairs*. The chairs must be well anchored with adequate metal stakes to withstand the concrete placement and consolidation. Once the chairs are secured, transverse bars are placed on the chairs as support for the longitudinal rebar. The longitudinal rebar is placed on the transverse bars and secured firmly using appropriate ties, typically every 4–6 ft. (1.2–1.8 m). The advantage of the hand placement method is that it allows for easy verification of rebar placement, including elevation and lap distances. Unfortunately, the manual method is slower and more labor intensive than mechanical methods. Figure 17.5 shows reinforcing steel before placement of concrete in a CRCP pavement.

FIGURE 17.5 Steel Reinforcement in CRCP Pavement.
Courtesy: Wouter Gulden, ACPA-SE Chapter.

Reinforcing steel can also be placed using the mechanical method. This involves placing rebar in a prepositioned location while the paver places and consolidates the fresh concrete around it. Great care should be taken to assure proper placement of the longitudinal reinforcement, since the precision of the mechanical systems could deviate by inches from intended locations. A double-lift construction method would help in placing the reinforcing steel more precisely; however, this method could be costly.

17.4   PCC SLAB CONSTRUCTION: SLIPFORMING

There are two methods of constructing PCC pavement slabs: using a slipform paver and using fixed forms. In slipform paving, the paving machine rides on treads over the area to be paved. The paver is guided using previously set stringline or laser sensors. Fresh concrete is deposited in front of the paving machine, which then spreads, shapes, consolidates, screeds, and float-finishes the concrete in one continuous operation. Great coordination between concrete production, delivery, and placement is needed to maintain adequate forward progress of the slipform paver.

In fixed-form paving, stationary metal forms are set, aligned, and staked rigidly on a solid foundation. Final compaction, preparation, and shaping of the subgrade or subbase are completed after the forms are set. Forms are cleaned and oiled to ensure easy release after the concrete hardens. Once concrete is deposited near its final position on the subgrade, spreading is completed by a mechanical spreader riding on top of the preset forms and the concrete. The spreading machine is followed by other machines that shape, consolidate, and float-finish the concrete. After the concrete has reached the required strength, the forms are removed. Concrete should be placed as close to its final location as possible. This is done using concrete-placing equipment. Concrete should not be moved excessively once deposited to minimize the potential for segregation. Concrete should not be placed in large piles and moved horizontally over long distances since the mortar fraction tends to flow ahead of the coarse aggregate and would increase segregation.

Consolidation is the process of compacting the fresh concrete by removing the entrapped air and causing it to mold without gaps and honeycomb around reinforcement and forms. Mechanical vibration is the most common technique for consolidation. This involves the use of internal or immersion-type vibrators often called *spud* or *poker vibrators*. A flexible shaft vibrator probe consists of a vibrating head connected to a driving motor by a flexible shaft. An unbalanced weight inside the head rotates at high speed, causing the head to rotate in a circular orbit. During vibration, the friction between aggregate particles is temporarily destroyed and the concrete behaves like a liquid; solids settle under the influence of gravity, and large entrapped air bubbles rise. The vibrated sphere of influence depends on the vibrator size and frequency of operation, and ranges between 3 and 14 inches for a vibrator head diameter between 0.75 and 3.0 inches, respectively. An insertion time of 5–15 seconds will usually provide adequate consolidation.

Improper vibration may be harmful to long-term durability of the concrete. Some effects of undervibration include honeycombing; an excessive amount of entrapped air voids that are called *bugholes*; sand streaks (which result when heavy bleeding washes mortar out from along the form, or due to segregation from striking reinforcement steel without adequate vibration); cold joints (from delay in placing adjacent concrete lifts or insufficient vibration of the initially placed concrete before the second lift is added); and placement lines or "pour" lines (darker lines between adjacent placement of concrete batches that are not knit together due to insufficient vibration).

When edge forms are used, they should be set accurately and firmly to the specified elevation. Slab edge forms are usually wood or metal. Wood forms should be well braced with steel stakes to maintain horizontal and vertical alignment, and they should not warp or bulge under the concrete pressure.

Rain protection of the concrete should be anticipated, and the contractor should be well prepared when it occurs during the concrete-placing operation. When rain occurs, all batching and

FIGURE 17.6 Slipform Paving.
Courtesy: Wouter Gulden, ACPA-SE Chapter.

placing operations should stop and the fresh or plastic concrete should be protected by plastic sheeting so the rain does not wash the cement from the surface or indent the surface. Resurfacing should proceed as needed after the rain stops and operations resume.

Screeding or strikeoff is the process of cutting off excess concrete to bring the surface in line with the desired slab elevation and grade. This is usually done using a template called a *straightedge*. This is moved across the concrete in a sawing motion with approximately a 1-inch surcharge ahead of the straightedge to fill in low areas. Sometimes a straight edge is combined with a vibrator (called a *vibratory screed*) to ease the consolidation and strikeoff effort.

Figure 17.6 shows the different stages of slipform paving.

17.5   FINISHING

Finishing operations include floating and texturing. A slight hardening or stiffening of the concrete surface is necessary before finishing can take place. Usually finishing can proceed when the bleed water sheen has evaporated and the concrete surface can withstand foot pressure with no more than a ¼-inch depression or indent.

Bullfloating or darbying is the process of smoothing the surface, leveling off high and low spots, and embedding large aggregate particles just below the slab surface. A long-handle bullfloat is typically used in pavement construction since the slabs are too wide to fully reach by a short-handle darby. For nonair-entrained concrete, a wood bullfloat is used. However, for air-entrained concrete, an aluminum or magnesium alloy tool should be used.

Consolidation, screeding, and bullfloating must be completed before excess bleed water collects on the surface of the concrete. Care should be taken to avoid overworking the concrete surface and preventing undesirable segregation and increasing the W/C ratio of the concrete surface by incorporating the bleed water. Increasing the surface W/C ratio has significant consequences on durability since it may reduce the strength and watertightness and increase the potential for crazing, dusting, and scaling.

Texturing is the process of increasing the roughness of the concrete surface to improve traction and reduce hydroplaning. Texturing of PCC pavements is done either by using a special broom or a tining machine, or by dragging a rough textured burlap or cloth along the surface. A tining machine drags a metal wire comb-like device along the transverse direction of the pavement. The tined grooves allow the water to escape from between the tires and the pavement surface, and reduce the potential for hydroplaning. Tining practices will vary by state agency; however, many require transverse grooves that are 3–5 mm (0.12–0.20 inches) deep, 3 mm (0.12 inches) wide, and spaced 12–20 mm (0.47–0.79 inches) apart (ACPA, 1995). At times, the area over the future joint locations is not textured in order to provide a good sawing and sealing surface.

Figure 17.7 shows some of the finishing operations.

17.6   JOINTING

After placing and finishing the concrete slab, joints are constructed to control cracking and to provide relief for concrete expansion caused by temperature and moisture changes. Joints are usually constructed using special saw-cutting equipment. If designed and constructed properly, joints in concrete pavements create vertical weakening planes in the concrete pavement to induce controlled cracks just below the saw cut. Properly constructed joints will be easier to maintain than uncontrolled random cracks. Once PCC joints have cured sufficiently, they should be cleaned and sealed with jointing compound to prevent noncompressible foreign debris from harming the concrete's contraction cycle. The FHWA HIPERPAVE software (FHWA, 2005) provides guidelines for timing joint construction based on concrete properties and environmental conditions.

FIGURE 17.7 Finishing Operations in PCC Pavement Construction.
Courtesy: Wouter Gulden, ACPA-SE Chapter.

The following guidelines from FHWA (1994) should be considered for proper joint construction procedures:

- Joint sawing should be timed correctly before stresses develop in the pavement that are large enough to cause cracking. These stresses are the result of restrained volumetric changes from both temperature and moisture changes in the new pavement.
- Joint construction should not begin until the new PCC pavement has gained enough strength to support the weight of the sawing equipment and operator, and to minimize excessive raveling due to the forces introduced by the cutting blade.

Figure 17.8 shows jointing operations in a PCC pavement.

17.7 CURING

Curing is the process of providing a satisfactory moisture and temperature condition that allows the concrete to develop its desirable properties. Proper curing will strongly affect the following properties: increase durability, strength, watertightness, abrasion resistance, volume stability, and resistance to cycles of freezing and thawing and to deicing chemicals. Pavement and other slabs are especially vulnerable to poor curing since strength development and resistance to freeze-thaw are especially affected at slab surfaces.

Curing methods and application timing significantly affect the development of concrete strength, permeability, and other durability-related characteristics. These curing procedures control the amount

FIGURE 17.8 Jointing.
Courtesy: Wouter Gulden, ACPA-SE Chapter.

and extent of moisture loss in the pavement, which significantly affect drying shrinkage stresses and cracking. Several curing methods are readily available for PCC pavement construction.

The most common curing method is the application of a liquid membrane. The objective of liquid membranes is to seal the surface and minimize moisture migration into the air. This in turn reduces the potential for drying shrinkage cracks. The two common methods used for applying liquid membranes are the sprayer bar and the handheld wand. The sprayer bar traverses the width of the finished concrete slab and applies the membrane through sprayer jets evenly spaced across the pavement width. The handheld wand is person controlled. The applied liquid membrane is designed to leave a white residue on the pavement to assist the operator to be assured of complete surface coverage. A single coverage for either method does not guarantee complete coverage. However, a double or triple application of liquid membrane can lead to an even-covered surface and lower the potential for moisture loss–related stresses and plastic shrinkage cracking. Figure 17.9 shows the application of curing material on a PCC pavement.

FIGURE 17.9 Application of Curing Compound to Minimize Evaporation during Curing in a New PCC Pavement.
Courtesy: Wouter Gulden, ACPA-SE Chapter.

Polyethylene sheeting is an effective method in retaining and trapping the moisture in PCC slabs. This minimizes evaporation, and thus drying shrinkage. The plastic sheeting must be well secured; otherwise, strong winds may cause the sheeting to become insecure and increase the potential for early-age damage.

In addition, because the sheeting cannot be placed until the PCC has hardened enough to sustain the disturbance from the sheeting application, a liquid membrane is usually applied first, with the sheeting placed later. Polyethylene sheeting also acts as a thermal insulator. This property can be beneficial if the construction site is subjected to rapid cooling following construction. However, it can be detrimental if used improperly. The insulation, in concert with the excess heat produced by fast-track PCC mixes, can result in too great a curing temperature, causing damage after placement.

17.8   TESTS FOR FRESH CONCRETE

17.8.1   CONSISTENCY

The slump test, ASTM C-143 (AASHTO T-119), is the most commonly accepted method used to measure the consistency of concrete (see Figure 17.10). The test is conducted by filling a metal conical mold 12 in. high with an 8-in. diameter base and a 4-in. diameter top. The slump cone is placed upright on a flat rigid surface, dampened, and filled in three layers of approximately equal volume. Each layer is rodded 25 times using a steel rod that is 5/8 in. in diameter and 24 in. long with a hemispherically shaped tip. Following the rodding, the last layer is screeded off, and the cone is slowly raised in approximately 5 ± 2 seconds. As the concrete subsides or settles to a lower elevation, the empty slump cone is inverted and gently placed next to the settled concrete (note: any additional vibration at this stage will induce additional subsidence in the slumped concrete). The "slump" is measured as the vertical distance the concrete settles, measured to the nearest 1/4 in. Since concrete loses slump with time, the entire slump test should be completed in 2.5 minutes. Note that if a portion of the molded concrete falls away or shears off, another test should be run on a different portion of the sampled concrete.

17.8.2   AIR CONTENT

A number of methods for measuring the air content of freshly mixed concrete are available. ASTM standards include the pressure method (C-231; AASHTO T-152), the volumetric method (C-173; AASHTO T-196), and the gravimetric method (C-138; AASHTO T-121). The pressure method

FIGURE 17.10 Slump Test.

FIGURE 17.11 Air Content Test.

(Figure 17.11) is based on Boyle's law, which relates pressure to volume. The air pressure meters are calibrated to read air content directly when a predetermined pressure is applied. The pressure meter method is the most commonly used method for determining air content in fresh concrete. The test requires less time than other methods, and the specific gravities for the concrete ingredients need not be known compared to the gravimetric method. The volumetric method is outlined in ASTM C-173 (AASHTO T-196) and requires removal of air from a known volume of concrete by agitating the concrete in an excess of water. The addition of alcohol accelerates the removal of air. The gravimetric method is based on determining the density (unit weight) of concrete. The measured density of concrete is subtracted from the theoretical density as determined from the absolute volumes of the ingredients, assuming no air is present (ASTM C-138 or AASHTO T-121).

17.8.3 DENSITY AND YIELD

The density (unit weight) and yield of freshly mixed concrete are determined in accordance with ASTM C-138 (AASHTO T-121). A calibrated known-volume bucket (usually ½ cu. ft.) is filled with the fresh concrete in approximately three equal layers and rodded 25 times per layer. The quotient of weight (or mass in SI units) and volume is the unit weight or density. The volumetric quantity of concrete produced per batch is called the *yield*, and is determined by summing the weights of all ingredients in the batch and dividing by the density of the fresh concrete. An approximate calculation of the volumetric air in concrete can be conducted by using the density and yield values provided the relative densities of the ingredients are known.

17.8.4 SAMPLING FRESH CONCRETE

It is critically important to obtain truly representative samples of freshly mixed concrete for quality control tests; otherwise, the test results will be misleading. Samples should be obtained and handled in accordance with ASTM C-172 (AASHTO T-141). It should be noted that a sample composed of two or more portions should not be taken from the very first or last portion of the batch discharge and should be protected from sunlight, wind, and other sources of rapid evaporation during sampling and testing.

QUESTIONS

1. What are the different stages of concrete pavement construction?
2. What are the two methods of paving concrete?
3. What is the importance of jointing?
4. What are the different types of steel provided, and why?
5. Why is curing important?

18 Maintenance and Rehabilitation of Pavements

Pavement Management Systems (PMS)

18.1 OVERVIEW

Maintenance and rehabilitation (M&R) of pavements should ideally be conducted through the use of pavement management, which can ensure the optimum use of tax dollars through the selection and use of the most cost-effective design, construction, and rehabilitation strategy. Pavement management involves all activities regarding the planning, design, construction, maintenance, and rehabilitation of pavements. It is based on the pavement management system (PMS), which consists of a set of tools or methods that help pavement managers to plan for constructing and maintaining highway or airport pavements in a serviceable condition over a given period of time.

Pavement information management systems (PIMS) or pavement management systems (PMS) were mentioned in the 1986 AASHTO *Design Guide*. This guide defined PMS, indicated its importance, and presented the steps needed to adopt and implement it. The guide also discussed the information needed, the importance of quantification of pavement performance and optimization in selecting strategies, different levels of PMS, and steps for improving PMS. The *Pavement Management Guide*, published by AASHTO in 2001, is the latest guide on PMS from AASHTO.

Pavement management is often considered as part of a bigger management scheme—such as asset management (by the U.S. Federal Highway Administration System)—and requires the consideration of a more holistic approach in the consideration, analysis, and implementation of highways. In the pavement management system, the implementation activities can be separated into three distinct parts: corrective maintenance, preventive maintenance, and rehabilitation. Corrective maintenance is provided in response to an existing problem, such as filling existing cracks; preventive maintenance is implemented in anticipation of a drop in the quality of the pavement, such as providing a thin overlay to extend the life of the pavement by another 5 years; and rehabilitation is adopted when a major structural improvement of the pavement is required, such as by recycling, placement of a new overlay, or total reconstruction.

PMS is implemented in two different levels—network and project. In the network level, PMS is utilized to select the best strategies for design, construction, and rehabilitation of all the pavements within an agency, to result in the best benefit-to-cost ratio within a given analysis period. In the project level, PMS is used to select the best design, construction, or rehabilitation alternative for a specific project within the network such that the project results in the maximum benefit-to-cost ratio over the given analysis period.

18.2 STEPS IN PMS

The information on the traffic, existing condition, environmental data, and construction history of the pavement network provides the basis of all subsequent activities in a PMS. Particular attention needs to be given to accurate but practical testing of existing pavements—the use of automated and nondestructive testing is gaining more and more favor. This information/data can be used to develop models for predicting the future condition of the pavements. This condition can be with respect to a conditioning index or residual life. Linked to this information is user input information on available M&R techniques, and their relative benefits and costs. The benefits and costs could be expressed simply in terms of extension of life or improvement in condition and money, or in terms of those factors as well as other fuel consumption of vehicles, user delays, and tire wear. The information collected and obtained by analysis up to this point can then be used to select the most appropriate M&R activity for any specific pavement at the most appropriate time so as to make the best use of the available budget. This step is carried out, in increasing order of sophistication, through ranking, prioritization, and optimization. This involves consideration of the "consequence" of adaptation of several alternative strategies on the condition of the different pavements. The optimization can be based on different criteria, for example, either on the concept of minimizing the total cost while keeping all pavements at or above a minimum condition, or on maximizing the total benefit, with the available budget in the PMS. Finally, mathematical modeling is used to select the most optimum strategy.

Therefore, the steps in PMS can be broadly divided into the following:

1. Collect information on pavements (distress survey).
2. Set up criteria for making decisions.
3. Identify alternative strategies.
4. Predict cost and performance of alternative strategies, and compare them.
5. Select and implement the most cost-effective strategy.

18.3   DIFFERENT PMS APPROACHES

Although the same in principle, there are different types of pavement management systems software, with appropriate models available to the user. For example, the World Bank has developed the High-way Development and Management System (HD-4) as a tool to evaluate pavement construction and maintenance strategies using technical and economical considerations. Its successful use depends on the calibration of pavement performance prediction models to local conditions for each agency.

MicroPaver, based on the work by the U.S. Army Corps of Engineers (n.d.), is a pavement management software which can be used for selection of cost-effective maintenance and repair (M&R) strategies for highways and airfield pavements. It allows the user to prepare a network inventory with condition ratings, develop pavement condition deterioration models, predict the condition of a pavement at a specific time and determine M&R needs, and evaluate the effect of different budget strategies. MicroPaver (also referred to as *Paver*) can be obtained from the University of Illinois at Urbana-Champaign, Technical Assistance Center (phone: 800-895-9345; e-mail: techctr@uiuc.edu). The key steps in the PMS are illustrated below, with reference to the MicroPaver approach.

A complete PMS consists of a series of steps, which starts with setting up a pavement inventory for the network that needs to be analyzed (for example, all of the pavements in a city or county or in an airport). The network is made up of several branches, each of which consists of several pavement sections. Each section should have specific information regarding its inventory, maintenance, and inspection information. Images of distresses could be stored and retrieved for each section. Input of work, traffic, and test (nondestructive and destructive) data could also be stored for each section. The results of a pavement inspection are then input for each "sample unit," several of which make up one section. Using inbuilt tables of distresses (which are also coded with two-digit numbers for speedy entry) specific to the type of the pavement being inspected, the type, severity, and quantity of the distresses observed during inspection could be input. Video inspection data could also be input. An example of a pavement survey form as used in MicroPaver is presented in Table 18.1.

The next step is the determination of the Present Condition Index (PCI) of the entire section, using the inbuilt condition calculator in the software (for an example, see Table 18.2). PCI is a

TABLE 18.1
Example of Pavement Survey Form

Distress	Description	Severity	Quantity	Units
1	Alligator cracking	L	6	Sq ft
1	Alligator cracking	M	30	Sq ft
4	Bumps/sags	L	28	Ft
7	Edge cracking	L	142	Ft
7	Edge cracking	M	15	Ft
10	Longitudinal/transverse cracking	L	26	Ft
10	Longitudinal/transverse cracking	M	27	Ft
10	Longitudinal/transverse cracking	H	28	Ft

numerical indicator that rates the surface condition of the pavement, and is used as a measure of the present condition of the pavement. It can be used as a basis for establishing maintenance and repair needs. The procedure for determination of PCI is explained in ASTM D-6433-03 for roads and in ASTM D-5340-04 for airfields. The procedure consists of the following: conducting a survey of distress (quantity and density); determining a "deduct value" for each distress and computing the total corrected damage value, according to guidelines given in the standard; and then subtracting the total corrected damage value from 100, to obtain the PCI at any time. The PCI can range from 0 to 100, corresponding to a condition of "failed" to "excellent," respectively.

MicroPaver uses a hierarchical structure with networks, branches, and sections in its inventory management. The results of an inspection can be input through an interface, and the online distress guide can be consulted to aid in the inspection and input. The distress type, quantity, and severity are combined to determine the PCI and then the condition of the pavement.

TABLE 18.2
Present Condition Index (PCI) of Several Sections in a Network

Age at Inspection	PCI	Model	Difference	Status	Network ID	Branch ID	Section ID	Surface	Rank	Inspection Date
9	61	55	6		1	IFARB	1	AAC	S	10/16/1992
1	100	95	5		1	IFARB	1	AAC	S	3/24/1984
10	64	49	15		1	IFARB	1	AAC	S	10/12/1993
8	72	60	12		1	IFARB	1	AAC	S	10/16/1991
7	80	65	15		1	IFARB	1	AAC	S	5/23/1990
6	79	70	9		1	IFARB	1	AAC	S	6/6/1989
4	89	80	9		1	IFARB	1	AAC	S	5/6/1987
3	94	85	9		1	IFARB	1	AAC	S	10/13/1986
2	95	90	5		1	IFARB	1	AAC	S	9/28/1985
1	100	95	5		1	IFARB	1	AAC	S	9/30/1984
11	50	44	6		1	IFARB	1	AAC	S	11/17/1994
1	94	95	1		1	IINTE	1	AAC	P	9/30/1984
8	44	60	16		1	IINTE	1	AAC	P	10/16/1991

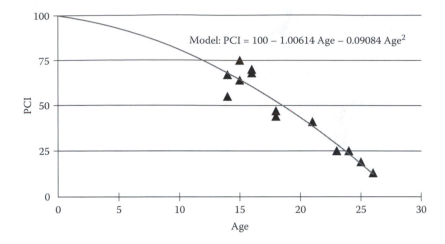

FIGURE 18.1 Change in PCI with Age of Pavement.

The results give the condition index and values along with distress summaries (and projected distresses). Based on minimum condition criteria or projected deterioration rates, an inspection schedule report can be generated. Finally, summary charts can be used to develop a graph and compare any two attributes of the pavement network. The software allows the user to query the whole database or for specific sections to generate specific reports. Geographical Information Systems (GIS) tools have been integrated into MicroPaver from its 5.1 version, which allows the user to link individual pavement sections to available GIS databases and display summaries and PCI reports in graphical format.

For the purpose of building predictive models relating age with condition/performance, Paver identifies and groups families of pavements which are constructed in the same way and are subjected to similar conditions such as traffic and environment. This process is called *family modeling*. Each pavement section is assigned to a family, and the family model is used to predict the section's future condition. Options of using unique knowledge about specific pavement sections, and viewing (such as outliers, equations, and coefficients of regression) and selecting the statistics (such as constraints) used to generate the models, are available. An existing model can be modified or updated with new data. The data are used to form predictive equations/models relating age with PCI, and then view the condition of the entire pavement network or any specific portion of it. The models (see an example in Figure 18.1) can be used to view the condition of any pavement section (condition analysis) at any given time, as well as to view the deterioration of the condition of the pavement with age—the results are provided in the condition analysis report.

Several forms of information can be viewed in this report, including the average condition of pavement for each year in the reporting period, and a histogram for each year with sections grouped into several PCI groups (which range from excellent to failed). *Critical PCI* is defined as the PCI value at which the rate of PCI loss increases with time, or the cost of applying localized preventive maintenance increases significantly.

18.3.1 CRITERIA FOR MAKING DECISIONS

MicroPaver provides a "work plan" tool that helps the user in planning, scheduling, budgeting, and analyzing alternative M&R activities, for specific sections and analysis periods. There are three ways in which the analysis can be conducted: (1) critical PCI methods, which optimize the M&R activity against a specific available budget or determine the budget needed to maintain a specified condition level; (2) minimum condition, which works by rationing M&R by pavement condition;

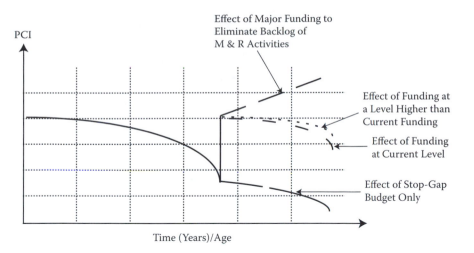

FIGURE 18.2 Conceptual Change in PCI with Time for Different Budget Scenarios.

and (3) consequence model, which works by measuring the impact of localized M&R action over the first year.

Inbuilt in the Paver database is a list of global maintenance work types, along with application interval, improvement in the condition every year (delta), and cost (which can be edited as required). Cost of major M&R is determined by the PCI at the time the work is performed. *Global M&R* is defined as activities applied to entire pavement sections with the primary objective of slowing the rate of deterioration. This policy is applied to pavements above the critical PCI. The work quantity is counted using "multipliers" listed for specific distress types and quantities. Paver allows the user to evaluate the effect of a selected budget, and compare it with that for unfunded M&R on the condition of a pavement (Figure 18.2). Options are also available to overlay the section condition plot with family condition plots, view the difference between the predicted condition and that with the M&R option, and the difference in condition from year to year.

18.4   DISTRESS SURVEY

Distress surveys involve the measurement and evaluation of type, extent, and intensity of various types of distresses—surface defects, rutting or distortion, cracking, and patching. Such surveys can be conducted by an individual walking along a section of the pavement or by using multiple cameras from a moving specialized vehicle. An example of a vehicle used for such surveys is shown in Figure 18.3. The vehicle is equipped with sensors and video cameras for continuous recording of different types of distresses such as cracks, which can be analyzed by pavement managers. Guidelines for identifying and quantifying different types of distresses are provided in ASTM D-6433 for roads and ASTM D-5340 for airfields.

Structural evaluation of pavements can be done in a number of different ways. In general, the deflection or the curvature of a pavement when subjected to a specific load is used for such measurement. Such evaluation can be first done with large spacing, followed by testing at closer locations to quantify the structural capacity, taking into consideration the variability of the results. The falling weight deflectometer can be used according to ASTM D-4694; the equipment is shown in Figure 18.4. The basic principle is the use of impulse load to measure the resulting deflection in the pavement—the higher the deflection for a specific load, the worse is the structural condition of the pavement.

FIGURE 18.3 Automatic Road Analyzer.

Skid resistance of pavements is measured as an evaluation of pavement safety. Skid resistance is usually measured in terms of friction factor, f, defined as F = F/L, where F is frictional resistance to motion in the plane of the interface, and L is the load acting perpendicular to the interface. Since the friction factor is dependent on many parameters such as tire type, speed, temperature, and water film on pavement thickness, standards are followed to measure it. Measurements can be done according to ASTM E-274 (using a full scale tire). In this test a locked tire is towed by a vehicle on a wet pavement (water is sprayed in front of the tire), and friction force between the tire and the pavement surface and the speed of the vehicle are recorded (Figure 18.5). The testing results in the determination of a skid number (SN), which is calculated as follows:

$$SN = \frac{F}{W} * 100$$

FIGURE 18.4 Falling Weight Deflectometer.

FIGURE 18.5 Skid Tester.
Courtesy: David Morrow, Dynatest Consulting Inc.

where:
   F = tractive horizontal force applied to the test tire at the tire-pavement contact patch, lbf
   W = dynamic vertical load on the test wheel, lbf

Note that the type of tire used in the test (ribbed or smooth) must be noted in the results by following the appropriate designation in the standard. In general, instrumentation in modern equipment automatically and continuously measures the SN for a specific test time period and averages it for reporting.

For pavements such as those in airports, under icy conditions, or in winter, spot measurement of frictional properties can also be conducted with the use of a specialized test vehicle, according to ASTM E-2101. This is important to detect otherwise hidden effects of contaminants on the pavement surface that can lead to unsafe conditions. In this test (valid only for icy conditions where the depth of loose snow does not exceed 1 in.), the steps consist of accelerating a vehicle over the pavement and applying brakes to lock all the wheels, and measuring the deceleration (which is related to the friction between the tire and the pavement surface) of the vehicle during braking.

For airports, the International Runway Friction Index (IRFI) is reported as a harmonized value of the friction characteristics of the pavement (ASTM E-2100). The parameter is calculated from friction numbers determined by local friction devices through established correlations, and can be used to monitor winter frictional characteristics of runways for maintenance.

Surface roughness can be measured using different types of profilometers, such as through a multiwheel > 23-ft-long profilometer (ASTM E-1274). The test consists of moving the profilometer over the pavement at less than 3 mph, and determining the roughness and rate of roughness of the pavement as the height of each continuous "scallop" or excursion of the surface records above and below the base reference level. Roughness can also be measured by vehicular response to pavement roughness (ASTM E-1082). In this type of device, the response of a spring mass system supported by a vehicle suspension system in response to pavement roughness is measured using sensors.

18.5   MAINTENANCE AND REHABILITATION OF ASPHALT PAVEMENTS

The choice between maintenance and rehabilitation is generally made on the basis of the existing surface and structural condition of the pavement. For example, if the surface condition is above a certain level, maintenance can be selected, whereas if it is below that level, both maintenance and rehabilitation can be considered. For the latter case, depending on the structural condition of the pavement,

TABLE 18.3
Maintenance and Rehabilitation Methods

Distress	Maintenance	Rehabilitation	Appropriate Recycling Process
Fatigue/alligator cracking		Reconstruction, thick hot mix asphalt (HMA) layer	Cold in-place recycling (CIR), full depth reclamation (FDR) with overlay
Bleeding	Chip seal		
Block cracking	Slurry seal, chip seal, sealing	Overlay	CIR, FDR with overlay
Corrugations	Thin overlay	Thick overlay	CIR with thin overlay
Joint cracks	Seal		
Polished aggregate	Chip seal, slurry seal, open graded friction course		
Potholes	Patching, full depth repair		CIR
Slippage cracks	Thin overlay on milled surface	Thick overlay	CIR with thin overlay
Thermal cracks	Thin overlay		Hot in-place recycling (HIR) with/without thin overlay
Rutting	Thin overlay	Thick overlay	CIR with thin overlay
Raveling	Chip seal, slurry seal, fog seal, sand seal, thin overlay		HIR with/without thin overlay

Note: For overlays on pavements with surface defects such as corrugations, the pavement must be milled to the depth of good mix prior to the application of the overlay.

either a significant maintenance activity (such as a thin overlay) or rehabilitation (reconstruction) can be selected. Table 18.3 shows a general guideline for the selection of different techniques.

18.5.1 MAINTENANCE

Proper maintenance of pavements is vital for the safety and comfort of the traveling public, and the overall economy of a nation. Furthermore, appropriate maintenance work results may result in significant enhancement of pavement life and/or lowering of rehabilitation cost in the future. The importance of proper maintenance has been recognized worldwide, and steps are being taken to develop better techniques for selecting appropriate techniques and new techniques. For example, in the United States since 1997, the FHWA, along with the Foundation of Pavement Preservation (FPP) and AASHTO, have vigorously started the adoption of good practices of pavement maintenance under the common theme of "pavement preservation" (www.fhwa.dot.gov/preservation, FHWA, n.d.; www.fp2.org, FPP, n.d.). A wealth of information can be accessed from the reference library section of the National Center for Pavement Preservation (NCPP, n.d.) at http://www.pavement-preservation.org/reference/. Some of the more important corrective and preventive maintenance activities are discussed below.

18.5.1.1 Primary Corrective Maintenance Activities

Crack *sealing* is the placement of specialized materials into working cracks (horizontal movement > 2 mm) to reduce the intrusion of incompressible materials and prevent intrusion of water into the underlying pavement. The sealant material must be an agency-approved product, selected on the basis of temperature and traffic, and applied properly (for example, with a backer-rod) on a clean and dry surface. The preparation equipment (such as a melter for hot applied sealant materials) must be in good working condition, and the operation must be completed in dry weather conditions with minimum temperature (typically 4°C and rising). Crack *filling* is the placement of materials into nonworking cracks (horizontal movement < 2 mm) to substantially reduce infiltration of water and reinforce adjacent pavement.

For pothole repairs, *pothole patches* must be made with good-quality HMA or cold mix, using a fine mix with adequate binder for workability, compaction, and durability, after cleaning, drying, and applying a tack coat in the existing potholes. The patching material could be of locally available cold mix, mix produced according to specifications, or propriety cold mix. The compatibility between aggregate and asphalt binder must be ensured for mixes. In the conventional method, a pothole is patched by placing the materials and compacting, say, with truck tires to keep at least a 3 to 6 mm crown on the pothole. Prior to placing the materials, the water and loose debris from the pothole must be removed. Cutting the pothole sides to nearly vertical surfaces, applying tack coat on the sides, and utilizing a small roller or plates for compaction would result in a better patching. A large variety of spray injection techniques are now becoming available for pothole patching.

18.5.1.2 Primary Preventive Maintenance Activities

Slurry seal: A slurry seal consists of a mixture of well-graded fine aggregates and mineral filler with dilute asphalt emulsion, often with additives or modifiers such as hydrated lime. The mix is generally produced by a slurry machine at the site and applied to the pavement with a squeegee. Slurry seal is effective in sealing minor cracks, reducing the potential of raveling, preventing further oxidation of the asphalt of the surface layer, and improving the friction properties of the pavement. Slurry seals with specific gradations are applied for specific pavements with different types of traffic (low, moderate, and heavy). The process requires a curing period of 2 to 6 hours, is not effective in pavements with heavy cracking, and should be used with caution in areas with superelevation where the equipment may not have good control over the flow of the slurry. Compaction by a light rubber-tired roller may be needed.

Chip seal: A chip seal consists of an application of asphalt binder or rapid-setting emulsion followed with an application of aggregate layer (Figure 18.6). If multiple layers are used (such as double or triple seal coats), finer gradations are used in each successive layer. Precoated aggregates could also be used, and one-size aggregates are often used. Compaction of the aggregate layer is required by a steel-wheeled or rubber-tired (preferred) roller. The minimum pavement temperature for the use of a chip seal is 15°C. Chip seals can be used as only wearing layers in pavements with light traffic. If aggregates are not properly embedded, windshield damage can result. A *rubberized* chip seal contains

FIGURE 18.6 Chip Seal.
Courtesy: Maureen Kestler, USDA Forest Services.

ground tire rubber in addition to the usual components of a chip seal. Better resistance
against climate- and traffic-induced stress is obtained in this seal, as well as resistance
against reflective and minor fatigue cracks. It is referred to as a stress-absorbing membrane (SAM) or stress-absorbing membrane interlayer (SAMI) when used underneath an
HMA overlay.

Sandwich seal: This is the application of a chip seal using two layers of aggregate with one
application of asphalt in between. Enhanced frictional properties can be obtained with this
method.

Cape seal: A cape seal (originally from Cape Province of South Africa) consists of a chip
seal covered with a slurry seal. It prevents the occurrence of loose stones in chip seals, and
could be used in pavements with high traffic volumes. It does require a curing period of 2
to 6 hours.

Fog seal: A fog seal consists of the application of a diluted asphalt emulsion (slow or medium
setting) without the application of any aggregate. It is used for preventing raveling and for
sealing surfaces to resist oxidation of the surface layer. The temperature should be above
16°C for application, and the layer needs some curing time before opening to traffic. The
pavement should be porous enough to absorb some of the asphalt emulsion, and until some
of the emulsion of the surface is worn away by traffic, friction may be reduced. A better
friction can be obtained by the use of *sand seal*, in which the sand is applied over a rapid- or
medium-setting emulsion layer.

Microsurfacing: Microsurfacing consists of the application of a mixture of high-quality aggregates and polymer-modified emulsion binder, with additives. The components are mixed in
a traveling pug mill, and areas with traffic do not require any compaction. Microsurfacing
is used to seal the pavement and prevent raveling. It can also be used to resist oxidation of
the surface layer, improve friction, and fill in minor ruts. This process uses relatively fast-
setting emulsions and requires mix design for determination of the asphalt content, and
sophisticated equipment for laydown.

Thin HMA overlay: An open graded friction course can be used. Generally such layers are ½
to 1 in. thick, with gap-graded aggregates, and increasingly with polymer-modified asphalt
binders. These layers are used to facilitate quick drainage of water and improve tire-
pavement contact and hence friction. Also known as *porous friction courses* (PFC), these
layers can contain ground tire rubber as well as high asphalt contents to improve adequate
durability. Generally these layers do not require any compaction, but do require sealing of
cracks, if any, as well as relatively low voids in the underlying layers.

Ultrathin friction course: This is a procedure in which an HMA layer made with polymer-
modified asphalt and gap-graded aggregate is placed on a heavy tack coat of polymer-
modified emulsion asphalt. Products such as Novachip are applied in one pass (binder,
aggregates, and screeding) by a single specialized piece of equipment. It is used for sealing
the surface and minor cracks, increasing friction, and reducing tire noise.

18.5.2 RECYCLING

The Asphalt Recycling & Reclaiming Association's (ARRA) publication *Asphalt Recycling Manual*
(ARRA, 2003; and see www.arra.org) is a good source of information on asphalt pavement recycling. The common types of recycling operations include hot mix recycling, hot in-place recycling
(HIR), cold in-place recycling (CIR), and full depth reclamation (FDR). Of these, hot mix recycling
is used very commonly for producing hot mix asphalt, which can be used as overlays in preventive
maintenance or as thick layers in rehabilitation. Hot in-place and cold in-place recycling are commonly used for preventive maintenance operations, whereas full depth reclamation is generally used
for rehabilitation work. The preventive maintenance applications are discussed first, followed by the
rehabilitation methods.

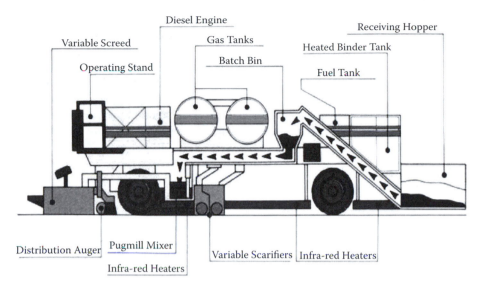

FIGURE 18.7 Schematic of Single-Pass Hot In-Place Recycling Equipment.
Courtesy: Wirtgen, GmbH.

18.5.2.1   Hot In-Place Recycling

Hot in-place recycling has been described as an on-site, in-place method that rehabilitates deteriorated asphalt pavements and thereby minimizes the use of new materials. Basically, this process consists of four steps: (1) softening of the asphalt pavement surface with heat; (2) scarification and/or mechanical removal of the surface material; (3) mixing of the material with recycling agent, asphalt binder, or new mix; and (4) laydown and paving of the recycled mix on the pavement surface. The primary purpose of hot in-place recycling is to correct surface distresses not caused by structural inadequacy, such as raveling, cracks, ruts and holes, and shoves and bumps. It may be performed as a single-pass operation or a multiple-pass operation. In a single-pass (Figure 18.7 and Figure 18.8)

FIGURE 18.8 Hot In-Place Recycling.

operation, the virgin materials are mixed with the restored reclaimed asphalt pavement (RAP) material in a single pass, whereas in the multistep process, a new wearing course is added after recompacting the RAP materials.

The advantages of hot in-place recycling are that elevations and overhead clearances are preserved, it is comparatively economical, and it needs less traffic control than the other rehabilitation techniques. This process can also be used to recoat stripped aggregates, reestablish crown and drainage, modify aggregate gradation and asphalt content, and improve surface frictional resistance. Hot in-place recycling is usually performed to a depth of 20 mm to 50 mm (3/4 to 2 in.), with 25 mm (1 in.) being a typical depth.

18.5.2.2 Cold Recycling

Cold recycling can be divided into two main parts—cold in-place recycling and central plant (cold mix) recycling. Cold in-place recycling can further be divided into two parts—cold in-place recycling (limited to the asphalt surface or/surface plus binder layer) and full depth reclamation (recycling of asphalt layer plus part of the granular base material). Cold milling is also used for obtaining materials for hot mix recycling (discussed later in this chapter).

18.5.2.2.1 Cold Milling

The advent of cold milling (cutting) has revolutionized the recycling of asphalt pavements. Cold milling has been defined as the method of automatically controlled removal of pavement to a desired depth with specially designed equipment, and restoration of the surface to a specified grade and slope, free of bumps, ruts, and other imperfections. The Asphalt Recycling & Reclaiming Association guideline specifications for cold milling require that the milling machine be power operated, self-propelled, and, self-sufficient in power, traction, and stability to remove a thickness of HMA surface to a specified depth. The alternative to cold milling is ripping and crushing operations with earthmoving equipment, scarifiers, grid rollers, or rippers, which are not commonly used today.

The modern cold-milling equipment has tungsten carbide teeth on drums, with variable cutting width for a variety of pavements/lanes, and excellent maneuverability for difficult milling situations like around utility inspection covers in city streets. Figure 18.9 shows a few examples of milling equipment. In many cases, such equipment has been integrated with a system of spraying recycling/ stabilizing materials, mixing, and compacting, in recycling equipment, as shown in Figure 18.10.

18.5.2.2.2 Cold In-Place Recycling

Cold in-place recycling (CIR) is defined as a rehabilitation technique in which the existing pavement materials are reused in place. The materials are mixed in-place without the application of heat. The reclaimed asphalt pavement (RAP) material is obtained by milling (cutting), planing, or crushing the existing pavement. Virgin aggregate or recycling agent or both are added to the RAP material, which is then laid and compacted. The use of cold in-place recycling can restore old pavements to their desired profile, eliminate existing wheel ruts, restore the crown and cross slope, and eliminate potholes, irregularities, and rough areas. It can also eliminate transverse, reflective, and longitudinal cracks. Some of the major reasons for the increased use of cold in-place recycling are the increased scarcity of materials, particularly gravel and crushed rock; the method's high production rate and potential of cost savings; minimum traffic disruption; the ability to retain the original profile; the reduction of environmental concerns; and a growing concern for depleting petroleum reserves.

The steps in cold in-place recycling consist of preparation of the construction area, milling of the existing pavement, the addition of recycling agent and virgin materials, laydown, compaction, and placement of the surface course.

The addition of new aggregates may not be necessary in some projects. At present, two different methods are used for cold in-place recycling: the single machine and the single-pass equipment.

Milling Drum

Close-up View of Milling Drum in Recycler WR 4200 Showing Teeth

Schematic of Changeable Tooth Holder, Lower Part Welded to Drum Upper Part Exchangeable

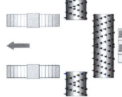

Closed Milling Width, 2.80 m

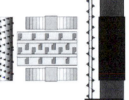

Maximum Milling Width, 4.2 m

Schematic of Variable Width Milling

Extension Cutters   Position of Foamed Asphalt Bar

Cutter   Twin-Shaft Pugmill

Milling (Cutting) Drums in WR 4200

Cutting Around Utility on City Street

FIGURE 18.9 Modern Cold-Milling Equipment.
Courtesy: Mike Marshall, Wirtgen GmbH.

The single machine or single-pass equipment is capable of breaking, pulverizing, and adding recycling agents in a single pass (Figure 18.11). The single-pass equipment train consists of a series of equipment, each capable of a particular operation. The usual components are a cold-milling machine, portable crusher, travel-plant mixer, and laydown machine.

Sometimes cold central plant recycling is conducted to make use of existing RAP stockpiles and/or make recycled mixes for later use. Sophisticated equipment is available to prepare such mixes, using one or more recycling additives, as shown in Figure 18.12.

18.5.2.2.3 Full Depth Reclamation

Full depth reclamation has been defined as a recycling method where all of the asphalt pavement section and a predetermined amount of underlying materials are treated to produce a stabilized

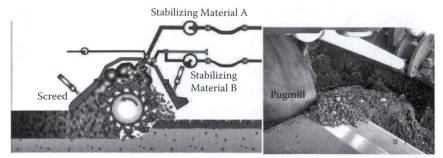

Integrated Cutting, Spraying, Mixing and Compaction Unit with Options for Using
Multiple Stabilizing Materials

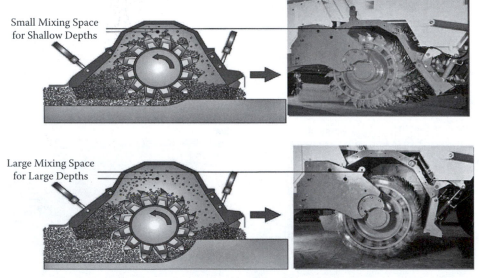

Variable Mixing Space

FIGURE 18.10 Combined Milling and Recycling Equipment.
Courtesy: Mike Marshall, Wirtgen GmbH.

base course. Different types of additives, such as asphalt emulsions and chemical agents such as
calcium chloride, Portland cement, fly ash, and lime, are added to obtain an improved base. The
five main steps in this process are pulverization, introduction of additive, shaping of the mixed
material, compaction, and application of a surface or a wearing course. If the in-place material
is not sufficient to provide the desired depth of the treated base, new materials may be imported
and included in the processing. This method of recycling is normally performed to a depth of
100 to 300 mm (4 to 12 in.). The major advantages and benefits of full depth reclamation are as
follows:

1. The structure of the pavement can be improved significantly without changing the geometry of the pavement and shoulder reconstruction.
2. It can restore old pavement to the desired profile, eliminate existing wheel ruts, restore
   crown and slope, and eliminate potholes, irregularities, and rough areas. Pavement-
   widening operations can also be accommodated in the process. A uniform pavement
   structure is obtained by this process.

Project Information

Traffic	Existing Pavement Structure	Distress in Existing Pavement	CIR Method Used
38,500 vehicles Per Day, 16% Trucks	150-1200 mm HMA, 200 mm Cement Treated Base, 300 mm Aggregate Subbase	Block Cracking, Wheel path potholes, Fatigue Cracking, Severe Aging of the Surface Mix	Recycle with 1.5% Cement and 2.5% Foamed Asphalt; Recycling Depth: 60 mm for shoulders, 100 mm for Mainline; Apply Fog Seal with Sand, with application of diluted emulsion at the rate of 0.008-0.0100 liters/sq m

Cracks in Existing Pavement

Asphalt Tanker
WM1000 Cement Slurry Mixer
WR 4200 Recycler

Recycling Train

WR 4200 Recycler

Recycling with Traffic in Adjacent Lane

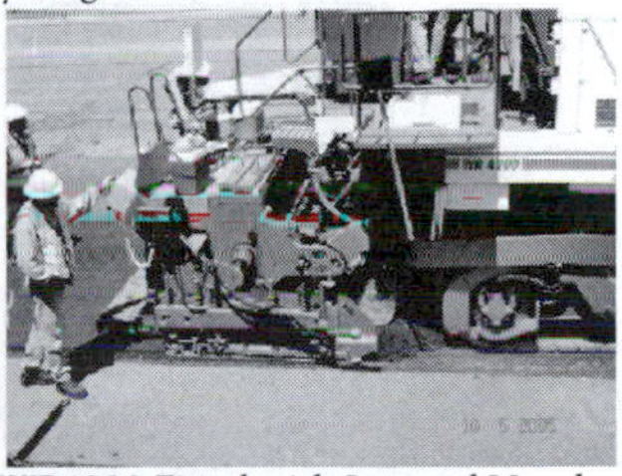

WR4200 Fitted with Integral Vogele AB500 Tamping and Vibrating Paving Screed

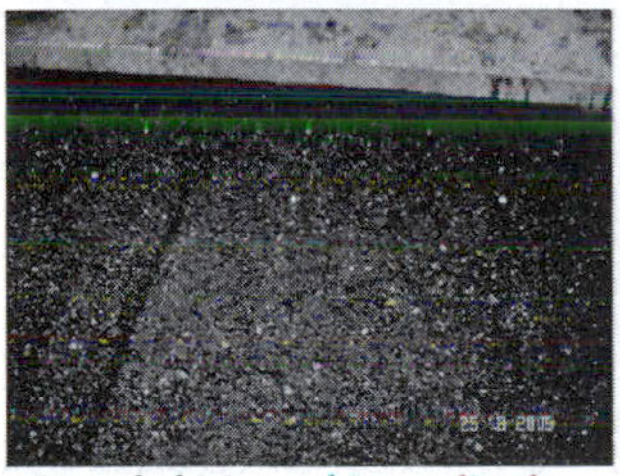

Recycled Material Immediately Behind the WR 4200 Paving Screed

Compaction with Steel Drum Roller

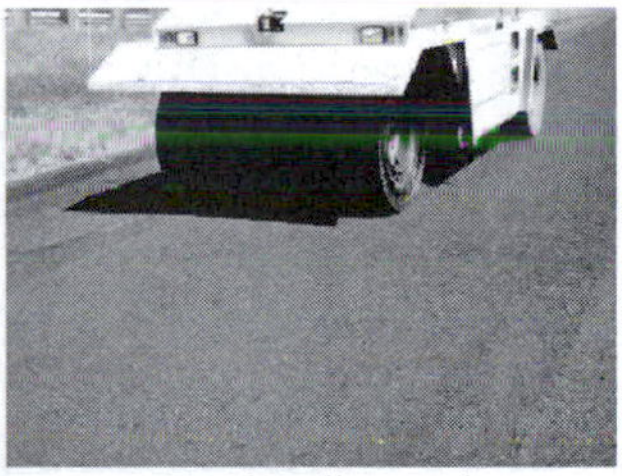

Finish Rolling with Rubber Tired Roller

Recycled Shoulder   Recycled No.1 Lane   Mill and Fill No.2 Lane   Recycled Shoulder

4 Hours After Completion of Recycled Lane 1.
Traffic Running on Recycled Mat with Fog/Sand Seal

FIGURE 18.11 Example of a Cold In-Place Recycling Project.
Courtesy: Mike Marshall, Wirtgen, GmbH.

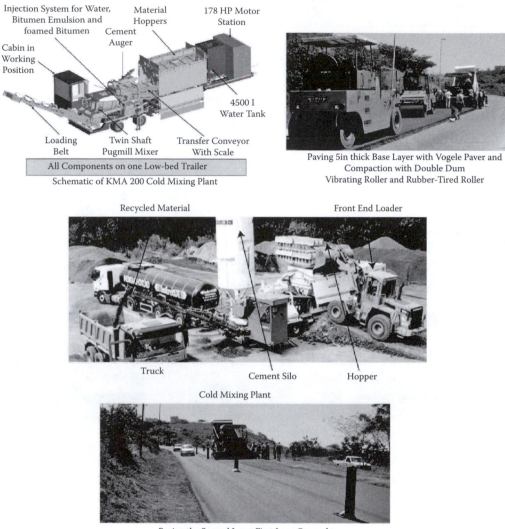

Cold mixing Plant-KMA 200
Components

Injection System for Water,   Material   178 HP Motor
Bitumen Emulsion and   Hoppers   Station
 foamed Bitumen   Cement
                       Auger
Cabin in
Working
Position

                                            4500 I
                                           Water Tank

    Loading   Twin Shaft   Transfer Conveyor
    Belt   Pugmill Mixer   With Scale

    All Components on one Low-bed Trailer

       Schematic of KMA 200 Cold Mixing Plant

Paving 5in thick Base Layer with Vogele Paver and
         Compaction with Double Dum
  Vibrating Roller and Rubber-Tired Roller

        Recycled Material   Front End Loader

          Truck   Cement Silo   Hopper

                    Cold Mixing Plant

          Paving the Second Lane: First Lane Opened to
              Traffic Immediately After Compaction

FIGURE 18.12 Example of a Central Plant Cold Recycling Project.
Courtesy: Mike Marshall, Wirtgen, GmbH.

3. It can eliminate alligator, transverse, longitudinal, and reflection cracking. Ride quality
   can be improved.
4. Frost susceptibility may be reduced.
5. The production cost is low, and only a thin overlay or chip seal surfacing is required on
   most projects.
6. Engineering costs are low.
7. Materials and energy are conserved, and air quality problems resulting from dust, fumes,
   and smoke are eliminated. The process is environmentally desirable, since the disposal
   problem is avoided.

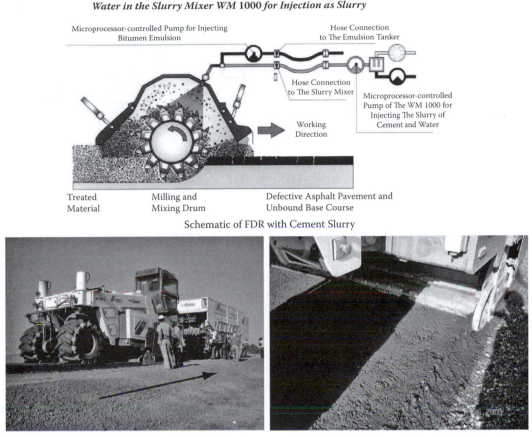

*Variant: Injection of Emulsion; Premixing Cement and
Water in the Slurry Mixer WM* **1000** *for Injection as Slurry*

Schematic of FDR with Cement Slurry

FDR Using WR 2500S Recycler and WM 100 Cement
Slurry Equipment

300 mm FDR with Cement Slurry with 5.8% Cement

FIGURE 18.13 Example of a Full Depth Reclamation Project with Cement Slurry.
Courtesy: Mike Marshall, Wirtgen, GmbH.

Full depth reclamation has been recommended for pavements with deep rutting, load-associated cracks, nonload-associated thermal cracks, reflection cracks, and maintenance patches such as spray, skin, pothole, and deep hot mix. It is particularly recommended for pavements having a base or subgrade problem. In this method, the first step is to rip, scarify, or pulverize or mill the existing pavement to a specified depth. The resulting material can be processed further for size reduction and mixed with recycling agents and new materials, if required. These include the multiple-step sequence, two-step sequence, single machine, and single-pass equipment train. Modern equipment is available for single-pass FDR, with the option of using water, emulsion, foamed asphalt, or cement slurry as the recycling additive. Figure 18.13 shows an example of an FDR process with cement slurry.

Foamed asphalt is being used increasingly in FDR. Foaming facilitates better dispersion of the asphalt into the materials to be recycled. A small amount of water is sprayed into hot asphalt as it is mixed with pulverized recycled pavement and soil. As the hot liquid and water mix, the liquid expands in a mini-explosion, creating a thin film of asphalt with about 10 times more coating potential. In modern single-pass equipment, foamed asphalt is created within the equipment in a separate foaming chamber and is directly added to the pulverized road material. An example of such equipment is shown in Figure 18.14.

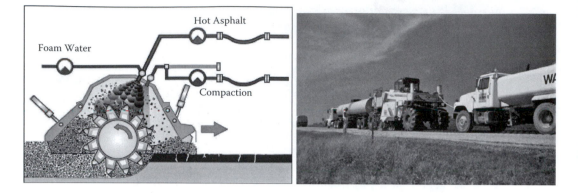

FIGURE 18.14 Schematic and Use of Foamed Asphalt Recycling Equipment.
Courtesy: Mike Marshall, Wirtgen, GmbH.

18.5.2.3   Hot Mix Recycling

Hot mix recycling has been defined as a method by which reclaimed asphalt pavement is combined with new aggregate and an asphalt cement or recycling agent to produce HMA. The RAP may be obtained by pavement milling with a rotary drum cold-milling machine or from a ripping/crushing operation.

The objective of the next step, crushing, is to reduce the RAP to the maximum acceptable particle size. One example of such a limit is that at least 95% of the RAP passes the 2 in. sieve. Cold-milling machines can crush the RAP in place, whereas in the ripping/crushing operation, front-end loaders are generally used to break up the pavement material so that it can be loaded into a truck for crushing at a central plant. The amount of aggregate degradation by cold milling is a function of the aggregate top size and gradation of the HMA pavement. For crushing in a central plant, different types of crushers are available, for example, compression crushers and impact crushers.

A "RAP breaker" or "lump breaker" is also used between the bin and the belt for size reduction. Impact crushers are most widely used in recycling. To produce a homogeneous RAP product, the RAP material is first blended thoroughly with a front-end loader or a bulldozer, and then crushed to downsize the top stone size in the RAP to one smaller than the top size in the HMA being produced (for example, 16 mm for a 19.5 mm top-size mix). This is to ensure that the asphalt aggregate blend is broken as much as possible and there is no oversize material.

RAP from different sources and containing different asphalt content and aggregates with different gradations should be stockpiled separately. RAP can be stockpiled either before or after processing, and a front-end loader or a radial stacker can be used for the purpose. The two major problems associated with stockpiling are consolidation and moisture retention. In the past, it was believed that low, horizontal stockpiles are better than high and conical stockpiles, which can result in reagglomeration of RAP. However, experience has actually proven that large, conical stockpiles are better and that RAP does not recompact in large piles. Actually, there is a tendency to form a crust over the 20–25 cm (8–10 in.) of pile depth. This crust can be easily broken by a front-end loader. Also, the crust tends to shed water and prevent the rest of the pile from recompacting. In the case of a low, horizontal stockpile, this 20–25 cm (8–10 in.) crust is also present.

Modifications are required in the batch plant to recycle RAP since attempts to introduce RAP directly with the virgin aggregates result in excessive smoke and material buildup problems in the dryer, hot elevator, and screen tower. The most widely used method for batch plant hot mix recycling is the "Maplewood method." Even though variations exist, basically a separate cold feed bin

FIGURE 18.15 Reclaimed Asphalt Pavement (RAP) Feeder in a Batch Plant.

introduces the RAP into the weigh hopper or the pugmill by a chute and belt conveyor (Figure 18.15). There the RAP is joined by the virgin material coming from the cold feed bins through the dryer and the screen decks. The temperature to which the virgin aggregates are superheated depends on the properties of the RAP material.

The RAP material cannot be processed in normal drum mix plants since excessive "blue smoke" is produced when the RAP comes in contact with the burner flame. The condition is further aggravated by the buildup of fine aggregates and asphalt binder on metal flights and end plates. It has been suggested that most of the smoke problem is caused by the light oils in soft grades of asphalt binder used to rejuvenate the aged asphalt in the RAP. Although the smoke problem could be solved by various processes such as lowering the HMA plant's production rate, decreasing the moisture content of the RAP, lowering the discharge temperature of the recycled mix, introducing additional combustion air, and decreasing the percentage of RAP, it was found that a more effective way to rectify the problem was to modify the drum mix plant.

Although variations exist in the process, basically the center entry method (Figure 18.16) is the most widely used method for hot mix recycling in a drum mix plant. In this method, the RAP is introduced into the drum downstream of the burner flame to mix with the superheated new aggregates. The hot virgin aggregates heat up the RAP material by conduction. The RAP is protected

FIGURE 18.16 RAP Feeder in a Drum Plant.

414Pavement Engineering: Principles and Practice

from coming in direct contact with the burner flame by a dense veil of aggregate added prior to the point where the RAP is added. It is very important to have the veil of virgin aggregate. Otherwise, overheating of RAP can result in blue smoke, and it may not be possible to use the design amount of RAP material. Sometimes special flight design, steel ring dams, or circular steel flame shields are utilized to force the RAP to mix with the virgin aggregates before being subjected to the high gas treatment. These techniques eliminate the "blue-smoke" problem.

Since the late 1980s, two new drum designs for more efficient heat transfer to RAP material during mixing have been developed. These are the double-barrel and triple-drum design. The double-barrel counterflow drum mix plant has more mixing space than a conventional drum mixer. The shell of the drum is used as the shaft of the coater. A 3.0–3.3-m (10–11-ft) diameter coater is created with an extremely large insulated mixing area. The virgin aggregate material is dried in the inner drum and superheated to 315–343°C (600–650°F) (when running 50% RAP). It then drops through the wall of the drum and meets with the RAP in the annular space. Approximately 1½ minutes of mixing time occurs in this outer shell. Since the outer shell does not rotate, easy access is available to add various other recycle components to the process as they become necessary and available. The heat of the inside barrel is transferred through the rotating shell to mixing in the annular space. The outer shell of the double barrel remains at approximately 49°C (120°F) at all times, leading to a very efficient plant. In this method the virgin and the RAP material are not exposed to the hot gases or to the steam of the drying process, and thus the light oils are not removed from the mix. In the outer section of the double barrel, due to the moisture removed from the RAP, a steam or inert atmosphere occurs, resulting in a much lower oxidation or short-term aging of the recycled HMA mix in the mixing chamber. Another benefit derived from this type of plant is the much longer life occurring with the bags in the baghouse due to the relatively lower temperature of the exhaust gases. As dust is discharged from the baghouse through a rotary airlock on the double-barrel plant, a screw conveyor is used to transfer the mix back into the outer shell. The holes through which the virgin aggregates are directed into the outer shell are also responsible for channelizing any smoke from the inner mixing section to the outer space. The pollutants go directly to the flame, where they are burnt. This results in reduced emission and blue smoke. The counterflow dryer design also leads to higher production rates with much lower fuel consumption. The triple-drum design also uses an outer shell; however, a stainless steel cylinder is used to enclose the combustion chamber. This cylinder (without any flight or steps of a regular drum) is believed to be effective in transferring heat to the RAP material through conduction and radiation. The virgin aggregate is introduced from the opposite end of the burner flame. The RAP material is introduced in the annular space formed by the outer shell. The superheated virgin aggregates fall into the annular space and mingle with the RAP material. The factors controlling the production limit in a drum mix plant are the moisture content and ambient temperature of the RAP and new aggregate. A practical limit of 30:70 (30% RAP and 70% new aggregate) has been recommended in the available literature, although research continues to enable the use of a higher percentage of RAP.

18.6 MAINTENANCE AND REHABILITATION OF CONCRETE PAVEMENTS

18.6.1 JOINT AND CRACK SEALING

Cracks and joints are sealed to prevent water and noncompressible materials from entering and causing potential damage. Different types of sealants are used for this purpose:

Hot-pour liquid sealants, which are heated and liquefied and then poured into cracks and joints. They can be opened to traffic after they have cooled and gained enough rigidity. They have a service life of 3–5 years.

Silicone sealants are silicone polymer compounds that are poured into joints at ambient temperatures, are commonly ready for traffic loads within 30–60 minutes of application, and have a service life of 8–12 years.

Compression seals are preformed rubber compounds (commonly neoprene materials) that are forced into a joint under compression. They form a seal by pushing against the walls of the joint, are immediately ready for traffic, and have a service life exceeding 15 years.

18.6.2 SLAB STABILIZATION

The purpose of slab stabilization is to fill voids that have been created beneath the slab due to a combination of pumping, erosion, or consolidation of the base. If left untreated, these voids could grow in size and could potentially contribute to faulting, corner breaks, or cracking. These voids are best filled by pumping grout (cement and fly ash slurry) through predrilled holes through the slab.

18.6.3 DIAMOND GRINDING

Diamond grinding is used to restore pavement surface friction or smoothness. The process uses gang-mounted diamond saw blades to grind off a thin surface layer approximately 2–20 mm thick. The undesirable roughness may have been caused by faulting, studded tire wear, or slab curling. Although not desirable for new construction, diamond grinding can also be used in newly constructed slabs to attain initial specified smoothness.

Patches constructed in rigid pavements are used to treat localized slab problems such as spalling, popouts, scaling, joint deterioration, corner breaks, and punchouts. A partial-depth patch is used if the pavement distress is limited in depth; otherwise, a full-depth patch should be used. HMA is commonly used for emergency patches. However, fast-setting cements are available and can be used to reduce the setting time to a few hours. When constructing permanent patches in PCC pavements, Portland cement or high early strength gain cement should be used.

Spalls may be caused by the infiltration of incompressible materials into joints when they contract and open during cold weather. During warm weather the pavement expands, closing the joints. However, the presence of incompressible materials in the joints will prevent the joints from closing and produce high compressive stresses along the joint faces or walls. As these compressive stresses build up, they may cause spalling of the concrete along the joints. Force may cause spalling at both the top and bottom of the pavements. Other causes of spalling include dowel bar misalignment, corrosion of metal joint inserts, reinforcing steel that has been placed too near the surface, D-cracking, alkali-silica reactions, and lack of consolidation of concrete near the joints.

Partial-depth patches are used to restore localized areas of slab damage that are confined to the upper one-third of slab depth. This is commonly used for moderate spalling and localized areas of severe scaling that do not exceed 3 inches deep and that cover an area less than 12 sq. ft.

Full-depth patches are used to restore localized areas of slab damage that extend beyond the upper one-third of slab depth or originate from the slab bottom. Full-depth patches are commonly used for repairing severe spalling, punchouts, corner breaks, severe slab cracking, and localized areas of severe scaling (ACPA, 1995).

Additional information on methods for determining the optimal timing for the application of preventive maintenance treatments in rigid pavements can be found in NCHRP Report 523 (Optimal Timing of Pavement Preventive Maintenance Treatment Applications; http://onlinepubs. trb.org/onlinepubs/nchrp/nchrp_rpt_523.pdf).

18.6.4 Load Transfer Devices

Rehabilitation of load transfer devices between two adjoining slabs involves cutting a slot perpendicular to the joint between the two slabs, inserting a dowel, and refilling the slot with a fast-setting polymer concrete. Combined with other pavement restoration methods, rehabilitation of load transfer devices has the potential to extend the service life of rigid pavements by 10–20 years, depending on the pavement condition at the time of the repair and subsequent pavement traffic loading and environmental conditions.

In 1992, the FHWA Special Project (SP-204; FHWA, 2006) was undertaken to encourage industry to develop equipment for economically constructing slots for retrofit load transfer. This equipment would make load transfer across faulted joints or working cracks a cost-effective maintenance and rehabilitation technique, and would extend the service life by providing positive load transfer for undoweled or transversely cracked JPCP or for working cracks in underreinforced JRCP.

18.6.5 Precast Panels for Repair and Rehabilitation

The use of precast concrete pavements is being evaluated for the application of rapid repair of localized failures in concrete pavements (such as full-depth repairs) and to rehabilitate long lengths of existing poorly performing asphalt and concrete pavements. Other benefits include better control over concrete batching, forming, and curing; increased performance and durability from post-tensioning; savings through reduced pavement thickness; and reduced construction time lane closures.

FHWA demonstration projects in Michigan and Colorado using precast paving for full-depth repairs of jointed concrete pavements were opened to traffic within 6 to 8 hours of lane closure. Precast 6 × 12 ft panels were used to repair deteriorated joints and slabs along sections of the interstate highway system. The repairs involved removal of deteriorated concrete, preparation of the base support, placement of a fast-setting bedding material, installation of precast panels, and installation of retrofitted dowel bars at the transverse joints. These projects are under evaluation to determine long-term performance.

In 2002, the Texas DOT successfully completed the first pilot project utilizing precast panels for pavement construction. Numerous lessons were used in this pilot project, including the following design and construction techniques: (1) use of longitudinal post-tensioning to tie together a series of precast panels to provide a jointless slab length of 250 ft, where each precast panel was also pretensioned in the transverse direction; (2) use of a thinner precast slab compared to the need for a thicker conventional jointed slab; (3) novel post-tensioning techniques; and (4) placement of the precast panels directly over an asphalt concrete base or leveling course.

Concrete Pavement Technology Program (CPTP) Task 52 is the continuation of the FHWA's SP-205 field demonstration program. The SP-205 is developing field-tested guidance on concrete pavement rehabilitation and repair techniques as well as strategies that emphasizes the do's and don'ts, and whys and whens, for concrete pavement restoration (CPR) and preventive maintenance of concrete pavements. The rehabilitation and maintenance strategies considered are full-depth patching, partial-depth patching, subsealing, joint resealing, retrofitted load transfer, and grinding and grooving. Periodic evaluation of the field test sites is being carried out under CPTP. About 40 sites are under evaluation.

18.6.6 Portland Cement Concrete Overlays

PCC overlays are increasingly being used as a rehabilitation technique for both existing PCC and hot mix asphalt pavements by potentially extending service life, increasing structural capacity, reducing maintenance requirements, and lowering life cycle costs. Improvements in PCC paving technology such as the use of zero-clearance pavers, fast-track paving concepts, and high early strength PCC mixtures have greatly increased the ability of PCC overlays to serve as a viable rehabilitation alternative. The FHWA in cooperation with the American Concrete Institute (ACI) has developed a synthesis report, "PCC Overlays State of the Technology Synthesis Report." This material presents the latest information on the design, construction, and performance of PCC overlays. It describes design and construction techniques for the four types of PCC overlays that are commonly used in highway pavement applications: bonded, unbonded, conventional whitetopping, and ultrathin whitetopping. Information is also provided on the selection of PCC overlays as a possible rehabilitation alternative for existing pavements.

Whitetopping is a Portland cement concrete overlay on existing asphalt concrete pavement. It can be used as a road surface course where other paving materials and methods have failed due to rutting or general deterioration. There are three types of whitetopping: conventional (thickness greater than 8 in.), thin (thicknesses over 4 but less than 8 in.), and ultrathin (2 to 4 in.). Ultrathin whitetopping (UTW) is a bonded, fiber-reinforced concrete overlay. UTW is designed for low-speed traffic areas or areas with a lot of stop-and-go traffic, such as street intersections, bus stops, or toll booths. Joint spacing is critical to a good performing UTW. The use of a short joint spacing is common in both directions of the slab. In effect, a mini-block paver system is formed. The limited experience with UTW indicates that joint spacing should be no more than 12 to 18 in. each way per inch of whitetopping thickness. For example, a 3-in. UTW surface should be jointed into 3×3 or 4×4 ft squares. Joints are sawed early to control surface cracking.

QUESTIONS

1. What are the different steps in a pavement management system?
2. What are the different types of mixes available for maintaining asphalt pavements?
3. List and briefly describe the different types of asphalt pavement recycling processes.
4. How does a maintenance engineer decide between partial-depth and full-depth repair in rigid pavements?
5. What materials should be used for partial-depth and full-depth repair in rigid pavements?
6. How does a designer select the appropriate technology for joint and crack sealing?

19 Airport Pavements

19.1 TYPES, IMPORTANCE, AND SPECIFICATIONS

Essentially there are three different pavement areas in an airport—runways, taxiways, and ramps or aprons. Runways (Figure 19.1) are used for takeoff and landing of aircrafts, taxiways are pavements that connect the aprons to the runways, and aprons are areas that are used for aircraft parking, loading, and unloading. Generally runways can be distinguished by white marks on them, while taxiways have yellow markings. Helipads, used for the landing and takeoff of helicopters, have a distinct *H* sign marked on the area.

Airport pavements must be constructed under strict guidelines and specifications, to avoid damage to aircrafts as well as injuries to passengers. Loose particles of damaged airport pavement mix or materials can cause foreign object damage (FOD). Loose particles can damage jet engines by getting sucked in, can damage propellers, and may become deadly projectiles.

An example of specifications for airport pavement hot mix asphalt (HMA) is the Federal Aviation Administration's (FAA) Item P401 (Plant Mix Bituminous Pavement; http://www.faa.gov/airports_airtraffic/airports/construction/pavement_design/; FAA, n.d.). It specifies materials, composition, construction methods, materials acceptance, contractor quality control, method measurement, basis of payment, as well as relevant ASTM and AASHTO standards. Additional guidance can be obtained from engineering briefs such as the FAA's Engineering Brief EB59A (http://www.faa.gov/airports_airtraffic/airports/construction/engineering_briefs/; FAA, n.d.), which is written for all pavements designed for gross aircraft weight of less than 60,000 lb and taxiways and aprons for gross aircraft weights of 60,000 lb or more.

The FAA provides guidance for the design, construction, and maintenance of airports through their specifications (http://www.faa.gov/airports_airtraffic/airports/; FAA, n.d.). AC150/5320-6D (FAA, 1995) is for airport pavement design and evaluation, whereas AC 150/5100-13A (FAA, 1999) is for nonprimary airports. The airport pavement design process has been developed from a nomograph/chart-based empirical procedure to spreadsheet-based design as well as the latest layered elastic pavement design [LEDFAA (FAA, 2004); and FAA Rigid and Flexible Iterative Elastic Layered Design, or FAARFIELD (FAA, 2008)]. AC 150-5380-6B (FAA, 2007) gives guidance on the maintenance of airport pavements. AC 150-5370-10C (FAA, 2007) gives standards for the construction of airports; it contains important updates on existing specifications. Examples of important engineering briefs are EB59A, Item P401 (Plant Mix Bituminous Pavements, or Superpave) (FAA, 2006), and EB60 (Semi Flexible Wearing Course for Apron Pavements) (FAA, 2001). The latter EB specified the use of a porous asphalt mix layer with voids filled with highly modified cement grout, which is used as an alternative to coal tar sealant for fuel resistance. Such layers (Densiphalt; see http://www.densiphalt.com; Densit, n.d.) have been used by Massport and the Port Authority of New York and New Jersey (PANYNJ) as well as in the Port of Baltimore in the United States. Densiphalt consists of open graded asphalt mix with 25–30% voids, limestone filler, and cellulose fiber. The mortar for the grout is composed of special cement with microsilica, superplaticizer, quartz sand, and water with a water-cement ratio of 0.27. For construction, the existing surface is removed to a depth of 2–3 in. and replaced with the open graded material. The special grout is then mixed with water and then applied on the surface of the open graded material, which removes the air voids; a curing compound is applied to control moisture release from the surface. EB No. 62 specifies the

FIGURE 19.1 Runways and Taxiways.
Courtesy: Massachusetts Port Authority.

use of polymer composite micro-overlay (applied as a slurry) for a fuel-resistant wearing surface as an alternative to a coal tar seal coat.

The method of establishing or changing a standard begins with the FAA headquarters office initiating the process, and preparing a draft, which gets reviewed by headquarter airport offices, the FAA regional offices, and the industry, and is finally accepted or rejected by the headquarters review committee.

19.2 STRUCTURAL DESIGN OF AIRPORT ASPHALT MIX PAVEMENTS

The design of airport pavements can be done with airfield pavement design software following the Advisory Circular AC 150/5320-6D (Airfield Pavement Design and Evaluation). The design process started with the basic shear failure–based California Bearing Ratio (CBR) method developed and modified by the U.S. Army Corps of Engineers in the 1960s and 1970s. The main drawback with this design method is its inaccurate representation of specific aircraft loads/gears and limited outputs with a lack of ability to conduct sensitivity analysis.

The *formula* for relating thickness of pavement to CBR is as follows:

$$t = \alpha \ (A_c)^{0.5} \ [-0.0481 - 1.1562 \ (\log CBR/P) - 0.6414 \ (\log CBR/P)^2 - 0.473 \ (\log CBR/P)^3]$$

where:

α = load repetition factor
A_c = tire contact area, in.^{2}
CBR = CBR of the layer being considered
P = tire pressure (psi) at depth t used in calculating the ESWL

The basis for this formula is equal deflection between the ESWL and multiple gear, assuming equal contact area.

Currently, a spreadsheet-based design (based on the CBR method, with gear configurations related through empirical data and theoretical concepts) is used, which has been made user-friendly with the use of Visual Basic macros. The design software is available from http://www.faa.gov/airports_airtraffic/airports/construction/design_software/ (FAA, n.d.).

This spreadsheet-based design allows for inputs for subgrade compaction requirements, layer conversion, multiple subbase layers, and charts on thickness versus CBR and thickness versus annual departures. For the aircraft mix, the design is based on "critical aircraft," and allows consideration of frost in design. Note that the design curves on the basis of which this spreadsheet software has been developed are provided in Advisory Circular AC 150/5320-6D in Figures 3-2 through 3-15. This design method is intended to result in pavements that would last for 20 years without needing any major maintenance if no major changes in forecasted traffic are encountered.

For asphalt pavements, there are 14 design curves that give the total pavement thickness as well as the thickness of the HMA surface for different types of aircrafts over a particular subgrade. These thicknesses are provided for both critical areas (with concentrated/departing traffic) and noncritical areas (with arriving traffic). The thickness of the HMA surface courses for noncritical areas is provided on the charts, whereas those for the base and subbase courses for noncritical areas can be considered as 0.9 times the thickness of the corresponding layers in the critical areas.

The software provides two options—the main module for designing airports with aircrafts heavier than 30,000 lb, and a submodule one for designing airports with aircrafts weighing less than 30,000 lb. The design for the main module is based on a pavement structure consisting of a pavement surface layer (HMA), a stabilized or unstabilized base layer, a stabilized or unstabilized subbase layer(s), and a subgrade.

There are 10 basic steps in the main module. For the proper completion of the design process, it is important that the steps are completed in the proper numerical order. (Please use Circular AC 150/5320-6D to refer to the figures and tables mentioned in the following steps.)

Step 1. Enter airport name and data: this step involves the input of the airport name, city and state, Airport Improvement Program (AIP) project number, as well as designer's name and comments, if any.

Step 2. Enter subgrade CBR and frost code: it has been suggested that the design CBR value should be equal to or less than 85% of all the subgrade CBR values, which corresponds to a design value of one standard deviation below the mean CBR value. Knowing the type of the soil (such as gravelly soil), the percentage finer than 0.02 mm by weight (say, 5%), and the soil classification (say, GW), one can determine the frost code (FG-1) from Table 2-4 in Advisory Circular AC 150/5320-6D. The frost code is used along with the frost penetration depth information to decide on a specific frost protection system for the pavement.

Note that for specific aircraft types and loads, requirements regarding the compaction of subgrade of cohesive and noncohesive soils are provided in Table 3-2 in Advisory Circular AC 150/5320-6D.

Step 3. Enter subbase information. For the subbase, the number of subbase layers, CBR, and frost code are required.

Step 4. Select default aggregate base. For FAA design (P209), crushed aggregate base is the default which can be substituted by either aggregate case (P208) only if the gross weight of the design aircraft is less than 60,000 lb and the thickness of the surface HMA layer is increased by 1 in. or with lime rock base (P211) material.

Step 5. Calculate frost penetration depth. This step is optional, to compare against the required frost protection depth. In this step, the required inputs are the Air Freezing Index (degree days °F) and the unit weight of the soil (in pcf). The frost penetration depth is calculated from the interpolation of the following data, developed on the assumption of a 12-in. (300-mm) thick rigid pavement or a 20-in. (510-mm) thick flexible pavement.

Frost Penetration Depth (In.)

Degree Days, °F	Soil Unit Weight (Pcf)			
	100	115	125	150
200	20.5	21.5	23.8	25.5
400	27.5	30.5	35	38.5
600	34	38	44.5	49
800	40	44.5	54	59
1000	45	51	62	69
2000	69.5	79	102	113
3000	92	105	140	156
4000	115	130	177	205
4500	125	145	197	225

Note that the Air Freezing Index is a measure of the combined duration and magnitude of below-freezing temperatures occurring during any given freezing season. The average daily temperature is used in the calculation of the Freezing Index. For example, if the average daily temperature is 12 degrees below freezing for 10 days, the Freezing Index would = 12 degrees × 10 days = 120 degree days. It is recommended that the design Air Freezing Index be based on the average of the three coldest winters in a 30-year period, if available, or the coldest winter observed in a 10-year period.

For economic reasons, the maximum depth of frost protection suggested is 72 in. There are different ways of providing frost protection to the pavement, depending on the frost code of the soil. Seasonal frosts can cause nonuniform heave and loss of soil strength during melting, resulting in possible loss of density, development of pavement roughness, restriction of drainage, and cracking and deterioration of the pavement surface. Whether such effects are possible or not depends on whether the subgrade soil is frost susceptible, there is free moisture to form ice lenses, and the freezing front is deep enough to penetrate the subgrade soil.

The frost susceptibility of the soil is affected primarily by the size and distribution of voids (where ice lenses can form), and is determined through empirical correlations from the soil classification (Unified Soil Classification System) and the percentage finer than 0.02 mm. The table relating these parameters to the frost codes is given as Table 2-4 in Advisory Circular AC 150/5320-6D. It can be assumed that sufficient water for creating ice lenses is present if the moisture content of the soil is at or greater than 70% of its saturation. It is advised that sufficient moisture for detrimental frost action is assumed to be present all the time.

The protection from frost action can be provided either by making sure that deformations from frost actions are limited or avoided, or by making sure that the pavement has adequate structural capacity during the frost melting (and ignoring the heave due to freezing). The most costly option is complete frost protection, in which the difference between the pavement depth designed from load-carrying consideration only and the depth of frost penetration (assuming the latter is greater than the former) is replaced with nonfrost-susceptible materials to protect the subgrade soil from frost. This could involve replacement of a significant amount of the subgrade soil. The complete frost protection method is recommended only for pavements on FG-3 and FG-4 soils, which are extremely variable in horizontal extent.

The limited subgrade frost protection is meant for limiting the effects of frost, and allowing a tolerable (based on experience) level of heave. In this case, 65% of the depth of frost penetration should be made up of nonfrost-susceptible materials. This design method is recommended for FG-4 soils except where the conditions require complete protection, as well as for FG-1, FG-2, and FG-3 soils when the functional requirements of the pavement permit a minor amount of frost heave. It is recommended that consideration be given to using transition sections where horizontal variability of frost heave potential permits.

In the third method, the reduced subgrade strength method, the CBR of the subgrade is reduced (on the basis of the frost group/code) so as to design a thicker pavement and hence have adequate strength during the thawing or melting of ice lenses. The reduced CBRs versus the frost group data are given in Table 3-1 in Advisory Circular AC 150/5320-6D. This method is recommended for pavements on FG-I, FG-2, and FG-3 subgrades which are uniform in horizontal extent or where the functional requirements of the pavement will permit some degree of frost heave. The method is also permitted for variable FG-1 through FG-3 subgrades for less sensitive pavements which are subject to slow-speed traffic and for which heave can be tolerated.

For airport pavements in permafrost regions, if the complete protection system is not found practical, the method of reduced subgrade strength can be used. The system of providing insulation panels to protect the permafrost can also be used (with a case-by-case approval from the FAA, as no design standard exists), with careful construction and possible restriction of aircraft loads to prevent failure of the insulation panels.

Steps 6–8. Enter aircraft mix. There are two directives regarding the thickness of the pavement over and above those that are determined through design. First, for different types of aircrafts (and different loads for each of those), minimum base course thickness has been specified in Table 3-4. Second, for number of departures in excess of 25,000, the total pavement thickness should be increased according to Table 3–5.

The design method is based on the determination of pavement thickness for each of the aircrafts in the list, with consideration of both load and number of departures, and the selection of the most critical aircraft, which requires the thickest pavement. If accepted by the designer, the thickness of the pavement required for the critical aircraft is considered as the design thickness from this point onward in the design process.

In the design process, it is assumed that 95% of the gross weight is carried by the main landing gears and 5% is carried by the nose gear of aircrafts. For the weights of civil aircrafts, FAA Advisory Circular AC 150/5300-13 can be consulted. The designs are based on the maximum anticipated takeoff weights with the assumption that this conservative approach would offset any unforeseen changes in operation and traffic, and also the ignoring of the arriving traffic.

The design curves on which the software is based provide thickness for four different sets of aircrafts—single-gear, dual-gear, dual-tandem-gear, and wide-body aircrafts. Once the critical or design aircraft is determined, each of the other aircrafts is converted to an equivalent design aircraft, first by converting into the equivalent gear, and then by converting to the equivalent annual departures, of the design aircraft. Each wide-body aircraft is treated as a 300,000-lb (136,100 kg) dual-tandem aircraft when computing equivalent annual departures. Finally, all the equivalent annual departures are summed up, and the pavement is designed for the design aircraft with the total equivalent annual departures. However, the heaviest aircraft must be considered to provide for adequate thickness of the surface layer, depth of compaction, and drainage structures.

Note that the lateral distribution of aircraft traffic along the path of travel (in the form of a normal distribution) or aircraft wander is taken into consideration in the design process with the help of the "departure-to-coverage ratio." The input is in terms of annual departures. Departures are converted to "coverages." Coverage is a measure of the number of maximum stress applications that occur on the surface of the pavement due to the applied traffic. One coverage occurs when all points on the pavement surface within the traffic lane have been subjected to one application of maximum stress, assuming the stress is equal under the full tire print. Assuming normal distribution of aircraft traffic along the travel path, each departure (also called *pass*) of an aircraft can be converted to coverages using a single pass-to-coverage ratio. These ratios are given in Table 5 in Appendix 2 of Advisory Circular AC 150/5320-6D. Annual departures are converted to coverages by multiplying by 20 and dividing that product by the pass-to-coverage ratio, which is tabulated with respect to 12 aircraft/gear types. There is also a figure in Appendix 2 of Advisory Circular AC 150/5320-6D (Figure 3) which shows plots of "load repetition Factor" versus coverages. After the pavement is designed for coverages, the thickness is multiplied by the appropriate load repletion factor to determine the final thickness for each specific type of aircraft (this step is done prior to the selection of the critical aircraft).

Step 9. Compute thickness for stabilized layers. This step requires the selection of a stabilized layer and input of the equivalency factor, for a base and/or the top layer of the subbase layer(s). Guidance regarding the selection of proper equivalency factors is provided in the software.

Step 10. Go to the design summary. In this step, the design calculations done in Step 8 are repeated with the information provided in Step 9, and a summary showing the different input parameters, designed parameters, as well as recommended guidelines is shown, along with options for printing, on the screen. Options are also available to see plots of total pavement thickness versus number of departures for the design aircraft and versus CBR of the subgrade.

A step-by-step example is provided below.

Example

Consider airport pavements to be built on an area with subgrade soil which has the following properties: percentage passing No. 200 sieve = 85%; liquid limit on minus #40 material = 35%; plastic limit on minus #40 material = 7%; percentage finer than 0.02 mm = 12%; and results from three CBR tests conducted in the laboratory = 6, 8, and 7. The unit weight of the soil is 115 pcf.

From a weather station very close to the airport site, the average of the three coldest winters in a 30-year period shows that the design Air Freezing Index is 1000 degree days.

The forecasted air traffic for the airport is as follows:

Aircraft	Gear Type	Annual Departures	Average Maximum Takeoff Weight (lb)
B727-200	Dual	2000	90,500
B737-200	Dual	5000	28,600
B747-200	Double dual tandem	1200	833,000
ABA300-B2	Dual tandem	1000	304,000

1. From the plasticity chart in Figure 2-2, the soil is classified as ML; from Table 2-3, the soil is described as sandy silt.
2. Design CBR = Average – 1 * standard deviation = 7 – 1 = 6.
3. From Table 2-4, the frost group/code = FG-4.

4. Consider sufficient water to be present in the subgrade to cause detrimental frost action.
5. Using Figure 2-7, the depth of frost penetration is 47 in.
6. Determine the required thickness of the pavement for each aircraft.

Aircraft	Gear Type	Maximum Takeoff Weight (lb)	Annual Departures	Required Thickness (in.)
B727-200	Dual	190,500	2000	34 in., Figure 3-3
B737-200	Dual	128,600	5000	29 in., Figure 3-3
B747-200	Double dual tandem	833,000	1200	45 in., Figure 3-7
ABA300-B2	Dual tandem	304,000	1000	34 in., Figure 3-5

7. Determine the design aircraft. From the above table, it can be seen that the B747-200 requires the thickest pavement, and hence it should be considered as the critical or design aircraft. Since the B747-200 has double-dual-tandem gear, all traffic should be grouped into the double-dual-tandem configuration.
8. Group forecast traffic into the landing gear of the design aircraft.

Aircraft	Gear Type	Annual Departures	Multiplied by Factor (Table Provided in ¶ 305a in AC150/5320–6D)	Equivalent Double-Dual-Tandem Gear Departure
B727-200	Dual	2000	0.6	1200
B737-200	Dual	5000	0.6	3000
B747-200*	Double dual tandem	1200	1.0	1200
ABA300-B2	Dual tandem	1000	1.0	1000

*Each wide-body aircraft is treated as a 300,000-lb dual-tandem aircraft when computing equivalent annual departures.

9. Convert aircraft to the equivalent annual departures of the design aircraft.

Aircraft	Gear Type	Equivalent Double-Dual-Tandem Gear Departure (A)	Wheel Load * (lb) (B)	Wheel Load of Design Aircraft (lb) (C)	Equivalent Annual Departure of the Design Aircraft $(10^{[(\log A) * (B/C)^{1/2}]})$
B727-200	Dual	1200	45,240	35,625	2950
B737-200	Dual	3000	30,543	35,625	1658
B747-200*	Double dual tandem	1200	35,625**	35,625	1200
ABA300-B2	Dual tandem	1000	36,100	35,625	1047
				Total	6855

*Main gears carry 95% of the gross load; divide the load carried by each gear by the number of wheels.

** Each wide-body aircraft is treated as a 300,000-lb dual-tandem aircraft when computing equivalent annual departures.

10. Final design. Final design should be made for 6855 annual departures for a B747-200 weighing 833,000 lb. See Figure 3-7 (this figure is reproduced in Figure 19.2 in this book) regarding design curves for a B747-200; designed total thickness is 47.5 in. Next, consider Figure 3-7 (here, Figure 19.2) again, with a subbase CBR of 20. The combined (base + surface) thickness is obtained as 18 in. From Figure 3-7, the thickness of the HMA surface required for critical areas is 5 in.

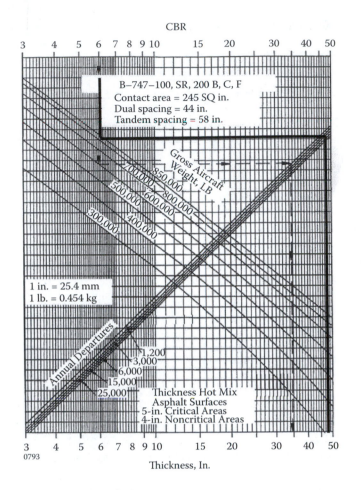

FIGURE 19.2 Design Curves for B747 Family of Aircraft.
Source: From Federal Aviation Administration, AC 150/5320-6D (Washington, DC: FAA).

Therefore, the thickness of the base course is $(18 - 5) = 13$ in. (if a stabilized course with a 1.4 layer equivalency number is used, the thickness is reduced to $13/1.4 = 9.5$ in.).

The thickness of the subbase is therefore $(47.5 - 18) = 29.5$ in. (if a stabilized course with a 1.6 layer equivalency number is used, the thickness is reduced to $29.5/1.6 = 18.5$ in.).

Since the stabilized base and subbase are required for new pavements designed to accommodate aircraft weighing 100,000 lb or more, this should be done. For the base course, consider a plant mix bituminous concrete layer with a layer equivalency of 1.4 (compared to a P209 crushed aggregate base). For the subbase course, consider the use of a P209 crushed aggregate base course with a layer equivalency of 1.6 (compared to a P-154 subbase course). Note that the frost penetration depth is 51 in. Therefore, to provide complete frost penetration, the thickness of the pavement section should be increased from $(5 + 9.5 + 18.5) = 33$ in. to 51 in. This means a layer of $(51 - 33) = 18$ in. of nonfrost-susceptible material must be placed below the designed pavement section.

The final design is as follows:

Layers	Thickness Required (in.)		
	Critical Areas	Noncritical Areas	Edge
HMA surface (P401)	5	4 (from Chart 7-3)	3.5 (0.7 * T, thickness in critical areas; minimum is 0.7T; Figure 3-1)
HMA base (P401)	9.5	9 (0.9T)	7 (0.7T)
P209 crushed aggregate subbase	18.5	17	17 (0.7T does not apply for the subbase)
To provide complete frost protection: nonfrost-susceptible material	18 in. below pavement section		
To provide limited frost protection: 65% of frost penetration depth must have nonfrost-susceptible material	0 in., none required		

The summary information for this design, as obtained from the spreadsheet program, is shown in Figure 19.3. The plots of thickness versus CBR of the subgrade and annual departure are shown in Figure 19.4.

19.2.1   DESIGN OF FLEXIBLE PAVEMENT OVER EXISTING PAVEMENT

For flexible pavements, two options are allowed—HMA overlay on existing flexible pavement and HMA overlay over existing rigid (PCC) pavement. Note that the FAA does not allow overlay pavement sections containing granular separation courses between the existing and new layers (called *sandwich pavement*) because of the possibility of the granular layers getting saturated with water, and the trapped water having a detrimental effect on the performance of the pavement.

19.2.1.1   HMA Overlay on Existing Flexible Pavement

For this option, the method used is known as the *thickness deficiency approach*, and it consists of the following steps.

1. Using the design curves or software presented above, determine the thickness required for a new pavement for the load and annual departures. Determine the thickness of all of the pavement layers.
2. Compare the required thickness of the new pavement with the existing pavement, and determine the thickness of the overlay. In this step, the existing base may be considered as the subbase and the existing surface may be considered as the base. A higher quality material layer may be converted to a lower quality material layer, using layer equivalencies provided in Tables 3-6 through 3-8 of Advisory Circular AC 150/5320-6D. While using the layer equivalencies, proper consideration should be given to the condition of the existing pavement—since defects such as surface cracking, a high degree of oxidation, and evidence of low stability would reduce the equivalency factors.

FLEXIBLE PAVEMENT DESIGN FOR		10/31/2005

Trial-1

City-1

Engineer - Airport Design Engineer

AC Method

AIP No.   1

Example problem

47.5"	**Total Thickness Required (inches)**

No thickness adjustments required

Stabilized Base/Subbase Are Required

Initial Pavement Cross Section	
5"	Pavement Surface Layer (P-401)
13"	Base Layer (P-209)
29.5"	Subbase #1 (P-154) CBR= 20
0"	Subbase #2 CBR= 0
0"	Subbase #3 CBR= 0

Stabilized or Modified Cross Section		Factors
5"	P-401 Plant Mix Bituminous Pavements	
9.5"	P-401, Plant Mix Bituminous Pavements	1.4
18.5"	P-209, Crushed Aggregate Base course	1.6
0"	Material as defined by user	
0"	Material as defined by user	

Frost Considerations		
115 lb/cf	Dry Unit Weight of Soil	
1000	Degree Days °F	
51"	Frost Penetration Depth	
6	Original CBR value of subgrade Soil	
6	CBR Value used for the Subgrade Soil	Non-Frost Code for Subgrade Soil
20	CBR Value used for subbase #1	Non-Frost code for Subbase #1
0	CBR Value used for subbase #2	Non-Frost code for Subbase #2
0	CBR Value used for subbase #3	No frost selection made for Subbase #3

Design Aircraft Information	

The Design Aircraft is a BOEING747 - 780,000 lbs — ()

833000 lbs	Gross Weight	**20** Design Life (years)
6,857	**Equivalent Annual Departures of a 300,000 lb Dual Tandem Gear - see Para. 305 AC 150/5320-6D	

Subgrade Compaction Requirements for Design Aircraft			
Non-Cohesive Soils		**Cohesive Soils**	
Compaction	Depth Required	Compaction	Depth Required
100%	0 - 23	95%	0 - 9"
95%	23 - 41"	90%	9 - 18"
90%	41 - 59"	85%	18 - 27"
85%	59 - 76"	80%	27 - 36"

See Appendix 5 to AC 150/5320-6D, Airport Design and Evaluation, for application of this software.

FIGURE 19.3 Summary Output.

Example

Consider an existing pavement with subgrade CBR = 6; the HMA surface course is 5 in. thick, the base course is 8 in. thick, and the subbase is 12 in. thick. The subbase CBR is 15. Frost action is negligible. Assume an existing pavement is to be strengthened to accommodate a dual-wheel aircraft weighing 150,000 lb and with an annual departure level of 6000.

HMA surface: 4 in.; base: 14 in.; subbase: 15 in.; total thickness: 33 in.

The total thickness required to protect the subgrade is 33 in.; the combined thickness of the surface and the base should be 18 in. to protect the subbase.

The existing pavement is 8 in. deficient in total pavement thickness. Assume that the HMA surface of the existing pavement can be substituted as a base layer with a layer equivalency of 1.5. If 4 in. of the existing HMA surface layer is considered as the base in this manner, the thickness of the new base becomes (8 + 4 * 1.5) = 14 in. This satisfies the base course requirement, leaving 1 in. of HMA surface. A 3-in. HMA overlay on top of this would satisfy the HMA surface thickness requirement.

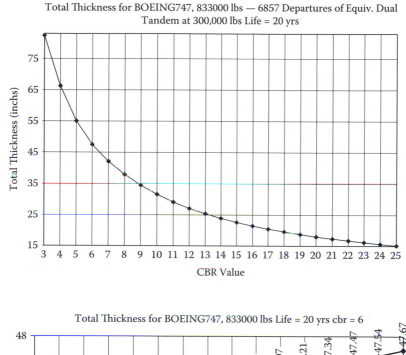

FIGURE 19.4 Change in Thickness with CBR and Departures.

Note that the CBR of the different layers can be determined by evaluation tests, such as those using nondestructive testing (NDT). Also, rounding off thicknesses of equivalent layers (derived using layer equivalencies) is not recommended.

19.2.1.2 HMA Overlay on Existing Rigid Pavement with or without Existing HMA Overlay

To design HMA overlay on existing rigid pavements, the thickness deficiency approach, considering the requirement for a new rigid pavement and the existing rigid pavement, is used. The following formula is used:

$t = 2.5 \, (Fh_d - C_b h_e)$ where the different parameters are explained below.

t = required HMA overlay thickness, in.

F = factor controlling degree of cracking in the base rigid pavement.

h_d = exact (not rounded off) thickness of new rigid pavement required, in., considering the modulus of subgrade reaction, k, of the existing subgrade and the flexural strength of the existing concrete.

C_b = condition factor indicating the structural integrity of the existing rigid pavement, with values ranging from 1 for pavements with slabs having nominal structural cracking to 0.75 for slabs with structural cracking. These two conditions are explained in Figures 4-4 and 4-5 in Advisory Circular AC 150/5320-6D. Note that the overlaying of slabs with severe structural cracking and hence C_b less than 0.75 is not recommended, as this could lead to severe reflective cracking. In such situations, it is advised that such slabs be replaced and load transfer along inadequate joints be restored (and increase C_b as a result), and then an overlay be applied. NDT can be used to determine an appropriate value of C_b, a single value is recommended to be used for an entire area, and the value should not be varied along a pavement feature.

h_e = thickness of existing rigid pavement, in.

The factor F is used to control or limit the amount of cracking expected in the concrete underneath the applied HMA overlay. The failure mechanism is assumed to be like this: the underlying rigid pavement cracks progressively with traffic load repetitions until the average size of the slab pieces reaches a critical value, beyond which shear failure occurs with a significant increase in deflection under traffic. F is considered to be a function of subgrade strength and traffic volume, and is recommended as 1.0; it can also be determined from Figure 4-3 in Advisory Circular AC 150/5320-6D.

To design a HMA overlay over an existing HMA layer over an existing rigid pavement, the same formula given above is to be used, assuming that there is no existing HMA layer. Then the new HMA overlay thickness should be determined by subtracting from the required overlay thickness the thickness of the existing HMA layer or part thereof (if the existing HMA layer is not in such a condition as to be considered in whole).

Example

Consider an existing 12-in. thick rigid pavement is to be strengthened to provide adequate design for 6000 departures of a dual-wheel aircraft weighing 150,000 lb. The flexural strength of the existing concrete is 700 psi, and the modulus of subgrade reaction is 300 pci. $C_b = 0.95$.

Using Figure 3-18 (from AC150/5320-6D), the required single-slab thickness is found to be 13.1 in. The F factor is considered to be 1.

$t = 2.5 \, (Fh_d - C_b h_e)$
$t = 2.5 \, (1 * 13.1 - 0.95 * 12)$
$t = 4.25$ in.

Now suppose that in the existing pavement there was a 2-in. thick HMA overlay on top of the rigid pavement. The calculation for the required new HMA overlay would have proceeded as above, ignoring the presence of the old overlay. Now that it is known that a 4.25-in. HMA overlay is needed, the thickness of the existing overlay must be subtracted from 4.25 in. to determine the required new overlay thickness. Based on NDT and/or engineering judgment, it is now seen that, because of its existing condition, the existing 2-in. HMA can at best be considered as an HMA layer of effective thickness of 1.25 in.; the new overlay that should be used is $(4.25 - 1.25) = 3$ in.

19.3 DESIGN OF CONCRETE PAVEMENTS

AC 150/5320-6D specifies the design procedure for rigid pavements using charts. A subbase of at least 4-in. thickness is required unless certain conditions regarding soils and drainage are satisfied (given in Table 3-10 in AC/5320-6D). Stabilized subbases are required for all pavements carrying aircrafts weighing more than 100,000 lb. For the subgrade, the compaction requirements are less strict compared to the requirements for flexible pavements. Guidance is provided for compaction of cohesive and noncohesive soils (such as 90% of maximum density for cohesive soils in the fill sections). The use of a nonrepetitive static plate load test (AASHTO T-222) is recommended for the determination of the modulus of subgrade reaction, k. Recommended values for subgrades for different thicknesses of subbase are provided (Figure 19.5).

The thickness of concrete slabs is determined using charts for a variety of aircrafts. The inputs required for using these charts are concrete flexural strength, subgrade modulus, gross weight of the design aircraft, and annual departure for the design aircraft. Figure 19.6 shows an example of using the design charts. The dashed line shows the sequence of input values. Note that the given

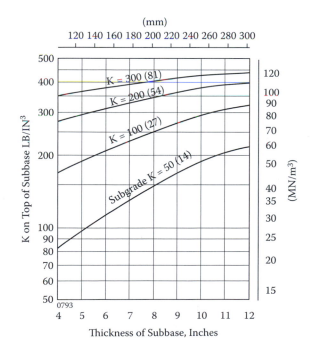

FIGURE 19.5 Effect of Stabilized Subbase on Subgrade Modulus.
Source: From Federal Aviation Administration, AC 150/5320-6D (Washington, DC: FAA).

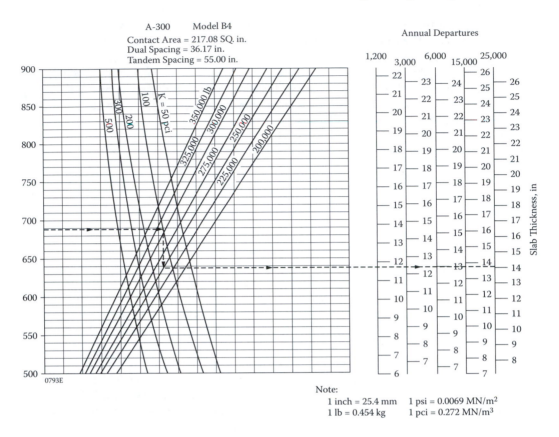

FIGURE 19.6 Example of Design Chart for Rigid Pavement.
Source: From Federal Aviation Administration, AC 150/5320-6D (Washington, DC: FAA).

charts are for jointed edge loading, and based on the consideration of loading, either tangent to or perpendicular to the joint.

19.4   DESIGN FOR AIRPORT PAVEMENTS WITH LIGHT AIRCRAFTS

Aircrafts weighing less than 30,000 lb are characterized as light aircrafts by the FAA, and airport pavements handling such aircrafts are designed on the basis of methods presented in Chapter 5 of AC150/5320-6D. The design of flexible pavements is conducted with the use of the gross aircraft weight and the CBR of the subgrade (Figure 19.7). Once the total thickness of the pavement is determined, the thickness obtained by considering a CBR of 20 is subtracted from it to get the thickness of the subbase. The thickness corresponding to a CBR of 20 is made up of the surfacing and the base. The minimum thickness of HMA over granular base is 2 in. Note that for airports handling aircrafts weighing less than 12,500 lb, highway HMA mixes used by the state department of transportation could be utilized. For rigid pavements, two thicknesses are specified: 5 in. for aircrafts weighing less than or equal to 12,500 lb, and 6 in. for aircrafts weighing more than 12,500 lb.

19.5   ADVANCED DESIGN METHODS

The FAA has developed and is in the process of releasing a suite of design methods/software (FAARFIELD; www.airporttech.tc.faa.gov/pavement/3dfem.asp; FAA, 2008) for both flexible and rigid pavements. These design methods have been evaluated and calibrated on the basis of

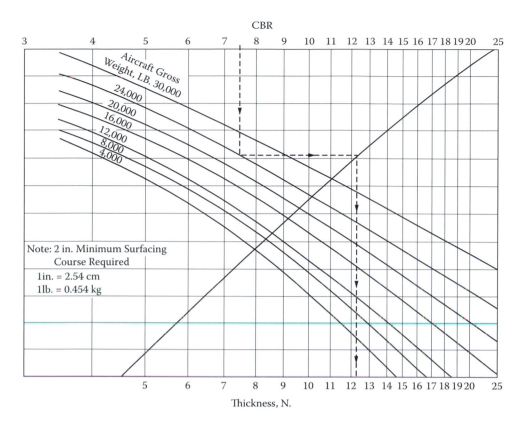

FIGURE 19.7 Design Curves for Flexible Pavement, Light Aircraft.
Source: From Federal Aviation Administration, AC 150/5320-6D (Washington, DC: FAA).

full-scale testing at the National Airport Pavement Test Facility (NAPTF; http://www.airporttech. tc.faa.gov/naptf/; FAA, 2006). The flexible pavement design method is the same as in the original layered elastic analysis program, LEDFAA (with LEAF as the main computational program), and utilizes subgrade strain and asphalt layer tensile strain as the two primary responses for design. For rigid pavement, the design method utilizes a 3D-Finite Element Method (FEM) model (NIKE-3D is the computational program) of an edge-loaded slab on multiple layers for design. Currently the trial version of LEDFAA is available for download.

19.5.1 Asphalt Pavements

For asphalt pavements, the primary reason for moving to a more sophisticated design from the CBR-based spreadsheet program is the necessity to design pavements for aircrafts with significantly higher loads and more complex gear configurations, compared to the aircrafts on which the original design curves were developed, as well as compared to the "wide-body aircrafts" for which adjustments had already been made to the original design curves/methods. Two notable examples of such aircraft are the Boeing B777 and the Airbus A380. The load and gear configurations of these aircrafts are shown in Figure 5.7 in Chapter 5. LEDFAA, being a M-E design, should be capable of conducting a more rational analysis and hence a more accurate design of pavements required by aircrafts such as the two mentioned above.

The failure criteria are based on two responses: compressive strain on top of the subgrade, to prevent rutting; and the tensile strain at the bottom of the HMA layer, to prevent bottom-up fatigue cracking. The material properties, for which default values are provided in the software, are stiffness modulus, E, Poisson's ratio, μ, and layer thickness, t. The main distinctive feature of this design process is the way of considering traffic and layer characteristics. There is no equivalent departure or design aircraft—the damage for each aircraft is calculated and added. The entire aircraft fleet mix must be input in the design.

The cumulative damage factor (CDF) is calculated as follows:

$$CDF = \sum \frac{n_i}{N_i}$$

where:

n_i = actual passes of individual aircraft
N_i = allowable passes of individual aircraft

With the CDF, 1 means the entire life of the pavement is used up.

The gear location and lateral distribution of the aircraft along the path of travel for each aircraft are considered, and the CDF is calculated for each 10-in.wide strip over a total 820-in. width. Miner's rule is used to sum the damage for each layer. Note that the vertical strain on top of the subgrade is computed at multiple points, with all of the wheels in the main gear contributing to the computed strain (for example, 16 wheels for the B747 and 20 wheels for the A380). The models are as follows.

19.5.1.1   For Vertical Strain ε_v on Top of the Subgrade

$$\text{Number or coverages to failure, } C = \left(\frac{0.004}{\varepsilon_v} \right)^{8.1}$$

where $C \leq 12,100$.

$$C = \left(\frac{0.002428}{\varepsilon_v} \right)^{14.21}$$

where $C > 12,100$.

19.5.1.2   For Horizontal Strain ε_h at the Bottom of the Surface Layer

$$\log_{10} C = 2.68 - 5 * \log_{10} \varepsilon_h - 2.665 * \log_{10} E_A$$

where E_A is the modulus of the asphalt mix surface layer.

FAARFIELD/FEDFAA is Windows-based software that allows the user to input a fleet of aircraft, for which an inbuilt library contains all relevant information such as gear layout, loads, and tire pressure. The pavement structure can be checked for the aircrafts for a design period, or, for a specific design period, the thickness required to be added can be determined. Note that the structural property of the layers used in this method is elastic modulus (E), and no longer CBR. The following expression can be used to convert one form to the other: E, psi = 1,500 (CBR).

19.5.2   RIGID PAVEMENTS

For rigid pavements, the CDF is calculated using the horizontal edge stress at the bottom of the Portland cement concrete layer. The edge and the interior stresses are determined using 3D finite element models. The steps in design consist of the following: (1) compute interior stress, (2) compute 75% of the free edge stress with gear oriented parallel to the slab edge, (3) compute 75% of the free edge stress with gear oriented perpendicular to the slab edge, and (4) select the highest of the three cases as the working stress for design.

The failure model is of the following form:

$$\frac{DF}{F_{cal}} = \left[\frac{F_s'bd}{(1-\alpha)(d-b)+F_s'b} \right] * \log C + \left[\frac{(1-\alpha)(ad-bc)+F_s'bc}{(1-\alpha)(d-b)+F_s'b} \right]$$

where:

DF = design factor, defined as $\frac{R}{\sigma}$

R = flexural strength of concrete

σ = concrete tensile strength

Fcal = stress calibration factor, 1.13

F's = stabilized base compensation factor, F's = 0.3 for stabilized base, 1 for nonstabilized base

$$\alpha = 0.8, a = 0.5878, b = 0.2523, c = 0.7409, d = 0.2465$$

Failure is defined as number of coverage for SC = 80 at any given value of $\frac{R}{\sigma}$.

19.6 NONDESTRUCTIVE TESTING AND REHABILITATION OF AIRFIELD PAVEMENTS

The purpose of using nondestructive testing and evaluation is to identify failure mechanisms in pavements and hence to develop the most cost-effective treatments/remedies. The process of evaluation should include a condition survey, a review of records, the development of a test plan, testing and evaluation, and the preparation of a report.

The records should provide information on user knowledge, as built drawings, aerial photos, maintenance information, and results from prior evaluations. The network to be evaluated can be defined on the basis of type of material (such as HMA or PCC) and its use (ramp, taxiway, or runway), thickness, age, and/or subsurface layers.

The condition survey is very important for checking surface conditions and for avoiding foreign object debris. It consists of working to identify distress—its quantity and severity—and combining the different data into a Pavement Condition Index (PCI), which is related to the condition of the pavement. The higher the PCI, the better is the condition. The data can be used to conduct condition analysis and develop pavement condition versus time models using software such as MicroPaver. This also helps in identifying pavements whose PCIs are close to or beyond critical PCI. Functional characteristics such as smoothness and surface friction also require testing.

When developing the testing plan for pavements, one can plan for destructive, semidestructive, or nondestructive testing. Destructive testing includes digging test pits; semidestructive testing can be coring or conducting a dynamic cone penetrometer (DCP); nondestructive testing can be tests such as falling weight deflectometer (FWD) or ground-penetrating radar (GPR).

Generally, pavement coring is done to obtain the thickness of the layers (to be used for analysis with GPR or FWD data) as well as to test HMA cores for their properties. The coring and removal of the upper layers can also provide access for DCP testing or for obtaining subbase, base, or subgrade soil samples.

Soil boring can be done with a drill rig, and can be done with split spoon sampling and the standard penetration test (SPT). For NDT, the DCP testing can be done to check the stiffness of the layers below the HMA layers, and the results can be converted to CBR (and to modulus, if required). The GPR can be used to determine the thickness of the different layers in the pavement, and detect buried structures or hidden conditions such as voids or trapped moisture.

The FWD is the most commonly used NDT for determination of the structural condition of the pavement. It applies an impulse load with a load pulse duration of approximately 25 ms. It represents a moving load more accurately than any other existing type of deflection-based NDT.

The Road Rater, a vibratory loading apparatus, is also used for testing. It can operate at 5–60 cycles per second, with loads of 1000–8000 lb.

The heavy weight deflectometer (HWD) is also used instead of the FWD to apply higher loads. However, note that if the layers are not sensitive to the load magnitude (force amplitude), the FWD can be used in place of the HWD.

The FWD/HWD applies a load and simultaneously measures the deflection at different locations. There are three relevant ASTM standards for deflection-measuring equipment. D4695-96 (Standard Guide for General Pavement Deflection Measurements) specifies the use of static, vibratory, and impulse load equipment; D4602-93 (Standard Guide for Nondestructive Testing of Pavements Using Cyclic Loading Dynamic Deflection Equipment; 2002) specifies the use of vibratory testing equipment; and D4694-96 (Standard Test Method for Deflections with a Falling Weight Type Impulse Load Device) specifies the use of impulse loading equipment.

The data from FWD testing can be used for both qualitative as well as quantitative evaluation of the pavement. The qualitative evaluation helps in understanding the variability of structural conditions within the network, whereas the quantitative evaluation provides the data needed as inputs for design (modulus, E). It can also provide information regarding voids in PCC and load transfer across joints in PCC pavements.

The impulse or dynamic stiffness modulus (ISM or IDM) provides an indication of the stiffness of the overall pavement. It is defined as follows:

$$I(D)SM = \frac{L}{d_0}$$

where L = applied load and d_0 = maximum deflection.

The I(D)SM can be plotted along the length of a runway, for example, to evaluate the variability of the structural condition in the initial stage of analysis. The more detailed analysis involves backcalculation of layer moduli, which can be used with a layered elastic design program. The backcalculation program BACKFAA can be downloaded from the FAA website (http://www.airporttech.tc.faa.gov/naptf/download/index1.asp; FAA, 2008).

For PCC pavements, closed-form solutions are used for the determination of stiffness from FWD test results. Test locations are at the center of the slab for layer properties, transverse joints for load transfer, slab corners for voids, and longitudinal joints for load transfer. Load transfer efficiency as well as voids can be evaluated.

The results of backcalculation analysis can be used in both the spreadsheet-based CBR design as well as LEDFAA. The backcalculated modulus of the subgrade can be used directly in LEDFAA, while it can be used to estimate CBR (through correlation, E = 1500 CBR) to be used in the spreadsheet design. In the pavement evaluation part, the primary inputs are traffic, pavement thickness and composition, pavement condition, layer properties, and subgrade strength.

Traffic, in the form of aircraft type, frequency, and weights, can be obtained from the airport master plan (forecast) as well as from the airport manager's office. The pavement thickness and composition data can be obtained from cores, test pits, as-built plans, and pavement management system reports.

The pavement condition inputs come from analysis of pavement condition data, which includes description and quantity of distresses and the PCI at different points along the feature. The layer properties include moduli values backcalculated from FWD, DCP, thicknesses from cores or GPR, and laboratory testing (split tensile test for PCC, and CBR on soils from boring).

For analysis of NDT data for estimation of layer properties for use in spreadsheet design, one can evaluate deflection basins, select analytical sections, and determine subgrade modulus—and then estimate CBR from correlation. The conventional FAA design method is based on CBR for flexible pavements and the Westergaard method for rigid pavements. For mechanistic methods, the layered elastic method is used for flexible pavements (which can be assumed as semi-infinite layers) and FEM is used for rigid pavements (which have joints).

The report of output of the pavement evaluation process includes allowable loads and the remaining life of a pavement feature (such as a runway). For the design aircraft, the allowable loads for different sections of the pavement are determined, and the lowest allowable load in that pavement is reported. Similarly, for the different sections the allowable number of load repetitions and hence the remaining lives are determined.

19.7 ACN-PCN

All airports in countries belonging to the International Civil Aviation Organization (ICAO) are obligated to report the strength of pavements (with bearing strengths greater than 12,500 lb) in terms of a standardized system, known as the ACN-PCN system. Details of this method and relevant calculations are provided in Advisory Circular AC 150/5335-5a (FAA, 2006).

ACN is defined as a number that expresses the relative effect of an airplane at a given weight on a pavement structure for a specified standard subgrade strength. PCN is a number that expresses the load-carrying capacity of a pavement for unrestricted operations.

The ACN-PCN system is structured so that a pavement with a particular PCN value can support, without weight restrictions, an airplane that has an ACN value equal to or less than the pavement's PCN value. This is possible because ACN and PCN values are computed using the same technical basis. Note that the ACN-PCN system is only intended as a method of reporting relative pavement strength so airport operators can evaluate the acceptable operations of airplanes, and is not meant for pavement design or evaluation.

The ACN-PCN system is based on four standard levels of subgrade strengths (modulus of subgrade reaction for PCC pavements, and CBR for flexible pavements). The ACN for each aircraft is provided by the aircraft manufacturer, and can be determined using the FAA's COMFAA software (www.airporttech.tc.faa.gov/naptf/download/index1.asp; FAA, 2008).

The ACN for rigid pavements is derived on the basis of the Westergaard solution for a loaded elastic plate on a Winkler foundation (interior load case), assuming a concrete working stress of 399 psi. For flexible pavements, the CBR method based on Boussinesq's solution for stresses and displacements in a homogeneous, isotropic elastic half-space is used.

To standardize the ACN calculation and to remove operational frequency from the relative rating scale, the ACN-PCN method specifies that ACN values be determined at a frequency of 10,000 coverages.

A single wheel load is calculated to define the landing gear–pavement interaction for each type of pavement section. The derived single wheel load implies equal stress to the pavement structure and eliminates the need to specify pavement thickness for comparative purposes. This is achieved by equating the thickness derived for a given airplane's landing gear to the thickness derived for a single wheel load at a standard tire pressure of 181 psi (1.25 MPa). The ACN is defined as two times the derived single wheel load (expressed in thousands of kilograms).

The determination of a pavement rating in terms of PCN is a process of determining the ACN for the selected critical (i.e., most demanding) airplane and reporting the ACN value as the PCN for the pavement structure.

The PCN can be determined in two ways—the airplane method (simpler) or the technical evaluation method (requires more time and resources).

The airplane method is useful when information on traffic and runway characteristics is limited and engineering analysis is neither possible nor desired. In the airplane method, the ACN for all airplanes currently permitted to use the pavement facility is determined and the largest ACN value is reported as the PCN. The assumption in using this method is that the pavement structure has the structural capacity to accommodate all airplanes in the traffic mixture and that each airplane is capable of operating on the pavement structure without restriction. Note that the pavement capacity can be significantly overestimated if an excessively damaging airplane, which uses the pavement on a very infrequent basis, is used to determine the PCN.

Example of the Use of the Airplane Method

Consider a flexible pavement runway with a subgrade strength (CBR) of 10 and traffic having the operating gross weights and ACNs shown below.

Aircraft	Operating Weight (lb)	Tire Pressure (psi)	% Gross Weight on Main Gear for ACN	ACN (from COMFAA)	Annual Departures
B737-300	130,000	195	90.86	32	5000
A300-B4	370,000	205	94.00	57	2000
B767-300ER	370,000	190	92.40	50	1500
B777-200	600,000	215	95.42	52	500

The PCN should be expressed as follows:

PCN (highest number from the table above)/F or R (depending on flexible or rigid)/code for CBR (given in Table 2-2)/code for tire pressure (given in Table 4-2) / U or T (depending on whether the airplane method or the technical evaluation method is used)

Therefore, in this example, the PCN is

$$PCN57/F/B/X/U$$

19.8 REHABILITATION OF PCC AIRPORT PAVEMENTS

The final step is to decide on the rehabilitation option, depending on whether the problem is functional or structural. In deciding on a specific method, the considerations should include those of economy, in terms of initial construction cost as well as life cycle cost (in terms of present worth), operational disruption, local capabilities, and reliability. Alternatives include mill and fill, perpetual pavement, overlay, break and seat, rubblization, recycling, or reconstruction.

The condition of the pavement section is measured in terms of serviceability, expressed as a Pavement Condition Index (PCI). The serviceability of a pavement goes down with time after construction, and there is a critical PCI (65 for primary airports and 55 for small airports—details given in the Government Accountability Office report GAO/RCED-98-226; Government Accountability Office, 1998) below which rehabilitation must be done to improve the condition of the pavement. If the condition of a pavement warrants rehabilitation, as mentioned earlier, an investigation must be conducted to determine the cause of the failure and select the best rehabilitation alternative. The investigation can be done on the basis of existing information, a visual condition survey (as part of the pavement management system), nondestructive testing, pavement coring and boring, traffic and pavement analysis, as well as drainage analysis.

There are two available options for rehabilitating a PCC airport pavement using HMA. The first is an HMA overlay over the existing PCC pavement, with saw cut and seal on top of the joints or providing a crack relief layer. The second is the application of an HMA overlay over a fractured or rubblized PCC pavement.

The sequence of work in the first option with the crack relief layer includes repairing the existing concrete slabs, installing edge drains, placing a crack relief layer, providing a HMA leveling course, and finally applying a HMA surface course. The work with the crack/break and seat approach includes removing the HMA overlay, if any; correcting drainage problems; cracking/breaking concrete; seating cracked pieces; removing and patching soft areas; sweeping pavement surfaces; applying a tack coat; and placing an HMA leveling and surface course. The modulus of cracked/broken and seated concrete can be between 500 and 800 ksi. The rubblization process includes removing HMA overlay, if any; correcting drainage problems; rubblizing PCC pavement (generally, the top half or two-thirds is rubblized); rolling rubblized concrete; applying a prime coat; and placing HMA leveling and overlay courses. The modulus of rubblized concrete can range from 200 to 500 ksi.

The break and seat process is considered to be more costly and more effective compared to the rubblization process.

19.9 CONSTRUCTION QUALITY CONTROL AND ACCEPTANCE TESTING

The FAA's Advisory Circular 150/5370-10A presents standards for specifying the construction of airports. The main features of the quality control and quality assurance (QC/QA) of HMA (which is specified under P401 specifications) are that the contractor has the full responsibility for all QC testing, and the engineer has the full responsibility for all QA testing, with the provision that in both QC and QA, the testing laboratory (or laboratories) meets the requirements of ASTM D-3666 (Standard Specification for Minimum Requirement of Agencies Testing and Inspecting Bituminous Materials). The contractor must employ technicians who are certified by the appropriate agency, and should submit the QC plan to the engineer.

The main scope of work for QC includes the following:

1. Asphalt content: a minimum of two tests per lot
2. Aggregate gradation: a minimum of three per lot
3. Moisture content of aggregate: a minimum of one per lot
4. Moisture content of mixtures: a minimum of one per lot
5. Temperature: a minimum of four times per lot to determine temperatures of the dryer, asphalt in the storage tank, mixture at the plant, and mixture at the site
6. In-place density: required to ensure that specified density is achieved
7. Sampling: at the discretion of the engineer
8. Control charts: for the production process
9. Documentation: includes daily reports

The main scope of work for QA involves the following:

1. Tests for stability, flow, and air voids: for each sublot and evaluation of acceptance on the basis of the percent within limits (PWL) specification; air voids are calculated on the basis of the maximum specific gravity measured in each sublot.
2. Mat and joint density: evaluation of acceptance for each lot on the basis of PWL; cores from mats are taken at a minimum of 1 ft distance away from the joint; location of method of determination of joint density (such as with cores taken directly at the joint) varies; evaluation of acceptance of joint density on the basis of PWL; one core is taken from each of four sublots in a lot.
3. Thickness: measured from cores taken for density measurement.

4. Smoothness: measured with a 12-ft straight edge perpendicular and parallel to the center-line at a minimum distance interval of 50 ft; compacted surface shall not vary more than 3/8 in. for base courses and ¼ in. for surface courses; remedial actions are required when more than 15% of all measurements within a 2000 sq yd lot exceed tolerance.
5. Grade: measured by running levels at intervals of 50 ft or less; compacted mat shall not vary from the gradeline elevations and cross sections shown in the plans by more than ½ in.; remedial actions are necessary, when more than 15% of all the measurements within a lot are outside the specified tolerance.

Example of Specifications

For plant-produced mix, variability of the following:

1. Stability (lb) $= \pm 270$
2. Flow (0.01 in.) $= \pm 1.5$
3. Air voids (%) $= \pm 0.65$

For in-place mixes (results from tests with cores, which are compared to results from tests on laboratory-compacted samples):

1. For mat density, contractor should target 98.5% (mat density variability is 1.3%).
2. For joint density, contractor should target 96.5% (joint density variability is 2.0 to 2.1%).
3. For air voids, mixes designed at 3.5% have the most tolerance for acceptance.

A few of the important items are as follows:

1. During production, check stability and flow on the job mix formula.
2. Obtain thickness from cores.
3. Resampling is done for mat density only.
4. Test section is required, whose dimensions are provided, along with specific instructions regarding how it should be tested ("300 feet long, 20–30 feet wide with a longitudinal joint included; treated as a lot with three sublots for acceptance and pay").

19.10 CONSTRUCTING, CHECKING, AND IMPROVING FRICTION/SKID RESISTANCE OF RUNWAYS

Checking the friction numbers of runways, and improving them if necessary, are critical for the safe operation of airports. Good friction is important for traction and good braking performance, and to avoid skidding and loss of control of aircrafts. Good friction is provided as surface finish, as well as through regular checking and maintenance operations. FAA Advisory Circular AC 150/5320-12C provides guidance on the measurement, construction, and maintenance of skid-resistant airport pavement surfaces.

For the design and selection of construction materials of pavement structures, drainage factors (such as slope) and micro- and macrotexture of the surface should be taken into consideration. Microtexture, contributed by fine-particle surfaces, provides friction at low speeds and when making contact between tire and residual rainwater; while macrotexture, provided by the roughness of the pavement surface as a whole, is more important at high speeds, and for providing drainage during rainstorms. Macrotexture can be improved by selecting suitable coarse aggregates (those with resistance to polish, a rough texture, and an angular shape). Skid resistance for asphalt mix pavements can be improved with the use of a 0.75–1.5-in. thick layer of permeable friction course, which has coarse gradation and large interconnected voids for the quick drainage of water. Chip seals and aggregate slurry seals can also be used for temporary enhancement of skid resistance. For Portland cement concrete (PCC) pavement, brush, broom, or burlap drag or wire combing/wire tining can be used to texture the surface while it is still in plastic state.

For both asphalt mix and PCC pavements, grooving can be done by the transverse saw cut method. For new asphalt mix pavements, grooving is recommended after 30 days of paving; while for PCC pavements, a vibrating ribbed plate or a ribbed roller can be used to cut grooves while it is in plastic state, or a saw cut can be used if it is in a hardened state. The FAA specifies the depth and width of grooves as 1.4 in. × 1.4 in., at ½ in. center-to-center spacing, for the entire runway, perpendicular to the direction of aircraft movement.

Over time, the skid resistance of surfaces deteriorates due to the polishing action of the aggregates as well as the accumulation of materials such as rubber from tires (especially in touchdown areas) in the void spaces of permeable friction course (PFCs) or surface grooves. The problem is made worse by the spilling of jet fuels and presence of ice and snow on the pavement surface. Therefore, regular friction surveys at specified intervals (depending on the number of jet aircrafts landing daily per runway end, specified by the FAA) should be conducted, using continuous friction-measuring equipment (CFME), preceded by visual inspection. In addition, periodic inspection of groove depth should be made, and corrective actions should be taken if 40% of the grooves are found to be equal to or less than 1/8 in. in depth and/or width for a distance of 1500 feet.

The friction surveys should be conducted 10 and 20 feet to the right of the runway centerline at speeds of 40 and 60 mph, using water from the CFME equipment to wet surfaces in front of the friction-measuring tires to provide water of at least 0.04 in. depth. Periodic inspections of water accumulation in depressed areas during rainstorms must be made, and corrective actions should be taken for water depths exceeding 1/8 in. The FAA provides friction numbers (Mu) for minimum, maintenance/planning, and new constructions, depending on the type of equipment used, for 40 and 60 mph measurements. For example, the three numbers for the Dynatest Runway Friction tester (Figure 19.8) are 0.5, 0.6, and 0.82, respectively, for 40 mph, and 0.41, 0.54, and 0.72 for 60 mph testing. The NASA grease smear method is used for evaluating the macrotexture of the pavement surface.

The rubber removal techniques may affect the structural integrity of the pavement, and hence the use of PFCs in airports with aircraft traffic has not been recommended by the FAA. Generally,

Water Nozzle

FIGURE 19.8 Friction Testing.
Courtesy: David Morrow, Dynatest Consulting Inc.

contaminants such as rubber are removed by using high-pressure water, approved chemicals, high-velocity impacts, or mechanical grindings, under dry conditions, until the friction number of the contaminated area has been raised to within 10% of that of the uncontaminated areas. Also, to check whether the techniques result in structural damage, a test section can be used for demonstration.

Painted areas in airport pavements can provide less friction, and the skid-resistance properties of such areas can be improved by adding silica sand or glass beads to paints.

19.11 ASPHALT MIXES FOR AIRPORT PAVEMENT

The mix design procedures discussed in Chapter 14 are applicable for airport pavements. In many cases (such as FAA), the Marshall design procedure is being used, although a change to the Superpave (as discussed in Chapter 14) system is expected in the near future. There are some specific issues regarding the mix design and construction of asphalt mix pavements for airports, which are discussed below.

19.11.1 FUEL-RESISTANT MIX

Jet fuel from parked aircrafts can damage asphalt mix on the surface. Specially formulated asphalts have been used to counter this problem. Mass loss (maximum allowable 1%, for example) of an asphalt mix when immersed in jet fuel for 24 hours (plain immersion or immersion and application of brushing) has been used to check the suitability of a material. Coal tar sealers have been applied in the past, although currently specialty products (tar free), such as Sealoflex, are being used. Fuel-resistant asphalts (often modified with polymer) have been used in many airports such as those in Boston and New York City (La Guardia), United States; Cairo, Egypt; Saint Maarten, the Netherlands; and Kuala Lumpur, Malaysia.

19.11.2 CONSTRUCTION AND MAINTENANCE OF LONGITUDINAL JOINTS

Damaged longitudinal joints are of very serious concern in airfield pavements. Loose materials from such areas can cause foreign object damage to aircrafts, leading to loss of life and equipment. The different good techniques discussed in Chapter 14 can be adopted for airfield pavement construction for obtaining good joints. Wherever possible, suitable-size aggregates (finer aggregates produce better joints) and echelon paving should be used. Typical distresses such as cracking and raveling in longitudinal joints happen in about 4–5 years after construction. Initially a crack occurs along the joint, which leads to secondary cracks. Such cracks should be sealed promptly before they become a problem. Hot poured rubberized asphalt is usually used for crack sealing.

19.11.3 TIME OF CONSTRUCTION

In many cases, airport pavement work is conducted at night under a tight schedule (to avoid busy hours of aircraft traffic). Proper management of construction work is crucial for a successful operation, and contractors with proven experience in handling airport jobs are preferred. Mixes and materials must be obtained from plants that are close by, and if required, mixes that require lower temperatures, and hence can be stored on site (such as warm mix asphalt), can be considered.

QUESTIONS

1. What are the different types of pavements in an airport?
2. Design an asphalt pavement runway to be built on an area with subgrade soil which has the following properties. Use the spreadsheet method.

Percentage passing No. 200 sieve = 80%; liquid limit on minus #40 material = 32%; plastic
limit on minus #40 material = 5%; percentage finer than 0.02 mm = 16%; results from
three CBR tests conducted in the laboratory = 5, 6, 5. The unit weight of the soil is 112
pcf.

From a weather station very close to the airport site, the average of the three coldest winters
in a 30-year period shows that the design Air Freezing Index is 1500 degree days.

The forecasted air traffic for the airport is as follows:

Aircraft	Gear Type	Average Annual Departures	Maximum Takeoff Weight (lb)
B737-200	Dual	5000	128,600
ABA300-B2	Dual tandem	1000	304,000

3. Design the above pavement using FAARFIELD/LEDFAA 2.
4. Name a very important form of maintenance work for airport pavement, and describe
   briefly how it is conducted.

20 Nondestructive Tests

Nondestructive tests used in pavement engineering can be broadly divided into several categories: nuclear equipment, deflection based, electromagnetic, and seismic equipment. Examples of such equipment are discussed in the following sections.

20.1 NUCLEAR GAGE

The basic operational principle of a nuclear gage is that a source (such as Cesium-137) of nuclear particles (such as photons) is inserted into (for example, 6 in.) or placed on the surface of the layer whose density is to be measured. As these particles are released into the layer, a detector on the equipment on the surface of the layer detects and counts the number of particles coming back to that point. As the density increases, the number of detected particles decreases. Working on this principle, the gage is calibrated using a block of known density, and then it is used for determining the density by using the calibration factor.

The properties that can be measured include the density of hot mix asphalt (HMA) or soil as well as the moisture content of soil. The mode of transmission of the particles from the source inside the layer to the detectors can be either direct (when the source is inserted into the layer) or backscatter (in which case the particles are reflected off layers, when the source is placed on the surface). A set of Geiger-Mueller tubes is used as detectors (two are used for a thin-layer density gage) for measuring density, whereas an He 3 tube is used as a detector (of neutrons) for moisture content. The available nuclear gages have sufficient capabilities of storing data that can be collected from multiple tests, and are generally powered by rechargeable NiCd batteries.

The relevant American Society for Testing and Materials (ASTM) standard is D2922. The equipment needs to be verified for calibration by measuring the density of a block of known density, ensuring that the reading is within ±2 pcf of the actual density of the material at each depth, and comparing the densities measured at any time to those measured earlier (say, an average of the last four). Factors that can influence this "standard count" include temperature, the time elapsed between measurements, the presence of any affecting material (such as a wall, other gage, or person) close to the gage, and operator error. In the backscatter technique the measurements are very sensitive to surface roughness, and special care must be taken to make the surface level with water or fine sand (note that this leveling would increase the density by 1–1.5%).

The ratio of the particle count obtained on a test material to that obtained from a standard block is known as the *count ratio*. The gage is calibrated by measuring the count ratio for blocks of materials such as aluminum, magnesium, granite, and limestone of known densities. Procedures are available for ensuring the short- and long-term electronic stabilities of the system, and correcting measurement anomalies due to factors such as a natural decrease in intensity of particles due to decay of the source material, test location close to a wall, varying material composition or surface features, and absorption of rays because of the elemental composition of the material. Typical equipment is shown in Figure 20.1.

FIGURE 20.1 Nuclear Gage Equipment Used on Hot Mix Asphalt (HMA).

20.2 FALLING WEIGHT DEFLECTOMETER (FWD)

A falling weight deflectometer (FWD) is extensively used for estimation of pavement layer moduli and for determination of the structural condition of pavements. The information obtained from FWD testing can be used in structural analysis to determine capacity, estimate expected performance life, and design a rehabilitation plan for pavements. Deflections prior to and after pavement rehabilitations are done to evaluate the effectiveness of specific rehabilitation methods. An FWD can also be used to test load transfer efficiency of joints within concrete pavements. American Society for Testing and Materials standards are available for the use of an FWD for pavement deflection-based testing (see *ASTM 4694, 4695-96*). Standard calibration procedures for load cells and deflection sensors are also available.

In this testing, an impulse load is generated by dropping a mass ranging from 6.7 kN to 120 kN through different heights onto a base plate through a set of rubber buffers (Figure 20.2). There are two sets of sensors—a load cell to measure the applied load and an array of geophones (which are located on the surface of the pavement) to measure the resulting deflection. FWD tests can be conducted at different load levels, although some protocols would require the deflections to be normalized to a 9,000-lb load and "corrected" for temperature (if different from the standard 77°F). For example,

$$\text{Adjusted Center Deflection} = D_0(1.598837 - 0.009211683 * T^{0.96}) \tag{1}$$

where T = temperature of the asphalt pavement surface (°F).

An example of FWD equipment is the Dynatest Model 8000 FWD, which has a load range of 7–120 kN. It is a fast (up to 60 test points per hour) and reliable tool that can be operated by a single person. This is self-sufficient equipment, with sensors for the measurement of air and surface temperature, and includes Windows-based ELMOD 4 software for data analysis.

20.2.1 DIRECT USE OF DEFLECTIONS

Using the deflections noted at the different sensors, a series of expressions can be used to determine relevant properties of the pavement, such as the following.

Array of Geophones on the Pavement Surface

FWD Unit on Trailer

Real Time Data Acquisition—GPS
Coordinates, Load and Deflections

FIGURE 20.2 Falling Weight Deflectometer (FWD).

The *area parameter* (which can be used together with the center deflection) represents the normalized area of a slice taken through the deflection basin from 0 to 0.9 m (0 to 3 feet).

$$\text{Area Parameter} = \frac{6(D_0 + 2D_{12} + 2D_{24} + D_{36})}{D_0}$$

where D_0, D_{12}, D_{24}, and D_{36} are the deflection readings (in inches) from sensors located 0, 12, 24, and 36 in. from the center of the FWD loading plate, respectively.

$$\text{Area Correction Factor} = 0.7865 + (1.4578 * 10^{-4} * T^{1.68})$$

where T = temperature of the asphalt pavement surface (°F).

The Surface Curvature Index (SCI) and shape factors are indicators of the relative stiffness of the upper pavement layers.

$$SCI = D_0 - D_2$$

$$Shape\ Factor = \frac{(D_0 - D_2)}{D_1}$$

where D_0 is the center deflection and D_1 and D_2 are the deflections at the first and second sensors, respectively (located at 8 and 12 in. from the load).

Subgrade modulus:

$$Subgrade\ Modulus = -466 + \left(9000 \times \frac{0.00762}{\left(\frac{D_{36}}{1000} \right)} \right)$$

where the subgrade modulus is given in psi, and D_{36} is the deflection reading (in inches) from the sensor located 36 in. from the center of the FWD loading plate.

A software program for calculating the above parameters (FWD) can be downloaded free from the Washington State Department of Transportation (DOT) website at http://www.wsdot.wa.gov/biz/mats/pavement/testing.htm#fwd (2007).

20.2.2 BACK-CALCULATION

Another approach to using the FWD data is to "back-calculate" moduli of the different layers of the pavement, using mechanistic principles. This process involves using layer thickness and Poisson's ratio (which must be known, estimated, or assumed with reasonable accuracy), assuming moduli of the different layers ("seed" moduli), and determining the different deflections at the locations of the FWD sensors. The process is repeated until the calculated deflection basin is found to be approximately identical to the measured deflection basin (within tolerances). An example of a criterion to stop the back-calculation process is that the summation of the "errors" (the difference between calculated and actual deflections) is less than 5%. Note that the modulus of the subgrade can be determined from the FWD deflections at the outer sensors without back-calculation.

Back-calculation can be done by different software, some of which is freely available. The Evercalc software is available free as part of the Everseries suite of software from the Washington State DOT website at http://www.wsdot.wa.gov/biz/mats/Apps/EPG.htm (2007). This is a mechanistic-based pavement analysis program that uses an iterative process of matching the measured surface deflections with the theoretical surface deflections calculated from assumed moduli. The program is capable of handling up to five layers and can be run with or without a rigid or stiff base. User-provided seed moduli are not required. The seed moduli can be estimated using the internal regression equations that are based on relationships between layer moduli, load, and various deflection basin parameters. Provisions are available for nonlinear material behavior. The main steps are indicated below. Note that other software (such as MODULUS, available from the Texas Transportation Institute at http://tti.tamu.edu; n.d.) have different features but similar steps.

Steps in Performing Back-Calculation with Evercalc Program

1. The Evercalc program is opened.
2. "File" is selected, and "General File" is chosen.
3. The file is opened, it is named, and the units are changed to U.S. units.
4. The appropriate number of layers is selected.

5. The number of sensors is chosen to be 9, and the plate radius is input as 5.9 in.
6. "Stiff Layer" and "Temp. Correction" are both unselected, and the "Sensor Weigh Factor" is chosen to be uniform (as an example).
7. The radial offsets for sensors 1 through 9 are selected to be 0, 8, 12, 18, 24, 36, 48, 60, and 72, respectively.
8. The layers are then filled:
   Select the appropriate layer IDs.
   For HMA, select Poisson's ratio as 0.35. For soil layers, select 0.4.
   Input the values of "Initial Modulus," "Min. Modulus," and "Max. Modulus."
9. The max iterations are chosen to be 5, and the "RMS Total %" and "Modulus Tot.%" are also chosen to be 5.
10. "Stress and Strain Location" is chosen, the general file is saved, and the main menu is displayed.
11. Next, the Deflection file is opened.
12. The appropriate pavement temperature is selected.
13. The appropriate layer thicknesses are selected.
14. The appropriate number of drops is selected.
15. The deflection data from the FWD are put into the section with the drops.
16. The file is saved (should be the same prefix as the general file), and the main menu is displayed.
17. The back-calculation is performed by choosing "File." "Perform Backcalculation" is pressed, and then "Interactive Mode" is chosen. The name of the general file is chosen, the "OK" button is pressed, and then the name of the corresponding deflection file is chosen. Next the output file is chosen, and then the summary file. This prompts the output, and "OK" is pressed at the bottom and a graph is displayed. This graph shows how closely the input into the program matches the calculated values.
18. The window is then closed, and "Print" is chosen in the top tool bar. "Print/View Output" is chosen, and the file is selected to view the summary.
19. The "Average RMS(%)" is viewed.
20. This whole process is repeated, with the modulus values of the layers changed, until the error is near 6% or less.
21. Tips: if the bottom of the calculated deflection basin deviates from the actual data, the modulus of the top layer needs to be changed; changing the middle two layers controls the middle of the deflection basin, and the top part of the deflection basin is controlled by changing the modulus of the bottom layer.

Numerous studies have been conducted on the accuracy of load and deflection measurements as well as on the refinement of the back-calculation procedures, resulting in continuous improvement of both equipment as well as analysis procedures. As a result, the FWD has become the principal nondestructive testing tool for pavement engineers. The use of back-calculated moduli using an FWD has been recommended for rehabilitation of pavements by the newly developed NCHRP-1-37A pavement design guide. Such use is absolutely necessary for determination of moduli of subsurface layers as well as for determination of moduli and consistency of new and innovative pavement materials. Examples of such use include the testing of full depth reclaimed pavements with different types of additives, such as foamed asphalt. Furthermore, the FWD is being used by numerous researchers in the determination of pavement layer properties in full-scale test sections and for the evaluation of subgrades under unusual conditions as well as the effect of environment and new materials on pavement properties.

One key requirement for the successful application of an FWD is the use of accurate layer thickness and condition data. Without the use of any nondestructive tool, the only solution is to take cores, in sufficient numbers, such that accurate estimates of layer thickness and conditions

FIGURE 20.3 Portable Falling Weight Deflectometer.
Courtesy: Michael Santi, Idaho DOT, and Maureen Kestler, USDA Forest Services.

can be made. However, taking cores defeats the whole idea of nondestructive testing and hence is not an attractive option. Nondestructive instruments such as ground-penetrating radar (GPR) have now become very valuable tools for pavement engineers for providing the data necessary for using an FWD.

20.3 PORTABLE FALLING WEIGHT DEFLECTOMETER

The portable FWD (Figure 20.3) is a single-person-use portable device that uses an accelerometer to determine the deflection due to a falling weight. The results are given as maximum deflection, and the bearing capacity modulus is calculated from it. This device can be used to estimate the stiffness of the upper layer of the pavement or to compare the stiffness and hence determine the effectiveness of compaction. The weight is of 10 kg falling through a height of 80 cm onto a plate of diameter ranging from 140 to 200 mm. The peak force is 25 kN, with impulse duration of 25–30 ms.

20.4 ROLLING WHEEL DEFLECTOMETER (RWD)

The RWD is a truck-based system that can be used to measure the deflection under moving dual tires of an 18-kip single-axle load for primarily flexible pavements. The equipment (Figure 20.4) consists of a semi trailer with four laser units in the undercarriage and close to the pavement surface. Of the four laser units, three (located forward and outside of the deflection basis) are used for measuring the unloaded pavement profile, while the fourth one (located between and under the dual tires) is used for determination of the deflected profile. The deflection is measured as the difference between the deflected and the undeflected profile. A reading is taken at every 0.1 in., and the data are averaged to reduce the random error of measurement. The tractor unit contains all of the required data acquisition system, and the equipment can also record digital images and videos of the test section during testing as well as GPS coordinates. The test data can be used for network-level analysis in pavement management as well as for delineation of segments for rehabilitation, and to establish limits of deflection for new pavements and load restrictions for spring thaw periods. Since it can measure the deflection while moving at highway speeds, there is no need for lane closures and traffic control during its operation, and as much as 200–300 lane miles can be tested in one day.

RWD Semi Tractor Trailer

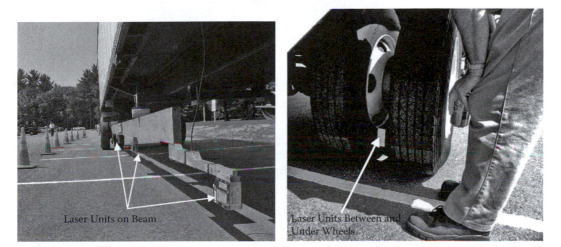

Laser Units on Beam

Laser Units Between and Under Wheels

FIGURE 20.4 Rolling Wheel Deflectometer (RWD).

20.5 GEOGAUGE (SOIL STIFFNESS GAUGE) FOR SOIL COMPACTION

The GeoGauge (Soil Stiffness Gauge; Figure 20.5) can be used for measuring soil layer stiffness as part of the soil compaction control process. This handheld equipment works by measuring the impedance at the surface of the soil. It measures the stress imparted to the surface and the resulting surface velocity as a function of time. The GeoGauge imparts very small displacements to the soil ($< 1.27 \times 10^{-6}$ m or $< .00005$") at 25 steady-state frequencies between 100 and 196 Hz. Stiffness is measured as a ratio of force to displacement, and can be used to determine the modulus. The stiffness is determined at each frequency, and the average of the 25 is displayed. The GeoGauge is supported on the soil by a ring-shaped foot through isolators. Shakers attached to the foot measure force and displacement. Moduli can be calculated with the use of Poisson's ratio. The equipment is powered by batteries and can store multiple readings, and the data can be downloaded to a computer. The equipment is calibrated by measuring force and displacement of a reference mass.

20.6 GROUND-PENETRATING RADAR (GPR)

The GPR works on the principles of radiating short electromagnetic pulses from an antenna and analyzing the amplitude and velocity of the reflected waves from pavement layers (Figure 20.6). The pavement thickness and density can be computed from these amplitudes and arrival times. The GPR can be used as a handheld device for use on short stretches of pavement, but for pavement thickness measurement of long stretches of test pavements generally a vehicle-mounted 1 GHz horn antenna GPR system is used (Figure 20.6). Data are collected at normal driving speeds, and passes can be made out in the wheelpaths and centerline of each lane in each direction. The equipment consists

FIGURE 20.5 GeoGauge Equipment.
Courtesy: Humboldt Mfg. Co.

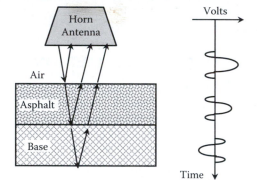

Measurement Setup   GPR Waveform   Sample Field Data

Antenna

FIGURE 20.6 Schematic of Operation and Ground-Penetrating Radar (GPR) Equipment.
Courtesy: Kenneth Maser, Infrasense, Inc.

of an electronic box and an antenna, together with a tether and a laptop computer with the software required for data acquisition and analyses. The data can then be analyzed to develop continuous thickness profiles, which are very useful for determination of pavement condition (along with other devices such as the FWD) for rehabilitation or estimation of remaining life, as well as for segmentation of a project for proper consideration of mix design and construction methods for rehabilitation or recycling operations.

20.7 PORTABLE SEISMIC PAVEMENT ANALYZER (PSPA)

The portable seismic pavement analyzer (PSPA; Figure 20.7) is a rapid nondestructive testing device that provides the modulus of the top pavement layer in real time. It is a portable unit that consists of a "source," two "receivers," and an electronics box, and comes with a tether and a laptop computer with the software that controls both data acquisition as well as analysis. The analysis procedure uses the surface wave energy to determine the variation in modulus with wavelength (strictly speaking, surface wave velocity with wavelength). The analysis method implemented in the PSPA is called the ultrasonic surface waves (USW) method, which is a simplified version of the spectral-analysis-of-surface-waves (SASW) method. Briefly, this method utilizes the surface wave energy to determine the variation in modulus with wavelength. For simplicity, the surface wave velocity is converted to modulus using mass density and Poisson's ratio.

PSPA with Data Acquisition System

Source (Foreground) and Two Receivers in PSPA Unit

Laptop Computer Shows Results of
Analysis within Seconds

FIGURE 20.7 Portable Seismic Pavement Analyzer (PSPA).

FIGURE 20.8 Free Free Resonant Column (FFRC) Test.

Up to a wavelength equal to the thickness of the top pavement layer, the moduli from the dispersion curve are equal to the actual moduli of the layer. As such, the modulus of the topmost layer can be directly estimated without a need for back-calculation.

20.8 FREE FREE RESONANT COLUMN (FFRC) TEST

The basic principle of this test is that the resonant frequency that is developed within a specimen of a certain material when it is subjected to an impulse load is dependent on its dimension and stiffness. Hence in this test, a specimen of known dimension is subjected to a tapping by hammer (Figure 20.8) fitted with a load cell, and an accelerometer is used to record the resulting signal at the other end of the specimen. The amplitude-frequency plot, displayed by a data acquisition system, which is connected to both load cell and the accelerometer, is used to determine the resonant frequency. The resonant frequency is then used to calculate the modulus, using the mass density (ASTM C-215).

20.9 ULTRASONIC TEST

In this test (Figure 20.9), a transducer-generated electric pulse is used to cause a mechanical vibration in one side of a specimen, and a transducer is used to sense the propagating wave at the other end of the specimen. The time of travel of the wave through the specimen is measured, and the modulus of the material is calculated using the bulk density and Poisson's ratio.

20.10 MAGNETIC INDUCTION TOMOGRAPHY (MIT)

The MIT-Scan equipment (Figure 20.10) uses magnetic tomography to detect and evaluate metal dowels placed in concrete pavements. The equipment consists of five sensors that emit an electromagnetic pulse and detects the induced magnetic field in the dowels. It is towed on rails across a concrete pavement slab, and both numerical and graphical data are output quickly from dedicated software in the attached computer. The device utilizes a number of sensors to capture redundant data, as well as filtering techniques to detect both the position and orientation of dowels inside the concrete.

FIGURE 20.9 Ultrasonic Test Equipment.
Courtesy: Karen O'Sullivan, WPI.

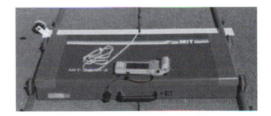

FIGURE 20.10 MIT-Scan Equipment on Rails on Concrete Slab.
Source: U.S. Federal Highway Administration, 2004.

QUESTION

1. Use any back-calculation software to estimate the moduli of the different layers in a pavement in a rocky fill area in Maine, which has the following results, from an FWD test. Make reasonable assumptions, as required.

   Layer 1: HMA surface and binder (consider as one layer) thickness = 4 in.; layer 2: granular base thickness = 12 in.; layer 3: subgrade

   FWD test details and data:

   Plate radius (in.): 5.9; number of sensors: 7

   Sensor offsets (in.): .0, 8.0, 12.0, 18.0, 24.0, 36.0, and 60.0

   Load (lbf): 8953.0

   Measured deflections (mils): 7.250, 5.420, 4.690, 4.000, 2.960, 2.230, and 1.690

① esti

21 Economic Analysis and Cost-Saving Concepts

21.1 ECONOMIC ANALYSIS

A proper economic analysis should be conducted for every pavement construction project. In order to conduct economic analysis, it is necessary to understand some key concepts and the techniques which help us to compare alternatives on the basis of economics and have reasonable estimates of different types of costs associated with pavement projects.

21.1.1 Engineering Economy

Economic analysis can be conducted at two levels—first, at the network level to determine the feasibility and scheduling of a project, and second, at the project level to achieve maximum economy within a specific selected project by comparing different alternatives. For the second case, all of the alternatives must be able to satisfy the project requirements, and should be considered over the same time period. The economic evaluation of the different alternatives must consider accurate (as far as possible) estimates of economic variables, and costs and benefits associated with the pavement. Life cycle cost (LCC) analysis provides a rational method of conducting such economic analysis.

21.1.2 Concept of Life Cycle Cost

The life cycle cost means the total cost that is incurred during the complete life cycle of the pavement. This includes construction, maintenance, and rehabilitation costs and considers return from salvage, if any. The utility of computing LCC is that it may be more prudent to spend more money to build a better pavement initially and spend comparatively less money in maintenance and rehabilitation than it is to save money by building a lower quality pavement, which will require more and/or more frequent maintenance and rehabilitation, and therefore will require spending significantly more money over the long run.

The techniques used for conducting economic analysis to compute LCC vary, and they depend on many factors such as the level at which the analysis is being conducted (network or project), the applicability of the techniques to the specific type of project (for example, public or privately owned), and the consideration of level of acceptance from the decision maker and the public (which, for example, may be more sensitive to initial construction costs). However, all applicable techniques must consider future costs and benefits, and the differences between the different types of alternative pavement types for the specific project.

21.1.3 Techniques

The different steps involved in LCC analysis are establishing alternatives, determining timings of activities, estimating costs, and computing the costs, respectively. Two of the more commonly used economic indicators are the net present value (NPV) or net present worth (NPW) and equivalent uniform annual cost (EUAC). NPV is the discounted monetary value of expected net benefits (i.e., benefits minus costs). In this method, monetary values are assigned to benefits and costs, which are discounted using a discount rate to determine their present values (PV), and then the present value

457

of the costs is subtracted from the present value of the benefits. NPV is calculated using the following formula:

$$NPV = \text{Initial Cost} + \sum_{k=1}^{N} \text{Rehabilitation Cost}_k \left[\frac{1}{(1+i)^{n_k}} \right]$$

where:

i = discount rate, the percentage figure representing the rate of interest that money can be assumed to earn over the period of time under analysis

n = year of expenditure

Note that the discounting (using a discount rate, i) reflects the time value of money, and it expresses the benefits and costs at different times in terms of a common unit of measurement. The real discount rate of 3–5% has been suggested for use with real dollar cost estimates.

In the EUAC method, the NPV of all discounted costs and benefits is expressed as uniform annual payments throughout the analysis period, and is useful for situations where budgets are established on an annual basis. EUAC is calculated as follows:

$$EUAC = NPV \left[\frac{i(1+i)^n}{(1+i)^n - 1} \right]$$

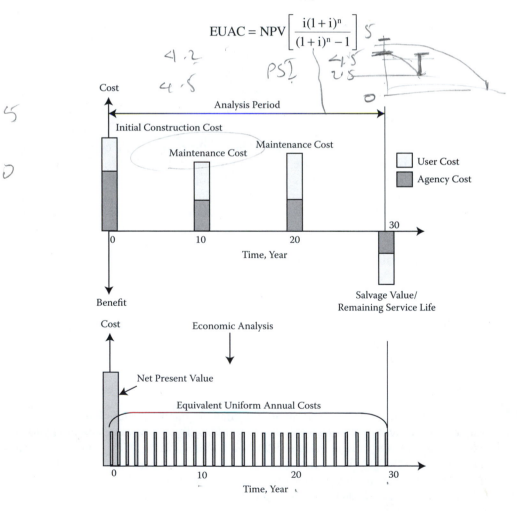

FIGURE 21.1 Concepts of Expenditure and Economic Analysis.

where:
    NPV = net present value
      i = discount rate
      n = year of expenditure

The concepts of NPV and EUAC are illustrated in Figure 21.1.

Example Problem

For an analysis period of 20 years, there are two alternative pavement designs with equal salvage values. The first alternative is to reconstruct the pavement when it reaches failure, while the second alternative is to improve the condition of the pavement at the 8th, 16th, and 24th years. The costs are as follows:

Alternative 1: initial cost: $18 million; cost to reconstruct after 20 years: $9 million.
Alternative 2: initial cost: $15 million; cost for improvement at 8th, 16th, and 24th years: $5 million each occurrence.

Which alternative should be selected?

SOLUTION

Use a discount rate of 4%.

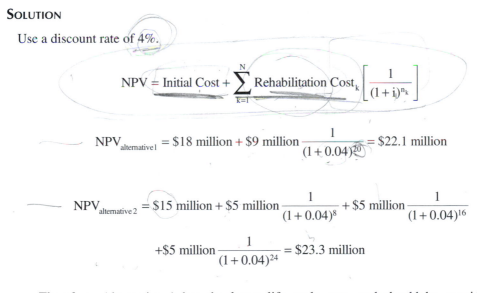

$$NPV = \text{Initial Cost} + \sum_{k=1}^{N} \text{Rehabilitation Cost}_k \left[\frac{1}{(1+i)^{n_k}} \right]$$

$$NPV_{\text{alternative 1}} = \$18 \text{ million} + \$9 \text{ million} \frac{1}{(1+0.04)^{20}} = \$22.1 \text{ million}$$

$$NPV_{\text{alternative 2}} = \$15 \text{ million} + \$5 \text{ million} \frac{1}{(1+0.04)^8} + \$5 \text{ million} \frac{1}{(1+0.04)^{16}}$$

$$+\$5 \text{ million} \frac{1}{(1+0.04)^{24}} = \$23.3 \text{ million}$$

Therefore, Alternative 1 has the lower life cycle cost, and should be considered for selection.

21.1.4 COSTS IN LIFE CYCLE COST ANALYSIS

The costs include those incurred by the pavement-managing entity (such as a department of transportation) as well as the users, because of the construction or maintenance/rehabilitation work.

Note that costs that are common to alternatives can be excluded from the LCC analysis. The same concepts are applicable for benefits as well.

The agency costs include those associated with construction, maintenance, and rehabilitation; engineering and administrative costs; as well as traffic control costs, and should take into account the salvage value at the end of the analysis period. In case the design life of one alternative exceeds the other, then the remaining service life (RSL) must be taken into consideration. The user costs, which are relatively difficult to estimate, include those associated with vehicle operation, travel time, and accidents. Guidance on evaluating travel time costs can be obtained from a U.S. Department of Transportation (USDOT) memorandum from http://ostpxweb.dot.gov/policy/Data/VOTrevision1_2-11-03.pdf.

21.1.5 PROBABILISTIC VERSUS DETERMINISTIC APPROACH

In the above discussions, the values of the outputs (such as NPV) were obtained by using equations, assuming unique values of the input, and this resulted in specific values of the output. This is the *deterministic* approach, in which the uncertainty associated with the inputs, such as future costs, discount rate, or year of rehabilitation, is not considered, and the values are considered to be constants. However, in real life, the input values are not constant—they are variable. In the more sophisticated *probabilistic* method, this variability of input values (such as initial costs and discount rate) is considered and the output is determined as a (normal) distribution (rather than a single value). Using a cumulative distribution, one can estimate the different costs (corresponding to different percentiles or probability) and also determine which alternative is more variable. The method uses Monte Carlo simulation, which picks random numbers from the distribution of the input variables, outputs the results also as a random variable, and then sums up all the results to provide a distribution of the output parameter. The use of this probabilistic approach enables the user to quantify the risk (risk analysis) associated with each option (due to the variability of input parameters) and hence make a more informed decision on the basis of risk the agency is willing to tolerate.

21.1.6 INFORMATION ON LIFE CYCLE COST ANALYSIS FOR PAVEMENTS

Since the early 1990s, the U.S. Federal Highway Administration (FHWA) has been actively sponsoring workshops on LCC and encouraging users to employ LCC analysis to evaluate alternatives for different projects. The use of LCC analysis has been made compulsory for all projects in excess of $25 million which have federal involvement in the United States. Simultaneously, the FHWA Office of Asset Management has been offering a very useful collection of tools for conducting LCC analysis—these tools include detailed and summary documents and spreadsheet-based software, complete with manuals. The materials can be downloaded free from the following website: http://www.fhwa.dot.gov/infrastructure/asstmgmt/lcca.cfm.

Information on discount rates can be obtained from the U.S. Office of Management and Budget (OMB) through its Circular No. A-94 at http://www.whitehouse.gov/omb/circulars/a094/a094.html#1.

Life cycle cost analysis software can also be obtained from the Asphalt Pavement Alliance (APA) website at http://www.asphaltalliance.com/library.asp?MENU=544.

21.2 COST-SAVING CONCEPTS

Over the years, cost-saving concepts in pavement design and construction have been developed on the bases of economic analysis as well as necessity. Two such concepts are perpetual pavement and the recycling of pavement materials. The concept of perpetual pavement is based on economic analysis—it provides a better pavement now with higher initial cost, to prevent excessive costs in the future. The concept of recycling has grown partly on the basis of necessity and partly on that of engineering economics. Periodic shortages of asphalt (as a result of shortage of gasoline) and ever increasing costs of asphalt and construction processes have made pavement agencies recycle more and more pavement materials over the years. In fact, recycling is a routine activity for most pavement agencies, and the last few decades have seen tremendous improvements in recycling equipment and processes all over the world.

21.2.1 Principles of Perpetual Pavements

Perpetual pavements refer to full-depth asphalt pavements that are engineered to resist typical distresses and survive for long periods of time, without requiring any major rehabilitation. The basic requirements of such a pavement are as follows:

1. The bottom must have high asphalt content, and the overall thickness should be large enough to prevent fatigue cracking (by having a large number of allowable repetitions to failure as well as by having a relatively small tensile strain).
2. The intermediate layer should be durable and stable—requiring strong aggregate structure as well as sufficient resistance against moisture damage.
3. The wearing course must possess sufficient resistance against rutting and wearing failures, and be impermeable, although, if needed, an open graded friction course should be used to improve driving conditions.

The concept of design of perpetual pavement is shown in Figure 21.2.

Over time, a perpetual pavement is expected to have distresses on the surface or within 2–3 in. of the surface, which can be repaired relatively easily and economically by replacing or improving the top layer only, for example through resurfacing or recycling. This is a significantly better option compared to a conventional pavement, in which fatigue cracks coming up from the bottom part would necessitate the replacement of the entire pavement structure that is up to the depth of the crack initiation. The time interval of resurfacing of perpetual pavements can be in the range of 12 years, giving them a relatively longer life of 25 years or more (with two resurfacings).

The structural and mix design of the asphalt-rich bottom layer (the base, which is in the zone of the maximum tensile strain and hence the zone of fatigue crack initiation) can be based on the "maximum limiting" strain of 60 µε, whereas that for preventing subgrade rutting can be based on the "maximum limiting" strain of 200 µε.

The materials used for constructing the different layers must be selected carefully, with proper consideration of the type(s) of distresses that can be expected in each specific layer. The foundation

FIGURE 21.2 Concept of Perpetual Pavement.
Source: Adapted from Newcomb et al. (2001).

Material (subgrade) must be well compacted and have adequate resistance against swelling or frost heaves, and should be capable of supporting necessary construction equipment. The base should have a relatively high content (to give a high density of the layer) of an asphalt whose high temperature grade should be the same as that of the surface layer, while the low temperature grade could be the same as that of the layer above this layer (the intermediate layer). Performance tests should include those for rutting, fatigue, as well as moisture damage. The intermediate layer should contain rut-resistant aggregate structure and an optimum content of asphalt of the same high temperature grade as that of the surface layer, and low temperature grade one grade lower than that of the surface layer. For aggregates, either large-size stones or smaller stones with stone-on-stone contact can be used. Performance tests should include those for rutting and moisture damage. The wearing (surface) layer should be made up of polish-resistant aggregates with stone-on-stone structure (gradation) and an optimum amount of asphalt with appropriate (or one grade higher) high and low temperature grade. This asphalt may contain polymers and/or fibers, as required for the mix, to achieve high density, impermeability, and rutting resistance. At a minimum, performance testing for rutting should be conducted.

Construction of a perpetual pavement should be done using proper equipment and quality control/assurance techniques. Key factors are the use of volumetric properties to check the quality of hot mix asphalt (HMA), and sophisticated tools such as laser for detecting (and hence preventing) the segregation of HMA, a nuclear gage for assuring proper density, ground-penetrating radar for ensuring adequate and uniform thickness, and profilometers for smoothness. Continuous access to an adequately staffed and equipped quality control laboratory is critical.

Once the perpetual pavement is constructed, it is necessary to monitor it for distresses, such that adequate steps can be taken when the critical type and level of distress are observed. Annual surveys of distress and ride quality are necessary. In general, the resurfacing can consist of the milling of 2–4 in. of the surface and replacing it with a similar or slightly better mix. Structural evaluation can be triggered by the observation of critical distresses on the surface, and can be conducted with the use of a falling weight deflectometer and ground-penetrating radar (and cores). If necessary, a slight increase in thickness can be made during resurfacing. Proper tack coat must be used to ensure bonding between the existing pavement and the new wearing layer.

Detailed information on perpetual pavement can be obtained from the APA webpage at http://www.asphaltalliance.com/library.asp?MENU=517.

21.2.2 ECONOMIC BENEFITS OF RECYCLING

When properly selected, all the different types of recycling methods are usually cheaper than the conventional rehabilitation methods, even though the relative savings will depend on the kind of recycling technique used. The primary savings in hot and cold mix asphalt pavement recycling come from savings in the cost of virgin asphalt cement, whereas the savings in hot in-place recycling come by the elimination of transportation costs and use of very little amount of virgin material. The major savings in the case of cold in-place recycling come by eliminating the need for a fuel or emission control system, since the process is done at ambient temperature; the elimination of transportation costs; and the addition of only a small percentage of virgin asphalt binder. Savings of up to 40%, 50%, 55%, and 67% can be achieved by using hot mix recycling, hot in-place recycling, cold in-place recycling, and full depth reclamation, respectively. These savings are achieved when one of the recycling methods is used in place of a conventional method or some other recycling method. For concrete pavements, existing materials can be rubblized and the layer can be compacted and used as a base for an asphalt pavement, or the existing concrete can be recycled with new aggregates for a new concrete pavement.

In addition to the material and construction cost savings, a significant amount of cost savings (in terms of user costs) can be realized by the reduced interruptions in traffic flow when compared with conventional rehabilitation techniques. Recycling can be used to rejuvenate a pavement or correct a mix deficiency and conserve material and energy—options not available with the conventional paving techniques. A conventional overlay may require upgrading shoulders to maintain a profile, raising guard rails to maintain the minimum safety standard, and restricting overlays below bridges to maintain underpass height. On the other hand, recycling can effectively be used to maintain the highway geometry and thus result in substantial overall savings as well.

Information on recycling pavements can be obtained from the FHWA's website at http://www.fhwa.dot.gov/pavement/recycling/.

QUESTIONS

1. What is life cycle cost analysis?
2. What are the different types of costs associated with a pavement?
3. Determine which of the alternatives presented below has the lower life cycle cost. Assume that both have the same life of 30 years and the same salvage value. Assume a discount rate of 4%.
   Alternative A: initial cost: $15 million; rehabilitation at the 10th and 20th years: each cost $4 million.
   Alternative B: initial cost: $17 million; rehabilitation at the 15th year: $4.5 million.
4. What is the basic concept of perpetual pavement?
5. Consider two alternatives with equal lives for constructing an asphalt pavement: option 1 involves milling and removing the existing 2-in. layer and replacing it with 4 in. of HMA; option 2 uses cold in-place recycling to recycle the top 1.5 in. of the existing pavement and then overlays it with 2 in. of new HMA. Obtain the costs of milling, recycling, and construction of new HMA from your local HMA industry (e.g., an asphalt association such as the National Asphalt Pavement Association [NAPA] and/or the Asphalt Recycling and Reclaiming Association [ARRA], or the local or state department of transportation), and select the best option.

22 Instrumentation in Asphalt and Concrete Pavement

Analysis of data from properly instrumented pavement test sections can provide invaluable information for the proper design and rehabilitation of pavements. Such data consist of environmental as well as response data from different layers, the more important ones of which are discussed in the following sections.

22.1 TEMPERATURE

A thermocouple is generally used for determination of temperature. Its principle of operation is based on the fact that the voltage drop across dissimilar metals which are placed in contact is a function of temperature. There can be different types of thermocouples. For example, in a T type, a twisted-stranded-shielded-soldered pair of thermocouple wire (constantan and copper) is used (Figure 22.1). The exposed end of the thermocouple is surrounded by copper tubing, which is attached to the cable insulation, by heat-shrinkable Teflon tubing.

22.2 SOIL MOISTURE CONTENT

Soil moisture content is measured using the time domain reflectometry (TDR) technique, which is based on the change in dielectric constant of a material with a change in moisture content. The dielectric constant of dry soil can range from three to eight (whereas that of free water is 81). A TDR probe is shown in Figure 22.2. This probe (CS615; Campbell Scientific) has two parallel conducting rods held on one end by an epoxy head, containing a bistable multivibrator. A suitable oscillation frequency of the vibrator is used for measurement of dielectric constant and to determine moisture content with the help of a data acquisition system.

22.3 FROST DEPTH

The depth of frost in soil is determined using the concept of measurement of soil resistivity, which differs widely from an unfrozen state (20,000–50,000 ohms normally) to a frozen state (from 500,000 up to several million ohms). Electrical resistance is measured between the conductors mounted along a cylindrical probe. The depth of frost penetration is measured from the resistance data collected from adjacent pairs of electrodes down the resistivity probe, noting the depth of change in resistivity. Copper rings in PVC rods (Figure 22.3) are generally used for measuring resistivity.

22.4 STRAIN IN ASPHALT OR CONCRETE PAVEMENT AND SOIL LAYERS

The horizontal strain in an asphalt or concrete pavement layer is determined on the basis of change in resistance of a body due to its elongation or shortening. The elongation or shortening is caused by stresses coming from the traffic load. Two features are important for the proper functioning of the strain gages. First, the strain gage must be embedded in material with high flexibility and low stiffness, and have sufficient durability against stresses and temperature fluctuations. Secondly, the strain gage must be securely anchored in the pavement to behave as "part" of the pavement. A Dynatest PAST II strain gage is shown in Figure 22.4 as an example. It has a resistance of 120 ohms

FIGURE 22.1 Thermocouples.
Courtesy: Lauren Swett.

FIGURE 22.2 Time Domain Reflectometry (TDR) Probe.
Courtesy: Lauren Swett.

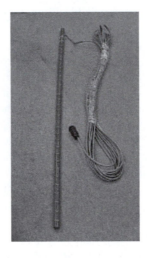

FIGURE 22.3 Frost Depth Measurement Instrument.
Courtesy: Lauren Swett.

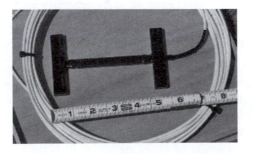

FIGURE 22.4 Asphalt or Concrete Strain Gage.
Courtesy: Lauren Swett.

FIGURE 22.5 Soil Strain Gage.
Courtesy: Lauren Swett.

and a gage factor of 2.0, and can endure a temperature range of –30°C to +150°C. The maximum strain it can record is 1500 microstrain. For soils, for measuring the vertical strain, generally linear variable deformation transducers (LVDT) such as those with a ±5 mm range are used. Figure 22.5 shows a Dynatest Soil and Stress Deformation Transducer (FTC-1).

22.5   STRESS IN SOIL LAYERS

Soil pressure cells are used for the measurement of vertical stress in different types of soils. The critical features include sensitivity and resolution of the pressure, and resistance to the damaging effects of loading and the environment. An example of a pressure cell (Figure 22.6) is the SOPT gage from Dynatest, which consists of a liquid-filled cell with an integrated pressure transducer within a thin membrane, all consisting of pure titanium. Measurable pressures range from 1.5 to 120 psi, for different size aggregates.

22.6   DATA ACQUISITION SYSTEMS

Data are obtained from in-place instruments through the use of appropriate data acquisition systems. Such data acquisitions systems are controlled by a suitable software (such as LABVIEW) through a computer system, and include appropriate modules or devices to interface between the instruments

FIGURE 22.6 Soil Pressure Cell.
Courtesy: Lauren Swett.

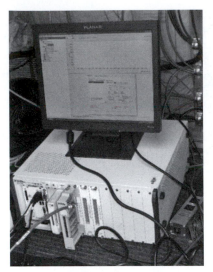

Data Acquisition System with
Module in Computer

Close-up View of the Module Showing
Connection from a Strain Gage

FIGURE 22.7 Data Acquisition Systems.

and the computer. In many cases, data loggers are used to collect and store data, which are then periodically transferred to a computer by using a suitable software and connection cable. As required, the software can be used to trigger data acquisition on the basis of some other signal, such as data coming from a weigh-in-motion (WIM) signal (as a truck passes by the instrumented section). The relevant information about the instruments (such as gage factor or calibration constant) is input into the software, which then makes all the relevant computations on the basis of the acquired data and stores the data in ASCII or spreadsheet format, or even displays data in graphical format. Various types of hardware and software are available—an example is shown in Figure 22.7.

QUESTIONS

1. List the various instruments used in pavements for environmental and load response-related data.
2. Conduct a literature search on Transportation Research Information System (TRIS) on the Internet to obtain at least one report on full-scale instrumentation in pavements. Prepare a review of the report with particular emphasis on the type of instruments, installation, and experience of the researchers.

23 Specialty Applications

23.1 ASPHALT MIXTURES

Over the years different types of asphalt mixtures have been developed in different parts of the world for specialty applications. Some examples of such mixes are presented below.

23.1.1 POLYMER MODIFIED ASPHALT

Polymers are used in asphalt to increase the range of temperature in which the asphalt mix can be used without any distress and to improve bonding of asphalt binder and aggregates. Elastomers and plastomers are the two major types of polymers used in asphalt. Block copolymers, random polymer, and natural and synthetic latex are examples of elastomers, which are used primarily for decreasing temperature susceptibility, increasing tensile strength at high strain, and improving the cohesion and adhesion of asphalt binder. For example, styrene butadiene rubber (SBR) is used for reducing the rutting potential of asphalt mixes at high temperatures. Plastomers consist of polyethylene and ethylene copolymers, and are used primarily to decrease temperature susceptibility, increase modulus, or increase tensile strength. For example, ethyl vinyl acetate (EVA) is used for modifying the strength, stiffness, and workability of the asphalt mix. Polymers can be either preblended with the liquid asphalt binders or blended with the mix at the HMA plant. A growing use of polymer-modified asphalt is in chip seal coats and other thin-layer maintenance applications, as an interlayer for resisting reflective cracking, as well as in other specialty mixes such as the open graded friction course (OGFC) and stone matrix asphalt (SMA). The use of polymers has become almost routine in many applications, and this has necessitated the development of special characterization tests (such as the direct tension test and the multiple stress creep and recovery test, or MSCR) and conditioning tests (such as the modified rotating German flask, or MGRF) for polymer-modified asphalt binders. Two issues that are specifically important for the good performance of polymer-modified asphalt mixes are storage stability and degradation. Storage stability involves the compatibility of the polymer and asphalt binder when stored for long time periods at elevated temperatures, and degradation refers to the deterioration of the properties of the polymer due to the presence of oxygen at high temperatures. For information on modified asphalts, the reader is advised to see the Association of Modified Asphalt Producers (AMAP) Web site (www.modifiedasphalt.org).

23.1.2 ASPHALT RUBBER MIXES

Crumb rubber, obtained from recycling vehicle tires, has emerged as a commonly used additive in asphalt binder in the United States. Started primarily as a way of recycling tires and avoiding filling up landfill space, the use of crumb rubber has grown steadily since the early 1990s, and continues to grow as different advantages of its use are being investigated. Generally, crumb rubber is used at about 15–25% of the total mass of the binder. It is crushed to a maximum size (generally 1 mm) before being added to asphalt binder at an elevated temperature (wet process), or to aggregate at an HMA plant (dry process). The reported benefits of using crumb rubber in asphalt mixes include greater flexibility, rougher surface texture, greater resistance to aging, and a decrease in draindown potential of asphalt binder. These qualities translate to greater fatigue life, better surface friction, improved elasticity, and ability to add a higher amount of asphalt binder to increase film thickness in mixes such as OGFC. The use of rubberized asphalt (or what is known as asphalt rubber in the United States) continues to grow in chip seals and other maintenance mixes, as an interlayer, as well as in structural HMA. The extra cost of the crumb rubber can be offset by the reduced thickness of layers and/or the enhanced life of the pavement. Swelling and degradation of crumb rubber in asphalt binders are the two important issues. A wealth of information is available from the Rubber Pavement Association Web site (www.rubberpavements.org).

23.1.3 STONE MATRIX ASPHALT (SMA)

Stone matrix asphalt is a binder-rich, aggregate, interlocked, rut-resistant, and durable asphalt mix that can be used for high-volume, heavy, as well as slow or standing traffic pavement, such as that in industrial loading areas, bus lanes, or container terminal areas. The key feature of SMA is its unique gradation, which consists of 20–30% passing the 4.75 mm sieve material, and hence it is of relatively coarse nature. Based on available guidelines, the gradation must be determined as part of the mix design process, through the use of the concept of "stone-on-stone contact." In principle, it refers to a situation in which the fine aggregates have been reduced to a point where the coarse aggregate particles are in contact with each other, resulting in a strong load-bearing coarse aggregate skeleton.

The mix consists of a relatively high amount of mineral filler (percentage passing the 0.075 mm sieve)—around 10%, for example, compared to around 5% for a typical dense graded mix. The high percentage of mineral filler acts as a stabilizer for the asphalt binder, which is also needed at a relatively high content ($\geq 6\%$ compared to around 5% for a dense graded mix). Fibers are also added to the binder to reduce the tendency of draindown of asphalt binder from the mix.

Because SMA derives its strength primarily from the coarse aggregate skeleton, aggregates from only strong parent rock types such as granite, basalt, or quartzite and with fractured faces are generally used. Also, the design gradation should be strictly followed (or adjusted appropriately) during production. Attention to quality control, especially aggregate gradation, is critical for the successful production of SMA. Similarly, the temperature of the mix should be carefully adjusted to allow good mixing, prevent draindown, and provide adequate temperature for laydown and compaction.

Design air voids are around 3%, while those in compacted mats should not exceed 5%. Fibers are generally added at 0.3 to 1.5% by mass of the total mix as stabilizers to prevent draindown of the mix. The fibers can be of mineral or cellulose type, and vary in length from 1 to 6 mm. For example, cellulose fibers are about 1 mm long and 0.05 mm thick.

The time of mixing in the plant should be increased by 5–15 seconds, compared to that for dense graded mixes. Mixing and compaction temperatures are dependent on many factors such as type of asphalt, but generally the compaction process should be completed before it cools down below 130°C. SMA layer thicknesses vary from 2.5 to 5 times the maximum aggregate size, and are generally put down in 38–50 mm thick layers. If the relatively high amount of mineral filler is not available as aggregates, lime or fly ash could be used, through mineral filler feeder, lime feeder, or cold feed bins. Fat spots or puddles of asphalt binder can show up on the surface if more than adequate binder has been used. In the field, compaction should start immediately after placement, using a > 10-ton static steel-wheeled roller. A vibratory roller can be used only sparingly, and a rubber-tired roller should be used only for finishing. Sand can be spread on finished surfaces to reduce skid hazards, if any.

Once the surface asphalt layer wears off under traffic, the surface texture of SMA provides good friction, and also helps in noise reduction, which can be as much as 2.5 dB. SMA costs 20–30% more than dense graded mixes because of factors such as higher quality of materials required, larger mixing time, and more intensive quality control.

23.1.4 PO ROUS FRICTION COURSE (PFC)

Porous friction courses (sometimes referred to as *open graded friction courses*) are layers made of asphalt mix with a high amount of interconnected air voids that drain out water quickly during a rainstorm, reducing glare, splash, and spray and the potential of hydroplaning; improving friction; and hence reducing the potential of accidents significantly. PFCs also contribute to the reduction of tire noise on pavements.

PFCs consist of very coarse "open/gap" gradation, in which materials passing certain sieve sizes are missing. The aggregates should be durable, rough textured, and resistant to abrasion and polish. Typically the gradation consists of 5–15% passing the 2.36 mm sieve. The asphalt content can range from 5 to 6.5%, with 5–7% polymer by weight of the binder and 0.15–0.5% fibers by weight of the mix. It has a relatively high amount of asphalt binder to provide adequate resistance against raveling and premature aging, often modified with polymer. The mix may also consist of a suitable stabilizer, such as fiber. Design as well as in-place voids in PFCs can range from 10 to 20%.

Fibers in the PFC can be added to the pugmill in packaged or loose form in a batch plant, or by blowing into a drum in a drum plant. Pelletized form fiber can be added to the pugmill or drum also. The mixing time needs to be increased, and storage times in silos should be limited to very short times to prevent draindown.

Haul distances should be limited, and trucks carrying PFCs should have a thorough coating of asphalt-release agent on their beds to prevent bonding of mix on the bed, as well as tarping to prevent cooling down and formation of crust on the surface of the mix.

To prevent trapping of water, PFC should be placed on impermeable asphalt pavement surfaces only. It should preferably not be used on a brand-new dense graded pavement where the in-place voids are still relatively high. Before the placement of the PFC, the underlying pavement can also be sealed using an appropriate amount of slow-setting emulsion tack coat (that can penetrate surface voids). The PFC layer should be open (daylighted) on the edges of the pavement to allow the free flow of water permeating through it.

Use of a materials transfer device is recommended, and the presence of a hot screed helps in preventing pulling of the mat under the paver. Generally a few passes of a static steel-wheeled roller is sufficient for properly seating the PFC (there is no need for compaction). Specific areas such as intersections or turning areas, ramps, and curbed areas may not be suitable for PFCs since they can ravel and shove, or prevent draining.

Since they are critical in draining water quickly, the voids in the PFCs should be prevented from getting filled up with materials such as sand from snow treatments. Also, since they are placed in thin lifts, such as those with 1/2- to 5/8-inch thickness, snow plows may not be used to prevent lifting and dislodging of the mix. Use of anti-icing chemicals is a better option, and pores clogged with debris can be cleaned with high-pressure cleaners.

23.1.5 WARM MIX ASPHALT (WMA)

Warm mix asphalt (WMA) is a concept that involves the use of an additive and/or the modification of the asphalt mix process to lower the viscosity of asphalt binder, and hence enables the production of asphalt mixes at a relatively lower temperature (generally, 10–75°C, depending on the technology used). There are four widely used WMA production processes. The WAM-Foam® technology involves the use of a two-component binder system called WAM-Foam® (warm asphalt mix foam). In this process, a soft-foamed binder and hard-foamed binder are added at different stages during plant production. In the Aspha-Min® process, a synthetic zeolite called Aspha-Min® is used during mixing at the plant to create a foaming effect in the binder. Sasobit, Fischer-Tropsch paraffin wax, and Asphaltan B®, a low-molecular-weight esterified wax, are mixed directly with the asphalt binder, whereas Evotherm is a nonproprietary product that includes additives to improve coating and workability.

In the WAM-Foam® process, the mixing is done in two stages. In the first stage, a soft binder is used to provide effective coating of aggregates at a lower temperature (typically 212–250°F, or 100–120°C). In the second phase a harder binder, in the form of a powder, foam, or emulsion, is added to bring the grade of the total binder to the level of a binder used for a conventional HMA. Aspha-Min® is a manufactured synthetic sodium aluminum silicate, better known as zeolite. These zeolites contain about 21% water by mass. When the zeolites are added to the aggregates at the same time as the binder (around 185–360°F, or 85–182°C), the water is released, resulting in the creation of foamed asphalt, which in turn improves the coating of the aggregates and workability of the mix. Zeolites (commonly used amount as 0.3% by mass of the mix) can be added directly to the pugmill of a batch plant or through a recycled asphalt pavement (RAP) feed or pneumatic feeder in a drum plant. Sasobit is a synthetic paraffin wax that melts around 210°F (99°C) and reduces the viscosity of the asphalt binder. This enables the production temperatures to be reduced by 18–54°F (10–30°C). Sasobit is added directly to the asphalt binder at the plant, generally at the rate of 1.5–3.0% by mass of the asphalt binder. Asphaltan B® is a mix of wax constituents and higher molecular weight hydro-carbons, which also reduces the viscosity of the asphalt. It is generally used at 2–4% by weight of the asphalt binder, and can be added directly to the binder or during mixing.

Evotherm is a chemical that is delivered in an emulsion with high asphalt residue. The water from this emulsion is given out when it is mixed with hot aggregates. The product is supposed to enable the production of HMA at a significantly lower temperature, and enhance workability, adhesion, and coating of the asphalt mixes.

The benefits of WMA over conventional HMA are many. Because of the lowering of the production temperature, less oxidative aging of the asphalt binder and hence the production of a superior asphalt mix can be expected. A lowering of temperature also means savings in fuel (burner) costs, as well as lowering of emissions. It also means that the construction season in many parts of the world which have colder climates in winter can be extended. Finally, it offers the potential of recycling larger amounts of RAP materials.

As the use of WMA grows in different parts of the world and in the United States, research and field trials keep on generating useful information on this mix, and guidelines for successful use of this product are being developed. Several concerns that are specific to this kind of mix are as follows: the use of the proper grade of asphalt binder, the need to ensure that aggregates are free of moisture (since they are heated to a lower temperature), the effect of low-viscosity binder in early stages of life of the pavement, and the effect, if any, of the WMA additives on the low-temperature properties of the asphalt mixes. Although WMA may need a lower amount of asphalt binder to compact (during mix design in the laboratory), current guidelines encourage the use of asphalt content as would be used for a conventional HMA. The warm mix asphalt homepage (warmmixasphalt.com) contains a growing list of publications related to WMA.

23.2   CONCRETE WHITETOPPING

Whitetopping refers to the placement of a layer of concrete over existing HMA pavements as a reha-bilitation operation. The thickness of ultrathin whitetopping ranges from 2 to 4 inches, and fiber-reinforced concrete is used in such cases. Before the placement of whitetopping, the existing HMA surface is textured for providing good bond between the new and old materials. The behavior of a pavement with bonded whitetopping is considered to be different from that of other pavements. The composite section results in much lower stress in the concrete section and hence allows the use of thinner section. Many successful applications have been noted, and research continues on the better design concepts of whitetopping layers (NCHRP, 2004). For more information please see the ACPA (www.acpa.com) and FHWA Web sites (FHWA, 2005).

QUESTIONS

1. What are the different types of polymers used in asphalt mixes?
2. How does asphalt rubber improve HMA pavements?
3. How is an SMA mix designed?
4. What are the main characteristics of porous friction courses?
5. What are the different processes of producing warm mix asphalt?
6. Using current cost data for fuel for dryer drums in HMA plants (from local industry), estimate the savings in dollars per ton if a warm mix asphalt is produced at 125°C instead of a hot mix asphalt at 150°C.

References

Abou-Ayyash, A. 1974. Mechanistic Behavior of Continuously Reinforced Concrete Pavement. PhD. thesis, University of Texas at Austin.

Abrams, D. A. 1918. Design of Concrete Mixtures, Lewis Institute, Structural Materials Research Laboratory, Bulletin No. 1, PCA LS001, http://www.portcement.org/pdf_files/LS001.pdf.

ACI 318.2002. Building Code Requirements for Structural Concrete. American Concrete Institute, Farmington Hills, MI.

ACI Committee 302. 2002. Concrete Floor and Slab Construction. American Concrete Institute, Farmington Hills, MI.

Acum, W. E. A. and L. Fox. 1951. Computation of Load Stresses in a Three Layer Elastic System. *Geotechnique* 2 (4): 293–300.

Ahlborn, G. 1972. *ELSYM5: Computer Program for Determining Stresses and Deformations in Five Layer Elastic System*. Berkeley: University of California.

Ahlvin, R. C. and H. H. Ulery. 1962. Tabulated Values for Determining Complete Pattern of Stresses, Strains and Deflections Beneath a Uniform Circular Load on a Homogeneous Half Space. *Highway Research Board Bulletin* 342:1–3.

Ahmed, N. V., R. L. Lytton, J. P. Mahoney, and O. T. Phillips. 1978. Texas Rehabilitation and Maintenance District Optimization Systems. Research Report No. 207–3. College Station: Texas Transportation Institute, Texas A&M University.

Allen, D. L. and R. C. Deen. 1986. A Computerized Analysis of Rutting Behavior of Flexible Pavements. Transportation Research Record No. 1095. Washington, D.C.: Transportation Research Board.

American Association of State and Highway Transportation Officials (AASHTO). 1986. *Guide for Design of Pavement Structures*. Vol. 2. Washington, D.C.: AASHTO.

American Association of State and Highway Transportation Officials (AASHTO). 1993. *AASHTO Guide for Design of Pavement Structures*. Washington, D.C.: AASHTO.

American Association of State and Highway Transportation Officials (AASHTO). 1998. *Supplement to the AASHTO Guide for Design of Pavement Structures Part II-Rigid Pavement Design and Rigid Pavement Joint Design*. Washington, D.C.: AASHTO.

American Association of State and Highway Transportation Officials (AASHTO). 2000. *MDM-SI-2, Model Drainage Manual, 2000 Metric Edition*. Washington, D.C.: AASHTO.

American Association of State and Highway Transportation Officials (AASHTO). N.d. [Home page], www.transportation.org.

American Concrete Institute. N.d. [Home page], http://www.aci-int.org.

American Concrete Institute, ACI Committee 211. 1991. *Standard Practice for Selecting Proportions for Normal, Heavyweight and Mass Concrete, ACI 211. 1–91*. Farmington Hills, MI: ACI.

American Concrete Institute, ACI Committee 214. 1977. *Recommended Practice for Evaluation of Strength Test Results for Concrete, ACI 214-77*. Reapproved 1997. Farmington Hills, MI: ACI.

American Concrete Institute, ACI Committee 318. 2002. *Building Code Requirements for Structural Concrete, ACI 318-02, and Commentary, ACI 318R-02*. Farmington Hills, MI: ACI.

American Concrete Pavement Association (ACPA). 1990. Guidelines for Unbonded Concrete Overlays. Technical Bulletin TB-005.0 D. Skokie, IL: ACPA.

American Concrete Pavement Association (ACPA). 1995. *Construction of Portland Cement Concrete Pavements*. National Highway Institute Course No. 13133. ASHTO/FHWA/Industry Joint Training. Federal Highway Administration, Department of Transportation. Washington, D.C.

American Concrete Pavement Association (ACPA). 1999. *Survey of States' Concrete Pavement Design and Construction Practices*. Skokie, IL: ACPA.

American Concrete Pavement Association (ACPA). 2002. Portland Cement Concrete Overlays: State of the Technology Synthesis. Product Code SP045P. Skokie, IL: ACPA.

American Concrete Pavement Association (ACPA). N.d. Whitetopping: State of the Practice. Product code EB210P. Skokie, IL: ACPA.

American Society for Testing and Materials (ASTM). 1989. *Annual Books of ASTM Standards, Road and Paving Materials: Traveled Surface Characteristics*. Vol. 04.03. Philadelphia: American Society for Testing and Materials.

American Society for Testing and Materials (ASTM). Various years. *Annual Book of ASTM Standards*. Vol. 04.08, *Soil and Rock*. Philadelphia: American Society for Testing and Materials.

American Society for Testing and Materials (ASTM). N.d. ASTM International. www.astm.org.

American Society of Civil Engineers (ASCE). 2004. Introduction to Mechanistic-Empirical Pavement Design, Reference Manual. Seminar, Baltimore, MD.

Anderson, D. A. and T. W. Kennedy. 1993. Development of SHRP Binder Specifications. *Journal of AAPT* 62:508.

Asphalt Emulsions Manufacturers Association (AEMA). 1997. *A Basic Asphalt Emulsion Manual*. 3rd ed., Manual Series No. 19. Annapolis, MD: AEMA.

Asphalt Institute. 1983. *Computer Program DAMA: User's Manual*. Lexington, KY: AI.

Asphalt Institute. 1984. *Mix Design Methods for Asphalt Concrete and Other Hot Mix Types*. Manual Series No. 2 (MS-2). May. Lexington, KY: AI.

Asphalt Institute. 1991. *Thickness Design: Asphalt Pavements for Highways and Streets*. Manual Series No. 1 (MS-1). Lexington, KY: AI.

Asphalt Pavement Alliance. N.d. Life Cycle Cost Analysis Software, www.asphaltalliance.com/library.asp? MENU=544.

Asphalt Pavement Alliance. N.d. Perpetual Pavements, http://www.asphaltalliance.com/library.asp?MENU =517.

Asphalt Recycling & Reclaiming Association (ARRA). 2003. Basic Asphalt Recycling Manual. Annapolis, MD: ARRA.

Association of Modified Asphalt Producers. N.d. [Home page], www.modified asphalt.org.

Ayres, M., Jr. 1997. Development of a Rational Probabilistic Approach for Flexible Pavement Analysis. Ph.D. diss., University of Maryland at College Park.

Bahia H. U. and D. A. Anderson. 1994. The Pressure Aging Vessel (PAV): A Test to Simulate Rheological Changes Due to Field Aging. ASTM Special Technical Publication 1241. Philadelphia: American Society for Testing and Materials.

Bahia H. U. and D. A. Anderson. 1995. The SHRP Binder Rheological Parameters: Why Are They Required and How Do They Compare to Conventional Properties? Preprint Paper No. 950793. Presented at the Transportation Research Board Meeting, Washington, D.C.

Baladi, G. 1989. Fatigue Life and Permanent Deformation Characteristic of Asphalt Concrete Mixes. Transportation Research Record No. 1227. Washington, D.C.: Transportation Research Board.

Barber, E. S. and C. L. Sawyer. 1952. Highway Subdrainage. *Proceedings, Highway Research Board*, 643–666.

Barenberg, E. J. and M. R. Thompson. 1991. *Calibrated Mechanistic Structural Analysis Procedures for Pavements, Phase 2 of NCHPR Project 1-26*. Washington, D.C.: National Cooperative Highway Research Program and Transportation Research Board.

Barksdale, R. D., Ed. 1991. *The Aggregate Handbook*. Washington, D.C.: National Stone Association.

Bhatti, M. A., J. A.Barlow, and J. W. Stoner. 1996. Modeling Damage to Rigid Pavements Caused by Subgrade Pumping. *Journal of Transportation Engineering* 122 (1, January–February): 12–21.

Bolander, Peter and Alan Yamada. Project Leader Dust Palliative Selection and Application Guide, Online document: The Forest Service, United States Department of Agriculture (USDA), http://www.fs.fed. us/eng/pubs/html/99771207/99771207.html.

Boltzmann, L. 1874. Zur Theorie der elastichen Nachwirkung, Sitz. Kgl. Akad. Wiss. Wien (Math-Naturwiss Klasse) 70, 275–306.

Boussinesq, J. 1885. *Application des Potentiels a L'Etude de L'Equilibre et du movement des Solides Elastiques*. Paris: Gauthier-Villars.

Bradbury, R. D. 1938. *Reinforced Concrete Pavements*. Washington, D.C.: Wire Reinforcement Institute.

Brademeyer, B. 1988. VESYS Modification. Final report, FHWA DTFH61-87-P-00441. Washington, D.C.: U.S. Department of Transportation.

British Standards Institution. N.d. [Home page], www.bsi-global.com.

Brown, E. R. and L. A. Cooley, Jr. 1999. *Developing Stone Matrix Asphalt Mixtures for Rut-Resistant Pavements*. Project D-09-08. NCHRP Report 425. Washington, D.C.: Transportation Research Board, National Center for Asphalt Technology, National Research Council.

Brown, Elton Ray. N.d. Class notes for CE 589, Pavement Construction. Civil Engineering Department, Auburn University.

Brown, Elton Ray. N.d. Class notes from CE 688, Asphalt Mix Design. Civil Engineering Department, Auburn University.

Burati, J. L., R. M. Weed, C. S. Hughes, and H. S. Hill. 2003. Optimal Procedures for Quality Assurance Specifications. FHWA-RD-02-095. Washington, D.C.: Federal Highway Administration.

Bureau of Reclamation. 1977. *Method for Determining the Quantity of Soluble Sulfate in Solid (Soil and Rock) and Water Samples*. Denver, CO: Bureau of Reclamation.

Burmister, D. M. 1943. The Theory of Stresses and Displacements in Layered Systems and Application to the Design of Airport Runways. In *Proceedings of the Highway Research Board*.

Burmister, D. M. 1962. Applications of Layered System Concepts and Principals to Interpretations and Evaluations of Asphalt Pavement Performances and to Design and Construction. In *Proceedings, International Conference on the Structural Design of Asphalt Pavements*, 441–453.

Butos, M., H. E. DeSolminihac, M. I. Darter, A, Caroca, and J. P. Covarrubias. 1998. Calibration of Jointed Plain Concrete Pavements Using Long-Term Pavement Performance. Transportation Research Record No. 1629. Washington, D.C.: Transportation Research Board.

Byrum, C. R. 1999. Development of a High Sped Profiler Based Slab Curvature Index for Jointed Concrete Pavements. PhD. dissertation, University of Michigan.

Byrum, C. R., W. Hansen, and S. D. Kohn. 1997. The Effect of PCC Strength and Other Parameters on the Performance of PCC Pavements. Sixth International Purdue Conference on Concrete Pavement Design and Materials for High Performance, Indianapolis, IN.

Carey, W. N. and P. E. Irick. 1960. The Pavement Serviceability-Performance Concept. Highway Research Bulletin 250, Highway Research Board, Washington, D.C.

Casagrande, A. and W. L. Shannon. 1952. Base Course Drainage for Airport Pavements. In *Proceedings of the American Society of Civil Engineers* 77, 792–814.

Cedergren, H. 1994. America's Pavements: World's Longest Bathtubs. *Civil Engineering* (September), 56–58.

Cedergren, H. R., J. A. Arman, and K. H. O'Brien. 1973. Development of Guidelines for the Design of Subsurface Drainage Systems for Highway Pavement Structural Sections. Report No. FHWA-RD-73-14. Washington, D.C.: U.S. Department of Transportation.

CERL. 2006. MicroPaver, U.S. Army Construction Engineering Research Laboratory. *Drainage of Highway and Airfield Pavements*

Chamberlain, E. J. 1987. A freeze thaw test to determine the frost susceptibility of soils. Special Report 87-1. Hanover, NH: USA Cold Regions Research and Engineering Laboratory.

Chang, H. S., R. L. Lytton, and H. S. Carpenter. 1976. Prediction of Thermal Reflection Cracks in West Texas. Report No. TTI-2-8-73-18-3. College Station: Texas Transportation Institute, Texas A&M University.

Choubane, Bouzid, Gregory A. Sholar, James A. Musselman, and Gale C. Page. 1999. A Ten-Year Performance Evaluation of Asphalt-Rubber Surface Mixes. Preprint Paper No. 00926. Presented at the Transportation Research Board Meeting in Washington, D.C.

Christensen, D. W. and D. A. Anderson. 1992. Dynamic Mechanical Test Data for Paving Grade Asphalt Cement. *Journal of AAPT* 61:117.

Christory, J. P. 1990. Assessment of PIARC Recommendations on the Combating of Pumping in Concrete Pavements. Sixth International Symposium on Concrete Roads, Madrid, Spain.

Coduto, Donald P. 1999. *Geotechnical Engineering: Principles and Practices*. Upper Saddle River, NJ: Prentice Hall.

Concrete Pavement Technology Program (CPTP). N.d. CPTP Status Report: Task 65 Engineering ETG Review Copy. Washington, D.C.: U.S. Department of Transportation.

Crovetti, J. A. and B. J. Dempsey. 1993. Hydraulic Requirements of Permeable Bases. Transportation Research Record No. 1425. Washington, D.C.: Transportation Research Board.

D'Arcy, H. 1856. *Les Fountaines Publiques de la Ville de Dijon*. Paris: Dalmont.

Darter, M. I. 1988. A Comparison Between Corps of Engineers and ERES Consultants, Inc. Rigid Pavement Design Procedures. Technical Report Prepared for the United States Air Force SAC Command, Savoy, IL.

Darter, M. I. and E. J. Barenberg. 1976. Zero-Maintenance Pavements: Results of Field Studies on the Performance Requirements and Capabilities of Conventional Pavement Systems Federal Highway Administration. FHWA-RD-76-105, Washington, D.C.

Darter, M. I. and E. J. Barenberg. 1977. Design of Zero-Maintenance Plain Jointed Concrete Pavement, Volume 1 – Development of Design Procedures. FHWA-RD-77-111. Washington, D.C.

Darter, M. I., K. T. Hall, and Cheng-Ming Kuo. 1994. *Appendices, Support under Concrete Pavements.* Washington, D.C.: National Cooperative Highway Research Program.

Darter, M. I., K. T. Hall, and C. Kuo. 1995. Support under Portland Cement Concrete Pavements. NCHRP Report 372. Washington, D.C.: National Cooperative Highway Research Program.

Darter, M. I., S. A. LaCourseiere, and S. A. Smiley. 1979. Performance of Continuously Reinforced Concrete Pavement in Illinois. Transportation Research Record No. 715. Washington, D.C.: Transportation Research Board.

Daugherty, R. L. and A. C. Ingersoll. 1954. *Fluid Mechanics with Engineering Applications.* New York: McGraw-Hill.

DeLong, D. L., M. G. F. Peutz, and A. R. Korswagen. 1973. Computer Program BISAR, Layered Systems under Normal and Tangential Surface Loads. External Report AMSR.0006.73. Amsterdam: Koninklijke/Shell Laboratorium Amsterdam.

Dempsey, B. J. 1982. Laboratory and Field Studies of Channeling and Pumping. Transportation Research Record No. 849, 1–12. Washington, D.C.: Transportation Research Board.

Dempsey, B. J. 1988. Core Flow Capacity Requirements of Geocomposite Fin-Drain Materials Used in Pavement Subdrainage. Transportation Research Record No. 1159. Washington, D.C.: Transportation Research Board.

Dempsey, B. J. 1993. Performance of Prefabricated Geocomposite Subdrainage System in an Airport Runway. Federal Aviation Administration Report DOT/FAA/RD-93/23. Washington, D.C.: FAA.

Densit. N.d. [Home page], www.densiphalt.com.

Elliott, R. P. and M. R. Thompson. 1985. Mechanistic Design Concepts for Conventional Flexible Pavements. Transportation Engineering Series No. 42. Urbana: University of Illinois.

Federal Aviation Administration (FAA). 1989. Airport Design. AC 150/5300-13, www.faa.gov/airports_airtraffic/airports/resources/advisory_circulars/index.cfm?template=Document_Listing&Keyword=150/5300-13&DocumentSelected=1.

Federal Aviation Administration (FAA). 1995. Airport Pavement Design and Evaluation. AC 150/5320-6D. Washington, D.C.: U.S. Department of Transportation.

Federal Aviation Administration (FAA). 1995. *LEDFAA User's Manual, AC 150/5320-16.* October. Washington, D.C.: U.S. Department of Transportation.

Federal Aviation Administration (FAA). 1999. Development of State Standards for Non-Primary Airports. AC 150/5100-13A, www.faa.gov/airports_airtraffic/airports/resources/advisory_circulars/index.cfm?template=Document_Listing.

Federal Aviation Administration (FAA). 2001. Engineering Brief # 60, http://www.faa.gov/airports_airtraffic/airports/construction/engineering_briefs/media/EB_60.pdf.

Federal Aviation Administration (FAA). 2004. LEDFAA: Layered Elastic Design, www.airporttech.tc.faa.gov/pavement/26ledfaa.asp.

Federal Aviation Administration (FAA). 2006. Engineering Brief # 59A, http://www.faa.gov/airports_airtraffic/airports/construction/engineering_briefs/media/EB_59a.pdf.

Federal Aviation Administration (FAA). 2006. Item P-401 Plant Mix Bituminous Pavements. Engineering Brief EB59A, May 12. Washington, D.C.: FAA.

Federal Aviation Administration (FAA). 2006. National Airport Pavement Test Facility, www.airporttech.tc.faa.gov/naptf/.

Federal Aviation Administration (FAA). 2006. Standardized Method of Reporting Airport Pavement Strength: PCN. AC 150/5335-5a, http://rgl.faa.gov/Regulatory_and_Guidance_Library/rgAdvisoryCircular.nsf/0/f4b2780536535ff7862571f7006848c2!OpenDocument&Click=.

Federal Aviation Administration (FAA). 2007. Guidelines and Procedures for Maintenance of Airport Pavements. AC No: 150/5380-6B, http://www.faa.gov/airports_airtraffic/airports/resources/advisory_circulars/media/150-5380-6B/150_5380_6b.pdf.

Federal Aviation Administration (FAA). 2007. Standards for Specifying Construction of Airports. AC No.: 150/5370-10C, http://www.faa.gov/airports_airtraffic/airports/resources/advisory_circulars/media/150-5370-10C/150_5370_10c.pdf.

Federal Aviation Administration (FAA). 2008. Advanced Airport Pavement Design Procedures: FAARFIELD— 3D Finite Element Based Design Procedure, www.airporttech.tc.faa.gov/pavement/3dfem.asp.

Federal Aviation Administration (FAA). 2008. [Software information page], http://www.airporttech.tc.faa.gov/naptf/download/index1.asp.

Federal Aviation Administration (FAA). N.d. [Home page], www.faa.gov.

Federal Aviation Administration (FAA). N.d. Airports, www.faa.gov/airports_airtraffic/airports/.

Federal Aviation Administration (FAA). N.d. Design Software, www.faa.gov/airports_airtraffic/airports/construction/design_software/.

Federal Aviation Administration (FAA). N.d. Engineering Briefs, www.faa.gov/airports_airtraffic/airports/construction/engineering_briefs/.

Federal Aviation Administration (FAA). N.d. Pavement Design and Construction, www.faa.gov/airports_airtraffic/airports/construction/pavement_design/.

Federal Highway Administration (FHWA). 1961. Design Charts for Open-Channel Flow. Hydraulic Design Series No. 3 (HDS 3). Washington, D.C.: U.S. Government Printing Office.

Federal Highway Administration (FHWA). 1989. Concrete Pavement Drainage Rehabilitation. FHWA Report, Experimental Project No. 12. Washington, D.C.: U.S. Department of Transportation.

Federal Highway Administration (FHWA). 1990. *Concrete Pavement Joints*. Technical Advisory 5040.30. Federal Highway Administration. Washington, D.C.

Federal Highway Administration (FHWA). 1990. Highway Subdrainage Design by Microcomputer: (DAMP). FHWA-IP-90-012. Washington, D.C.: U.S. Department of Transportation.

Federal Highway Administration (FHWA). 1992. Drainage Pavement System Participant Notebook. FHWA-SA-92-008, Demonstration Project No. 87. Washington, D.C.: U.S. Department of Transportation.

Federal Highway Administration (FHWA). 1994. Pavement Deflection Analysis. NHI Course No. 13127. Participant Workbook., Publication No. FHWA-HI-94-021.

Federal Highway Administration (FHWA). 1995. Geotextile Engineering Manual, Course Text. Publication No. FHWA-HI-89-050. Washington, D.C.: U.S. Department of Transportation.

Federal Highway Administration (FHWA). 1998. Geosynthetic Design and Construction Guidelines. FHWA-HI-95-038, NHI Course No. 13213, April. Washington, D.C.: U.S. Department of Transportation.

Federal Highway Administration (FHWA). 1998. FHWA. Life Cycle Cost Analysis in Pavement Design Demonstration Project No. 115. FHWA-SA-98-040. Washington, D.C.: U.S. Department of Transportation.

Federal Highway Administration (FHWA). 1999. Pavement Subsurface Drainage Design. FHWA-HI-99-028, NHI Course No. 131026, April. Washington, D.C.: U.S. Department of Transportation.

Federal Highway Administration (FHWA). 2001. Drainage Requirements in Pavements (DRIP). Developed by Applied Research Associates for the Federal Highway Administration, Contract No. DTFH61-00-F-00199. Washington, D.C.: U.S. Department of Transportation.

Federal Highway Administration (FHWA). 2003. Distress Identification Manual for the Long Term Pavement Performance Program. FHWA-RD-03-031, June. Washington, D.C.: U.S. Department of Transportation.

Federal Highway Administration (FHWA). 2004. Concrete Pavement Technology Update, www.fhwa.dot.gov/pavement/concretecptu101.cfm.

Federal Highway Administration (FHWA). 2005. Computer Based Guidelines for Concrete Pavements, Vol. 2: Design and Construction Guidelines and HIPERPAVEII User's Manual. Publication No. FHWA-HRT-04-122. Washington, D.C.: U.S. Department of Transportation.

Federal Highway Administration (FHWA). 2005. Ultra-Thin White Topping (UTW) Project, http://www.fhwa.dot.gov/pavement/utwweb/.

Federal Highway Administration (FHWA). 2006. Retrofit Load Transfer: Sharing the Load. SP-204, www.fhwa.dot.gov/Pavement/concrete/sp204.cfm.

Federal Highway Administration (FHWA). 2008. Recycling: Current Projects and Activities, www.fhwa.dot.gov/pavement/recycling/.

Federal Highway Administration (FHWA). N.d. [Home page], www.fhwa.gov.

Federal Highway Administration (FHWA). N.d. Asset Management: Life-Cycle Cost Analysis, www.fhwa.dot.gov/infrastructure/asstmgmt/lcca.cfm.

Federal Highway Administration (FHWA). N.d. Pavements: LTPPBind, www.fhwa.dot.gov/pavement/ltpp/ltppbind.cfm.

Federal Highway Administration (FHWA). N.d. Preservation, www.fhwa.dot.gov/preservation.

Federal Highway Administration (FHWA), Office of Asset Management. 2002. *Life-Cycle Cost Analysis Primer*. August. Washington, D.C.: US Department of Transportation.

Finn, F., C. L. Saraf, and R. Kulkarni. 1986. PDMAP, Development of Pavement Structural Subsystems, NCHRP Report 291, Transportation Research Board, Washington, D.C.

Finn, F. N., K. Nair, and C. Monismith. 1973. Minimizing Premature Cracking of Asphalt Concrete Pavements, NCHRP Report 195, June. Washington, D.C.: National Cooperative Highway Research Program and National Research Council.

Foley, G., S. Cropley, and G. Giummarra, 1996. *Road Dust Control Techniques—Evaluation of Chemical Dust Suppressants' Performance*, Special Report 54, ARRB Transport Research, Victoria, Australia.

Fonseca, O. A. and Witczak, M. W. 1996. A Prediction Methodology for the Dynamic Modulus of In-Place Aged Asphalt Mixtures. In *Proceedings, Association of Asphalt Paving Technologists*, Vol. 65.

Foster, C. R. and Ahlvin, R. G. 1954. Stresses and Deflections Induced by a Circular Load. In *Proceedings of Highway Research Board*.

Foundation for Pavement Preservation. N.d. Welcome to the Foundation for Pavement Preservation Web Site! www.fp2.org.

Fox, L. 1948. Computation of Traffic Stresses in a Simple Road Structure. Road Research Technical Paper No. 9. Crowthorne, UK: Department of Scientific and Industrial Research, Road Research Laboratory.

Fredlund, D. G. and A. Xing. 1994. Equations for the soil–water characteristic curve. *Canadian Geotechnical Journal* 31:521–532.

Friberg, B. F. 1938. Design of Dowels in Transverse Joints in Concrete Pavements. *Transactions, ASCE* 64 (2): 1809–1828

Fuller, W. B. and S. E. Thompson. 1907. The Laws of Proportioning Concrete. *Transactions, ASCE* 59:67–143.

Garber, Nicholas J. and Lester A. Hoel. 2002. *Traffic and Highway Engineering*. 3rd ed. Clifton Park, NY: Thomson Learning.

Gisi, A. J. and S. S. Bandy. 1980. Swell Prediction of Natural Soils in Kansas. Memorandum. Topeka: Kansas Department of Transportation.

Guide for Mechanistic-Empirical Design of New and Rehabilitated Pavement Structures. 2004. Final Report. Part 1. Introduction, Part 2. Design Inputs. National Cooperative Highway Research Program, Transportation Research Board, National Research Council, Washington, D.C., Copy No. 88.

Haas, R., F. Meyer, G. Assaf, and H. Lee. 1987. A Comprehensive Study of Cold Climate Airfield Pavement Cracking. In *Proceedings of the AAPT*, Vol. 56.

Hajek, J. J. and R. C. G. Haas. 1972. Predicting Low-Temperature Cracking Frequency of Asphalt Concrete Pavement. Transportation Research Record No. 407. Washington, D.C.: Transportation Research Board.

Halladay, Michael. 1998. The Strategic Highway Research Program: An Investment That Has Paid Off. *Public Roads* 61 (5, March–April 1998), http://www.tfhrc.gov/pubrds/marapr98/shrp.htm.

Halstead, W. J. 1983. Criteria for Use of Asphalt Friction Surfaces. NCHRP Synthesis 104. Washington, D.C.: Transportation Research Board.

Harichandran, R. S., M. S. Yeh, and G. Y. Balidi. 1989. MICH-PAVE User's Manual. Final Report. FHWA-MI-RD-89-032. East Lansing, MI: Department of Civil and Environmental Engineering, Michigan State University.

Harwood D. W., D. J. Torbie, K. R. Richard, W. D. Glauz, and L. Elefteriadou. 2003. Review of Truck Characterization as Factors in Roadway Design. NCHRP Report 505. Washington, D.C.: Transportation Research Board.

Hayhoe, Gordon F. 2004. LEAF—A New Layered Elastic Computational Program for FAA Pavement Design and Evaluation Procedures. Paper presented at the FAA Administration Technology Transfer conference, April 2002, http://209.85.173.104/search?q=cache:-www.airporttech.tc.faa.gov/NAPTF/Download/BAKFAA/Documentation/Conf%20Paper%20P-26.doc.

Heinrichs, K. W., M. J. Liu, M. I. Darter, S. H. Carpenter, and A. M. Ioannides. 1989. Rigid Pavement Analysis and Design. FHWA-RD-88-068. June. Washington, D.C.: Federal Highway Administration, Research, Development, and Technology.

Heitzman M. S. 1992. *State of the Practice: Design and Construction Of Asphalt Paving Materials with Crumb Rubber Modifier*. FHWA SA-92-022. Washington, D.C.: U.S. Department of Transportation.

Heukelom, W. 1973. An Improved Method of Characterizing Asphaltic Bitumens with the Aid of Their Mechanical Properties. In *Proceedings, Association of Asphalt Paving Technologists*, Vol. 42.

Highway Community Exchange, Federal Highway Administration (FHWA). N.d. Highway Community Exchange Online Forum, http://knowledge.fhwa.dot.gov/cops/hcx.nsf/HowToUseThisDiscussionSite Display?openform.

Highway Community Exchange, Federal Highway Administration (FHWA). N.d. NCHRP 1-37A (Mechanistic-Empirical) Pavement Design Guide, http://knowledge.fhwa.dot.gov/cops%5Chcx.nsf/home?openform&Group=NCHRP%201-37A%20(Mechanistic-Empirical)%20Pavement%20Design%20Guide.

Hines, M. L. 1993. *Asphalt Cement Properties Improved by Styrelf: Laboratory and Field Data*. Wichita, KS: Koch Materials Company.

Hoerner, T. E., M. I. Darter, L. Khazanovich, L. Titus-Glover, and K. L. Smith. 2000. *Improved Prediction Models for PCC Pavement Performance-Related Specifications*. Vol. 1, Final Report. Federal Highway Administration.

Holtz, Robert D. and William D. Kovacs. 1981. *An Introduction to Geotechnical Engineering*. Upper Saddle River, NJ: Prentice Hall.

Huang, Y. H. 1993. *Pavement Analysis and Design*. Englewood Cliffs, NJ: Prentice Hall.

Huber, G. A. 1994. Weather Database for the Superpave Mix Design System. Strategic Highway Research Program Report SHRP 648A. Washington, D.C.: Transportation Research Board, National Research Council.

Ioannides, A. M., C. M. Davis, and C. M. Weber. 1999. Westergaard Curling Solution Reconsidered. Transportation Research Record No. 1684. Washington, D.C.: Transportation Research Board.

Ioannides, A. M. and M. I. Hammons. 1996. Westergaard Type Solution for Edge Load Transfer. Transportation Research Record No. 1525. Washington, D.C.: Transportation Research Board.

Ioannides, A. M., M. R. Thompson, and E. J. Barenberg. 1985. Westergaard Solutions Reconsidered. Transportation Research Board No. 1043. Washington, D.C.: Transportation Research Board.

Janssen, Donald J., and Mark B. Snyder. 1994. Resistance of Concrete to Freezing and Thawing. SHRP-C-391, Strategic Highway Research Program, Washington D.C., http://gulliver.trb.org/publications/shrp/CHRP-C-391.pdf.

Jones. A. 1962. Tables of Stresses in Three Layer Elastic Systems: Stress Distribution in Earth Masses. Bulletin 342. Washington, D.C.: Highway Research Board.

Jones, David. 1999. Holistic Approach to Research into Dust and Dust Control on Unsealed Roads, Transportation Research Board, Proceedings of the Seventh International Conference on Low-Volume Roads, TRR No. 1652, Volume 2.

Jones, G. M., M. I. Darter, and G. Littlefield. 1968. Thermal Expansion-Contraction of Asphaltic Concrete. *Proceedings, AAPT* 37:56–100.

Kallas, B. F. and V. P. Puzinauskas. 1972. Flexural Fatigue Tests on Asphalt Paving Mixtures: Fatigue of Compacted Bituminous Aggregate Mixtures. STP 508. Philadelphia: American Society for Testing and Materials.

Kaloush, K. E. and M. W. Witczak. 2000. Development of a Permanent to Elastic Strain Ratio Model for Asphalt Mixtures: Development of the 2002 Guide for the Design of New and Rehabilitated Pavement Structures. NCJRP 1-37A Inter Team Technical Report, September. College Park: University of Maryland.

Kerali, H. G. R. 2000. *Highway Development and Management (HDM-4)*, Vol. 1: *Overview of HDM-4*. Paris and Washington, D.C.: PIARC (World Road Association) and the World Bank.

King, Gayle N., Harold W. Muncy, and Jean B. Prudhomme. 1986. Polymer Modification: Binder Effect on Mix Properties. In *Proceedings of the AAPT*, Vol. 55.

Koerner, Robert M. and Bao-Lin Hwu. 1991. Prefabricated Highway Edge Drains. Transportation Research Record No. 1329. Washington, D.C.: Transportation Research Board.

Kosmatka, Steven H., Beatrix Kekhoff, and William Panarese. 2002. *Design and Control of Concrete Mixtures*. 14th ed. Skokie, IL: Portland Concrete Association

Larson, G. and B. J. Dempsey. 1997. Enhanced Integrated Climatic Model, Version 2.0. Final Report, Contract DTFA MN/DOT 72114. Urbana: Department of Civil Engineering, University of Illinois.

Leahy, R. B. 1989. Permanent Deformation Characteristics of Asphalt Concrete. Ph.D. diss., University of Maryland at College Park.

Lee, Y. H. and M. I Darter. 1995. Development of Performance Prediction Models for Illinois Continuously Reinforced Concrete Pavements. TRR 1505. Washington, D.C.: Transportation Research Board.

Likos, William J. and Ning Lu. 2001. Automated Measurement of Total Suction Characteristics in the High Suction Range: Application to the Assessment of Swelling Potential. Transportation Research Record No. 1755. Washington, D.C.: Transportation Research Board.

Little, D. N, M. R. Thompson, R. L. Terrel, J. A. Epps, and E. J. Barenberg. 1987. Soil Stabilization for Roadways and Airfield. Final Report. Panama City, FL: Engineering and Services Laboratory, Air Force Engineering and Services Center, Tyndall Air Force Base.

Lytton, R. L., R. L. Boggess, and J. W. Spotts. 1976. Characteristics of Expansive Clay Roughness of Pavements. Transportation Research Record No. 568, 9–23. Washington, D.C.: Transportation Research Board.

Lytton, R. L., D. T. Phillips, and C. V. Shanmugham. 1982. The Texas Rehabilitation and Maintenance System. Paper presented at the Fifth International Conference on the Structural Design of Asphalt Pavements, Delft, The Netherlands, August 23–26.

Lytton, R. L., D. E. Pufahl, C. H. Michalak, H. S. Liang, and B. J. Dempsey. 1990. An Integrated Model of the Climatic Effects on Pavements. Texas Transportation Institute, Texas A&M University, Report No. FHWA-RD-90-033. McLean, VA: Federal Highway Administration.

Lytton, R., J. Uzan, E. Fernando, R. Roque, D. Hiltunen, and S. Stoffels. 1993. Development and Validation of Performance Prediction Models and Specifications for Asphalt Binders and Mixes, Strategic Highway Research Program, Report No. SHRP-A-357. Washington, D.C.: U.S. Department of Transportation.

Majidzadeh, K. 1981. Implementation of a Pavement Design System. Final Report, Research Project EES 579. Columbus: Ohio State University.

Majidzadeh, K. 1988. A Mechanistic Approach to Rigid Pavement Design. In *Concrete Pavements*. London: Elsevier.

Majidzadeh, K. E. M., D. Kauffmann, V. Ramsamooj, and A. T. Chan. 1970. Analysis of Fatigue and Fracture of Bituminous Paving Mixtures. Report No. 2546. Washington, D.C.: U.S. Bureau of Public Roads, Research and Development.

Maupin, G. W., Jr. and J. R. Freeman, Jr. 1976. Sample Procedure for Fatigue Characterization of Bituminous Concrete. FHWA-RD-76-102, June. Washington, D.C.: U.S. Department of Transportation.

May, R. W. and M. W. Witczak. 1992. An Automated Asphalt Concrete Mix Analysis System. In *Proceedings of the AAPT*, Vol. 61, Charleston, SC.

McCall, J. T. 1958. Probability of Fatigue Failure of Plain Concrete. In *Proceedings of the American Concrete Institute*, Vol. 55, No. 13.

McCullough, B. F., J. C. M Ma, and C. S. Noble. 1980. Limiting Criteria for the Design of Continuously Reinforced Concrete Pavements. Transportation Research Report No. 756. Washington, D.C.: Transportation Research Board.

McGennis, R. B., S. Shuler, and H. U. Bahia. 1994. Background of Superpave Asphalt Binder Test Methods. FHWA-SA-94-069. Washington, D.C.: U.S. Department of Transportation.

McLeod, N. W. 1970. Influence of Hardness of Asphalt Cement on Low Temperature Transverse Pavement Cracking. Proceedings, Canadian Good Roads Association.

McLeod, N. W. 1976. Asphalt Cements: Pen-Vis Number and Its Application to Moduli of Stiffness. *American Society of Testing and Materials Journal of Testing and Evaluation* 4 (4, July).

Mega Concrete. N.d. [Home page, in Polish], www.mega-sa.com.pl.

Mindess, Sidney and J. Francis Young. 1981. *Concrete*. Englewood Cliffs, NJ: Prentice-Hall.

Miner, M. A. 1945. Cumulative Damage in Fatigue. *Transactions, ASME* 67:A159–A164.

Minnesota Department of Transportation. 2000–2006. PaveCool: Asphalt Pavement Cooling Tool, www.mrr. dot.state.mn.us/research/MnROAD_Project/restools/cooltool.asp.

Mitchell, J. K. 1993. *Fundamentals of Soil Behavior*. 2nd ed. New York: Wiley International Sciences.

Moavenzadeh, F., J. E. Soussou, H. K. Findakly, and B. Brademeyer. 1974. Synthesis for Rational Design of Flexible Pavements Part 3: Operating Instructions and Program Documentation. Reports for FH 11-776. Washington, D.C.: U.S. Department of Transportation.

Molenaar, A. A. A. 1983. Structural Performance and Design of Flexible Road Constructions and Asphalt Concrete Overlay. Ph.D. Diss., Delft University of Technology, Netherlands.

Monismith, C. L., D. A. Kasianchuk, and J. A. Epps. 1967. Asphalt Mixture Behavior in Repeated Flexural: A Study on In-Service Pavement Near Morro Bay, California. IER Report TE67-4. Berkeley: University of California.

Monismith, C. L., J. A. Epps, D. A. Kasianchuk, and D. B. McLean. 1972. Asphalt Mixture Behavior in Repeated Flexural. Report No. TE 70-5. Berkeley: University of California, Institute of Transportation and Traffic Engineering.

Moulton, L. K. 1979. Design of Subsurface Drainage Systems for Control of Groundwater. Transportation Research Record No. 733, Transportation Research Board, Washington, D.C.

Moulton, L. K. 1980. Highway Subsurface Design. FHWA-TS-80-224. Washington, D.C.: U.S. Department of Transportation.

Najjar, Yacoub, Imad Basheer, Ali Hossam, and Richard McReynolds. 2000. Swelling Potential of Kansas Soils: Modeling and Validation Using ANN Reliability Approach. Paper presented at the seventy-ninth annual meeting of the Transportation Research Board, January 9–13.

National Asphalt Pavement Association (NAPA). 1994. Guidelines for Materials, Production, and Placement of Stone Matrix Asphalt (SMA). IS 118. Lanham, MD: NAPA.

National Asphalt Pavement Association (NAPA). 2003. Constructing Quality Hot Mix Asphalt Pavements: A Troubleshooting Guide and Construction Reference. QIP 112. Lanham, MD: NAPA.

National Center for Pavement Preservation (NCPP). N.d. NCPP Reference Library, www.pavementpreservation.org/reference/.

National Climatic Data Center (NCDC). N.d. Heavy Rainfall Frequencies for the U.S., http://ncdc.noaa.gov/oa/documentlibrary/rainfall.html.

National Cooperative Highway Research Program (NCHRP). 1990. Calibrated Mechanistic Structural Analysis Procedures for Pavements. Vol. 1, Final Report. Washington, D.C.: National Cooperative Highway Research Program.

National Cooperative Highway Research Program (NCHRP). 1994. Long-Term Performance of Geosynthetics in Drainage Applications. NCHRP Report 367. Washington, D.C.: Transportation Research Board, National Research Council, National Cooperative Highway Research Program.

National Cooperative Highway Research Program (NCHRP). 2004. Thin and Ultra-Thin Whitetopping. NCHRP Synthesis 338. Washington, D.C.: Transportation Research Board, National Research Council, National Cooperative Highway Research Program.

National Cooperative Highway Research Program (NCHRP). N.d. Design Guide: Mechanistic-Empirical Design of New & Rehabilitated Pavement Structures, www.trb.org/mepdg.

National Lime Association. 1991. Lime Stabilization Construction Manual. Bulletin 326. Arlington, VA: National Lime Association.

National Research Council. 1994. Sensitivity Analyses for Selected Pavement Distress. SHRP-P-393. Washington, D.C.: National Research Council.

National Slag Association. N.d. [Home page], www.nationalslag.org.

National Stone, Sand & Gravel Association (NSSGA). N.d. [Home page], www.nssga.org.

Nevile, A. M. 1981. *Properties of Concrete*. Pitman, 3rd ed.

Newcomb, David E., Mark Buncher, and Ira J. Huddleston. 2001. Concepts of Perpetual Pavements in Perpetual Bituminous Pavements. Transportation Research Circular, No. 503, December. Washington, D.C.: Transportation Research Board, National Research Council.

Occupational Safety and Health Administration (OSHA). N.d. [Home page], www.osha.gov/comp-links.html.

Occupational Safety and Health Administration (OSHA). N.d. OSHA Hazard Communication Standards, 29 CFR 1910.1200, www.ilpi.com/msds/osha/1910_1200.html.

Office of Management and Budget. 1992. (Transmittal Memo No. 64). Circular No. A-194, revised, October 29, www.whitehouse.gov/omb/circulars/a094/a094.html#1.

Packard, R. G., and S. D. Tayabji. 1985. New PCA Thickness Design Procedure for Concrete Highway and Street Pavements. In *Proceedings of the Third International on Concrete Pavement Design and Rehabilitation*. West Lafayette, IN: Purdue University.

Paris, P. C. and F. Erdogan. 1963. A Critical Analysis of Crack Propagation Laws. *Transactions of the American Society of Mechanical Engineers, Journal of Basic Engineering, Series D* 85 (3): 538.

Paris, P. C., M. P. Gomez, and W. P. Anderson. 1961. A Rational Theory of Fatigue. *Trend in Engineering* 13 (7): 9.

Parker, Frazier. N.d. Class notes for CE 584, Soil Stabilization. Civil Engineering Department, Auburn University.

PCA, 1991. Design and Construction of Joins for Concrete Highways. Concrete Paving Technology, Portland Cement Association.

PDMAP (Probabilistic Design Method for Asphalt Pavements), F. Finn, C. L. Saraf, and R. Kulkarni. 1986. Development of Pavement Structural Subsystems. NCHRP Report 291. Washington, D.C.: Transportation Research Board.

Peattie, K. R. 1962. Stress and Strain Factors for Three-Layer Elastic Systems: Stress Distributions in Earth Masses. Bulletin No. 342. Washington, D.C.: Highway Research Board.

Peattie, K. R. and A. Jones. 1962. Surface Deflection of Road Structures. In *Proceedings: Symposium on Road Tests for Pavement Design*, Lisbon, Portugal.

Peutz M. G. F., H. P. M. Van Kempen, and A. Jones. 1968. Layered Systems under Normal Surface Loads. *Highway Research Record*, no. 228: 34–45.

Pfeiffer, J. P. and P. M. Van Doormaal. 1936. The Rheological Properties of Asphaltic Bitumens. *Journal of Institute of Petroleum Technologists* 22:P 414.

Phillips, D. T., C. V. Shanmugham, S. Sathaye, and R. L. Lytton. 1980. Rehabilitation and Maintenance System: District Time Optimization. Research Report No. 239-3. College Station: Texas Transportation Institute, Texas A&M University.

Phillips, D. T., F. Ghasemi-Tari, and R. L. Lytton. 1980. Rehabilitation and Maintenance System: State Optimal Fund Allocation—Program I. Research Report No. 239-4. College Station:, Texas Transportation Institute, Texas A&M University.

PIARC (World Road Association) Technical Committee on Flexible Roads and on Surface Characteristics. 1993. Porous Asphalt. Publication No. 08.01B. Paris: PIARC.

Portland Cement Association (PCA). 1992. *Soil-Cement Laboratory Handbook*. Skokie, IL: PCA.

Portland Cement Association (PCA). 2002. *Design and Control of Concrete Mixtures*. Engineering Bulletin 001, 14th ed. Skokie, IL: PCA.

Powell, W. D., J. F. Potter, H. C. Mayhew, and M. E. Nunn. 1984. The Structural Design of Bituminous Pavements, TRRL Laboratory Report 1132. Transportation and Road Research Laboratory: U.K.

Powers, T. C. 1932. Studies of Workability of Concrete. *Journal of the American Concrete Institute*, Farmington Hills, MI, 28, 419.

Prithvi, S. Kandhal, Frazier Parker, and Rajib B. Mallick. 1999. Aggregate Tests for Hot Mix Asphalt: State of the Practice. Transportation Research Circular No. 479. Washington, D.C.: Transportation Research Board, National Research Council.

Raad, L. 1982. Pumping Mechanisms of Foundation Soils under Rigid Pavements. Transportation Research Record No. 849, 29–37. Washington, D.C.: Transportation Research Board.

Raad, L. and J. L. Figueroa. 1980. Load Response of Transportation Support Systems. *J. Transp. Eng., ASCE*, 106(1), 111–128.

Rauhut, J. B., R. L. Lytton, and M. I. Darter. 1984. Pavement Damage Functions and Load Equivalence Factors. FHWA-RD-84-01. Washington, D.C.: U.S. Department of Transportation.

Rauhut, J. B, R. L. Lytton, and M. I. Darter. 1984. Pavement Damage Functions for Cost Allocations. Vol. 2, FHWA-RD-84-019, June. Washington, D.C.: U.S. Department of Transportation.

Read, John and David Whiteoak. 2003. Shell Bitumen. The Shell Bitumen Handbook. 5th ed. London: Shell.

Ridgeway, H. H. 1976. Infiltration of Water through the Pavement Surface. Transportation Research Record No. 616, 98-100. Washington, D.C.: Transportation Research Board.

Roberts, Freddy L., Prithvi Kandhal, E. Ray Brown, Dah-Yinn Lee, and Thomas W. Kennedy. 1996. Hot Mix Asphalt Materials, Mixture Design, and Construction. Lanham, MD: National Asphalt Pavement Association (NAPA) Education Foundation.

Rose, G. and D. Bennett. 1994. Benefits from Research Investment: Case of Australian Accelerated Loading Facility Pavement Research Program. Transportation Research Record No. 1455. Washington, D.C.: Transportation Research Board.

Rubber Pavements Association. N.d. [Home page], www.rubberpavements.org.

Saarenketo, Timo, Pauli Kolisoja, Kalevi Luiro, Pekka Maijala, and Nuutti Vuorimies. 2000. Percostation for Real Time Monitoring Moisture Variations, Frost Depth and Spring Thaw Weakening. Transportation Research Report No. 00833. Washington, D.C.: Transportation Research Board.

Salsilli, R. A., E. J. Barenberg, and M. I. Darter. 1993. Calibrated Mechanistic Design Procedure to Prevent Transverse Cracking of Jointed Plain Concrete Pavements. In *Proceedings of the 5th International Conference on Concrete Pavement Design and Rehabilitation*, Purdue University, West Lafayette, IN.

Sayers, M. W., T. D. Gillespie, and W. D. O. Paterson. 1986a. Guidelines for Conducting and Calibrating Road Roughness Measurements. Technical Paper 46. Washington, D.C.: World Bank.

Schapery, R. A. 1986. Time-Dependent Fracture: Continuation Aspects of Crack Growth. In *Encyclopedia of Materials Science and Engineering*, ed. M. B. Bever. Elmsford, NY: Pergamon Press.

Schwartz, D. R. 1987. D-Cracking of Concrete Pavements. NCHRP Synthesis of Highway Practice No. 134. Washington, D.C.: National Cooperative Highway Research Program.

Seed, H. B. 1959. A Modern Approach to Soil Compaction. In *Proceedings of the Eleventh California Street and Highway Conference*, reprint No. 69, 93. Berkeley: Institute of Transportation and Traffic Engineering, University of California.

Seiler, W. J. Expedient Stress Analysis of Jointed Concrete Pavement Loaded by Aircraft with Multiwheel Gear. Transportation Research Report No. 1370, 29–38. Washington, D.C.: Transportation Research Board.

Shahin, M. Y. N.d. MicroPaver. Washington, D.C.: U.S. Army Corps of Engineers.

Shell. 1951. *Shell Pavement Design Manual: Asphalt Pavements and Overlays for Road Traffic*. London: Shell.

Shell. 1978. *Shell Pavement Design Manual: Asphalt Pavements and Overlays for Road Traffic*. London: Shell International Petroleum.

Sivaneswaran, N., Linda M. Pierce, and Joe P. Mahoney. 2001. *Evercalc® Pavement Backcalculation Program*, Version 5.20. March. Seattle: Materials Laboratory, Washington State Department of Transportation.

Smith, K. D., T. P. Wilson, M. I. Darter, and P. A. Okamoto. 1992. Analysis of Concrete Pavements Subjected to Early Loadings. Transportation Research Report No. 1370. Washington, D.C.: Transportation Research Board.

Smith, K. L., K. D. Smith, L. D. Evans, T. E. Hoerner, and M. I. Darter. 1997. Smoothness Specifications for Pavements. Final Report, NCHRP 1-31. Washington, D.C.: Transportation Research Board.

Solaimanian, M. and P. Bolzan. 1993. Strategic Highway Research Program Report SHRP-A-637: Analysis of the Integrated Model of Climate Effects on Pavements. Washington, D.C.: Transportation Research Board, National Research Council.

Solaimanian, M. and T. W. Kennedy. 1993. Predicting Maximum Pavement Surface Temperature Using Maximum Air Temperature and Hourly Solar Radiation. Transportation Research Record No. 1417. Washington, D.C.: Transportation Research Board, National Research Council.

Spangler, E. B. and W. J. Kelly. 1964. GMR Road Profilometer—A Method for Measuring Road Profiles, Research Publication GMR-452, General Motor Corporation.

Strategic Highway Research Program (SHRP). 1993. Distress Identification Manual for Long Term Pavement Performance Project. SHRP-P-338. Washington, D.C.: Strategic Highway Research Program.

Strickland, D. *Shell Pavement Design Software for Windows, 2000*. Wythenshawe, UK: Shell Bitumen.

Tabatabie, A. M. and E. J. Barenberg. 1980. Structural Analysis of Concrete Pavement Systems. *ASCE, Transportation Engineering Journal* 106 (5): 493–506.

Tangella, S., C. Rao, J. Craus, J. A. Deacon, and C. L. Monismith. 1990. Summary Report on Fatigue Response of Asphalt Mixtures. Interim Report No. TM-UCB-A-003A-89-3, Project A-0030A Strategic Highway Research Program. Berkeley: University of California, Institute of Transportation Studies.

Tayabji, S. D. and B. E. Colley. 1986. Analysis of Jointed Concrete Pavements. FHWA/RD-86/041. Washington, D.C.: U.S. Department of Transportation.

Tayabji, S. D., P. J. Stephanos, and D. G. Zollinger. 1995. Nationwide Field Investigation of Continuously Reinforced Concrete Pavements. Transportation Research Report No. 1482. Washington, D.C.: Transportation Research Board.

Taylor, H. F. W. 1997. *Cement Chemistry*. London: Thomas Telford.

Texas Transportation Institute. N.d. MODULUS, http://tti.tamu.edu.

Thompson, M. R. 1987. ILLI-PAVE Based Full-Depth Asphalt Concrete Pavement Design Procedure. In *Proceedings of the Sixth International Conference on Structural Design of Asphalt Pavements*, University of Michigan, Ann Arbor.

Timoshenko, Stephen and J. N. Goodier. 1951. *Theory of Elasticity*. 2nd ed. New York: McGraw-Hill. (Originally published in 1934.)

Tolman, F. and F. Gorkum. 1996. Mechanical Durability of Porous Asphalt. Paper presented at Eurobitume conference.

Transportation Research Board. N.d. Endurance Limit of Hot Mix Asphalt Mixtures to Prevent Fatigue Cracking in Flexible Pavements, www.trb.org/TRBNet/ProjectDisplay.asp?ProjectID=972.

Transportation Research Board. N.d. Models for Predicting Top-Down Cracking of Hot-Mix Asphalt Layers. NCHRP 01-42A, www.trb.org/trbnet/projectdisplay.asp?projectid=228.

Tsend, K. H. and R. L. Lytton. 1986. Prediction of Permanente Deformation in Flexible Pavement Materials. ASTM Symposium, Implication of Aggregates in the Design, Construction and Performance of Flexible Pavements, New Orleans, LA.

Ullidtz, P. 1987. *Pavement Analysis*. New York: Elsevier. *Engineering Manual, Military Construction* University, College Station, Texas (1980).

U.S. Army Corps of Engineers, Construction Engineering Research Laboratory. N.d. MicroPaver, www.cecer.army.mil/paver/Downloads.htm.

U.S. Department of Transportation (USDOT). 1992. Traffic Monitoring Guide. FHWA-PL-92-017, October. Washington, D.C.: U.S. Department of Transportation.

U.S. Department of Transportation (USDOT). 2003. Memorandum: Revised Departmental Guidance, http://ostpxweb.dot.gov/policy/Data/VOTrevision1_2-11-03.pdf, 2.

U.S. General Accounting Office. 1998. Airfield Pavement: Keeping Nation's Runways in Good Condition Could Require Substantially Higher Spending. GAO/RCED-98-226. Washington, D.C.: U.S. GAO.

U.S. Geological Survey. N.d. Rainplot USGS Web Site, http://ent.ucf.edu/research/usgs.

Uzan, J., S. Frydman, and G. Wiseman. 1984. Roughness of Airfield Pavement on Expansive Clay. In *Proceedings, Fifth International Conference on Expansive Soils*, Adelaide, South Australia, May 21–23.

Van der Poel, C. 1954. A General System Describing the Viscoelastic Properties of Bitumens and Its Relation to Routine Test Data. *Journal of Applied Chemistry* 4 (May): 221–236.

van Rooijen, Ronald C., Arian H. de Bondt, and Ronald L. Corun. 2004. Performance Evaluation of Jet Fuel Resistant Polymer-Modified Asphalt for Airport Pavements. Paper presented at the FAA Worldwide Airport Technology Transfer Conference.

Van Wijk, A. J., J. Larralde, C. W. Lovell, and W. F. Chen. 1989. Pumping Prediction Model for Highway Concrete Pavements. *ASCE, Journal of Transportation Engineering* 115 (2): 161–175.

Velasco, M. O., and R. L. Lytton. 1981. Pavement Roughness on Expansive Clay. Transportation Research Record No. 790, 78–87. Washington, D.C.: U.S. Department of Transportation.

Verstraeten, J., J. E. Romain, and V. Veverka. 1977. The Belgian Road Research Center's Overall Approach to Asphalt Pavement Structural Design. In *Proceedings of the Fourth International Conference on Structural Design of Asphalt Pavements*, Vol. 1, August.

Vesic, A. S. and S. K. Saxena. 1969. Analysis of Structural Behavior of Road Test Rigid Pavements. Highway Research Board No. 291. Washington, D.C.: Highway Research Board.

Von Quintas, H. L., J. A. Sherocman, C. S. Hughes, and T. W. Kennedy. 1991. Asphalt-Aggregate Mixture Analysis System: AAMAS. NCHRP Report No. 338, March. Washington, D.C.: National Cooperative Highway Research Program, National Research Council.

Wardle, L. J. 1976. Program CIRCLY: A Computer Program for the Analysis of Multiple Complex Circular Loads on Layered Anisotropic Media: User's Manual Geomechanics Computer Program No. 2. Melbourne, Australia: CSIRO Division of Applied Geomechanics.

Warren, H. and W. L. Dieckmann. 1963. Numerical Computation of Stresses and Strains in a Multiple-Layer Asphalt Pavement System. Unpublished report. Richmond, CA: Chevron Research Corporation.

Washington State Department of Transportation. 1995. *Everstress: Version 5*. Seattle: Washington State Department of Transportation.

Washington State Department of Transportation, State Materials Laboratory. 2007. Pavement Design Tools, www.wsdot.wa.gov/biz/mats/pavement/pave_tools.htm.

Washington State Department of Transportation, State Materials Laboratory. 2007. Pavement Testing: Falling Weight Deflectometer, www.wsdot.wa.gov/biz/mats/pavement/testing.htm#fwd.

Washington State Department of Transportation, State Materials Laboratory. 2007. WSDOT Pavement Guide, www.wsdot.wa.gov/biz/mats/Apps/EPG.htm.

Westergaard, H. M. 1926a. Stresses in Concrete Pavements Computed by Theoretical Analysis. *Public Roads* 7:25–35.

Westergaard, H. M. 1926b. Analysis of Stresses in Concrete Pavement Due to Variations of Temperature. *Proceedings, Highway Research Board* 6:201–215.

Westergaard, H. M. 1927. Theory of Concrete Pavement Design. Proceedings, Highway Research Board, Part 1, 175–181.

Westergaard, H. M. 1939. Stresses in Concrete Runways of Airports. In *Proceedings of the 19th Annual Meeting of the Highway Research Board*. Washington, D.C.: National Research Council.

Westergaard, H. M. 1948. *New Formulas for Stresses in Concrete Pavements in Airfields*. Vol. 113. Washington, D.C.: American Society of Civil Engineers Transactions.

Winkler, E. 1867. Die Lehre von der Elastizitat und Festigkeit [The theory of elasticity and stiffness]. Prague: H. Dominicus.

Witczak, M. W., H. Von Quintus, K. Kaloush, and T. Pellinen. 2000. Simple Performance Test: Test Results and Recommendations. Superpave Support and Performance Models Management. NCHRP 9-19, Task C Report, November. Tempe: Arizona State University.

Wood, S. L. 1991. Evaluation of Long-Term Properties of Concrete. *ACI Materials Journal* 88 (6, November-December): 630–642.

World Bank. 1995. *HDM III: The Highway Design and Maintenance Standards Model*. Washington, D.C.: World Bank.

Wu, C. L., J. W. Mack, P. A. Okamoto, and R. G. Packard. 1993. Prediction of Faulting of Joints in Concrete Pavements. In *Proceedings, Fifth International Conference on Concrete Pavement Design and Rehabilitation*, Vol. 2., Purdue University, West Lafayette, IN, April.

Yao, Z. 1990. Design Theory and Procedure of Concrete Pavements in China. 2nd International Workshop on the Theoretical Design of Concrete Pavements, Siguenza, Spain.

Yoder, E. J. and M. W. Witczak. 1975. *Principles of Pavement Design*, 2nd ed. New York: John Wiley.

Yu, H. T., M. I. Darter, K. D. Smith, J. Jiang, and L. Khazanovich. 1998. Performance of Concrete Pavements, Volume III – Improving Concrete Performance. Report No. FHWA-RD-95-111 Washington, D.C., Federal Highway Administration.

Conversion Factors

Multiply	By	To Obtain
Foot (ft)	30.48	Centimeter (cm)
Gallon (gal)	3.785	Liter (L)
Horsepower (hp)	745.7	Watt (W)
Inch (in.)	2.540	Centimeter (cm)
Acre	0.4	Hectare (ha)
Pound (lb)	0.454	Kilogram (kg)
Pound-force (lb-f)	4.448	Newton (N)
Pounds per square inch (psi)	6.895	Kilo Pascal (kPa)
Stokes	104	m^2/s
Mile (mi)	1.609	Kilometer (km)
Bar	105	Pascal (Pa)

$1 \ Pa = 1 \ N/m^2$.

10^{-6} = prefix: micro-.

10^{-3} = prefix: milli-.

Index